MUON-CATALYZED FUSION

AIP CONFERENCE PROCEEDINGS 181

RITA G. LERNER
SERIES EDITOR

MUON-CATALYZED FUSION

SANIBEL ISLAND, FL 1988

EDITORS:

STEVEN E. JONES
BRIGHAM YOUNG UNIVERSITY

JOHANN RAFELSKI
UNIVERSITY OF ARIZONA

HENDRIK J. MONKHORST
UNIVERSITY OF FLORIDA

AMERICAN INSTITUTE OF PHYSICS   NEW YORK 1989

L.C. Catalog Card No. 88-83636
ISBN 0-88318-381-1
DOE CONF 8805170

Printed in the United States of America.

CONTENTS

INTRODUCTION

CHAPTER 1: EXPERIMENTAL STATUS AND RESULTS

Invited Papers:

Contributed Papers:

CHAPTER 2: THEORETICAL MUOCHEMISTRY

Invited Papers:

Contributed Papers:

Chapter 5: Muon Production

Invited Papers:

Contributed Papers:

Chapter 6: Summary and Conclusions

Preface

Our quest towards the understanding of muon-catalyzed fusion has led to a beautiful picture of exotic atomic, molecular, and nuclear processes, unexpected resonances, and extremely rapid reactions. In addition, surprisingly large fusion yields have stimulated speculation regarding possible future applications of cold fusion.

It was in this highly charged atmosphere that we assembled on Sanibel Island in Florida, May 1–6, 1988 for the Muon-Catalyzed Fusion 1988 meeting. About 100 participants gathered from many countries, including Australia, Austria, Canada, Germany, India, Israel, Italy, Japan, the Netherlands, the Soviet Union, Sweden, Switzerland, the United Kingdom, and the United States. The meeting was sparked by discussions of new experimental results and theoretical concepts, as well as novel engineering ideas. Many of us met previously at the first MuCF meeting in Jackson Hole in 1984, which was followed by meetings in Tokyo (1986) and Leningrad (1987). We are all looking forward to the next rendezvous in Europe in 1989.

The papers assembled here represent the output of the workshop and our progress in the understanding of muon-catalyzed fusion and its implications. We thank all who participated in the meeting and have made contributions to these proceedings. We have arranged the contents into six chapters: Experimental Status and Results; Theoretical Muochemistry; Properties of Muomolecules; Muon Sticking and Muon Regeneration; Muon Production; Summary and Conclusions. There were also poster sessions and we have included the most significant among these contributions.

Our particular appreciation goes to Professor Hendrik Monkhorst, the Chairman of the Organizing committee, and his staff of the University of Florida at Gainesville for an exceptionally well-organized meeting in a most beautiful setting. We acknowledge the University of Florida and Brigham Young University for their essential financial support.

The great accomplishment of this meeting was the stimulation and continuation of the dialogue between the theory and experiment aiming to resolve the numerous open questions in muon-catalyzed fusion.

Steven E. Jones
Johann Rafelski
Hendrik J. Monkhorst

THE POLITICAL ECONOMY OF MUON-CATALYZED FUSION RESEARCH

Ryszard Gajewski
U. S. Department of Energy, Washington, DC 20545

Introduction

We in the United States like to view ourselves as a nation of immigrants. That aspect of our history has had a number of social, political, and economic consequences, many of which can be felt to this day. Of that multitude of socioeconomic phenomena, all stemming from the immigrant heritage of our nation, I have found one particularly relevant to muon-catalyzed fusion research. Let me explain.

The very nature of settling new territory implies a basic injustice: earlier settlers have an incomparably higher opportunity to succeed than those who come later. The Pilgrims who landed on Plymouth Rock had to struggle for survival against the harsh environment, but those who eventually survived made out well. Even those early colonists who followed them, could, and did, take advantage of the enormous opportunities offered by the vast unexplored wealth of the new continent.

By the time the wave of modern immigration hit Ellis Island, the situation had changed drastically. Not only were most of the obvious opportunities already exploited by the early settlers, but an air of xenophobia had begun to creep in. "What are all those newcomers, those primitive Irish, tricky Jews, lazy Italians, stupid Poles (and so on...) up to, anyway?" Even without the xenophobia (which we may disapprove of, but which is a well-recognized social phenomenon), the fact is that the new immigrants were entering a well-functioning society with a developed set of values, institutions, and yes, an established distribution of wealth.

It is important at this point to know that the following truth had been realized. For a new immigrant to be successful, it wasn't enough to be as good as his "established" counterpart. "To succeed" the Irish, the Jewish, etc. parents were telling their children, "...you must not only study as hard as your Yankee schoolmate, you must do better, much better." Many listened, and many succeeded beyond anybody's wildest dreams. That they succeeded is not so much a tribute to our people's intrinsic goodness, as it is to a system that allows—in fact, thrives on—a free competition of ideas, and rewards initiative and excellence.

Muon-catalyzed fusion is a latecomer to the scene of energy research in general, and fusion research in particular. It has to compete with several well-established fusion programs. The competition takes place in the socioeconomic arena (competition for funding), but in the final analysis, its outcome will be determined by technical considerations. It would be naive to neglect the socioeconomic aspects—the established fusion programs are enormous enterprises with a powerful infrastructure. That is why, in order to succeed, it is not enough for muon-catalyzed fusion to offer a promise comparable to that of the other approaches. Neither will the classical incantation, "More research is needed!" make this latecomer succeed. Like an immigrant's child, in order to succeed in a big way we must prove that we have something clearly superior to offer, something the society simply cannot ignore. And let's not complain that magnetic fusion was given decades of lavish (by muon-catalyzed fusion standards) support to sort out the various magnetic configurations and their characteristic instabilities. They were here early, we are latecomers. Who said life was fair?

Let's also bear in mind that not all immigrants went on to become millionaires. Some failed, but many went on to live honorable lives in businesses, sciences, and professions; some even made it into the civil service. Likewise, research on the physics of muonic molecule formation, alpha-sticking, etc. can go on for its value to the understanding of basic phenomena in nature. As such it will have to compete for support with other basic research activities in atomic and nuclear physics.

On the other hand, should evidence of a clear technical promise of muon-catalyzed fusion emerge superior to that of the other approaches, I have no doubt that such a new opening will be recognized very quickly. Once that happens, the sky will be the limit to the research opportunities and technological prospects of muon-catalyzed fusion.

What will it take to get there? Most everyone in this room knows the answers: significantly increase the number of fusions per muon without going to heroic measures such as super dense DT mixtures at super high pressures; or significantly decrease the cost of producing a muon. Either of these steps, or a combination of both, could lead to technologically attractive systems.

As is the case in other areas of research, muon-catalyzed fusion involves increasingly more and more international cooperation. We welcome this not only because it lessens the burden of research costs borne by any one country, but because it documents the universal striving of the human mind to better understand nature and better utilize that knowledge. A workshop such as this one provides the most useful vehicle for international cooperation, as a forum for discussions at the working scientist's level. Such discussions, even if they do not lead to the heaven of 1,000 fusions per muon, may help answer the question: does heaven exist?

I can see that I have drifted from political economy to theology, a clear indication that it is time to quit! I hope that all of you have a most fruitful and rewarding meeting.

Chapter 1
Experimental Status and Results

SURVEY OF EXPERIMENTAL RESULTS IN MUON-CATALYZED FUSION

Steven E. Jones
Department of Physics and Astronomy
Brigham Young University

INTRODUCTION

Muon-catalyzed fusion research is motivated both by a curiosity about nature and by the possibility of applications. After all, exoergic nuclear fusion is readily induced by negative muons. And muon-catalyzed fusion is indeed curious: we have uncovered many surprises in a rich tapestry of exotic atomic and molecular processes, unexpected resonances, and extremely rapid nuclear interactions. It is remarkable that a fundamentally nuclear process can be affected by changing the temperature and composition of the environment. This phenomenon demonstrates the subtle interplay of atomic and nuclear physics inherent in muon-catalyzed fusion (μcf).

Theoretical and experimental efforts have also dovetailed to expand our understanding of muon catalysis. The theoretical breakthroughs ten years ago achieved by Leonid Ponomarev and his colleagues motivated experiments involving μcf in mixtures of deuterium and tritium. Observed temperature, density and d/t ratio effects in turn led to refinements in the theory. We can say that much progress has been made, but that many areas remained unresolved, some even virtually unexplored.

Overall, we can look back over the past decade of research and conclude that muon-catalyzed fusion yields have significantly exceeded expectations, leading to renewed speculation regarding applications. To guide our discussion of recent progress in μcf research, let us consider a straightforward yet profound equation:

$$1/Y = \lambda_o/\lambda_c + W. \qquad (1)$$

where

Y = yield, the number of fusions per muon (average);

λ_o = muon-decay rate (0.455 per microsecond);

λ_c = muon-catalysis cycling rate (1/time between fusion neutrons); and

W = the probability of muon loss per catalysis cycle, for any cause.

It is informative to interpret this governing equation as a sum of probabilities:

1/Yield =	Probability of muon decay during any stage of the catalysis cycle	+ Probability of muon-scavenging due to dead-end processes	(1b)

Clearly, to increase to fusion yield one would try to increase the catalysis cycling rate λ_c, and minimize muon losses W. We will here review what we have learned about these important parameters, then examine the current fusion yields vis-a-vis energy applications for μcf.

THE MUON CATALYSIS CYCLING RATE

Figure 1 displays a subset of data obtained at the Los Alamos Meson Physics Facility since 1982 regarding the observed (unnormalized) muon catalysis cycling rate. (See Ref. [1].) We see that λ_c depends on the density of the deuterium-tritium mixture as well as on its temperature and composition.

Why is this so? Coordinated theoretical and experimental studies have led to a picture of the μcf cycle which is portrayed in somewhat simplified form in Figure 2. Important reaction rates and muon-loss probabilities are labeled on this diagram. Note that C_p, C_d, C_t, and C_{He} represent the atomic fractions of the three isotopes of hydrogen (p,d,t) and helium present in the reaction chamber. The cycling rate can be broken down into component terms according to the prescription:

$$\lambda_c^{-1} \simeq \left[\frac{q_{1s} C_d}{\lambda_{dt} C_t} + \frac{0.75}{\lambda_{10} C_t} + \frac{1}{\lambda_{dt\mu} C_d} \right] \phi^{-1}. \tag{2}$$

for temperature $\lesssim 500K$, and where ϕ = the density of the target mixture.

The parameter q_{1s} merits further discussion. It represents the probability that the dμ atom will reach the ground state before the muon is transfered to a triton to form a tμ atom, an energetically favorable reaction. The dμ $\rightarrow$ tμ transfer reaction is faster for a smaller q_{1s}. But experiments [1,2,3] show that q_{1s} is larger than predicted [4] and decreases more slowly with increasing tritium fraction and density than expected. This transfer reaction is a relatively slow one (requiring typically a few nanoseconds), so it is relevant to understand why q_{1s} is as large as it is seen to be, and how it could be

Figure 1.

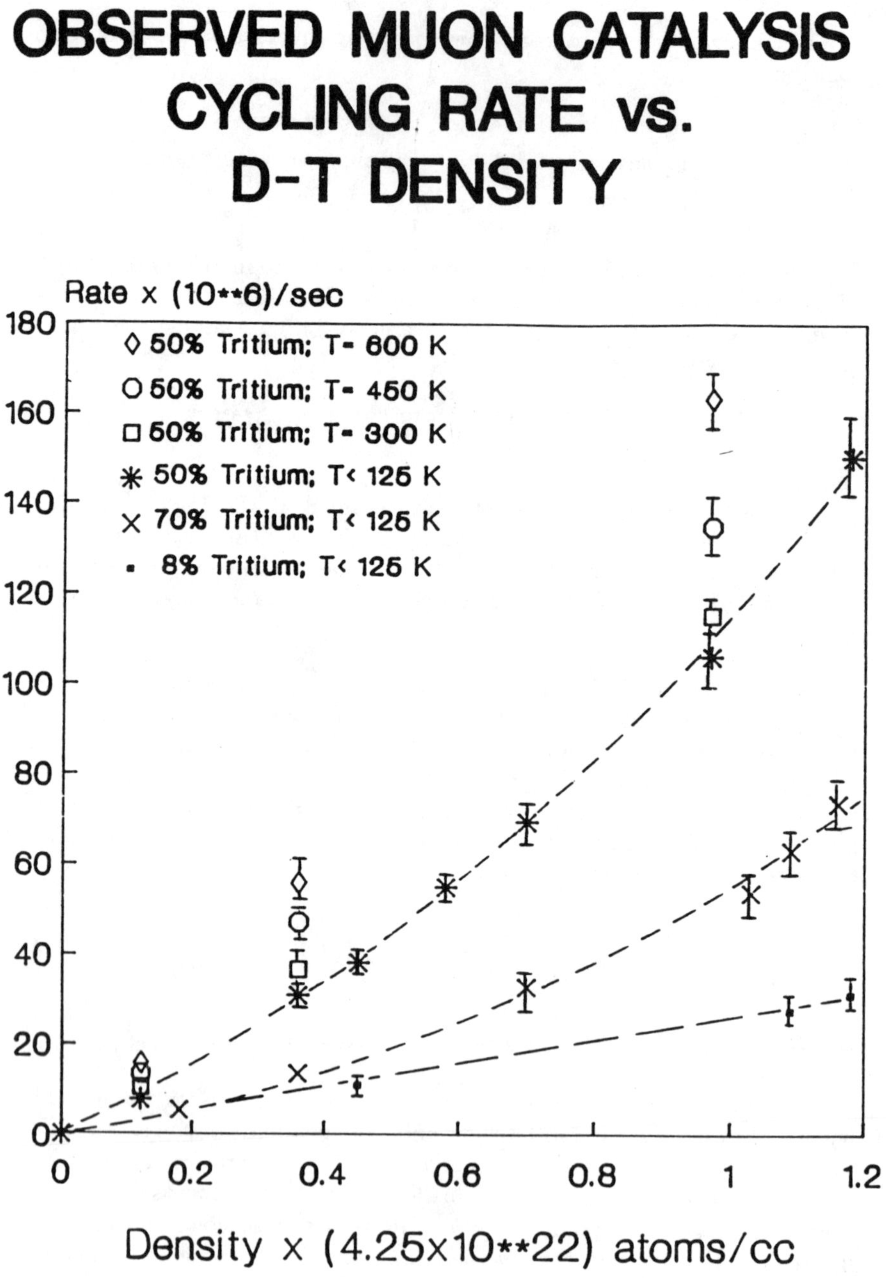
OBSERVED MUON CATALYSIS
CYCLING RATE vs.
D-T DENSITY
Rate x (10**6)/sec
◊ 50% Tritium; T= 600 K
O 50% Tritium; T= 450 K
□ 50% Tritium; T= 300 K
✻ 50% Tritium; T< 125 K
X 70% Tritium; T< 125 K
▪ 8% Tritium; T< 125 K
180
160
140
120
100
80
60
40
20
0
0
0.2
0.4
0.6
0.8
1
1.2
Density x (4.25x10**22) atoms/cc

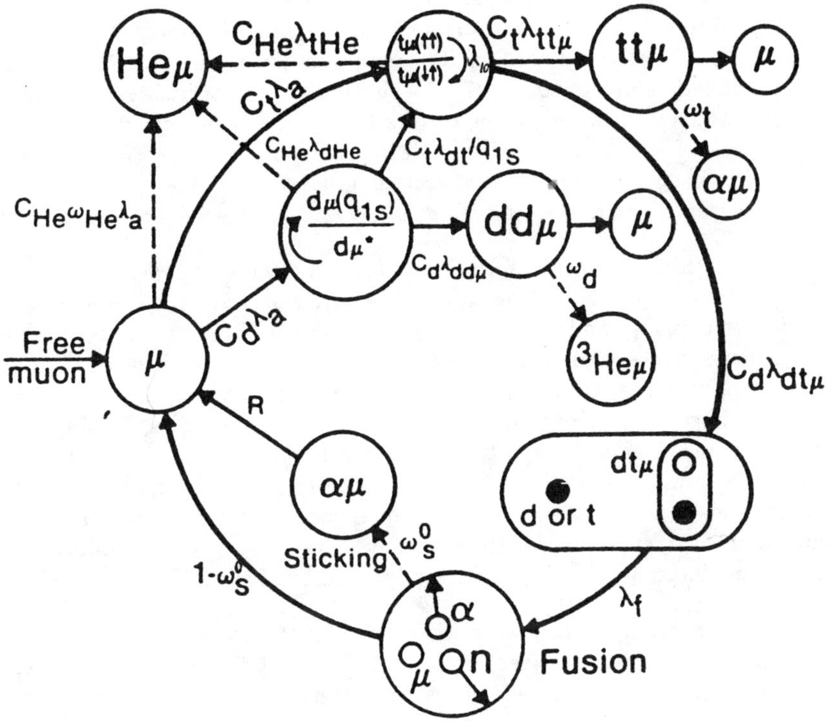

Figure 2. Scheme of the muon-catalyzed fusion cycle, showing reactions which occur when negative muons stop in a mixture of the hydrogen isotopes (p, d, and t) and helium (He) with respective fractions C_p, C_d, C_t, and C_{He}. Reaction rates are labeled with λ, and muon loss probabilities are labeled with ω.

reduced. A method to measure q_{1s} involving competition between the transfer reaction and helium scavenging (see Figure 2) has been described by Mel Leon [5]. Understanding the density, temperature and d/t ratio dependences of q_{1s} is a near-term goal of experiments.

Like q_{1s}, the hyperfine-quenching rate λ_{10} has proven stubbornly difficult to measure. A few years ago, results obtained at the Swiss Institute for Nuclear Research (now PSI) indicated that λ_{10} varied from (642 ± 27) /μs at 30 K to (317 ± 13) /μs at 300K.[6] These values were significantly less than the predicted value of 900/ μs [7] and showed a very surprising temperature dependence [6]. However, these results have since been retracted [2]. New results from LAMPF presented by Alan Anderson at this meeting show that λ_{10} is significantly larger than expected: greater than 1000 /μs. This may help to explain why hyperfine effects have been so elusive in μcf experiments.

Meanwhile, the rate of formation of dtμ molecules has been found [1,8] to depend strongly on temperature and density and on whether a tμ atom collides with a D_2 or a DT molecule (see Figure 2). These effects are reflected in the dependences seen in the muon catalysis cycling rate (Figure 1) and are consistent with the model of resonant dtμ formation developed by Ponamarev and collaborators.[9] Progress in measuring and understanding dtμ-formation has been rapid and gratifying for both theorists and experimentalists. It should be remembered that Ponomarev's predictions of fast, resonant dtμ formation were largely responsible for the renaissance of μcf research activity during the last few years. Furthermore, ideas on enhancing the dtμ-formation rate using lasers have recently been advanced by Hiroshi Takahashi [10].

We can conclude that reaction rates and the overall catalysis cycling rate are susceptible to further exploration, but that the rates are sufficiently fast to permit many hundreds of fusions during the muon lifetime. We turn our attention therefore to the question of muon-capture losses.

MUON-CAPTURE LOSSES (W)

Various ways in which muons may be lost from the catalysis cycle are shown in Figure 2. The muon may be captured and retained by a helium nucleus synthesized during dtμ, $dd\mu$, or ttμ fusion, with sticking probabilities ω_s, ω_d, and ω_t, respectively. In addition, small amounts (typically less than 1%) of protium are present, resulting in pdμ and ptμ fusion, with muon sticking probabilities ω_{pd} and ω_{pt}. The muon may also be scavenged by ambient helium in the hydrogen-isotope mixture, as indicated in Figure 2. All of these processes contribute to W, the total muon-loss probability per cycle [1]:

$$W \simeq \frac{q_{1s} C_d}{\lambda_{dt} C_t + \lambda_{dd\mu} C_d} (0.58 \lambda_{dd\mu} C_d \omega_d + \lambda_{pd\mu} C_p \omega_{pd} + \lambda_{d\mathrm{He}} C_{\mathrm{He}}) + \frac{1}{\lambda_{dt\mu} C_d} (\lambda_{tt\mu} C_t \omega_t + \lambda_{pt\mu} C_p \omega_{pt} + \lambda_{t\mathrm{He}} C_{\mathrm{He}}) + C_{\mathrm{He}} \omega_{\mathrm{He}} + \omega_s^{\mathrm{eff}}, \quad (3)$$

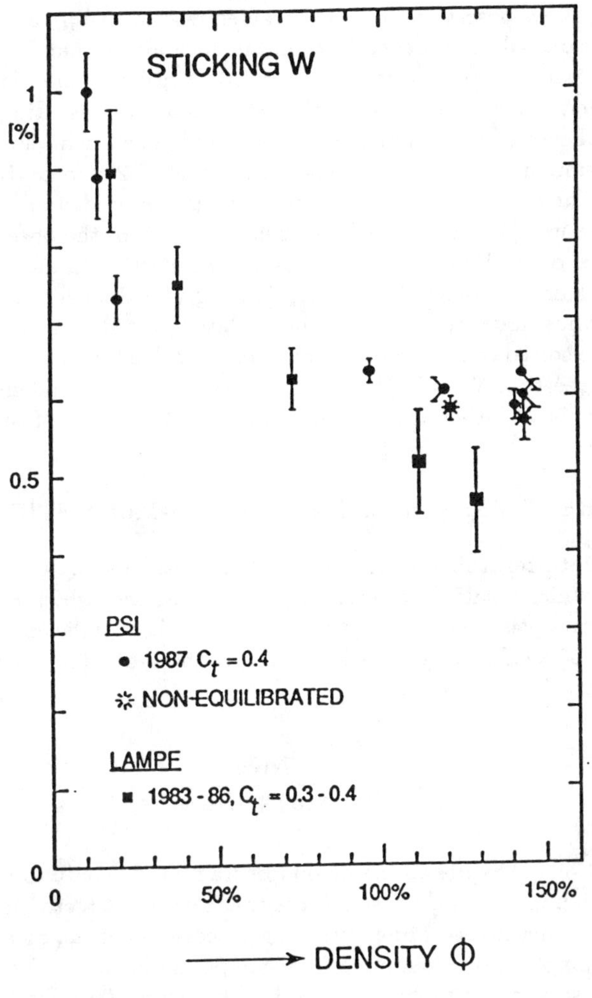

Figure 3. Data from LAMPF juxtaposed with the latest (1987) published results from PSI [11] show that the raw sticking W decreases with increasing density. Evidently, one or more muon-loss processes is density-dependent.

Experimentally measured values of W as a function of density are displayed in Figure 3. Results from LAMPF [1] and PSI [11] regarding the "raw sticking" W are in remarkable agreement and point to a rather striking density dependence.

What causes the obvious density-dependence of W? Looking closely at equation (4), we observe that some parameters such as $\lambda_{dt\mu}$ and possibly q_{1s} are significantly density-dependent, and in such a way that W will decrease with increasing target density, as observed. Until all muon-loss terms are fully understood, we cannot be certain whether ω_s, representing alpha-muon sticking following muon-induced d-t fusion and subsequent slowing down of the alpha-muon ion, is density-dependent or not [12]. In particular, if one assumes that q_{1s} is strongly density-dependent as predicted by Menshikov and Ponomarev [4], then one can account for much of the observed density-dependence of W. However, analysis of the LAMPF data has in fact shown only a weak density-dependence of q_{1s}, leaving a residual density-dependence in ω_s.[1] Thus, until q_{1s} and other interrelated parameters of equations (3) and (4) are sorted out completely, we cannot resolve this question. However, we can agree that W is indeed significantly density-dependent (Figure 3). After all, it is W rather than ω_s alone which influences the fusion yield (see equation 1).

DIRECT MEASUREMENT OF ALPHA-MUON STICKING

The data regarding W (Figure 3) were extracted by observing fusion neutrons, which results in a sensitivity to all processes which remove muons from the catalysis cycle. To measure ω_s alone, it is sufficient to count the number N of each of the charged products of the d-t fusion reaction, namely the α^{++} and $(\alpha\mu)^+$ ions:

$$\omega_s = \frac{N(\alpha\mu)}{N(\alpha) + N(\alpha\mu)} \tag{4}$$

Equation (4) expresses the muon loss fraction due only to $\alpha - \mu$ capture and retention following d-t fusion (note that ions are detected in coincidence with 14 MeV neutrons). Muon-stripping processes affect ω_s measured in this way, but complications stemming from competing dd and tt fusion channels, and muon scavenging by helium or other impurities (see Figure 2), can be excluded. Moreover, the ratio of equation (4) does not depend on absolute detector calibrations, a feature which reduces some systematic errors.

The experimental layout is portrayed in Figure 4 and is described in detail elsewhere (see ref. 13 and contributions by Michael Paciotti and John Davies, et al., in this volume). We recorded both the energy and the arrival time (relative to a fusion neutron) of each ion detected at the surface barrier

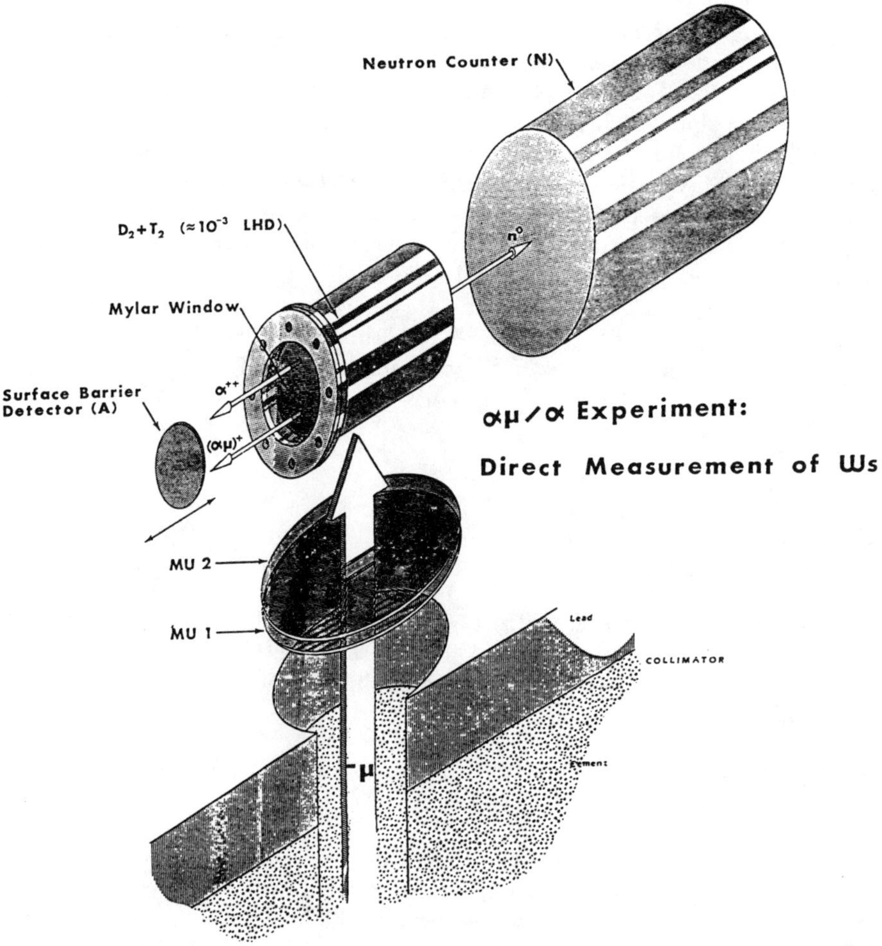

Figure 4. Scheme of the LAMPF/RAL experiment to measure alpha-muon sticking directly by detecting alpha and $(\alpha\mu)^+$ ions in coincidence with fusion neutrons.

detector. Figure 5 displays an energy spectrum of ions detected in a 10 nanosecond, neutron-coincidence time window chosen to select $(\alpha\mu)^+$ ions. A comparison with the same spectrum generated using Monte Carlo methods allows us to clearly identify both alpha and alpha-muon signals.

We are pleased to announce at this conference our first results obtained using this method. Data taken with a 60% deuterium - 40% tritium mixture at a pressure of 640 Torr are shown in Figure 5. At this pressure, we observe both alpha and alpha-muon signals. We need to correct the observed number of alpha particles for the fraction whose energy falls below the detection threshold (0.7 MeV) before reaching the detector (about half in this case, depending on the set position of the surface-barrier detector). Also, the number of detected alpha-muon ions must be corrected for the fraction lost due to stripping in the mylar window (8.3%) and in the gas (7.8%) before reaching the detector. These corrections are determined with the use of two separate Monte Carlo codes, which agree to within about 3%. The result for the initial sticking probability ω_s^o for data taken at LAMPF at a pressure of 640 Torr is:

$$\omega_s^o = (1.2 \pm 0.2 \pm 0.1 \quad \text{systematic})\% \quad (640 \quad \text{Torr, preliminary}) \qquad (5)$$

In order to gather better statistics, we separately collected alpha-muon ions at 1800 Torr and alpha ions at 490 Torr. This method is described in detail in Mike Paciotti's contribution in this volume. The result obtained in this way is:

$$\omega_s^o = (0.80 \pm 0.15 \pm 0.12 \quad \text{systematic})\% \quad (490 \quad \text{and} \quad 1800 \quad \text{Torr, preliminary}) \qquad (6)$$

While the statistics are better in this case, the estimated systematic error is large due mainly to uncertainty in the scaling of the fusion yield with the change in density [14]. It could well be that the fusion yield scales other than linearly with increasing density as assumed to obtain (6). This effect could then pull results (5) and (6) of the two approaches into better agreement. The measurements will be refined in future experiments at Rutherford Laboratory.[15]

We can already draw two important conclusions from these new results. First, the measured initial alpha-muon sticking probability appears to be in reasonable agreement with published theoretical calculations [16,17]:

$$\omega_s^o = (0.88 \pm 0.05)\% \quad (\text{theoretical}). \qquad (7)$$

Figure 5. Energy spectrum of ions detected at LAMPF in a time window chosen to select $(\alpha\mu)^+$ ions. Juxtaposition with the same spectrum generated by simulating the experiment using Monte Carlo methods (dashed line) allows us to clearly identify α^{++} and $(\alpha\mu)^+$ signals in the data.

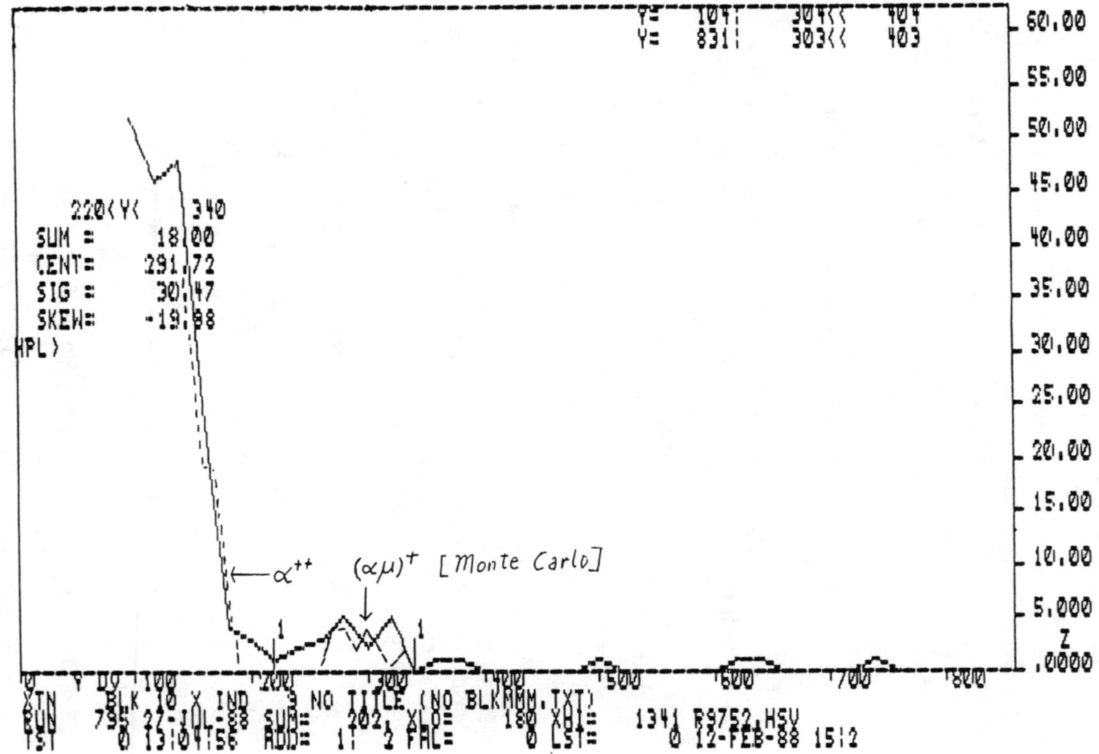

(See Mel Leon's paper in this volume for further discussion of recent of theoretical work on alpha-muon sticking.) Secondly, the directly measured value of sticking evidently rules out the possibility that sticking is very small (less than say 0.25%; see ref. [18]).

Before the measurement was made, one could argue that some process not included in equation (3) was making W large, while the sticking term ω_s^o was actually small. It is now clear that the initial sticking is large, around 1%, and that muon stripping as the alpha-muon ion slows down in the gas (density-dependent regeneration R) reduces this value very significantly:

$$\omega_s = \omega_s^o(1 - R \quad [density - dependent]) \tag{8}$$

Figure 6 displays the predicted [16] density-dependences of R and ω_s along with the (weighted) average value of ω_s based on measurements taken at LAMPF [1], PSI [8,19], and KEK [20] in liquid d-t mixtures. The separate measurements all agree within experimental errors. Note that the high-density value for sticking ω_s is about half the initial-sticking value. Significantly, we find that there remains some discrepancy between the experimental and theoretical values for sticking ω_s at liquid hydrogen density.

We have made considerable progress in understanding the alpha-muon sticking probability and related muon-loss mechanisms, but more work is clearly needed. It is also clear that sticking is the major bottleneck in the muon catalysis cycle, probably limiting the yield to a few hundred fusions per muon even at high densities.

EXPECTED FUSION YIELDS AND CONCLUSIONS

Since the catalysis cycling rate λ_c increases whereas the overall muon loss probability W decreases with increasing d-t density, we expect from equation (1) that the fusion yield will grow rapidly with increasing density. This is indeed the case, as demonstrated in Figure 7. In fact, the observed yield exceeds theoretical expectations of a few years ago by a comfortable margin.

But is it enough? Yuri Petrov has shown [21] that a hybrid reactor using μcf in conjunction with fission processes could generate power commercially when the μcf yield reaches about 150 fusions per muon. Figure 7 shows that this level has been reached in experiments. However, I suspect that fusion-fission hybrid reactors will remain unattractive as long as uranium remains inexpensive, particularly since hybrids partake of many of the problems of conventional fission reactors.

To produce power commercially using μcf alone would probably require

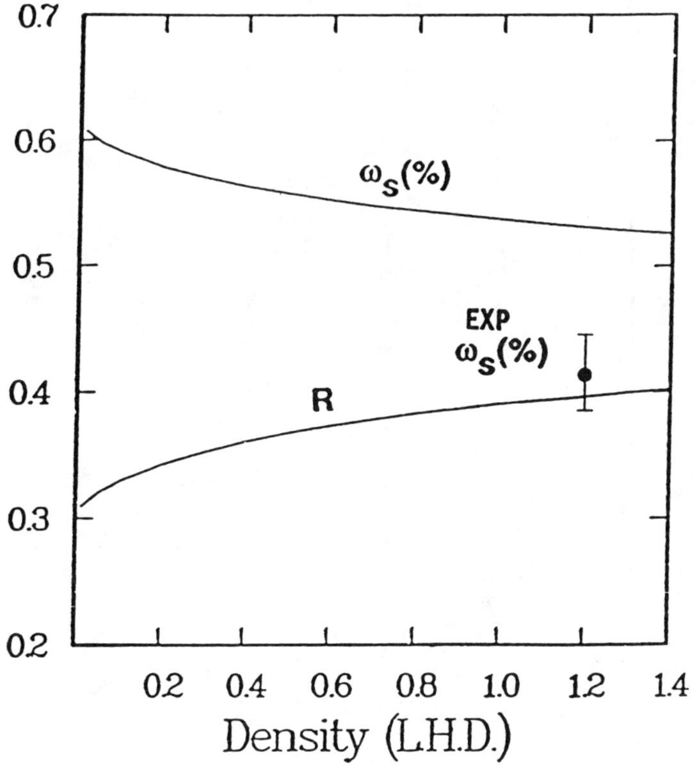

Figure 6. Calculated alpha-muon sticking ω_s assuming an initial sticking of 0.88% and regeneration R versus density [16], along with the observed sticking in liquid d-t mixtures (averaged from LAMPF, PSI and KEK experiments).

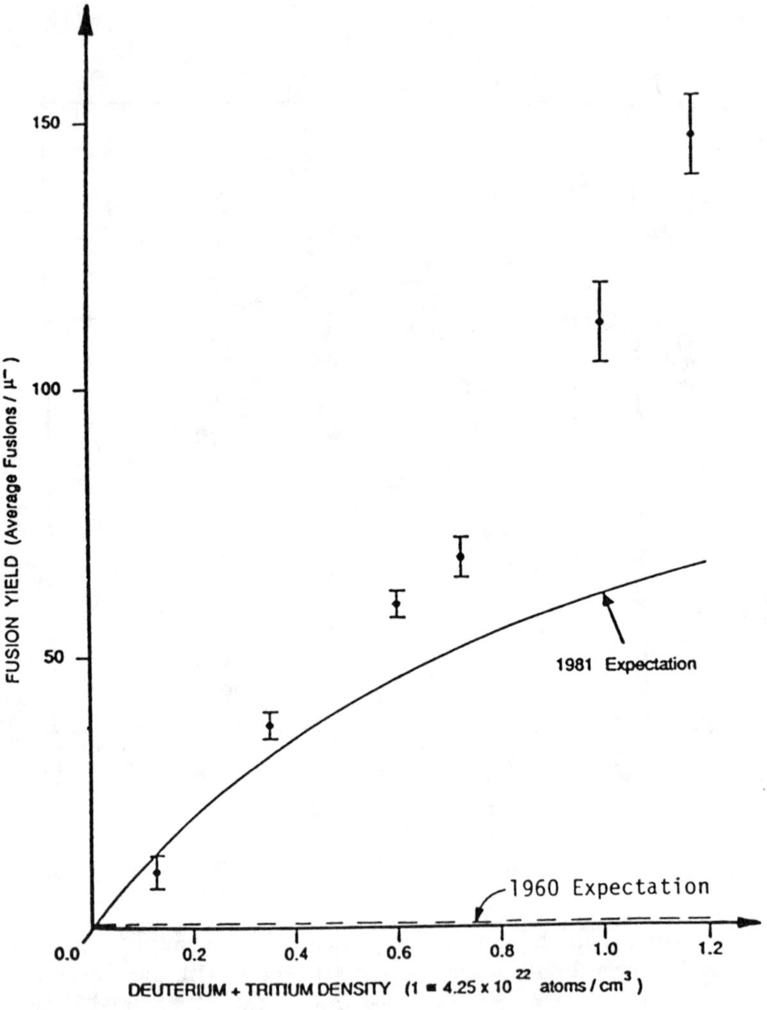

Figure 7. The average number of muon-catalyzed d-t fusion cycles ovserved by the BYU/Idaho/Los Alamos collabration as a function of density, for cold ($T < 100K$), equimolar deuterium-tritium mixtures. Note that the observed yield exceeds theoretical predictions at high densities.

an order-of-magnitude increase in the yield per muon, assuming current technology for muon production. Such a jump seems unlikely now because of the barrier imposed by alpha-muon sticking. However, some clever "imagineering" concepts were advanced at the workshop. For instance, Kulsrud and Tajima proposed a design in which alpha-muon ions would be repeatedly accelerated through solid d-t ice cells, to shake loose the muon (see paper by Kulsrud and Tajima in this volume). This concept appears to replace the sticking bottleneck with a very challenging engineering problem. In a paper presented at the workshop by Ponomarev, L. I. Men'shikov proposes that a cool plasma could greatly reduce the sticking coefficient [22]. Maintaining a sufficiently large $dt\mu$-molecular formation rate under such conditions would, they warn, be challenging.

In my opinion, the energy cost of producing muons must be very substantially reduced before energy production by means of μcf could be seriously considered. However, the field is young and active, and I think that such speculations are basically healthy as we vigorously strive to understand the beautiful phenomenon of muon-catalyzed fusion.

ACKNOWLEDGEMENTS

Discussions with numerous μcf colleagues have contributed to this summary. I particularly acknowledge valuable input from Alan Anderson, Antonio Bertin, Gus Caffrey, Jim Cohen, John Davies, Mel Leon, Ken Nagamine, Michael Paciotti, Claude Petitjean, Yuri Petrov, Leonid Ponomarev, Jan Rafelski, Antonio Vitale, and Alexi Vorobyov.

This research is supported by the Advanced Energy Projects Division of the U.S. Department of Energy.

REFERENCES

1. S. E. Jones, et al., Phys. Rev. Lett. 56, 588 (1986).
2. P. Kammel, et al., Muon Cat. Fusion 3, 483 (1988).
3. D. V. Balin, et al., Muon Cat. Fusion 2, 163 (1988).
4. L. I. Menshikov and L. I. Ponomarev, Pis'ma Zh. Eksp. Teor. Fiz.39, 542 (1984) [Sov. Phys. JETP Lett.39, 663 (1984), and 42, 13 (1985)].
5. M. Leon et al., Muon Cat. Fusion 2, 231, (1988).
6. W. H. Breunlich et al., Phys. Rev. Lett. 53, 1137 (1984).
7. A. V. Matveenko and L. I. Ponomarev, Zh. Eksp. Teor. Fiz. 59, 1593 (1970) [Sov. Phys. JETP 32, 871 (1971)]; V. S. Melezhik, Muon Cat. Fusion 1, 205 (1987).
8. W. H. Breunlich et al., Phys. Rev. Lett. 58 (1987) 137.
9. S. S. Gerstein and L. I. Ponomarev, Phys. Lett. 72B, 80 (1977); L. I. Menshikov and L. I. Ponomarev, Phys. Lett. 167B, 141 (1986).
10. H. Takahashi, Muon Cat. Fusion 2, 295 (1988).
11. C. Petitjean et al., Muon Cat. Fusion 2, 37 (1988).
12. L. N. Somov, et al., Muon Cat. Fusion 3, 465 (1988).
13. S. E. Jones, et al., Muon Cat. Fusion 1, 21 (1987); R. Gajewski and S. E. Jones, Muon Cat. Fusion 2 (1988).
14. M. A. Paciotti, et al., this volume.
15. J. D. Davies, et al., this volume.
16. J. S. Cohen, Muon Cat. Fusion 1, 179 (1987).
17. L. I. Ponomarev, Muon Cat. Fusion 3, 629 (1988).
18. H. Rafelski et al., Muon Cat. Fusion 1, 315 (1987).
19. H. Bossey et al., Phys. Rev. Lett. 59, 2864 (1987).
20. K. Nagamine et al., Muon Cat. Fusion 2, 731 (1988).
21. Yu. V. Petrov, Nature 285, 466 (1980); and Muon Cat. Fusion 3, 525 (1988).
22. L. I. Men'shikov, "Muon Catalysis Processes in Dense Low-Temperature Plasmas," Preprint IAE-4589/2, Moscow, 1988.

THE MEASUREMENT OF ddμ-MOLECULE FORMATION RATE AT HIGH DEUTERIUM PRESSURE (0.4-1.5 KBAR)

V.M.Bystritsky, V.P.Dzhelepov, V.V.Filchenkov, A.I.Gilev, V.B.Granovsky, Han Don Ir, N.Ilieva-Sokolieva, A.D.Konin, L.Marczis, D.G.Merkulov, A.I.Rudenko, A.B.Selikov, L.N.Somov, V.A.Stolupin, V.G.Zinov

Joint Institute for Nuclear Research,
101000 Moscow, USSR

ABSTRACT

The ddμ-molecule formation rate has been measured in gaseous deuterium at pressures of 390, 825 and 1490 bar. The measurements have been performed on the muon channel of the JINR phasotron with the use of a gas target and a full-absorption neutron detector.

The present work is devoted to the experimental study of muon catalysis (μCF) processes

$$d\mu + d \xrightarrow{\lambda_{dd\mu}\varphi} dd\mu \begin{cases} \xrightarrow{\beta\,\omega_d} {}^3He\mu + n & (1a) \\ \xrightarrow{\beta(1-\omega_d)} {}^3He + n + \mu & (1b) \\ \xrightarrow{1-\beta} t + p + \mu & (2) \end{cases}$$

in gaseous deuterium of high pressure (P $\sim$1 kbar). We have carried out a few runs on the muon channel of the JINR phasotron at 390, 825 and 1490 bar in a wide temperature range of T = 20-300 K. Now we have only fulfilled the analysis for T=300 K. The preliminary results are reported here.

The lay-out of the experimental set-up is given in Fig. 1. The main parts of it are the high pressure deuterium target and the full-absorption

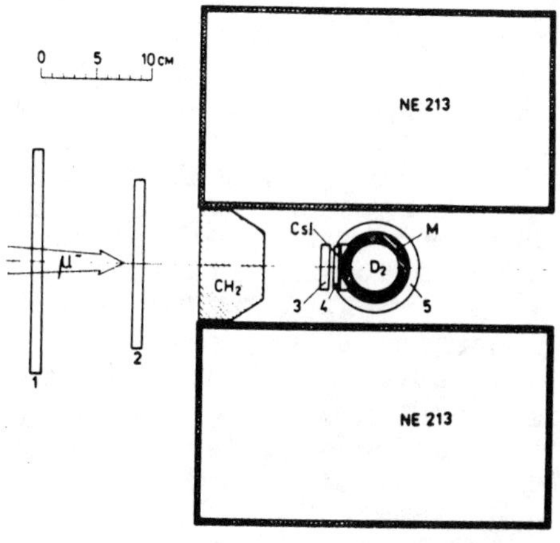

Fig. 1. The lay-out of the set-up for the experimental study of ddμ-molecule formation process.

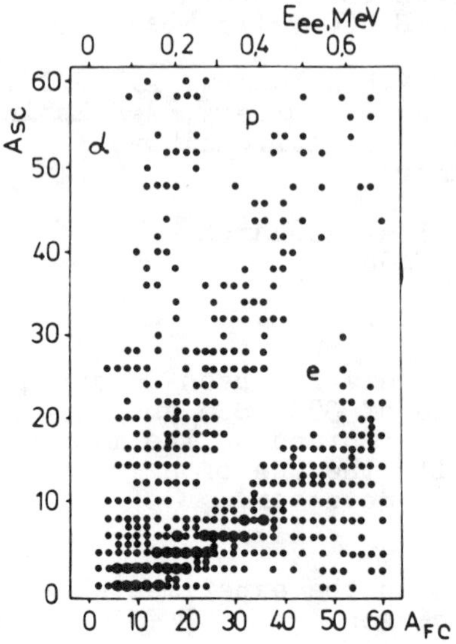

Fig. 2. The amplitude distribution of the signals of the fast component (FC) and the slow component (SC) detected by the neutron detector.

neutron spectrometer with the NE-213 scintillator for the registration of neutrons with an initial energy of E_n = 2.45 MeV from reactions (1a, b). The target [1] was made of special steel with the internal dimensions ⌀ 42x100 mm, the thickness of the walls was 9 mm. It was filled with pure deuterium (the hydrogen contamination was $\leq$ 0.3%) which was purified of the impurities with $Z > 1$ at a level of $\sim$ 0.1 ppm.

The detailed scheme of the neutron spectrometer and its parameters are presented in ref. [2]. The neutron spectrometer consists of two identical detectors symmetrically positioned around the target. The cuvettes of the scintillators were made of stainless steel with the dimensions ⌀ 310x160 mm. The internal surface of the cuvettes was covered with teflon which is characterised by its high reflection coefficient $\varkappa$ = 0.95. The energy resolution of the neutron detector was equal to $\sigma_{FWHM} = 0.09\ (1+1/\sqrt{E_{ee}(MeV)}\ )$ and the threshold energy of reliable n-γ separation was E_{ee}^{thr} = 50 keV (E_{ee} - equivalent electron energy). Fig. 2 shows a two-dimensional distribution of the signal amplitudes corresponding to the fast component (FC) and to the slow component (SC) of the light pulse characterising the quality of n-γ separation obtained in the expouser of deuterium at P = 1490 bar. This distribution is specially plotted without taking into account the selection criterion ($t_e > t_n$) in order to show the α-region caused by the registration of high energy background neutrons.

The spectrometer was calibrated using quasimonoenergetic neutrons selected from a ^{238}Pu-Be source by the TOF method and γ-sources. The detector efficiency ε and the pulse height distributions were calculated by means of a Monte-Carlo code. For a distance of L = 55 mm from the target center to each of the detectors, as in the case of our experiment (solid angle Ω = 66%), a registration efficiency of $\varepsilon \simeq 40\%$ (at $E_{ee}^{thr} = 100$ keV) for the neutrons with $E_n = 2.45$ MeV was obtained. In the first measurements only one detector was used and an efficiency $\varepsilon \simeq 20\%$ was calculated.

As shown in Refs. [3] the use of a high efficiency detector to register the products of muon catalysis reactions enables one to successfully employ the multiplicity of the μCF processes by registering the consecutive events of the μCF reactions, and allows in principle avoiding the quantity ε in the analysis. As shown in Refs.[3], the products of μCF reactions are detected with an "observed cycle rate", which for reaction (1) has the form

$$\lambda = \lambda_o + (\varepsilon + \omega_d - \varepsilon\omega_d)\bar{\lambda}_{dd} + \bar{\lambda}_x , \qquad (3)$$

where $\lambda_o = 0.455\ \mu s^{-1}$ is the muon decay rate, $\bar{\lambda}_{dd} = \lambda_{dd\mu}\beta\varphi$, λ_x is the partial disappearance rate of $d\mu$-atoms due to muon transfer to impurities. The expressions for the yields of the first, second, ... and of the k-th detected neutrons have the form:

$$\eta_{(1)} = \varepsilon\bar{\lambda}_{dd}/\lambda , \quad \eta_{(2)} = \eta_{(1)}^{2}(1-\omega_d), \ldots \eta_{(K)} = \eta_{(1)}^{K}(1-\omega_d)^{K-1} \qquad (4)$$

(without taking into account the losses due to the apparatus dead time). From formulas (3) and (4) it follows that the necessary condition for effective registration of neutrons from a few cycles of reactions (1a, b) is $\varepsilon\bar{\lambda}_{dd} \gtrsim \lambda_o$, which is satisfied in our experiment.

Counters 1, 2, 3 (plastic scintillators of size 200x200x15 mm, 150x150x 10 mm and 40x80x5 mm respectively) and 4 (CsI(Tl), 40x80x1 mm) were used for monitoring the muons of the beam coming to the target. Counter 5 (plastic scintillator of a cylindrical form) served to identify muon stops in the target and register the electrons from their further decay. The signal of a muon stop in the target

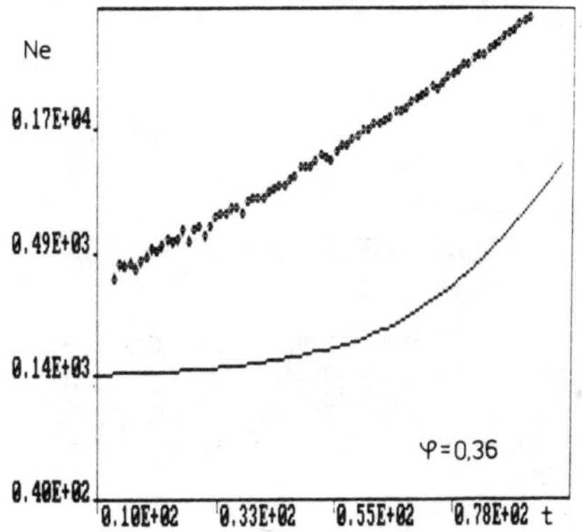

Fig. 3. The electron time distribution measured in the expouser with deuterium at P = 390 kbar (φ = 0.36).

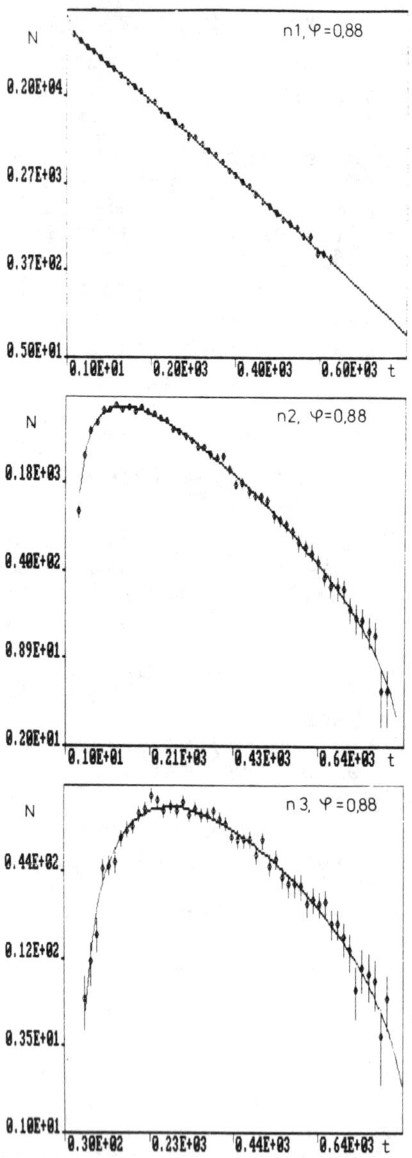

Fig. 4. The time distributions of the 1st, 2nd, 3rd detected neutrons obtained in the run at P = 1490 bar.

(1234,$\overline{5}$) triggered the gate with a duration of 10 μs, if in the previous 5 μs there was no muon signal (counter 1). During this time the consecutive signals (from one to eight) from the neutron detector were analysed in time and amplitudes of FC and SC. The necessary conditions for the analysis and for the recording of events were the detection of only one muon decay electron by counter 5 and the absence of a signal from counter 1 (the second muon) in the time interval of the gate.

Data processing of the events registered by the neutron detector consisted in n-γ (electron) separation and plotting the time distribution of the first, second and third neutrons with the criterion $t_e > t_n$. The electrons from a μ-decay registered by counter 5 were analysed and recorded separately. The time spectrum of electrons registered in expouser at P = 390 bar (φ = 0.36) is shown in Fig. 3. The points in this figure correspond to the experimental data for deuterium and the line is the background obtained in the run with an empty target. For the time $t_e > 0.5$ s the main sources of the electron background were accidental coincidences and muon stops in the light component (Al) of the target wall material. In expousers with φ = 0.63 and φ = 0.88 the relative level of the background was lower than for $\varphi = 0.36$. From the analysis of the electron time distributions a rate of muon transfer to impurities (nytrogen) $\lambda_x = (0.05 \pm 0.01)\varphi\ \mu s^{-1}$ was obtained.

The experimental time distributions of the first, second and third detected neutrons measured in expouser with $\varphi = 0.88$ is shown in Fig. 4 (time scale - 10 ns/channel, time bin - 0.16 μs). These spectra (and also that with $\varphi = 0.36$ and $\varphi = 0.63$) were analysed using expressions:

$$\begin{aligned} dn_{(1)}/dt &= K_1 \exp(-\lambda t), \\ dn_{(2)}/dt &= K_2 (t-\Delta) \exp[-\lambda(t-\Delta)], \\ dn_{(3)}/dt &= K_3 (t-2\Delta)^2 \exp[-\lambda(t-2\Delta)], \end{aligned} \qquad (5)$$

where K_1-K_3 are normalization parameters and $\Delta = 0.28$ μs is the apparatus dead time. The values of λ and $\varepsilon \bar{\lambda}_{dd}$ were obtained from the analysis of time spectra. The lines in Fig. 4 are time spectra calculated according to formula (5) with the optimal values of $\varepsilon \bar{\lambda}_{dd}$ and λ. In the following steps formula (3) for λ and the experimental values of $\omega_d = 0.126 \pm 0.004$ and $\beta = 0.582 \pm 0.017$ found in ref. [4] were used.

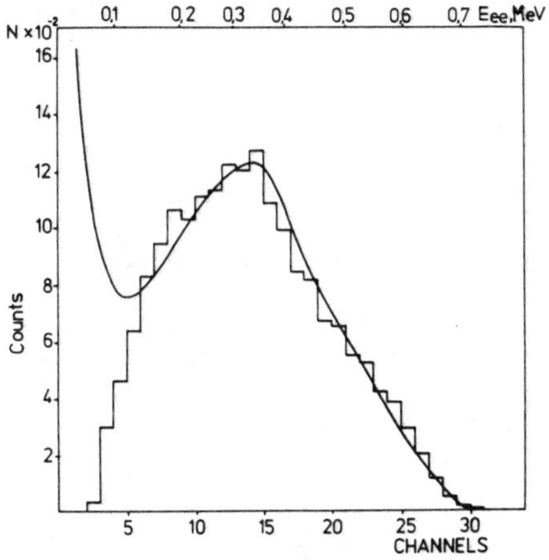

Fig. 5. The amplitude distribution (FC) for the first detected neutrons measured in the run with $\varphi = 0.88$.

The values of $\lambda_{dd\mu}(\varphi)$ were determined in two ways. In the first case we used the value λ_x obtained from the analysis of the electron spectra without using the neutron detection efficiency ε. In the second case we used the calculated value of ε. In Fig. 5 the pulse height spectrum of neutron events (FC) recorded in the run with $\varphi = 0.88$ is shown together with the results of the Monte-Carlo calculations (solid line). As seen there is a good agreement between calculations and experimental results.

The results obtained for $\lambda_{dd\mu}$ by the two methods agree within the error limits. Normalised to the liquid hydrogen density $\varphi = 1$, they are:

$$\lambda_{dd\mu}(\varphi = 0.88) = 2.53 \pm 0.33\ \mu s^{-1}$$
$$\lambda_{dd\mu}(\varphi = 0.63) = 2.48 \pm 0.32\ \mu s^{-1}, \qquad (6)$$
$$\lambda_{dd\mu}(\varphi = 0.36) = 2.40 \pm 0.31\ \mu s^{-1}.$$

These values are in agreement with the theoretical calculations of ref. [5] and with the experimental results for $\lambda_{dd\mu}$ obtained in ref. [4,6] at smaller deuterium densities.

As seen from the results (6) there is no dependence of $\lambda_{dd\mu}$ on density. It should be pointed out that we have obtained the relative dependence of $\lambda_{dd\mu}$ on density with a rather good accuracy. For this aim we have analysed the values of $(\lambda - \lambda_0)$ for different φ values taking into account the small correction due to the known dependence $\varepsilon(\varphi)$. It follows from this analysis that the value of $\lambda_{dd\mu}$ does not depend on deuterium density for $\varphi = 0.36$–0.88 within an accuracy of 3%.

REFERENCES

1. V.M.Bystritsky et al. Preprint JINR, 13-87-704, Dubna, 1987.
2. V.P.Dzhelepov, V.G.Zinov et al. Preprint JINR, 13-87-476, Dubna, 1987.
3. V.G.Zinov, L.N.Somov, V.V.Filchenkov. Preprint JINR, P15-82-478, Dubna, 1982 and Nucl.Instr.and Meth., 228, 174 (1984).
   V.M.Bystritsky, A.Gula, J.Wozniak. Atomkernenergie, Kerntechnik, 45, 197 (1984).
4. D.V.Balin, A.A.Vorobiev et al. Phys.Lett., 141B, 173 (1984).
5. L.I.Menshikov et al. Zh.Eksp.and Teor.Fiz. 92, 1173 (1987).
6. S.E.Jones et al. Phys.Rev.Lett., 56, 588 (1986).

STATUS OF MUON CATALYZED FUSION EXPERIMENTS AT UT-MSL/KEK

K. Nagamine
Meson Science Laboratory, Faculty of Science,
University of Tokyo, Bunkyou-ku, Tokyo, Japan
Metal Physics Laboratory, Institute of Physical & Chemical Research
Wakoh-shi, Saitama, Japan

T. Matsuzaki and K. Ishida
Metal Physics Laboratory, Institute of Physical & Chemical Research
Wakoh-shi, Saitama, Japan

Y. Hirata
Department of Nuclear Engineering, Faculty of Engineering
University of Tokyo, Bunkyo-ku, Tokyo, Japan

Y. Watanabe, Y. Miyake and R. Kadono
Meson Science Laboratory, Faculty of Science,
University of Tokyo, Bunkyo-ku, Tokyo, Japan

ABSTRACT

Since 1986, the experimental studies on muon catalyzed fusion (μCF) have been extensively in progress utilizing sharply pulsed negative muon available at the Meson Science Laboratory of the University of Tokyo (UT-MSL) located at KEK in Japan. Two major results have already been achieved; a) the first observation of radiative transition photons in the boundless decay from the ($d^4He\mu$) molecule during the process of muon transfer reaction from ($d\mu$) to He impurity in liquid D_2 ;b) X-ray measurement on the alpha sticking probability (ω_s) in the μCF for high C_T and liquid D_2/T_2, placing an upper limit on the ω_s. New directions of experiments are now in preparation based upon these two achievements ; a) search for the intermoleculor transition in ($dHe\mu$) and possible fusion process in ($d^3He\mu$) molecules and b) precise X-ray experiments on the ω_s in ($dt\mu$) μCF and its C_T dependence.

I. INTRODUCTION

Various new types of muon experiments have been carried out at the Meson Science Laboratory of the University of Tokyo (UT-MSL) located at the National Laboratory for High Energy Physics (KEK) since the establishment of the experimental facility for sharply pulsed (50ns pulse width and 50ms pulse separation (20 Hz)) muon beam[1,2,3]. At this UT-MSL/KEK, an instantaneously intense (10^4 μ^-/pulse in 5x5 cm^2) backward μ^- beam can be obtained with a substantially low e^- as well as π^- background by using the superconducting muon channel[1,2].

The most important features of the use of pulsed μ^- beam related to the μCF phenomena can be summarized as follows: a) a phenomena occuring at the long time delay after the μ^- beam arrival can easily be observed with the help of the substantially reduced background ; b) under the existence of a huge white-noise background, one can measure a weak signal of the muon beam associated events; c) various types of resonance stadies can be realized by using a high power pulsed laser or pulsed rf source which can only be produced with a low duty cycle.

Since 1986, two types of the μCF experiments have been performed at the UT-MSL/KEK by using these excellent features[4,5,6]. So far, most of the experiments have been done by the method of low energy photon measurements; a) measurement of the radiative transition photons in the decay process of the $(d^4He\mu)$ molecule during the process of the $(d\mu)$ to $(\alpha\mu)$ transfer process in liquid D_2 with 430 ppm He impurity, where a clean signal was seen for the characteristic energy spectrum of the radiative transition photons as well as its time dependence with the help of the feature a); the first trial was done for the X-ray measurements from the $(\alpha\mu)^+$ atoms formed in the α-sticking process in the μCF for high C_T(30%) liquid D_2/T_2, where a weak signal was attempted to be detected under the huge bremsstrahlung background from tritium β-decay with the help of the feature b). The importance of these experiments is clear ; the former experiment can be considered as the first experiment to observe directly the photons from the muon molecular ion (mesomolecule), while the latter experiment should be taken as the first direct measurement of the α-sticking phenomena in the μCF for the D_2/T_2 system where the muon catalyzed fusions are actually taking place more than 100 times.

During the one year since the μCF 87 at Leningrad, some progresses have been done for these two experiments including data analysis and discussions. Also, extensive preparatory works have been in process for the second X-ray experiment for the ω_S in high C_T and liquid D_2/T_2.

Present status of these works is summarized in the followings as well as the physics considerations related to the results of these experimental studies with the perspectives toward the future experiments.

2. X-RAY MEASUREMENT DIRECTLY OBSERVING MUON-MOLECULE

Many experimental observations on μCF phenomena have been successfully explained by considering the formations of the muon molecular ions, such as $(dt\mu)^+$, $(dd\mu)^+$, $(dHe\mu)^{++}$, etc. However, other than a rather indirect method of the temperature dependence measurements of the fusion neutron yield, there have been no direct observations of the existence of the muon molecule such as X-ray spectroscopy of the intramolecular transition, etc. Such a unhappy situation is mainly due to the unique energy region associated with the muon molecule e.g. 90 eV for (J=0, υ=0) to (J=1, υ=0) transition in $(dt\mu)$, to which neither conventional photon counting nor laser resonance technique can not be applied. On the other hand, without any doubts a direct spectroscopy of the muon molecule should be the most important next step to obtain a complete understanding of the μCF phenomena.

With respect to the molecular spectroscopy, we have recently measured the radiative transition photons from the bound state of $(d^4He\mu)_{J=1}$ to the unbound state of $(\alpha\mu)_{1s}$ + d in the process of μ transfer from $(d\mu)$ to the impurity ^4He atom in liquid D_2 (bound-free decay)[5,6]. In this case, because of the deep and unbound character of the ground state of the $(d\,^4He\mu)$, the transition photon was turned out to be rather easily detectable.

2.1 Observation of the Radiative Transition Photons in Boundless Decay of $(d^4He\mu)$ Molecule

In order to explain the anomalously high μ^- transfer rate from hydrogen isotops to the He impurity, Leningrad group led by Popov has developped theoretical studies on the transfer mechanism through the formation of the $(dHe\mu)$ molecule[7,8,9]. According to this model, the following processes are expected to take place (see also Fig. 1a). First, instead of the direct exchange reaction of $(d\mu) + {}^4He \rightarrow ({}^4He\mu) + d$, the molecular ion is formed through $(d\mu) + {}^4He \rightarrow [(d^4He\mu)^* e^-]^+ + e^-$ where the molecule $(d^4He\mu)$ is preferentially formed at the J=1 state of the $(2p\sigma)$ excited states through a dipole transition. The unstable $(2p\sigma)$ molecular state can subsequently deexcite through either a radiative E1 transition process to the dissociated ground state like $(d^4He\mu)^{++} \rightarrow ({}^4He\mu)^+ + d$ or Auger process like $[(d^4He\mu)^* e^-]^+ \rightarrow (d^4He\mu)^{++} + e^-$. According to the theoretical calculation[8], the radiative transition occurs at the rate of $10^{12} s^{-1}$, while the Auger contribution is estimated to be less than 15%. Also, the $1 \rightarrow 0$ intramolecular Auger process is forbidden so that the radiative transition occurs mainly from the J=1 state of the $(2p\sigma)$ state. In this case, the characteristic photon spectrum was predicted to be seen[8]; a unique peak energy with a broad and asymmetric shape (see, Fig. 1b).

At UT-MSL/KEK, the experiment was done for the 13.6 cc liquid D_2 target with 430 ppm He impurity which was dissolved by pressurizing the liquid D_2 surface with 2 atm He gas. The experimental arrangement was almost similar to the original X-ray experiment[4] on the α-sticking probability shown in the following section, except the details of the target chamber; adopting a larger size target chamber with inner chamber surface of a 50 μm Ta sheet for the reported experiment[5,6] and 100 μm Ag sheet for the refined experiment.

The observed results are shown in Fig. 1c and Fig.1d for the time integrated energy spectrum from 0.28 μs to 7.5 μs and for the energy integrated time spectrum (around 6.85 keV),respectively. As can be seen clearly in Fig.1c, a characteristic asymmetric and broad photon peak was observed with a central energy of 6.85(4)keV and a width of 0.74(4) keV which agress quite well with the theoretical predictions[8]. As shown in Fig.1d, the peak has a faster decay than the muon life time with a decay time constant of 0.98(5) μs. By a comparison between the predicted energy spectrum and the observed one, one can conclude that the radiative transition photon was observed. The present result nicely confirms the model of muon transfer mechamism through the formation of the muon molecule, providing us the first result of the transfer rate λ_{dHe} in a liquid phase. At the same time, the result could be considered as the first experiment to "see" the muon molecule.

2.2 Some Extended Studies on the (dHeμ) Molecules

After the first successful experiment on the radiative transition photon measurement for the bound-free decay of the (dHeμ) molecule, several new thoughts have been brought-in not only from us but also from the conventional molecular spectroscopy people, promoting us to perform the refined experiments.

The first step was taken to search for a multi-peak structure in the photon energy spectrum as is familiarly seen in the bound-free decay of the (electronic) excimer molecule[10]. Since the radiative E1 transition takes place between the bound state of (dHeμ) and the continuum state made of ($\alpha\mu$) and d, the scattering wave character depending upon the distance between ($\alpha\mu$) and d may appear as a correction term to the main E1 transition peak seen in Fig.1b producing a multi-peak structure. The refined spectrum measurements were done by replacing Ta inner layer sheet of the liquid D_2 chamber to Ag one, producing none of the electronic X-ray background from 4 keV to 22 keV. After careful measurements, any significant peak structure has not been found between 4 keV and 8 keV.

With respect to this problem, preliminary theoretical studies have been carried out by Hara et.al. based upon non-adiabatic treatment for the wave function of the muon molecle[11]. They could predict an existence of the second in addition to the main peak at 6.85 keV. Further refined experiments to see the second peak with the 10^{-3} intensity of the main peak is now in progress.

As for the rate of the μ^- transfer λ_{dHe}, there are two methods of determination. Namely, (1) the measurement of the photon intensity per μ^- ; $N_X = \kappa_X C_{He} \phi \lambda_{dHe} / (\lambda_0 + C_{He} \phi \lambda_{dHe})$, where κ_X is a fraction of the radiative transition and is assumed to be 0.85, and (2) the measurement of the decay time constant of the photon intensity ; $N_X(t)$ is proportional to $\exp[-(\lambda_0 + C_{He} \phi \lambda_{dHe})t]$. As reported already , the inconsistency was found between the values obtained by the two methods; 13.1(12) x $10^8 s^{-1}$ by the decay rate, while 6.7(13) x 10^8 s^{-1} by the total intensity. Here, the correction to the decay rate due to the (ddμ) molecule formation is small;less than 0.05 x $10^4 s^{-1}$ compared to the observed total decay rate other than λ_0 (0.52(5) x 10^6 s^{-1}). The same inconsistency was also found in the refined experiment with the Ag inner layer. Further refined experiment for the absolute photon yield determination is now in progress.

When this inconsistency is true, probable explanation might be an existence of the other decay route(s) from the J=1 state of the (2pσ) state in ($d^4He\mu$) in addition to the radiative E1 transition mentioned above. Since the main E1 transfer rate is estimated to be $10^{12} s^{-1}$, the rate of the other decay route might be of this order. Possible decay process might be, in addition to the process causing multiple energy peak structure , an enhanced decay processes which are normally considered to be very slow; e.g. Auger process, the intramolecular transition like J=1 to J=0, etc.

As an extension of the present work, similar X-ray studies are in progress for (d^{3}Heμ) system. In addition to the spectroscopy-type interests, there is a special interest in this system. If the intermolecular transition between J=1 to J=0 states takes place, there might be a chance for us to see the μCF in the (d^{3}Heμ) molecule. In this case, 18.4 MeV protons can be emitted during the μCF. The fusion rate at J=0 state was estimated by Kravtsov et.al. yielding 100 s^{-1} for (d^{3}Heμ). Recently, by using a correct non-adiabatic treatment of the three body system including nuclear effect, Kamimura[12] calculated the fusion rate at the J=0 state of the (d^{3}Heμ) to be 3(1) x $10^8 s^{-1}$. Therefore, if the intermolecular transition is allowed by any reasons, we could really expect to see the μCF in (d^{3}Heμ) system.

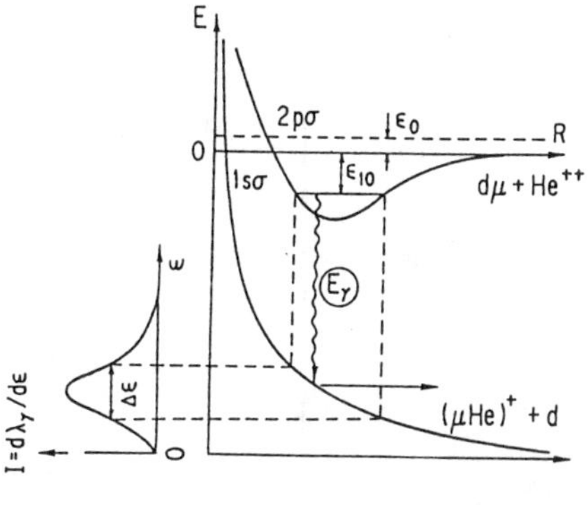

Fig. 1a Schematic picture of the bound-free decay process in the μ^- transfer from (dμ) to (Heμ) proposed by Popov et.al.[7].

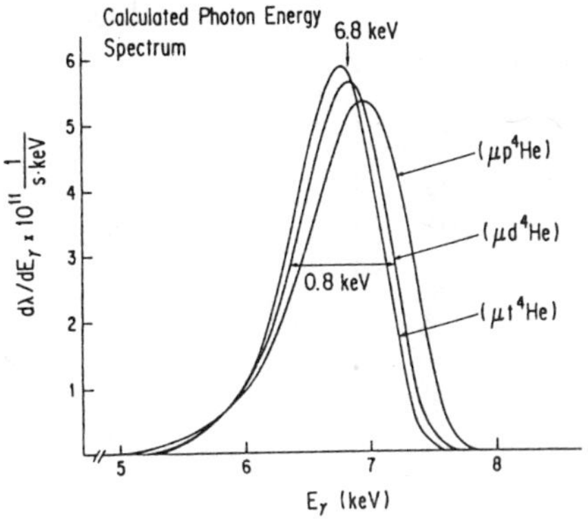

Fig. 1b Expected photon energy spectrum accompanying the bound -free decay from the (dHeμ) molecule calculated by Kravtsov et.al.[8].

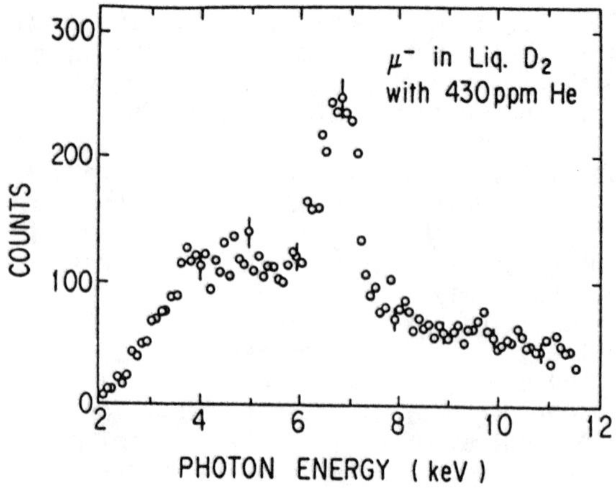

Fig. 1c Observed photon energy spectrum for the liquid D_2 with 430 ppm He impurity. The spectrum was obtained in the refined experiment using a target chamber with Ag inner-layer.

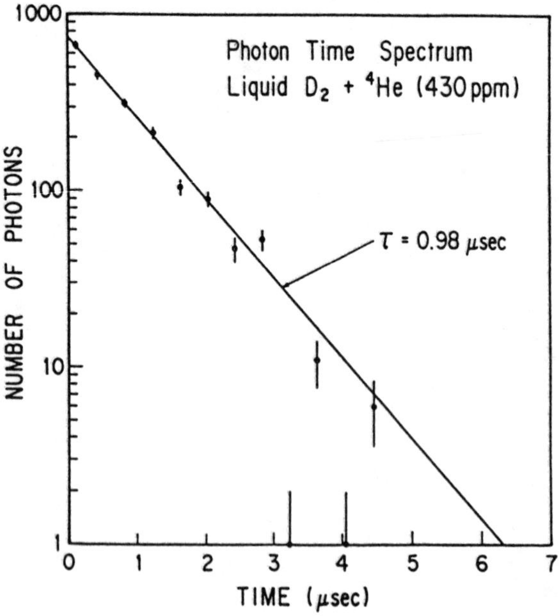

Fig. 1d Time spectrum of the photon peak seen in Fig. 1c. Constant background was subtracted.

3. X-RAY MEASUREMENT DIRECTLY OBSERVING ALPHA-STICKING PROBABILITY IN (dtμ)-μCF

It is still true that the most important parameter to be determined in the (dtμ)-μCF might be the α-sticking probability ω_s for high density and high C_T systems. The value of ω_s strictly places an upper limit for the energy production capability. The conventional neutron method observing a loss probability is an indirect method for the determination of the ω_s; the value can only be obtained as a small difference between the two large values (neutron yield and λ_c/λ_0) to which various additional corrections must be done for the loss processes other than the ω_s.

As a direct measurement of the ω_s, the two new methods have been proposed: the measurement of the $(\mu\alpha)^+$ ions by using either ionization chamber[13] or SSD[14] (ion method) and the measurement of muonic X-ray from $(\mu\alpha)$[4,15] (X-ray method). From a feasibility consideration, we can not apply the ion method to a high density system of the D_2/T_2 mixture.

The X-ray method, which is practically the only method to be applied to the direct ω_s measurement for a high density D_2/T_2 mixture, needs a theoretical relation connecting the observed X-ray intensity to the sticking probability ω_s. Recently, theoretical studies have been carried out on this subject by Cohen[16] and Markushin[17]. For instance, Cohen predicted a X-ray intencity per fusion (I_X/χ) being 0.24 for the D_2/T_2 with 1.2ϕ density where the calculation was started with an initial sticking probability of 0.848 obtained theortically by Bogdanova et.al.[18]. In these studies, starting from the predicted $(\alpha\mu)$ atomic states right after initial sticking, all the possible processes have been considered concerning the X-ray emission during slowing-down of $(\alpha\mu)^+$, such as μ^- stripping (reactivation), excitation, deexcitation, etc.

The X-ray measurement for high C_T and high density D_2/T_2 can not easily be done by the conventional detection system ; a huge bremsstrahlung background up to 17 keV due to tritium β-decay can easily smash away the weak K_α X-ray (8.30 keV) from the sticked $(\alpha\mu)$ atom. The X-ray experiment at SIN was done only for the C_T of 2×10^{-4}. The pulsed muons available at UT-MSL/KEK is particularly effective to overcome this difficalty as mentioned in the earlier section as well as the preliminary report[4,6].

3.1 The 1986 UT-MSL/KEK Experiment of the X-ray Measurement on ω_s for 30 % C_T Liquid D_2/T_2

In this subsection, we will present a brief description of the previous experiment as well as a correction to the content reported in the previous publication[4,6].

First, let us summarize the experimental method of the 1986 X-ray measurement whose details can be found in the previous report[4]. In our experiment, the main concern was focussed on the determination of the α-sticking probability in liquid and high C_T D_2/T_2 mixture. The C_T was taken as 30 % which is known to have the maximum λ_c at low tempaeratures[19,20]. The basic structure of the D_2/T_2 target chamber can be found in Fig.1 of the previously published report [4]. The D_2/T_2 gas mixture is contained in the doubly closed container, an innermost part of which is a gas container (1 litre) which is connected to a liquid container (1 cc) made of cupro-nickel (without any inner-layers in the 1986 experiment). The liquid container is indirectly cooled by the He flow cryostat through a thermal conductive copper rod. The X-ray from the liquid D_2/T_2 can penetrate towards the Si(Li) detector through the two 0.5 mm thick Be windows. During our experiment, the liquid container was maintained at the temperature between liquefaction temperature T_L and T_L-2 K.

The experiment was carried out by using the 60 MeV/c backward μ^- beam. The μ^- beam was confined to a spot of 2 cm diameter with the help of the longitudinal magnetic field of the SHC (35 kG). The essential part of the experimental arrangement is reproduced in Fig. 2. In order to obtain numbers of the μ^- stopping in the liquid D_2/T_2 target, the 16 channel plastic counter telescopes for the digital type of the μe detections are installed at the backward direction from the D_2/T_2 target chamber. Its efficiency and solid angle under the confinement field for decay e^- was determined by measuring muonic X-ray from liquid N_2 target under the exactly same geometry. The Si(Li) detector (5 mm thickness) for the low energy X-ray measurement as well as the NE213 liquid scintillation counter (2" dia. and 2" length) for 14 MeV fusion neutrons are placed inside the room-temperature gap (10 cm wide and 90 cm high) of the SHC in the direction perpendicular to the beam axis.

Data taking was made for the following spectra; (a) time spectrum of decay electrons, (b) two parameter (energy and time) correlation spectrum for the X-ray signals and (c) three parameter (energy, timing and n-γ descrimination) correlation spectrum for the neutron signals.

Observed energy spectrum from the X-ray detector is reproduced in Fig. 3 in a time-divided way with reference to the time of muon arrival (note that the incorrect time ranges in our previous publication[4,6] are corrected here). a) the "early" spectrum is essentially due to bremsstrahlung observed at the off-beam condition, where the 8.0 keV peak is electronic Cu K_α X-ray due to tritium beta-rays hitting the cupro-nickel wall of the liquid container; b) the "prompt" spectrum has an additional contribution from a low energy tail due to γ-ray from the multiple scattering of decay electrons; c) the "delayed" part which is the most essential part for the α-sticking effect, is made of a featureless spectrum with the bremsstrahlung background, which can be removed by using either "early" or "very delayed" spectrum. The corrected "delayed" spectrum is shown in Fig. 3(d). The corrected delayed spectrum was used as an object for the fitting analysis to find out the peak associated with the delayed X-ray due to the α-sticking. The fitting was made by assuming a Gaussian shaped peak superimposed on a linear background structure. Here, the FWHM of the Gaussian peak was taken to be either 0.25 keV corresponding to the detector resolution confirmed by the observed FWHM for the Cu K_α X-ray peak or 0.70 keV corresponding to the expected FWHM when a Doppler effect fully contributes to the muonic K_α line of $(\alpha\mu)$. We adopted results from both of these two fitting values corresponding to the two assumptions for the X-ray line-width. As shown in the previous report[4], in terms of X-ray yield per muon, the results are 0.035(22) and 0.055(45) corresponding to the FWHM of 0.25 keV and that of 0.70 keV, respectively. By using the decay time conotant of the fusion neutron λ_n obtained in the same measurement (1.20 (9) μs^{-1}) and the cycling rate λ_c adopted from the neutron experiment[19,20] (112 μs), we can obtain the values of the X-ray yield per fusion I_x/χ as 0.031 (20) % and 0.049 (40) % for the two values of FWHM.

The obtained X-ray intensity per fusion can be converted to the value of the initial sticking ω_s^0 through the theoretical calculation made by Cohen. As far as our analysis proceeds in such a way, we must consider only the case of full doppler broadening, since the energy spectrum calculated by the Cohen's model is more or less close to the one expected for full doppler broadining. The result is summalized as follows ; using the Cohen's value of I_x/χ (photon yield per fusion) versus initial sticking ω_s (0.25 % versus 0.865 %) we can convert the observed I_x/χ (0.049(40)%) to the initial sticking ω_s^0 (0.19(15)%). By using the stripping probability R obtained by Cohen[16], the ω_s^0 can be converted to the effective sticking ω_s^{eff} (0.11(9)%), which should be compared to the value obtained by the neutron method.

The value of ω_s mentioned in the preliminary report[4,6] was obtained by assuming $I_x/\chi = \kappa\omega_s^{eff}$, where the κ was taken to be 0.25. According to the Cohen's calculation , the number equivalent to κ $(I_x/\chi/\omega_s^0(1\text{-R}))$ is much larger as mentioned above.

As mentioned in the previous report[4,6], the ω_s^{eff} was obtained in our neutron measurement which was simultaneously done during the X-ray experiment; $\omega_s^{eff}(n) = 0.46(7)$ %. The result is significantly different from the value obtained by the X-ray method. In order to explain this discrepancy, we must consider the hidden process where the μ^- is lost from the $(\mu\alpha)$-state without emitting K_α X-ray. Possible mechanism is described in the following section.

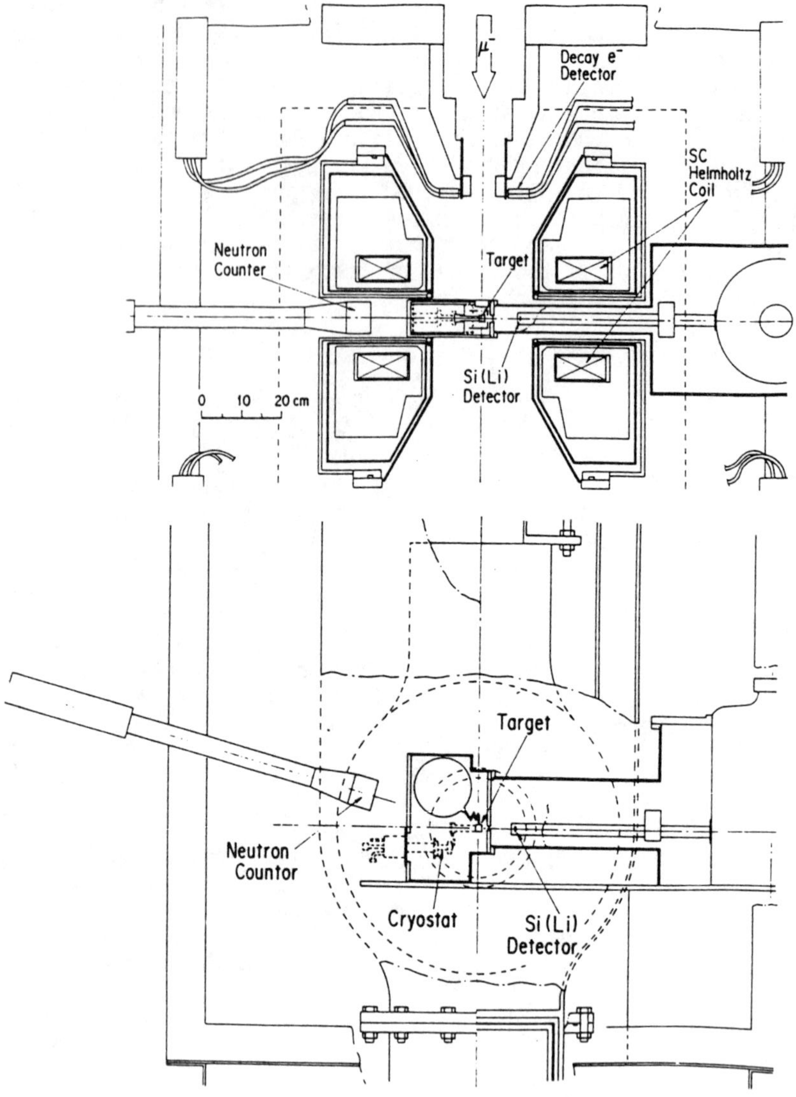

Fig. 2 Experimental arrangement for the X-ray and neutron measurement for liquid D_2/T_2 target at the 1986 UT-MSL/KEK Experiment.

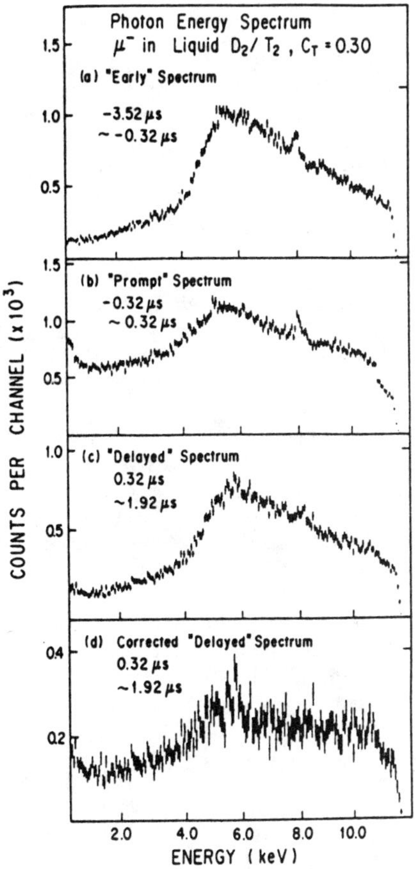

Fig. 3 Time-divided energy spectrum observed by the X-ray detector for the μCF in the liquid D_2/T_2 and delayed spectrum after bremsstrahlung background subtraction (d).

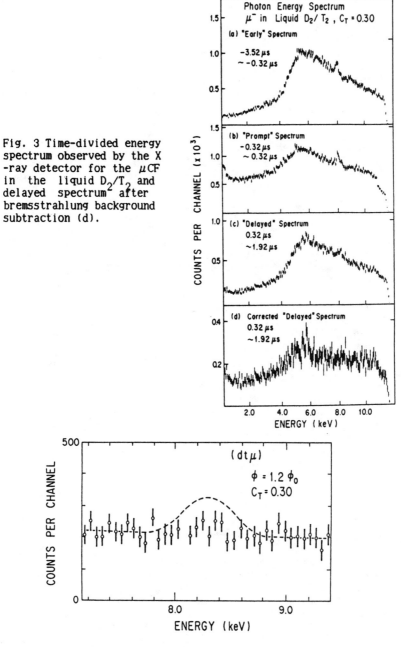

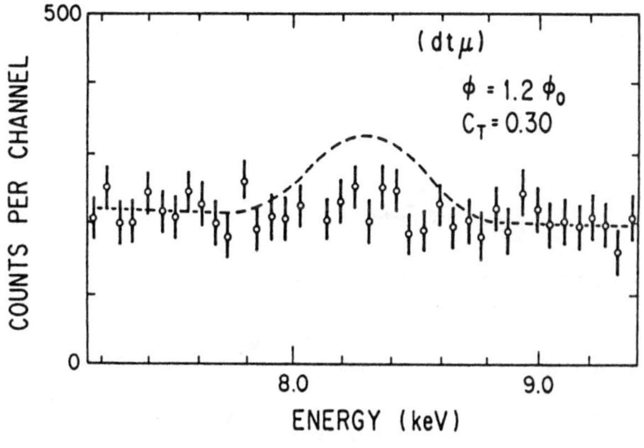

Fig. 4 Corrected "delayed" spectrum in the X-ray measurement. The curve is the expected energy spectrum which was calculated by us based upon the theory by Cohen[16] and which included Doppler broadening and counter resolution.

3.2 Considerations on the ω_s in dtμ-μCF and Towards the 1988 UT-MSL/KEK Experiment

In Fig. 6, the effective ω_s so far obtained in various types of experiments are summarized.[5] At the present stage, the following observation might be possible : a) there might be a slight density dependence when C_T is relatively high ; b) the observed small ω_s at high C_T can not be directly connected to the value obtained by the X-ray method at SIN, instead it seems to be reasonable to consider there exists a C_T-dependence in the ω_s at high density.

The small value of the ω_s derived in the present experiments needs a reasonable explanation. The C_T dependencce, if it is true at all, can be considered in the following way. Firstly, as suggested by Rafelski et.al[21], there might be an enhanced formations of $(\alpha\mu)_{2s}$ metastable state through the deexcitation of the 2p state by resonantly exciting the molecular level of the surrounding D_2 or T_2 (competing with Stark mixing effect). Secondly, the metastable $(\alpha\mu)_{2s}$ can be easily converted to $(t\mu)_{1s}$ by the change exchange collision, while $(d\mu)_{1s}$ can not be formed because of an energy-mismatch . If these processes occur simultaneously, we could expect the reduced ω_s value at high C_T and at high density.

On the other hand, the ϕ dependence observed in the neutron method cannot be explained by the theoretical considerations on the slowing-down kinematics of the $(\mu\alpha)^+$ atoms[16,17]. There seems to be a unexplained mechanism to enhance μ^- stripping from the $(\mu\alpha)^+$ atoms. Probably, the proposed C_T dependence might be correlated with this ϕ-dependence phenomena at high density of the D_2/T_2 mixture. In any case, precise measurements of the X-ray method for the ω_s and its ϕ- as well as C_T - dependence are urgently required.

At the UT-MSL/KEK pulsed muon facility, we are now extensively preparing the second X-ray experiment in the coming fall of 1988. The most important goal of the coming experiment is to gain high statistics for the X-ray intensity, at least to the level of 0.1 % in terms of the accuracy of ω_s. At the same time, the measurements will be extended to the C_T-dependence. Before the start of the run, various improvements had already been realized during these two years: a) the notorious Cu electronic K_α X-ray peak is completely reduced by adopting an Ag inner layer at the liquid target chamber; b) the X-ray background which existed after removing bremsstrahlung background was found to be coming from the fast neutrons and was now reduced to 1/3 of the value at the 1986 experiment by adopting the optimized shielding around X-ray detector; c) the D_2/T_2 target will be prepared at the JAERI which is located 50 km north of the KEK so that a flexibility of the experimental scheduling will be realized.

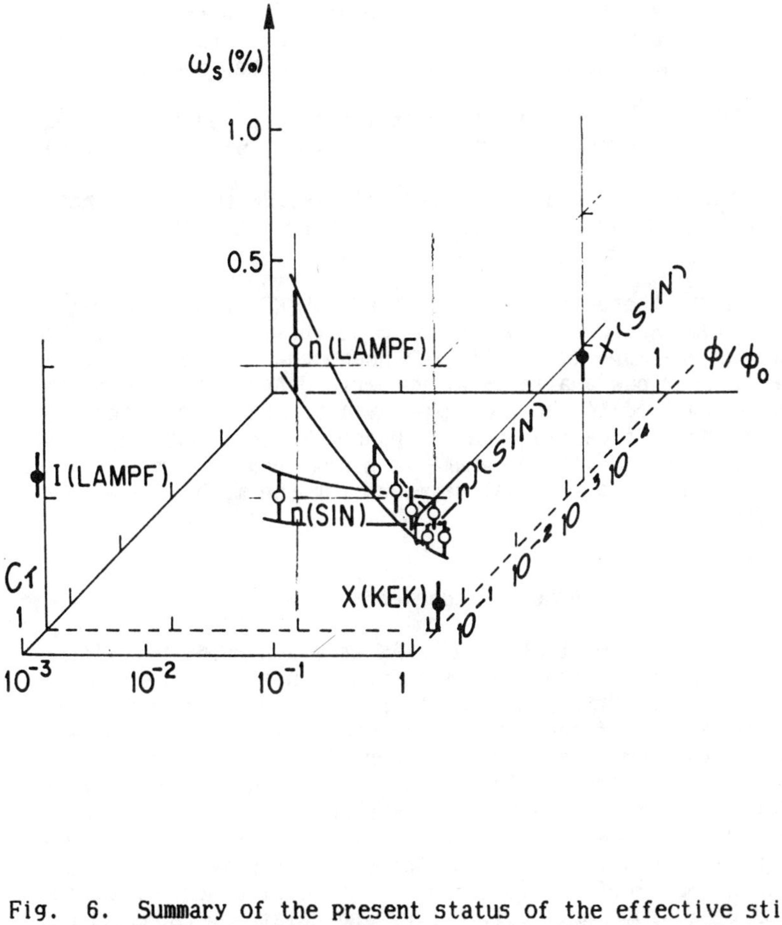

Fig. 6. Summary of the present status of the effective sticking probability for the μCF studies on the D_2/T_2 system. The neutron measurements are shown with white circles and the non-neutron measurements are shown with filled circles where I means the ion detection method and X means X-ray detection method.

4. CONCLUSION

Instead of summarizing the present manuscript, we would like to describe a list of possible data outputs during the 1988 experiments at UT-MSL/KEK.

1. The precise data of the life time and photon intensity for the radiative transition in the $(d^4He\mu)$ system, to clarify the existence of the hidden decay process from the muon-molecule state.

2. The radiative photon measurement for the $(d^3He\mu)$ system as well as the fusion-proton measurement for the μCF in $(d^3He\mu)$ system.

3. Satisfactory X-ray data for the $(dt\mu)$-μCF in liquid phase with the C_T of 0.30 and hopefully the corresponding data for the other C_T's.

Throughout these experiments at UT-MSL/KEK reported here, the pulsed muon source is now known to be very powerful tool to study various new aspects of the μCF phenomena. The more intense pulsed muon beam now available at Rutherford Appleton Laboratory will soon be applied to the μCF experiments. In Japan, the new accelerator project called Japanese Hadron Project (JHP) is now under serious considerations where we might expect drastically intense and high -quality negative muons for the much more advanced μCF studies.

Acknowledgement

The 1986 experiment at UT-MSL/KEK were achieved in collaboration with the following members to whom sincere thanks should be given ; Dr. K. Nishiyama, Prof. S.E. Jones and Dr. H. R. Maltrud and a part of the $(dHe\mu)$ experiment was assisted by Mr. H. Kurihara to whom we also give sincere thanks. We acknowledge the following persons for their kind helps and encouragements throughout the μCF experiments at UT-MSL/KEK: Prof. T. Yamazaki and Prof. H. Miyazawa and members of UT-MSL; Dr. T. Nishikawa, Prof. H. Sasaki, Prof. K. Katoh and other related members of KEK; Prof. A. Arima, Prof. T. Tominaga and associated members of the University of Tokyo; Dr. G.T. Garvey and related persons at Los Alamos National Laboratory. Helpful discussions with Dr. M. Leon are also acknowledged.

This work was supported in part by the Grant in Aid for Special Project Research on Meson Science of the Japanese Ministry of Education, Science and Culture of Japan.

REFERENCES

1. K. Nagamine, Hyp. Int. 8 787 (1981).
2. UT-MSL Newsletter 1-6 (1981-1986), eds. K. Nagamine and T. Yamazaki, unpublished, available from UT-MSL
3. Collected Papers on Muon Science Research at Meson Science Laboratory, University of Tokyo, (February, 1986), eds. K. Nagaimne and T. Yamazaki, unpublished, available from UT-MSL.
4. K. Nagamine, T. Matsuzaki, K. Ishida, Y. Hirata, Y. Watanabe, R. Kadono, Y. Miyake, K. Nishiyama, S.E. Jones and H.R. Maltrud, Muon Catalyzed Fusion 1, 137 (1987); K. Nagamine, Atomic Physics 10, 221 (1987)
5. T. Matsuzaki, K. Ishida, K. Nagamine, Y. Hirata and R. Kadono, Muon Catalyzed Fusion 2, 217 (1988)
6. K. Nagamine, T. Matsuzaki and K. Ishida, to be published in the Proceedings of the International Symposium on Muon and Pion Intersections with Matter (Dubna, July, 1987).
7. Yu. A. Aristov, A.V. Kravtsov, N.P. Popov, G.E. Solyakin, N.F. Truskova and M.P. Faifman, Yad Fiz. 33(1981)1066.
8. A.V. Kravtsov, N.P. Popov, G.E. Solyakin, Yu.A. Aristov, M.P. Faifman and N.F. Truskova, Phys. lett. 83A 379 (1984).
9. A.V. Kravtsov, A.I. Mikhailov and N.P. Popov, J. Phys. B: At. Mol. Phys. 19(1986)2579.
10. T. Kondow, privat, communication (1987).
11. S. Hara and T. Ishihara, private communication (1988).
12. M. Kamimura, contribution to this workshop.
13. D.V. Balin, A.I. Ilyin, E.M. Maev, V.P. Maleev, A.A. Markov, V.I. Medvedev, G.E. Petrov, L.B. Petrov, G.G. Semenchuk, Yu.V. Smirenin, A.A. Vorobyov and An.A. Vorobyov, Muon Catalyzed Fusion, 1, 127(1987)
14. S.E. Jones and M. Paciotti, contribution to this workshop.
15. H. Bossy, H. Daniel, F.J. Hartmann, W. Neumann, H.S. Plendl, G. Schmidt, T.von Egidy, W.H. Breunlich, M. Cargnelli, P. Kammel, J. Marton, N. Naegle, A. Scrinzi, J. Werner, J. Zmeskal and C. Petitjean, Muon Catalyzed Fusion 1, 115 (1987)
16. J.S. Cohen, Muon Catalyzed Fusion 1, 179 (1987); Phys. Rev. Lett. 58, 1407 (1987)
17. V.E. Markushin, Muon Catalyzed Fusion, 2, 395 (1988)
18. L.N. Bogdanova, L. Bracci, G. Fiorentini, S.S. Gerstein, V.E. Markushin, V.S. Melezhik, L.I. Menshikov and L.I. Ponomarev, Nucl. Phys. A454 653 (1985).
19. S.E. Jones et al., Phys. Rev. Lett. 56 588 (1986).
20. W.H. Breunlich et al., Phys. Rev. Lett. 58 329 (1987).
21. H. Rafelski, B. Mueller, J. Rafelski, D. Trautmann, R.D. Viollier and M. Daros Muon Catalyzed Fusion 1, 315 (1987)
22. Y. Akaishi, private communication.

FIRST DIRECT MEASUREMENT of $\alpha - \mu$ STICKING in $dt - \mu$CF

M.A. Paciotti, O.K. Baker, J.N. Bradbury, J.S. Cohen, M. Leon,
H.R. Maltrud, L.L. Sturgess,
Los Alamos National Laboratory, Los Alamos, N.M.,

S.E. Jones, P. Li, L.M. Rees, E.V. Sheely, J.K. Shurtleff, S. F. Taylor,
Brigham Young University, Provo, Utah,

A.N. Anderson,
Idaho Research, Boise, Idaho,

A.J. Caffrey, J.M. Zabriskie
Idaho National Engineering Laboratory, Idaho Falls, Idaho

F.D. Brooks, W.A. Cilliers, J.D. Davies, J.B.A. England, G.J. Pyle, G.T.A. Squier,
University of Birmingham, Chilton, England

A. Bertin, M. Bruschi, M. Piccinini, A. Vitale, A. Zoccoli,
University of Bologna, Bologna, Italy

V.R. Bom, C.W.E. van Eijk, H. de Haan,
Delft University of Technology, Holland, and

G.H. Eaton,
Rutherford-Appleton Laboratory,
Didcot, Oxfordshire.

ABSTRACT

Both $(\alpha\mu)^+$ and α particles have been observed in coincidence with fusion neutrons in a gaseous $D - T$ target at 2.8×10^{-3} liquid-hydrogen density. The initial muon sticking probability in muon-catalyzed $d - t$ fusion, measured directly for the first time, is $(0.80 \pm 0.15 \pm 0.12$ systematic)% in agreement with 'standard' theoretical calculations. However, this measured value does not support those theories that invoke special mechanisms to alter the initial sticking value.

INTRODUCTION

There has been a need for some time for a direct measurement of the $\alpha - \mu$ sticking probability in muon-catalyzed $d - t$ fusion.[1] The muon loss due to this sticking phenomenon is the most severe limitation to the ultimate fusion yield χ. The sticking probability, ω_s, has been inferred from the total muon loss rate after detailed corrections[2,3,4,5], and from x-ray measurements[6,7] plus cascade calculations[8,9]. Vorobyov expects that the LNPI direct ionization chamber method, so successful in $dd - \mu CF$, will work for measuring sticking in $dt - \mu CF$;[10] so far an upper limit of 1% comes from this work.[11]

The present paper describes the first direct measurement of the $\alpha-\mu$ sticking probability, using a low density ($\approx 10^{-3}$ lhd) $D-T$ mixture.[12] As such, it comes very close to measuring the initial sticking probability in $dt-\mu CF$. Extension of the this work is presently underway at Rutherford-Appleton Laboratory (RAL).[13]

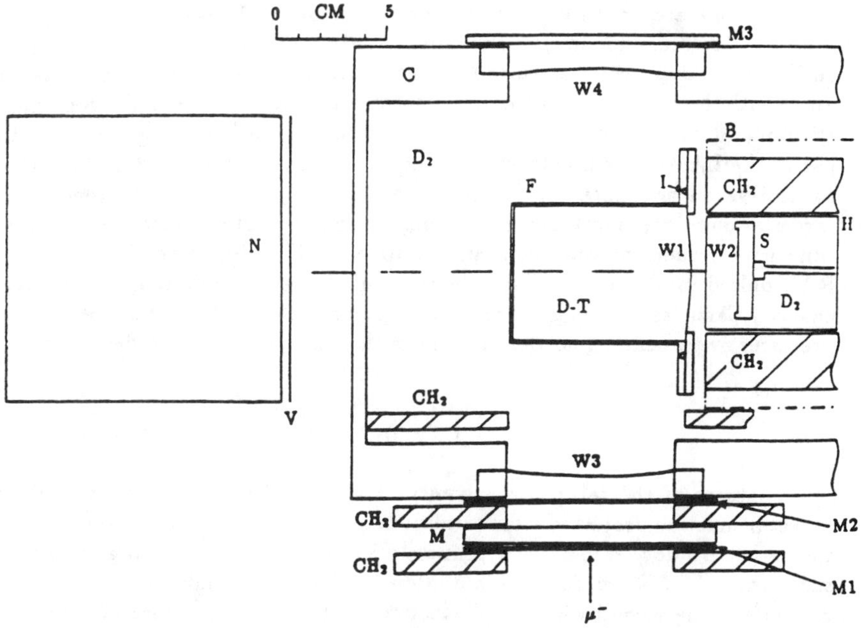

Fig. 1.

F:	Beryllium target flask, 6.4 cm diameter, 7.6 cm long; cylindrical walls are 1.5 mm thick; contains $D-T, C_t = 0.4$.
W1,W2:	1.5 micron mylar windows aluminized 1600 $\mathring{A}$ on T_2 side.
I:	Pure indium 'O'-ring sealing the target window.
S:	Silicon surface barrier detector, 1000 mm^2, 100 microns thick.
M1,M2,M3:	Muon telescope counters, each 1.6 mm thick
N:	Neutron counter, Bicron liquid scintillator (BC-501), 12.7 *cm* in diameter and 12.7 cm in depth; 1.6 mm veto counter in front.
B:	Dotted line indicates poletip of 1 kG permanent magnetic field.
C:	Secondary container for D_2, 12 l volume, Al walls, Lucite lid, all seals Viton 'O'-rings.
W3,W4:	Heat treated Al beam entrance and exit windows, 0.13 mm thick.
H:	Detector housing, sealed, cooled, and moveable; contains D_2.
M:	Moderator for slowing down 60-MeV/c μ^- beam.
V:	Charged particle veto counter.

The CONCEPT of the LAMPF EXPERIMENT

The $(\alpha\mu)^+$ ions produced by "sticking" events and the α-particles formed in the remaining 99^+ % majority of events are detected in coincidence with the 14.1 MeV neutron and are easily separable by range in the low pressure $D-T$ gas; the density is dictated by the very limited ranges of these ions. The fact that the ions are produced at 180° from the neutrons is useful for background rejection. Triggers due to (μ^-, pn) and $(\mu^-, p2n)$ captures in the target flask are substantially suppressed since the charged particle is not similarly correlated in angle with the neutron[14]. Beryllium, chosen for its good properties in containing tritium and its very low muon-capture probability, is the best target material for use in the LAMPF beam structure. The full μ^- beam is directed on the target to achieve an adequate event rate; many muons are therefore present in the target at one time, and there is then almost no possibility to measure the fusion time with respect to the muon arrival time. (This will be possible using the RAL pulsed muon beam.) Even though high-Z materials do not exhibit many charged particles in coincidence with neutrons, the high capture probability produces overwhelming singles rates in both neutron and silicon detectors.

LAYOUT

Features of the setup are shown in Fig. 1. Most important, the silicon detector must be protected from tritium beta radiation. T_2 diffuses out through the target window W1, limited primarily by the aluminum coating, at a measured rate of 2.5% per day and is diluted by the large volume of the secondary container. The second window W2 then keeps this dilute mixture at a distance from the detector where the intervening D_2 region is guarded by a magnetic field. The detector housing is sealed except for a long pressure-equilibrating capillary, necessary when the detector housing is cooled; cooling to −7° C improved the timing resolution from 3.5 to 3.0 ns. Target filling is challenging since the windows cannot support much differential pressure; as a molecular sieve cold trap cleans the incoming premixed $D-T$ gas, D_2 is admitted to the secondary container at a rate that maintains low differential pressure. Backgrounds are measured in an identical apparatus filled entirely with D_2 and normalized to incoming muons. A ^{233}U source, insertable between W1 and W2, gave identical energy and timing calibrations for each apparatus.

The character of the LAMPF experiment is revealed by typical rates; both peak rates during the LAMPF pulse and average rates are given in Table I. (During this particular run period, a thin production target caused the rates to be reduced by a factor of 3 below normal.) The neutron rate n is taken after pulse-shape discrimination[15]; the α rate includes noise; $\alpha\cdot(\gamma+n)$ is the trigger rate formed with a 150 ns coincidence width. Clearly the fusion process in not observable in either the neutron or the α-singles rate.

TABLE I. Typical rates.

RATE	*M1M2M3*	$\gamma + n$	n	α	$\alpha \cdot (\gamma + n)$
Peak/s	2.6×10^6	1.5×10^4	560	1440	2.8
Average/s	1.4×10^4	810	30	80	0.15

SYSTEMATIC EFFECTS

The factor 4 ratio between $(\alpha\mu)^+$ and α ranges was utilized by two different schemes; in an earlier experiment both ions were detected concurrently, while for the latest data, optimum detection of each ion required different fill pressures. Table II compares merits and systematic effects for each method. At 650 Torr the experiment is severely rate-limited and (at LAMPF) background-limited. We could not have been certain that $(\alpha\mu)^+$ had been seen without a higher density run. Using the dual pressure scheme, a strong $(\alpha\mu)^+$ signal is seen. However, the usefulness of the data may be limited by the systematic uncertainty of yield χ scaling with density ϕ. This scheme presents a good opportunity to measure stripping effects.

TABLE II. Systematic effects.

$(\alpha\mu)^+$ and α detected concurrently :

1) Single fill, $\phi = 1.0 \times 10^{-3}$ *lhd*, 650 Torr.
2) α's are collected from only 1/2 of target volume nearest the window.
3) μ^- stopping distribution must be well known.
4) α ranges must be known very well.
5) Gas impurity only affects the fusion yield χ , not the sticking result.
6) Slightly lower stripping since $(\alpha\mu)^+$ energies are higher.
7) No $t\mu$, $d\mu$ diffusion effect.
8) C_t only alters the yield χ.

Separate $D-T$ fills for $(\alpha\mu)^+$ and α observation :

1) Two fills, $\phi = 7.6 \times 10^{-4}$ (490 *Torr*)α and $\phi = 2.8 \times 10^{-3}$(1800 *Torr*)$(\alpha\mu)^+$.
2) Both ions are collected from the full target volume.
3) Assume muon stopping rate and distribution scales with density ϕ. The μ^- stopping distribution only comes into stripping and diffusion corrections.
4) Most uncertainty in range cancels out.
5) Gas impurity necessitates an important systematic correction.
6) Stripping is higher but still well known.
7) $t\mu$ and $d\mu$ may diffuse to the walls at the low density.
8) C_t change due to tritium diffusing out the window will alter the fusion yield at each pressure and thereby confuse the normalization.

A Monte Carlo code is useful for evaluating these systematic effects, but the experiment is not so complicated that the code is essential to the analysis. Figure 2 displays the distributions of fusion events in the dual-pressure scheme, reflecting principally the muon stopping distribution, detector solid angles, and particle range. α-particles originating near the back of the flask fall below the 0.7 MeV threshold and are not detected. The code finds a 93% active volume, whereas the $(\alpha\mu)^+$ active volume is 100%. According to the prescription, we could have used about 460 Torr to avoid this correction, but we favored instead a slightly higher density to obtain a higher χ. The overall efficiency is small (0.08% per stopped μ^-), independent of density, and, aside from the volume correction, cancels out of the sticking result.

Variation in the muon stopping intensity within the target volume was measured by counting the ^{56}Mn activation[16] of thin iron foils placed inside the non-tritiated target. Compared with the central maximum, the intensity falls to 1/2 along the axis near the window and the back wall. The predicted stop rate in hydrogen, based on these foils, was 4.7×10^{-4} μ^- stopped per incident μ^- at 490 Torr.

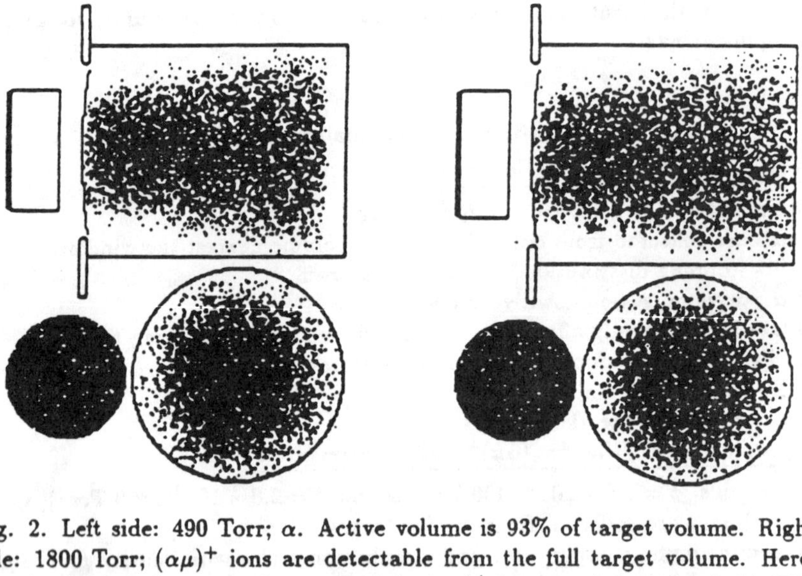

Fig. 2. Left side: 490 Torr; α. Active volume is 93% of target volume. Right side: 1800 Torr; $(\alpha\mu)^+$ ions are detectable from the full target volume. Here, the α's , having about 1/4 the range of the $(\alpha\mu)^+$ ions, are below threshold. The α locus at the detector and the projection along the target axis of the fusion distribution are shown below.

The Monte Carlo uses the Bichsel range code,[17] which was verified by degrading five α energies from the ^{233}U source through various gas thickness obtained by moving the detector. These checks were done between data runs utilizing the same windows and fill gas as fusion particles. Good consistency was obtained, and we conclude that the calculated range of a 3.5-MeV α degraded to the 0.7 MeV threshold was verified to about 2 mm.

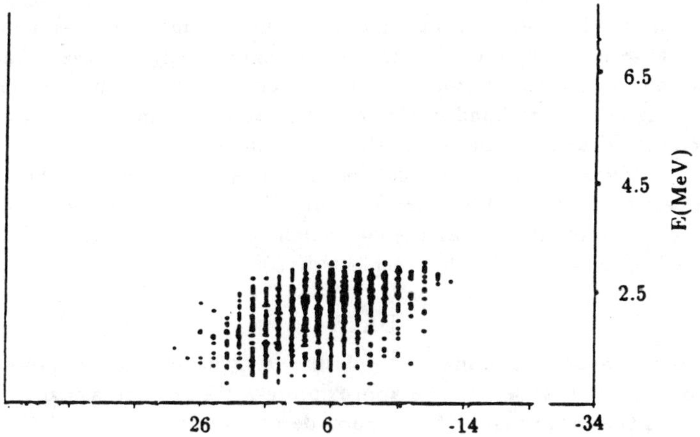

Fig. 3. Monte Carlo prediction for $(\alpha\mu)^+$ at 1800 Torr. Time vs energy correlation is evident. The higher energy $(\alpha\mu)^+$'s arrive early.

Corrections for stripping are easily and accurately given by the Monte Carlo using the energy-dependent stripping cross sections for $D-T$ gas, aluminum, mylar, and D_2 gas.[18] Table III lists stripping for the several materials assuming an average energy at each position. It can be used to assess the relative effect of the gas and of the windows, which are approximately equal.

TABLE III. Stripping probabilities.

STRIPPING MATERIAL	*STRIPPING PROBABILITY*
3.9 *cm avg.* $D-T$ *pathlength at* 1800 *Torr*	6.5%
1600 Å Al coating on target window	1.3%
1.5 *micron mylar target window*	4.0%
1.0 *cm* D_2 *at* 1800 *Torr*	1.0%
1600 Å Al coating on detector window	1.1%
1.5 *micron mylar detector window*	3.9%
1.5 *cm* D_2 *at* 1800 *Torr*	2.1%

However note that many of the stripped $(\alpha\mu)^+$ still have enough energy to be detected above threshold as α particles, particularly if the stripping occurs in the detector window or the last 1.5-cm D_2 region. The net result after the code has considered all of the above (stopping distribution, solid angle, energy loss, stripping at proper energies, and detection threshold) is that only 16% of the initially produced $(\alpha\mu)^+$ are unobserved. This effective stripping, $R_{eff} = 0.16$, is quite small considering that more efficient strippers such as aluminum and mylar have been introduced. Stripping is significantly less than that seen when the ion is allowed to stop fully in a medium[8], and convinces us that initial sticking is close at hand in this direct method. Verification of stripping calculations could be accomplished in these experiments, for example, by testing the effect of additional mylar. Reduced backgrounds at RAL will also be helpful. Fusion α particles generated between W1 and W2 could simulate $(\alpha\mu)^+$ events; however, C_t is so low there that the correction to ω_s^o is estimated to be much less than 1% of ω_s^o and is therefore neglected.

DATA

The pulse-shape discrimination[15] picture is given in Fig. 4 showing the location of the cut that selects the approximately 4% neutron signal from the remaining γ's that arise mainly from muon-decay electrons.

The α-data are presented first since the signal is so prominent (Fig. 5). Only the neutrons have been selected, and prompt muons have been rejected. The time difference between the α-ion and the fusion neutron is plotted along the abscissa while the ion energy is plotted along the ordinate. The box drawn shows the region where the α's are expected from Monte Carlo predictions. The position of the box along the time axis cannot be known from measurement, so the α-data themselves are used as a guide to positioning the box. $(\alpha\mu)^+$ ions are also expected in this plot, but with such low rate that background masks them.

The $(\alpha\mu)^+$ spectrum for all data taken at 1800 Torr is shown in Fig. 6, where the axes are the same as in Fig. 5. A quick comparison with its companion background plot shows a strong signal, but of course not as clean as the α data. The coincident background, which extends from threshold to well above the maximum $(\alpha\mu)^+$ energy, comes from from μ^- capture in the beryllium target flask; protons, deuterons, tritons, and alphas, are emitted in coincidence with neutrons. Non-coincident background, in the wings of the time distribution, originates from a variety of sources producing singles, including scattered muons which cannot be completely rejected and products from μ^- capture in beryllium, polyethylene, and aluminum.

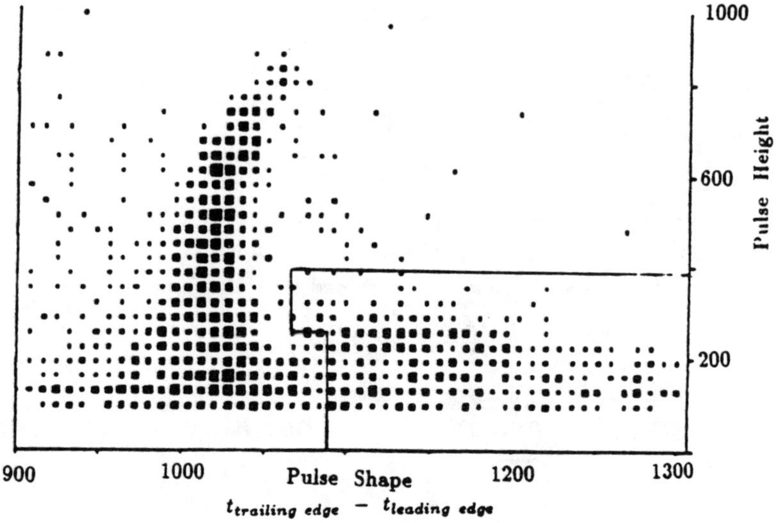

Fig. 4. Neutron separation by pulse shape.

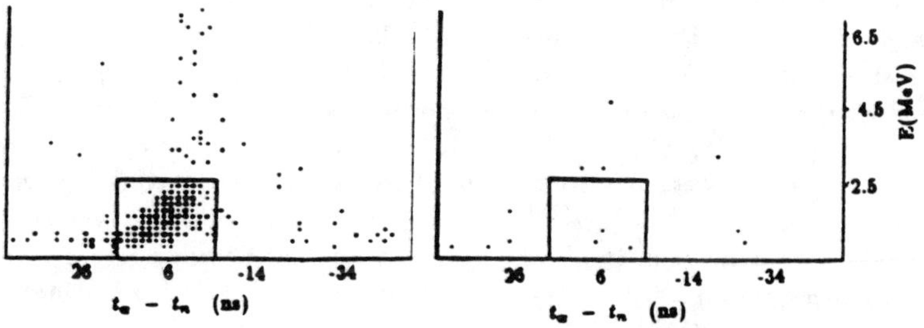

Fig. 5. α data on the left. Background on the right for 1/7 number of incident muons.

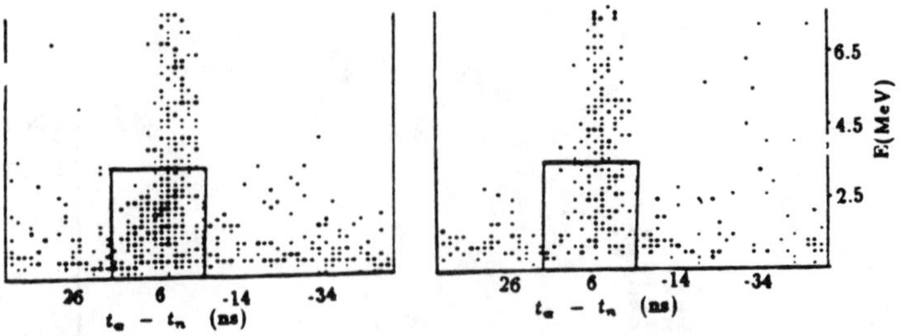

Fig. 6. The $(\alpha\mu)^+$ spectrum for all data taken at 1800 Torr. Background is shown on the right for 3/4 number of incident muons.

Table IV lists the raw numbers obtained from the data.

TABLE IV. Raw Data.

PRESSURE	*REGION*	*COUNTS*	*M1.M2.M3*
1800 *Torr*	$(\alpha\mu)^+$	115	16.7×10^9
1800 *Torr*	*Background*	90	29.1×10^9
490 *Torr*	α	295	7.2×10^9
490 *Torr*	*Background*	3	1.05×10^9

PURITY OF $D-T$ GAS

An anticipated source of trouble was impurities in the $D - T$ mixture. (See table II.) The mylar window precludes a high-temperature bakeout of the target, and there just is not enough gas to overpower small fixed amounts of contaminants as there is for the high-pressure targets.[12] The procedure used was to fill first to 1800 Torr and to collect $(\alpha\mu)^+$ data, then to bleed the gas pressure down to 490 Torr and collect α data. In this way, the same gas was used for the measurement and for the normalization. Time-dependent evolution of contaminants from the walls by tritium is not prevented by this procedure, but none was evident. A second 1800-Torr data run produced 30% more $(\alpha\mu)^+$ than did the first 1800-Torr fill when normalized to entering muons while no changes were evident in the background. It was not possible to accomplish the bleed procedure for this fill, and so confident normalization is lacking for more than 1/2 of the $(\alpha\mu)^+$ counts in Fig. 6. The direction of the observed effect is consistent with a cleaner mixture being obtained for the second fill following tritium scouring of the surfaces occuring during the first fill. Contamination of the D_2 in the secondary container could also contaminate the target gas by diffusion through the window; we expect less such impurity for a second fill. Although the increase is not statistically conclusive, its direction reinforces our belief that it is due to a change in gas properties. Consequently, only the first set of runs at 1800 Torr were used in deriving the sticking probability, the data appearing in Table IV. If impurity did cause the change in χ, we estimate about 600 ppm (with the Z of nitrogen) was present.

RESULTS

The initial sticking probability is given by

$$\omega_s^o = \frac{N_{\alpha\mu}}{N_\alpha + N_{\alpha\mu}}. \qquad (1)$$

Here

$$N_{\alpha\mu} = \frac{n_{\alpha\mu}}{N\phi^2(1 - R_{eff})}, \qquad (2a)$$

$$N_\alpha = \frac{n_\alpha}{N\phi^2\ 0.93} \qquad (2b)$$

where n_α ($n_{\alpha\mu}$) is the number of doubly (singly) charged ions detected, R_{eff} is the effective stripping, and 0.93 the active volume for α. N is the number of incident muons in each case (normalization to the high-energy coincident background was shown to be equivalent). We assume for now that both the number of μ^- stopped and the yield χ are proportional to ϕ. (See the next section.) The result is $\omega_s^o = (0.80 \pm 0.14)\%$ (statistical error).

ω_s^o measured in the 650-Torr experiment is $(0.91 \pm 0.3)\%$ (statistical error).[19] The error is larger since fewer $(\alpha\mu)^+$ were observed in the presence of substantial background.

SYSTEMATIC UNCERTAINTY

We now examine systematic uncertainties for items in Table II listed under "*Separate D − T fills for* $(\alpha\mu)^+$ *and* α". The $t\mu$ and $d\mu$ diffusion distances are limited by the muon lifetime at such low pressures and could become somewhat longer due to diffusion while the atoms are still epithermal. We estimate this distance to be about 1 cm at 490 Torr, somewhat smaller than the dimensions of the target. No correction is made at this time; generalization of Cohen's solution to the time-dependent Boltzmann equation[20] to get the spatial extent would be appropriate for a good estimate.

The most important uncertainty is a consequence of the use of two different densities in making the measurements. At our densities thermalization times equal or exceed the muon lifetime. Therefore most of the molecular formation will occur in the epithermal (transient) region where $\lambda_{dt\mu}$ is rapidly changing, and the yield χ may then not scale with density as assumed above. Table V attempts to outline the scope of this problem. Thermalization times, as well as average temperature and remaining triplet fraction at a relevant time of 2 μs (all from ref. 20) tell us that at the lowest density the $t\mu$ atoms remain hot, and the triplet quenching is not complete. The muon lifetime then selects rather different slices of the epithermal transient for each density. (See survey of experimental low-density transients in refs. 21 and 22 and calculated transient effects in ref. 23)

TABLE V. Density Effect.

QUANTITY	490 *Torr*	1800 *Torr*
Density ϕ	7.6×10^{-4}	2.8×10^{-3}
Thermalization time	6 μs	1.6 μs
Average temperature at 2 μs	540K	340K
$t\mu$ *triplet fraction at* 2 μs	0.33	0.04
Average $\lambda_{dt\mu}$	1.5×10^8	1.3×10^8
Extrapolated q_{1s} *from ref.*26	0.77	0.66

Next, let's see how large the effect is likely to be. Turning to Leon[24] to estimate singlet and triplet formation rates for our mixture at the two average temperatures, we find a higher rate for the less-thermalized, low-density case. The rates given in the Table V do not include screening effects,[25] but these will not alter the relative magnitudes. Nor should extension to the theory of direct molecular formation[23] seriously alter the relative magnitudes since the molecular-formation rates as a function of temperature still have similar shapes.

Other quantities remaining for discussion are q_{1s} and λ_{10}. An extrapolation of the Menshikov and Ponomarev q_{1s} is presented in Table V for each density[26];

here the strongest density dependence is predicted at low density. The direction of the q_{1s} density dependence is to offset the increased formation rate expected at the low density. In this simplified treatment, the 0.33 triplet fraction remaining in the 490 Torr sample has little effect on the conclusion; in the event that λ_{10} is actually larger than calculated[27], the triplet fraction could approach zero without largely affecting the $\lambda_{dt\mu}$ listed in the table.

Some comments are in order: 1) Expected epithermal enhancement of $\lambda_{dt\mu}$ makes the q_{1s} density dependence more important in determining yield at $C_t = 0.4$. Hence the RAL experiment may find a larger optimum C_t where the higher yield would be welcome. 2) λ_{10} effects would likely enter in a important way if a proper evaluation over the complete epithermal peak were done. 3) Thermalization could be more rapid than reported. 4) We should pay attention to the plunging cycle rates at low density,[4,5] keeping in mind that the lowest-density points (ϕ=1%) are already heavily into the epithermal region. But simply taking $\chi = \phi\lambda_c/\lambda_o = 0.08/\mu^-$ (450 Torr) with λ_c at $45/\mu s$ gives a predicted rate for *our* experiment that agrees with the absolute number of muon stops in the mixture found by foil activation. Unfortunately the foil test disagrees badly (by a factor of 2) with the number of stops estimated from beam properties. We might otherwise have hoped to bracket the extent of the epithermal enhancement at low densities (subject to assumptions about gas purity). 5) We anticipate that the RAL pulsed-beam data will add to the understanding of this low density region and answer some of the questions raised by the LAMPF experiment.

The initial C_t of 0.4 falls with time, and ionization chamber measurements indicate C_t dropped to 0.38 at the end of the data collection on the first 1800 Torr fill and to 0.37 at the end of the 490-Torr data collection. This small change is significant only if the optimum C_t is much higher so that 0.4 is on the rising edge of the λ_c vs C_t curve instead of at the plateau where we intended it to be.

Variations in ω_s^o depending on cuts used have been evaluated in the thesis of Li[19]; the rms scatter amounts to about 0.06% which we add to the statistical error. Based on the reliability of the above assumptions, we believe it unlikely that the density effect will alter ω_s^o by more than 15%. Accordingly a systematic uncertainty is quoted with the result: $\omega_s^o = (0.80 \pm 0.15 \pm 0.12$ systematic)% Subsequent experiments or calculations on the density effect can be used to correct this result.

The 650-Torr sticking measurement, as outlined in Table II, does not contain uncertainly due to the density effect, and its consistent value of (0.91 ± 0.3)% (statistical error), adds confidence to our conclusion.

CONCLUSION

In conclusion, we report a measurement of the initial $\alpha - \mu$ sticking probability in muon catalyzed $d - t$ fusion at low density. ω_s^o, measured directly for the first time, is $(0.80 \pm 0.15 \pm 0.12$ systematic)% in agreement with 'standard' theoretical calculations[28,29,30] and not supporting those theories that invoke special mechanisms to alter the initial sticking.[31,32,33] The reported value contains only a 16% correction to the observed sticking due to stripping. Additional experiments are underway at the Rutherford-Appleton Laboratory where a high-Z target in the pulsed muon beam produces lower backgrounds. Direct normalization to neutron singles, transient observation, and reactivation tests should be possible.

ACKNOWLEDGEMENTS

We wish to thank R. H. Sherman of the Los Alamos National Laboratory for making a test of the cleanliness in tritium of the thin mylar coated with aluminum. Little impurity was seen in his Raman spectrometer, giving us confidence to proceed.

This work is supported by the U. S. Department of Energy Division of Advanced Energy Projects.

REFERENCES

1. Panel discussion on the Future of μCF, C. Petitjean Chairman, Muon Cat. Fusion 1, 391 (1987)
2. S.E. Jones *et al.*, Phys. Rev. Lett. 56, 588, (1986).
3. C. Petitjean *et al.*, Muon Cat. Fusion 1, 89 (1987).
4. C. Petitjean *et al.*, Muon Cat. Fusion 2 37 (1988).
5. See both W. H. Breunlich and C. Petitjean, μCF Workshop, Sanibel Island (1988).
6. H. Bossy *et al.*, Phys. Rev. Lett. 59, 2864, (1987).
7. K. Nagamine *et al.*, Muon Cat. Fusion 1, 137 (1987).
8. J.S. Cohen, Phys. Rev. Lett. 58, 1407 (1987).
9. V. E. Markushin, Muon Cat. Fusion 3, 395 (1988).
10. A. A. Vorobyov, Muon Cat. Fusion 2 17 (1988).
11. G. Semenchuk, private communication (1986).
12. S. E. Jones *et al.*, Muon Cat. Fusion 1, 121 (1987).
13. J. Davies, μCF Workshop, Sanibel Island (1988).
14. N. C. Mukhopadhyay, Phys. Reports 30, 98, (1977).
15. A. J. Caffrey *et al.*, Muon Cat. Fusion 1, 53 (1987).
16. G. Heusser and T. Kirsten, Nucl Phys. A195, 369 (1972).
17. H. Bichsel, Private communication.
18. J.S. Cohen, Phys. Rev. A37, 2343 (1988).
19. P. Li, Thesis, Brigham Young University, unpublished (1988).
20. J.S. Cohen, Phys. Rev. A34, 2719 (1986).
21. W. H. Breunlich *et al.*, Muon Cat. Fusion 1, 67 (1987).
22. C. Petitjean *et al.*, Muon Cat. Fusion 2 37 (1988).
23. J. S. Cohen and M. Leon, Phys. Rev. Lett. 55, 52 (1985).

24. M. Leon, Phys. Rev. Lett. 52, 605(1984) and corrected figures in 52, 1655 (1984) (Caution is advised due to the emergence of the sub-threshold resonances as strong contributors to the molecular formation).
25. J. S. Cohen and R. L. Martin, Phys. Rev. Lett. 53, 738 (1984).
26. L. I. Menshikov and L. I. Ponomarev, Pisma Zh. Eksp. Teor. Fiz. 39, 542 (1984) [Sov. Phys. JETP Lett. 39, 663 (1984)].
27. M. Leon, μCF Workshop, Sanibel Island (1988).
28. L. N. Bogdanova *et al.*, Nucl Phys. *A454*, 653 (1986).
29. D. Ceperley and B. J. Alder, Phys. Rev. *A31*, 1999 (1985).
30. Chi-Yu Hu, Phys. Rev. *A34*, 2536 (1986).
31. J. Rafelski and B. Müller, Phys. Lett. 164*B*, 223 (1985).
32. M. Danos, B. Müller, and J. Rafelski, Muon Cat. Fusion 3 (1988).
33. M. Danos, L.C. Biedenharn, A. Stahlhofen, , μCF Workshop, Sanibel Island (1988).

μCF THOUGHTS FROM BIRMINGHAM AND THE RUTHERFORD APPLETON LABORATORY

J.D. Davies, F.D. Brooks, W.A. Cilliers,
J.B.A. England, G.J. Pyle, G.T.A. Squier,
University of Birmingham, U.K.

A. Bertin, M. Bruschi, M. Piccinini, A. Vitale, A. Zoccoli,
University of Bologna, Bologna, Italy

S.E. Jones, P. Li, L.M. Rees, E.V. Sheeley,
J.K. Shurtleff, S.F. Taylor
Brigham Young University, Provo, Utah, USA

G.H. Eaton
Rutherford Appleton Laboratory, U.K.

B. Alper
UKAEA Culham Laboratory, U.K.

V.R. Bom, C.W.E. van Eijk, H. de Haan
Delft University of Technology, Holland

A.N. Anderson
Idaho Research, Boise, Idaho, USA

A.J. Caffrey, J. Zabriskie
Idaho National Engineering Laboratory, Idaho Falls, Idaho, USA

M.A. Paciotti, O.K. Baker, J.N. Bradbury, J.S. Cohen, M. Leon,
H.R. Maltrud, L.N. Sturgess
Los Alamos National Laboratory, Los Alamos, N.M. 87545, USA

ABSTRACT

This paper gives some ideas to be learnt from magnetic confinement fusion and briefly describes the pulsed muon beam at ISIS, progress with the measurement of W_s(the μα sticking coefficient), future beam plans and possible experiments.

INTRODUCTION

μCF has had considerable progress; its continuation requires new ideas to be widely discussed at workshops like this. Thought must be given to collective and critical planning, exploiting resources, expanding funding and to acquainting those in authority or having influence, non readers of the specialised science publications, of μCF. Inter alia, there have been general articles in Nature, the New Scientist and Scientific American. Amusingly it was the latter that led to articles in the Economist, the (British) Times and BBC radio.

IDEAS FROM MAGNETIC CONFINEMENT FUSION

This subject was investigated last November at JET (Joint European Torus and 20km from ISIS) in a STOA Workshop of the European Parliament and EEC. Significant points included there no longer being an urgent need for fusion. Although magnetic confinement research has made considerable expensive progress, it has technical problems and a goal in the distant, costly future. Nevertheless it should not be abandoned but slowed and savings made. However research is not viable at less than the current European spend. The USA STARFIRE had reached similar conclusions. Mention was made of ITER - the start of a joint European, Japanese, USA, USSR project. The lesson for μCF is obvious - we must plan a world programme. Invited as an alternative method of fusion, μCF was presented as interesting physics with basic parameters made with relatively small scale, low cost equipment divorced from the large, expensive apparatus, the accelerator, which was provided anyway for other purposes. Progress was rapid with 'a low cost, long shot' possibility of energy production. Support for ISIS experiments by Culham, the British fusion centre, was mentioned.

In the future 550 fusions/μ could provide energy break-even, heat → heat. If ≥ 20% of the power output is required to drive the complex then the financial cost of such becomes too high. Therefore, in some sense, economic break-even requires 2500 fusions/μ .

THE ISIS PULSED MUON BEAM

The synchrotron sends protons in 50Hz pulses to make neutrons by spallation and fast fission at a distant uranium target: each pulse contains 2 bunches, 330ns apart. Recently the complex has run for appreciable periods at 750 Mev and 90 μa. The surface/cloud muon beam[1,2] is taken off a thin, intermediate transmission target. The raison-d'être of ISIS, so far, is to appear as a pulsed reactor; so the aims are long, steady runs and very low backgrounds. There is a very good cave around the muon target from which leak-paths are enthusiastically sought and blocked. Improved exit collimation from the cave to protect downstream magnets has also decreased external backgrounds.

The performance of the muon beam is given in table 1; ± 55kv on the velocity separator and a simple collimator reduced the electron contamination by x50. With little advertising for its first full year, the beam has been oversubscribed by x2½, mainly μSR experiments. A problem with having no switchyard is interchanging experiments.

A DIRECT MEASURE OF W_s, THE μα STICKING COEFFICIENT

This continues the LAMPF[3] determination of the ratio of $\frac{\mu\alpha - n}{\alpha - n}$

coincidences with a small, low density gas target described at the workshop. That experiment had a large 'α'-n background from the > 99.9% of the μ^- that stopped in the target walls with general accelerator background contributing to the high neutron count rate. Data were taken at 490mm Hg target pressure to see α-n coincidences and at 1800mm for μα-n since the event rate increased approximately as the square of the D-T density.

These backgrounds could be avoided by using a pulsed beam, target walls having high Z and delaying counting by ½ μs. As the atomic number of an elemental stopping medium is increased the effective muon life-time falls from $\tau=2$ μs for Z ~ 6 to 100ns for Z ~ 48 where it then flattens.

Figure 1 shows the apparatus modified for use at RAL. The target flask is made from silver since this is the structural material having τ_μ (effective) ≤ 100ns and of the lowest Z to reduce showering. There are 2 NE213 neutron counters to improve timing and the Si surface barrier α-detector was thinned to 50μ to reduce background.

The neutron backgrounds were examined as functions of energy and time after the muon pulse. That which was non muon-induced became softer with increasing time and came principally from the muon production target. The muon-induced background had an energy spectrum of time-independent shape but magnitude decaying with the τ_μ ~ 1μs characteristic of μ^--Al interactions. So the beam-pipe end-flange and target outer vessel were covered with 3mm of Pb. For E_n > 3 Mev this was still the dominant background and tests indicated an origin in the downstream beam-pipe - this will be lined with copper in future.

There is a tremendous flux of particles and energy during the beam burst - principally from the e/π contamination. Inhibiting each of the signal paths inside the LINK n/γ discriminator protected the neutron channel. Little background can get above the 0.6 Mev α counter threshold and pile-up was measured at 1½%. However the very many sub-threshold pulses during the burst cause a 50ns time walk and jitter in the pre-amp. If this cannot be inhibited then a thin scintillator will be used to take advantage of a photo-multiplier's stability.

μα-n events at 1520mm Hg target pressure and α-n coincidences at 765 and 490mm were clearly seen in the raw, on-line data as were fusion neutron singles at the same densities. The neutron data should provide the cross normalisation between high and low pressure required for this and the LAMPF experiments and also explore this region of epithermal production.

THE FUTURE

A grant application is being prepared for a kicker magnet to spatially separate the muon bunches; this will double the number of completed μSR experiments, as they use only a fraction of the available beam, and considerably improve their frequency range. The undeflected beam position would then be available for more permanent experiments such as μCF.

There are space and plans for a purpose-built muon beam on the other side of the proton beam and sharing the production target; a switch-yard would reduce background and permit the beam to alternate between 'permanent' experiments. Table 1 also gives the parameters of a 'conventional' decay channel at ISIS with a superconducting solenoid. The total fluxes of decay and cloud beams are comparable. As μCF experiments use only a small fraction of the μ^- stopping volume then the cloud beam would give much higher event rates and better signal to noise because of its smaller $\frac{\Delta p}{p}$ and spot size. The major advantage of the decay channel is the negligible π^- and e^- contamination (the latter only with an ultra-thin, solenoid window) n.b. a π^- gives 140 Mev of background but a large fraction of the muon mass goes into the neutrino. For the present beam a recently installed double collimator should enable higher separator volts to reduce the e^- contamination by much more than x50. Next year we aim to considerably reduce the π^- contamination with a thin degrader at an upstream focus at the expense of $\sim$ 25% increase in $\frac{\Delta p}{p}$. The choice of beam is open.

SOME POSSIBLE EXPERIMENTS

These are mentioned to illustrate the power of the facility.

μ^3He and μ^4He cascade x-rays

(A) Scavenging by He from excited (μ-hydrogen isotope) atoms can be distinguished from that via the ground or meso-molecular state; the former will dominate during the burst and the latter afterwards. Suppression of e/π contamination from the target and charged particles from the detector - GSPC or Compton suppressed SiLi - by a solenoid magnet will be required.

(B) The state of the 'stuck' μα, following D-T fusion, as it slows may be followed from the changes in yield of several spectral lines as functions of density.

Exploration of epithermal production and low temperature plasma.

The 'μt' intermediate state is created with 19ev and its thermalisation is sufficiently slowed at low densities that it spends significant time with energies equivalent to the required high temperatures. Neutron singles will explore $10 \rightarrow \frac{1}{2}\,\rho_{STP}$ and α-n coincidences densities below $1\rho_{STP}$.

Low temperature plasma.

This can be achieved with the pulsed, Θ-pinch of a solid D-T filament.

Hyperfine studies

A pulsed CO laser could induce $(\mu t)_s \rightarrow (\mu t)_t$ with polarized $(\mu t)_t$ being detected by transverse μSR.

REFERENCES

1. G. Eaton et al., NIM A269, 483 (1988).
2. F.D. Brooks et al., Muon Catalyzed Fusion 2, 85 (1988).
3. See M.A. Paciotti, these proceedings.

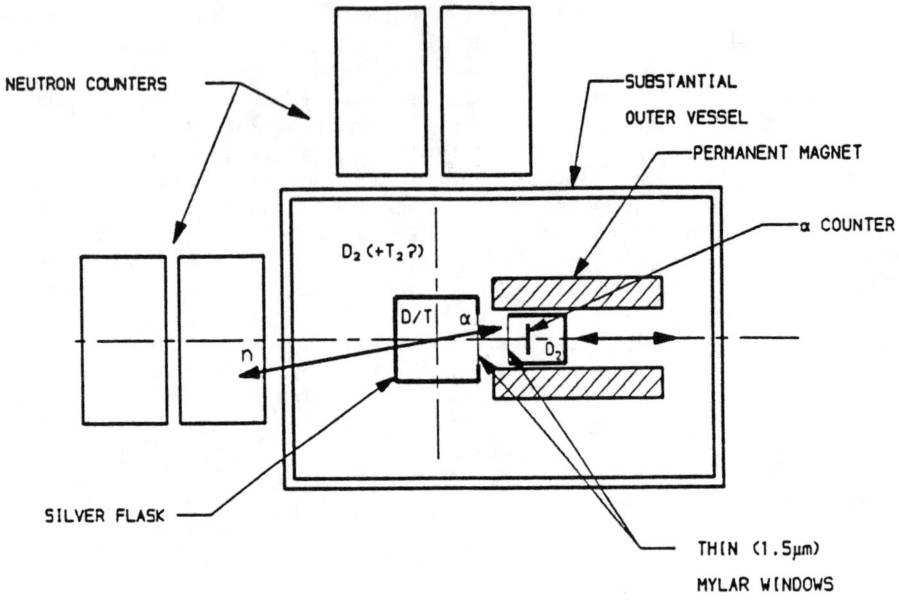

Fig. 1. Apparatus for measuring W_s.

Table 1

Column 1 - Performance of the ISIS surface/cloud pulsed muon beam
Column 2 - Calculate performance of a decay beam using an existing superconductor solenoid.
Both from a 1cm production target and 90 μa protons.

		Surface/Cloud		Decay
Beam spot		$2x3cm^2$	FWHM	$6x6cm^2$
$\Delta p/p$		5%(cloud)		12%
Intensity	29	$\sim 1.5\ 10^6 \mu^+/s$		negligible
at MeV/c	40	$6\ \ 10^4 \mu^-/s$		30% less
	60	$1.5\ 10^5 \mu^-/s$		equal
	100	"cloud" 30% less than "decay"		
π contam.		between 3:1 and		very low
e contam.		1:1 π or e : μ		low (?)
Polarized μ		at 29 Mev/c only		yes

INVESTIGATION OF Q_{1s}

A. N. Anderson

Introduction.

The probability Q_{1s} of a muon reaching the ground state of a $d\mu$ atom before transferring to a triton is an important parameter in the muon-catalyzed fusion cycle. It has proved very difficult to extract from experimental data because it varies strongly with both tritium concentration C_t, and density ϕ, and is combined with several other parameters in the steady-state neutron production (cycling) rate, λ_c, for which, for this discussion, we assume the following form:

$$\frac{\phi}{\lambda_c} = \frac{C_d Q_{1s}}{C_t \lambda_{dt}} + \frac{3}{4\chi} + \frac{1/4 + 3/4\frac{C_t\lambda_{10}}{\chi}}{C_{D_2}\lambda^0_{dt\mu-d} + C_{DT}\lambda^0_{dt\mu-t}} \tag{1}$$

with

$$\chi = C_t\lambda_{10} + C_{D_2}\lambda^1_{dt\mu-d} + C_{DT}\lambda^1_{dt\mu-t}$$

where λ_{dt} is the rate of μ transfer to t from the $d\mu$ ground state, λ_{10} is the rate for $t\mu$ triplet-to-singlet hyperfine quenching, and $\lambda^x_{dt\mu-z}$ is the mesomolecular formation rate from hyperfine state x with spectator atom z.

Leon [1] has predicted that at temperatures below $200K$, all of the resonant mesomolecular formation rates $\lambda_{dt\mu}$ except $\lambda^0_{dt\mu-d}$ are zero, so if we restrict our attention to runs at $100K$, the above expression simplifies to

$$\phi\tau_c = \frac{C_d}{C_t}Q_{1s}\tau_{dt} + \frac{3}{4C_t}\tau_{10} + \frac{1}{C_{D_2}}\tau^0_{dt\mu-d} \tag{2}$$

where terms $1/\lambda$ have been replaced by the equivalent expression, the time constant τ.

Consider the following experimental data from LAMPF, all at $\phi = .72$ LHD:

C_t	$\lambda_c(\mu s^{-1})$
.04	$9.6 \pm .2$
.10	$22 \pm .8$
.25	57 ± 1.0
.40	72 ± 1.1
.70	$32 \pm .8$

First, note that at $C_t = .7$, τ_c is relatively insensitive to Q_{1s}, so that we can set reasonably good limits on $\tau^0_{dt\mu-d}$ by assuming Qos is monotonic decreasing, that is, by limiting $Q_{1s}(.7)$ to the interval $0 \leq Q_{1s}(.7) \leq Q_{1s}(.4)$. Applying these two limits in turn generates two instances of equation (2), which can be solved to eliminate $\tau^0_{dt\mu-d}$, leaving only linear expressions relating $Q_{1s}\tau_{dt}$ and τ_{10}.

This analysis still requires an accurate estimate of *cd*. Although the data were taken at $100K$, the mixtures had been cooled rapidly from room temperature, and might be expected to be still at high-temperature equilibrium. The analysis was performed assuming equilibration at both $100K$ and $300K$, and shows little dependence on equilibration. The figures here all assume room-temperature equilibration.

Figure 1. shows the resulting allowed regions for $Q_{1s}\tau_{dt}$ as a function of τ_{10}, and the insensitivity of lds to both $Q0s$ and lq.

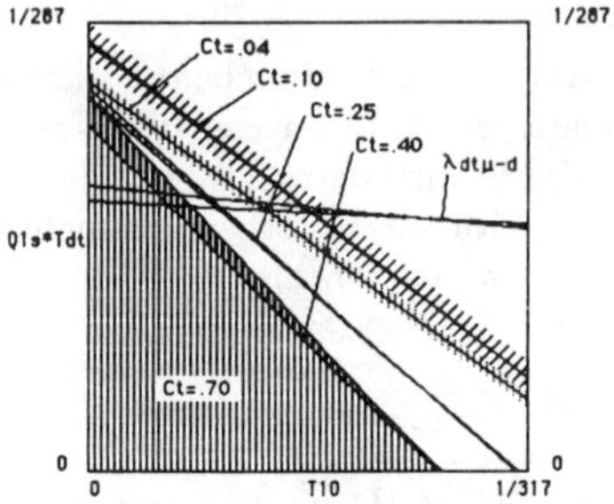

Figure 1. Regions for $Q_{1s}\tau_{dt}$ and $\lambda^0_{dt\mu-d}$ allowed as a function of λ_{10} at ϕ = .72LHD and T = $100K$. Times in μs. (equilib. at $300K$)

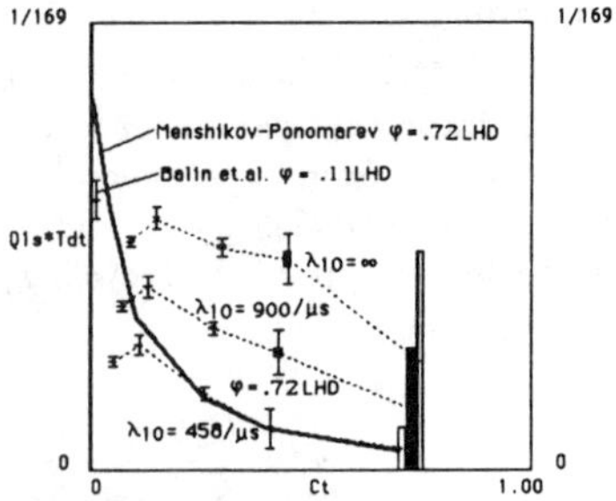

Figure 2. $Q_{1s}\tau_{dt} vs. C_t$ for various values of λ_{10}'s

The upper boundary of each C_t region corresponds to $Q_{1s}(.7) = Q_{1s}(.4)$ and the lower corresponds to $Q_{1s}(.7) = 0$. The regions for $C_t = .04$ and $C_t = .1$ are tightly constrained and appear as single lines. The hash marks around those regions represent the statistical error for the region.

Figure 2. shows the dependence of Q_{1s} on C_t for three choices of λ_{10}, corresponding to the best fit for the theoretical curve of Menshikov and Ponomarev (M-P) [2] (458), the theoretical value of λ_{10}, (900) [3], and the extreme (∞) suggested by the lack of transients in λ_c at high density observed in the LAMPF data. Each data point in Figure 2. consists of a rectangle indicating the relevant slice of the corresponding region in Figure 1., with error bars added accordingly.

Figure 2. also shows the experimental value $\lambda_{dt} = 2{*}0 \pm 20\mu s^{-1}$ claimed at $C_t = .0124$ by Balin et.al. [4] Although measured at a density of 84 *atm* and at room temperature, this number should vary only slightly with density or temperature.

Note that although the Balin value claims to be a measure of λ_{dt} alone, its extraction made no allowances for either epithermal effects or hyperfine quenching, either of which could affect their result.

The curves of Figure 2. indicate a sensitivity which in principle ought to allow the ambiguity in λ_{10} and Q_{1s} to be resolved. At first glance, the M-P fit at $\lambda_{10} = 447\mu s^{-1}$ seems like a good possibility, since it comes close to the value of Balin. Since Q_{1s} should be at least monotonic decreasing, it is obvious that the point at $C_t = .04$ has some systematic error, so perhaps it should be discarded. However, in order to fit M-P , λ_c for that point would need to be $\approx 5\mu s^{-1}$, or only *half* its observed value. It should be pointed out that the point at $C_t = .04$ is from a much earlier run than the others and thus subject to different systematics.

A value of $447\mu s^{-1}$ for λ_{10} should cause an *increasing* transient in λ_c, (i.e. a buildup) which, although difficult to see at high density, should be apparent at low density. Such a transient has not been seen [5], but that could be because it is combined with the *decreasing* epithermal transient which has been observed.

The existence of epithermal components raises another complication with this analysis. Although at high density the epithermal transients due to initial capture or excited-state transfer to t's are suppressed, there might still be an epithermal $\lambda_{dt\mu-t}$ contribution to λ_c from d to t transfer, which implies that the assumption that only $\lambda^0_{dt\mu-d}$ is nonzero may be invalid. Even a slight component of $\lambda^0_{dt\mu-t}$ can have an enormous effect on estimates of Q_{1s}, as shown by Figs. 4 and 5, in which a 5% contribution from $\lambda^0_{dt\mu-t}$ is assumed. Previous global fits to the LAMPF data have indeed suggested that $\lambda_{dt\mu-t}$ stays slightly nonzero for the entire interval $0K \leq T \leq 200K$.

How sensitive the above analysis can be to the value of $\lambda_{dt\mu-t}$ can be seen in Figures 3. and 4., which show the region where the M-P curve gives the best fit, assuming $\lambda_{dt\mu-t} = 0$, and $\lambda_{dt\mu-t} = .05\ \lambda_{dt\mu-d}$, respectively. In these figures the width of the 'ladder' represents the error bar for the fit. The patterned boxes in the ladder correspond to the value predicted by M-P for the C_t region with the same pattern.

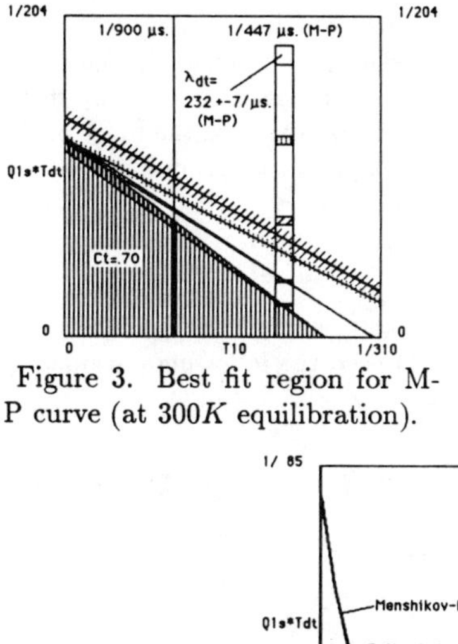

Figure 3. Best fit region for M-P curve (at $300K$ equilibration).

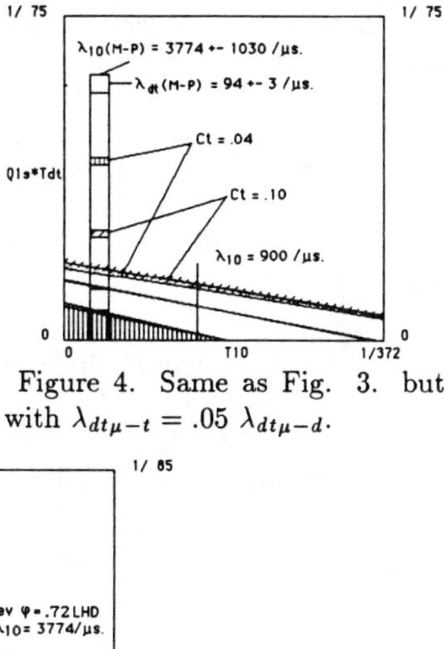

Figure 4. Same as Fig. 3. but with $\lambda_{dt\mu-t} = .05\ \lambda_{dt\mu-d}$.

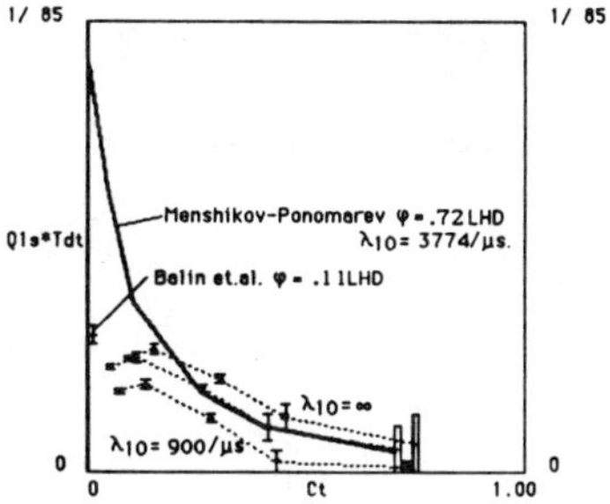

Figure 5. Same as Fig. 2. but with $\lambda_{dt\mu-t} = .05\ \lambda_{dt\mu-d}$.

The fit in Figure 4. no longer matches the data at $C_t = .1$, and no longer extrapolates the the Balin value for λ_{dt}, but notice that λ_{10} is forced to be large for any solution. Notice also in Figure 5. that the values for $C_t = .04$ have moved up relative to $C_t = .1$, thus becoming more plausible.

There seems to be no combination of parameters which yields an unambiguous good fit to these data for the Q_{1s} predicted by M-P , and the main reason is that under no circumstances does the measured Q_{1s} increase steeply enough at low C_t.

If one decides to reject M-P and accept the point at $C_t = .04$, then $\lambda_{10} = \text{inf}$ at least approximately approaches the value of Balin, and is a consistent, if not unique, solution to the problem. At this juncture, it is worthwhile to see if such a form for Q_{1s} is both reasonable and globally consistent.

Form of Q_{1s}.

Assuming the *du* deexcitation to be representable as a simple Markov chain, Q_{1s} can be expressed as the product of probabilities p_n that a $d\mu$ that has reached state n deexcites without transferring to a t. Such a probability can be written

$$p_n^{-1} = 1 + \frac{\phi C_t T_n}{\phi K_n + R_n} = \begin{cases} 1 + C_t A_n & \phi K \gg R \\ 1 + \phi C_t B_n & \phi K \ll R \end{cases}$$

where T is the transfer rate, and K and R are sums over destination states of rates for collisional deexcitation and ratiative deexcitation respectively. Thus Q_{1s}^{-1} can be a rather complicated polynomial in C_t and ϕC_t. For our purposes, however, it is only necessary to require that the mathematical form of Q_{1s} be able to approximate the more likely shapes expected. An S-curve, for example requires at least quadratic terms in ϕ and ϕC_t, whereas the data are likely to be insensitive to any higher terms.

Global Fit.

The non-linear least-squares fit to equation (1) used the following parametric forms:

$$\begin{aligned}
\lambda_{dt} &= \mathbf{LDT} + \mathbf{LDX}t \\
Q_{1s}^{-1} &= e^{\mathbf{Q1S}C_t}(1 + C_t\mathbf{QC} + C_t^2\mathbf{QC2} + \phi C_t\mathbf{QCD} + \phi C_t^2\mathbf{QC2D} + \phi^2 C_t^2\mathbf{QC2D2} \\
ln(\lambda^0_{dt\mu-d}) &= \sum_{i=0}^{4} \mathbf{LD}_i t^i + \mathbf{LDE}e^{\mathbf{LDL}t}(1 + \phi\mathbf{LDP}) \\
ln(\lambda^0_{dt\mu-t}) &= \sum_{i=0}^{4} \mathbf{LT}_i t^i \\
\lambda_{10} &= \mathbf{L10}(1 + \mathbf{L1T}t) \\
\lambda^1_{dt\mu-d} &= \mathbf{L1D} \\
\lambda^1_{dt\mu-t} &= 0
\end{aligned}$$

where t its the temperature of the run. Some parameters are forced positive by expressing them as exponentials of internal parameters. In the case of $\lambda^0_{dt\mu-d}$, this resulted in forcing an exponential rather than a linear dependence on ϕ. Figures 6 and 7 show that the resulting fit is still approximately linear, however.

Other parameters which could be allowed to vary include constant percentage and absolute errors added to each run (**SIGP** and **SIGA**), and a set of fudge factors which could be applied to groups of runs (**FDG**$_i$ and **GRP**$_i$) to allow for discrepancies due to many systematic variations from one group to another. Currently there are three groups defined which differ by as much as 10%. The data for $C_t = .04$, for example, are from a different group than the others.

Another parameter specified the average degree of molecular equilibration between room temperature and run temperature.

Results of Fit.

Figures 6 through 12 show the results for two preliminary attempts at fitting the model to approximately 280 LAMPF data runs. The first fit (S-crv) leaves most of the parameters unconstrained, the second constrains Q_{1s} to be M-P. The apparently reasonable χ^2's are misleading, not only because of the **FDG** and **SIGP** parameters, but by the distribution of the residual errors, which is highly nonuniform (see Figs. 8 and 9). The primary question remaining is whether the fit can best be improved by changes in the model or by renewed efforts to identify and eliminate systematic errors in the data.

Figures 6 and 7 show lists of parameters and their values for the two configurations. An '*' beside a parameter indicates it was allowed to vary in the fit. Some other parameters (e.g. **FDG**$_i$) are frozen at values found in previous fits. The curves in these figures show a family of lines for Q_{1s} and $\lambda^0_{dt\mu-d}$ corresponding to densities from 0.1 (dotted line) to 1 LHD (solid line).

Figures 8 and 9 show the distribution of residual error from the fits over C_t and temperature. The errors are also separated into groups of high ($\phi > .7$) and low density. The magnitude and sign of residuals is indicated by the length and direction of the diagonal line originating at each point. Scale for the residual errors is indicated by the small circle of radius 1σ in the corner of each figure. Note that the greatest errors are concentrated at high density.

Conclusions.

There seems to be no way to reconcile the shape of the M-P Q_{1s} with a large value for λ_{10}. There also seems to be no way to extract may of the contributing parameters to λ_c in isolation. A comprehensive global fit may allow an unambiguous analysis, but much work remains to be done before that can be determined.

References.

[1] M. Leon "Resonant Mesonic Molecule Formation in Muon-Catalyzed D-T Fusion" Phys. Rev. Lett. 52 (April (1984) 1655.
[2] Menshikov and Ponomarev " " Sov Phys JETP Lett. 39 663 (1984)
[3] A. V. Matveenko and Ponomarev " " Sov. Phys JETP 32 871 (1971)
[4] D. V. Balin *et. al.* "ANALYSIS OF dt-μ CF DATA" Muon Catalyzed Fusion 2 (1988)
[5] Ackerbauer "Muon Catalyzed Fusion at Low Tritium Concentration. Kinetics and First Experimental Results" Muon Catalyzed Fusion 2 (?)

S-crv: 18x277 system: tlim= 0-1000,philim=0.00-1.50,runs 0-1000
chisq=1.027894

ldt=	410.8828	ldx=*	0.0002	q1s=*	0.0028	qcd=*	0.0000
qc2d=*	2.4299	qc2d=*	3.2490	qc=*	0.0000	qc2=*	4.0477
lde=*	-500.4183	ldl=*	0.0009	ldp=*	-0.0010	ld0=*	505.8199
ld1=*	-0.4391	ld2=*	0.0002	ld3=*	-0.0000	lt0=*	4.1324
lt1=*	-0.0075	lt2=*	0.0001	lt3=*	-0.0000	lt4=	0.0000
l10=	1066.4023	l1t=	0.0000	fdg1=	0.9147	fdg2=	0.9795
fdg3=	1.0481	sigp=	0.0600	siga=	0.0000	grp1=	263.0000
grp2=	600.0000	k46=	0.0150	l1d=	0.0000	emx=	2.0000

* parameter was allowed to vary.

Config:S-crv phi=0.00-0.70 c=0.50-1.00
T= 260- 340 run=0-1000

Chisq=1.027894, 277 runs, 18 vars

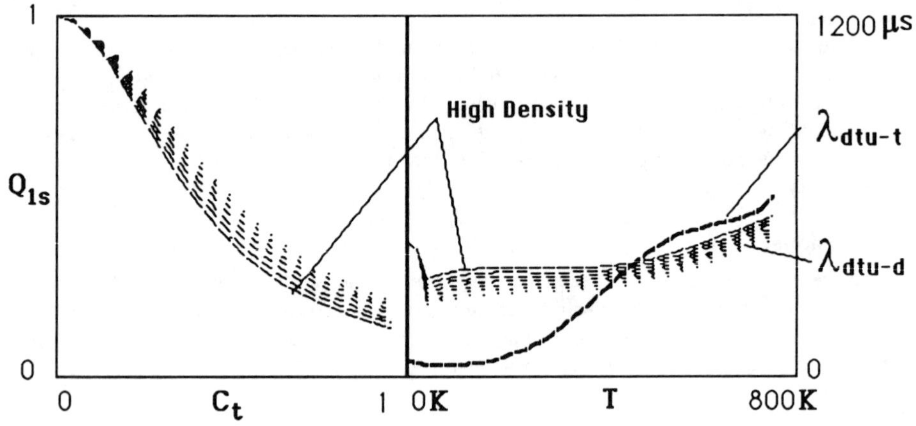

Figure 6. Parameters and plots for fit to cycling rates for 277 LAMPF runs. Q1s parameters are unconstrained.

M-P: 15x280 system: tlim= 0-1000,philim=0.00-1.50,runs 0-1000
chisq=1.368885

ldt=*	462.6848	ldx=*	-0.0001	q1s=	0.0021	qcd=	5.0000
qc2d=	2.4292	qc2d=	3.2480	qc=	5.0000	qc2=	4.0498
lde=*	-277.1476	ldl=*	0.0012	ldp=*	-0.0024	ld0=*	282.4969
ld1=*	-0.3377	ld2=*	0.0002	ld3=*	-0.0000	lt0=*	3.7676
lt1=*	-0.0058	lt2=*	0.0001	lt3=*	-0.0000	lt4=	0.0000
l10=*	533.7095	l1t=	0.0000	fdg1=*	0.9146	fdg2=	0.9795
fdg3=	1.0481	sigp=	0.0600	siga=	0.0000	grp1=	263.0000
grp2=	600.0000	k46=	0.0150	l1d=	0.0000	emx=	2.0000

* parameter was allowed to vary.

Config:M-P phi=0.00-0.70 c=0.00-0.40
T= 260- 340 run=0-1000

Chisq=1.368885, 280 runs, 15 vars

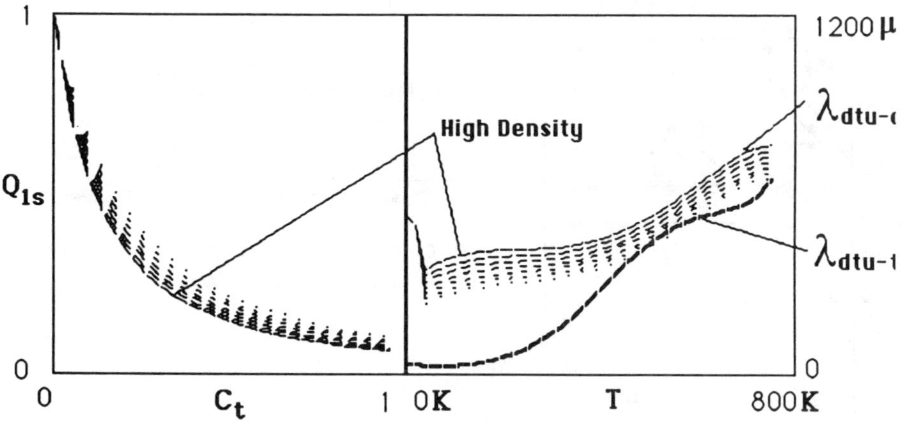

Figure 7. Parameters and plots for fit to cycling rates for 280 LAMPF runs. Q1s is constrained to values predicted by Menshikov-Ponomarev (M-P)

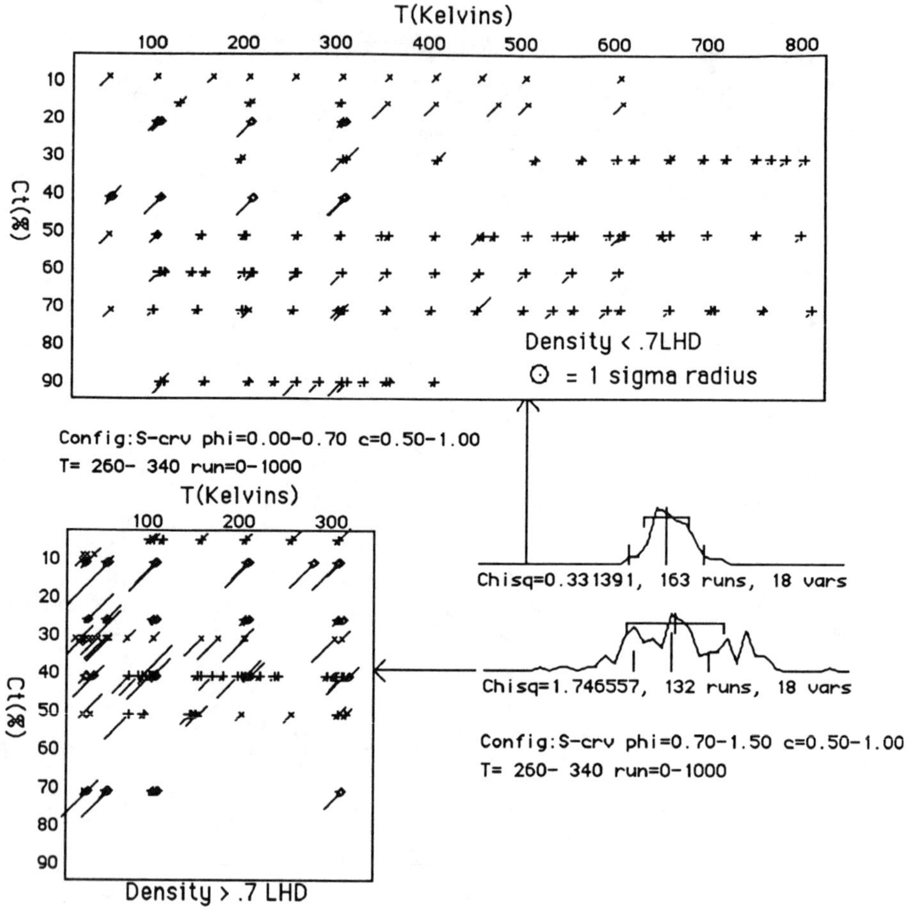

Figure 8. Map of residual errors to fit parameters shown in Figure 6. Length and direction of diagonals indicates magnitude and sign of residual error.

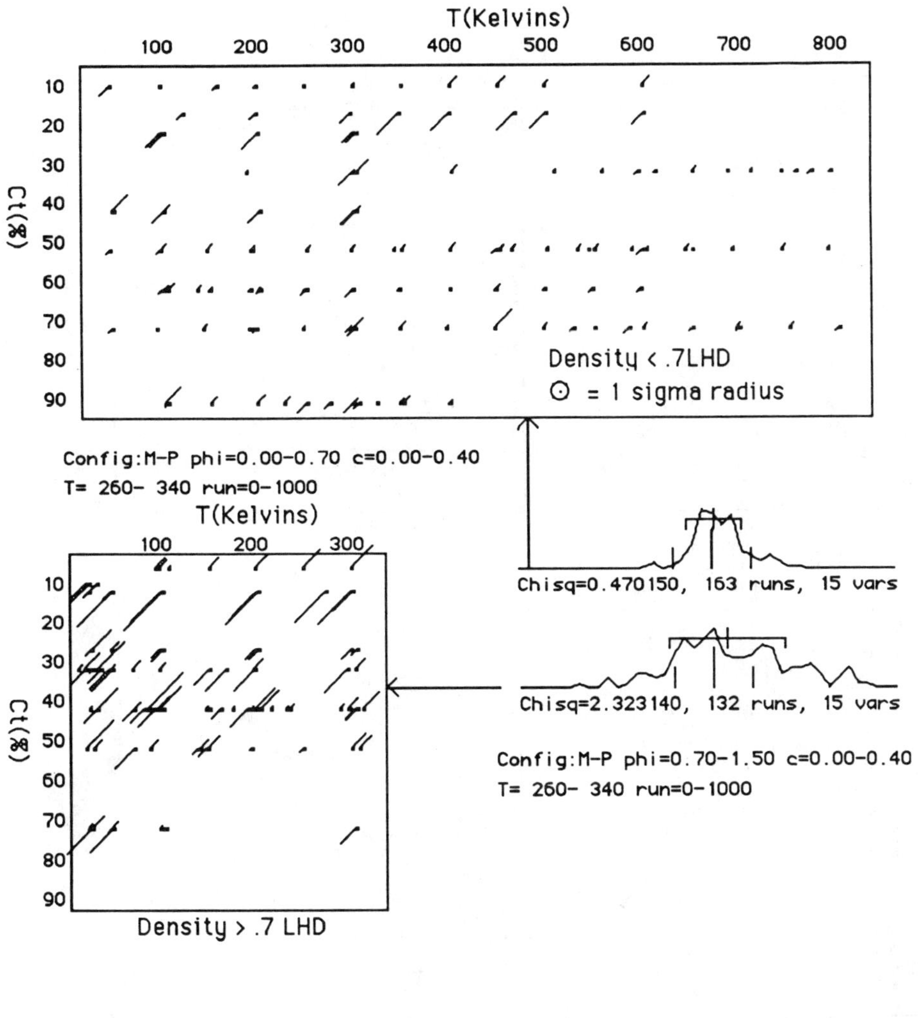

Figure 9. Map of residual errors to fit parameters shown in Figure 7. (Q1s constrained). Length and direction of diagonal lines indicates magnitude and sign of residual error.

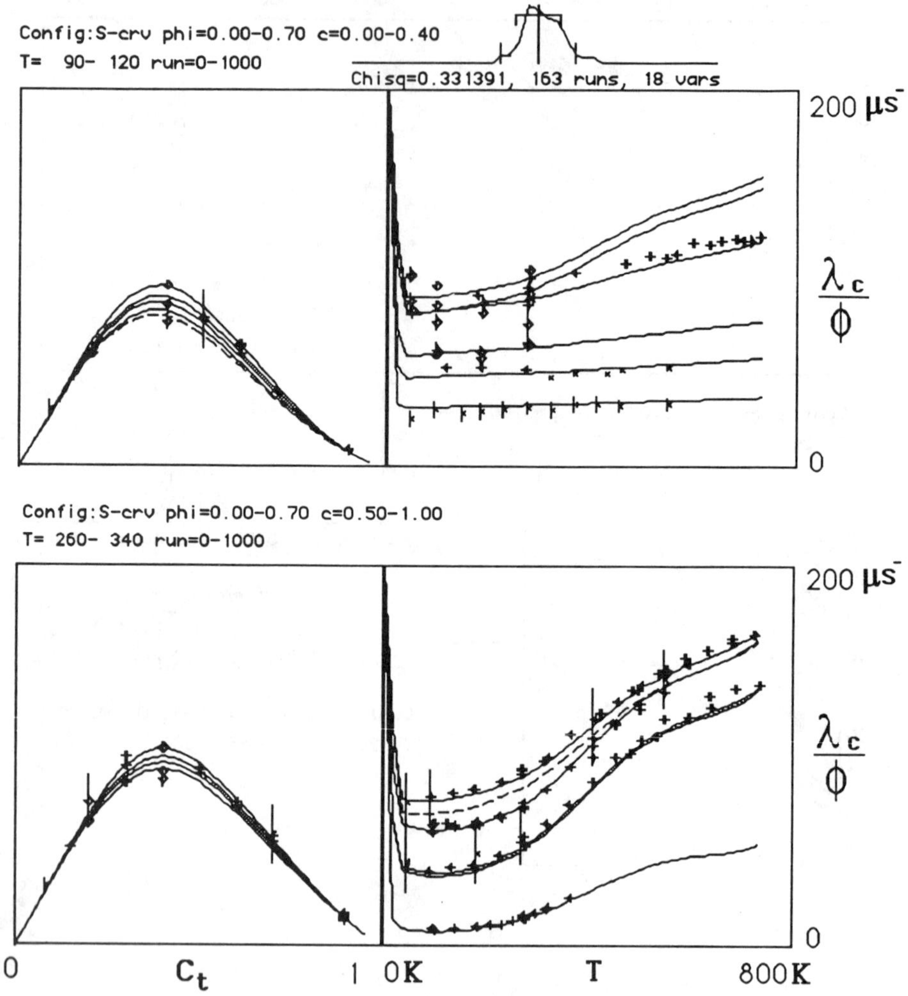

Figure 10. Selected coparison of data with fitted curves for unconstrained fit shown in Figure 3.

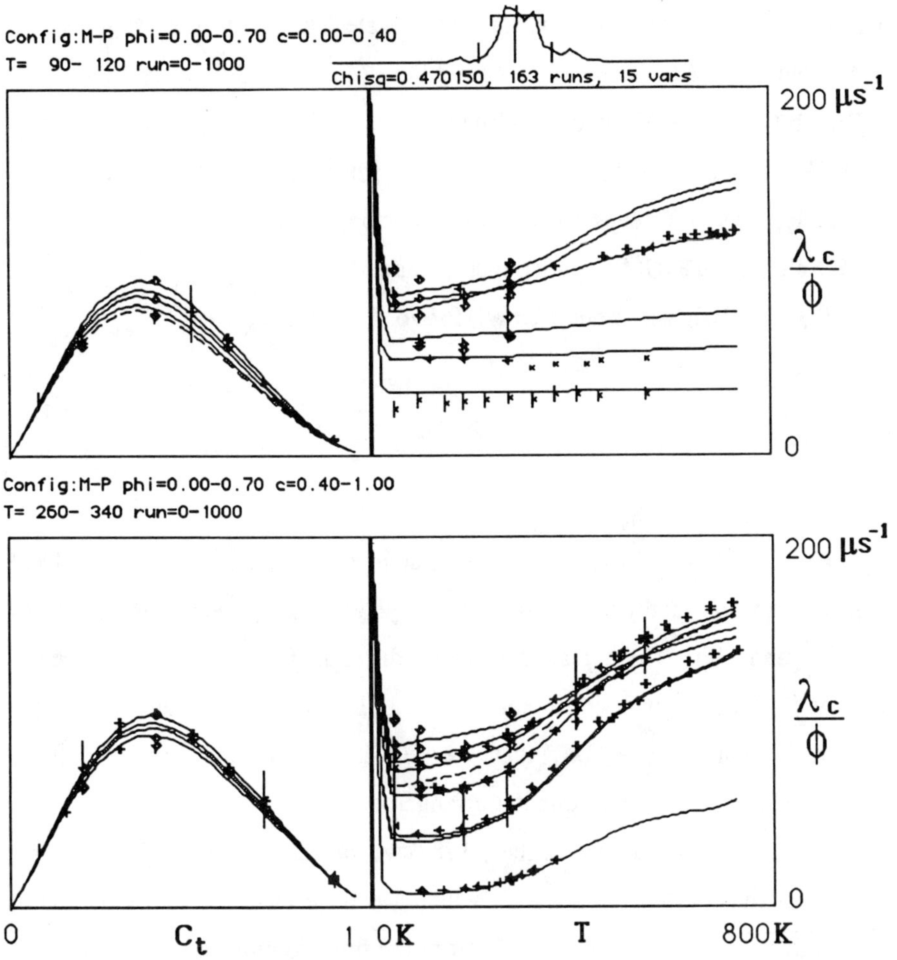

Figure 11. Selected comparison of data with fitted curves for M-P contstrained curve shown in Figure 4.

PROGRESS REPORT ON MUON CATALYZED FUSION STUDIES IN $H_2 + D_2$ AND HD GASEOUS TARGETS

K.A. Aniol, D.J. Margaziotis, California State University, Los Angeles
A.J. Noble, S. Stanislaus, C.J. Virtue, D.F. Measday, University of British Columbia, Vancouver, Canada
D. Horvath, Central Research Institute for Physics, Budapest, Hungary
B.C. Robertson, Queens University, Kingston, Canada
M. Salomon, TRIUMF, Vancouver, Canada
S.E. Jones, Brigham Young Univ., Provo, Utah U.S.A.

Last year at the Gatchina Conference we reported (1) results on $dd\mu$ and $pd\mu$ fusion product yields from gaseous targets of $H_2 + D_2$ and HD. These results were so different from theoretical expectations in the case of fusion neutrons from HD and fusion gamma rays from both HD and $H_2 + D_2$, that we felt it prudent to remeasure these processes in a new experimental arrangement.

In December, 1987 we ran at TRIUMF. We will report on that portion of the data which we have analyzed since our latest run, that is, on gamma ray yields from the muonic molecule $pd\mu$. Table 1 compares the experimental conditions for the April 1985 data (reported at Gatchina) and our December 1987 run.

TABLE 1
COMPARISON OF EXPERIMENTAL CONDITIONS FOR TWO SETS OF MEASUREMENTS.
50 ATM GAS TARGETS (20° C)ρ = 0.058 LH_2 DENSITY

	April 1985	December 1987
Targets		
	50%H_2 + 50%D_2	83%H_2 + 17%D_2
	25%H_2 + 50%HD + 25%D_2	56%H_2 + 38%HD + 6%D_2
	88%H_2D + G%H_2 + 6%D_2	48%H_2 + 52%HD + 6%D_2
Channel		
	M20A	M13
	8KHz μ–	56KHz π–
	0.5KHz e-	120KHz e-
		8 KHz μ^-
HD Source		
	homemade	purchased

Other differences: A new target was constructed for December, 1987. Because of the large π^-/e^- contamination in M13 a different trigger was used. This entailed different logic and hence both the data acquisition and data analysis computer codes were rewritten from the earlier (April 1985) runs.

Our original interest was simply to measure $\lambda_{dd\mu-p}$ and compare it to $\lambda_{dd\mu-d}$. We intended to use the fusion gamma ray-from

$p + d \rightarrow {}^3He + \gamma$ as a monitor. Since the $pd\mu$ molecule has no loosley bound states it is believed to be formed via a non-resonant mechanism.[2] which is calculated to have a negligible temperature dependence over our range of temperatures. Moreover, since the formation of the $pd\mu$ molecule entails the release of approximately 90 eV of energy we expected molecular structure to be insignificant. That is

$$d\mu + H_2 \rightarrow pd\mu + X \tag{1}$$

and

$$d\mu + HD \rightarrow pd\mu + X \tag{2}$$

should proceed at the same rates. We expected no difference in fusion gamma ray yields (#gamma rays/stopped muon) between cases (1) and (2).

Contrary to these expectations the $(H_2 + D_2)$ target gave twice as many gamma rays as the HD target and a temperature dependence was noted.[1] These features of the earlier experiment are corroborated by our recent measurement.

A sample gamma ray spectrum is shown in Fig.1. From such spectra the ratio of fusion gamma rays to decay electrons can be extracted. The yields are shown in figure 2. Very much the same conclusions can be drawn here.

(i) $H_2 + D_2$ gives a larger yield of fusion gammas than H_2 + HD.

(ii) The fusion yield from the equilibrated mix seems to be smaller than from the original mix.

Notice that figure 2 gives the yield in arbitrary units. The absolute yield will be discussed shortly.

Let us consider now the conventional picture from ref. 3 of $pd\mu$ formation and subsequent fusion. This will help us contrast more sharply the difference between theory and experiment. One notices that the formation of the J = 2 hyperfine state of $pd\mu$ robs fusion gamma rays. Fusion occurs only in the remaining three states J = 1, 1′, and 0.

The $(d\mu)_{J=1/2}$ state populates the $(pd\mu)_{J=1,0}$ states and $(d\mu)_{J=3/2}$, atom populates the $(pd\mu)_{J=1,2}$ states. The enhancement of the J = 1, 1′, 0, $pd\mu$ states, and subsequent enhancement of the gamma ray yield, by the transition $(d\mu)_{3/2} \xrightarrow{\phi C_d \lambda_d} (d\mu)_{1/2}$ is called the Wolfenstein-Gershtein effect.

One notices that depending on the population of these hyperfine $(pd\mu)_J$ states the gamma ray yield can vary widely. Using the population as depicted in ref. 3 and an effective value of $\lambda_d = 42 \times 10^6/s$ [5] (i.e. we will neglect the $(d\mu)_{1/2} \rightarrow (d\mu)_{3/2}$ *transition*) we can calculate the expected gamma ray yield ratios. These are tabulated in table 2, along with the measured values.

Table 2

Expected and Measured Yield Ratios of Fusion Gamma Rays from $(pd\mu))\rightarrow {}^3He\,\mu+\gamma$,

Experiment		Expected	Measured (overall temperatures
4/1/85	$\dfrac{Y_\gamma(.5H_2+.5D_2)}{Y_\gamma(HD)}$	1.0	$2.0 \rightarrow 2.2$
12/87	$\dfrac{Y\gamma(C_d=.17,H_2+D_2)}{Y\gamma(C_d = 0.3,H_2 + HD)}$	0.9	1.70 ± 0.26

An investigation shows several points where the difference between HD and H_2+D_2 might arise.

(i) Suppose we want to retain $\lambda_{pd\mu-d} = \lambda_{pd\mu-p}$. Then the distribution of $pd\mu$ hyperfine states is a possible parameter to play with. If λ_d depended on the D_2 concentration and not the atomic concentration, i.e., the process for atomic hyperfine transitions is

$$(d\mu)_{3/2} + D_2 \rightarrow (d\mu)_{1/2} + X$$

Then the largest ratio we could expect is

$$\frac{Y_\gamma\,(H_2 + .17D_2, \lambda_d \rightarrow \infty)}{Y_\gamma\,(H_2 + .52HD\,,\,\lambda_d = 0)} = 1.86$$

However, this is not realistic since λ_d has a known value.

(ii) Keep $\lambda_d(D_2) = 4.2\ x\ 10^6\ s^{-1}$, $\lambda_d(HD) = 0$ then at 50 atm, $(\phi = 0.058\phi_o)$ we get Table 3.

However, this possibility does not seem very strong. At $300°K$ the effective rate for $(d\mu)_{3/12} \rightarrow (d\mu)_{1/2}$ is less since $\lambda\ _d(1/2\rightarrow 3/2)$ also occurs. This reverse transfer would push the expected ratios down.

Table 3

Gamma Ray Yield Ratios if $\lambda d(D_2) =$ 42 x $10^6 S^{-1}$ $\lambda d(HD) = 0$

Experiment	Expected	Measured
$\frac{Y\gamma(H_2 + .50D_2)}{Y\gamma(HD)}$	1.54	$2 \rightarrow 2.2$
$\frac{Y\gamma(H_2 + .17D_2)}{Y\gamma(H_2 + .52HD)}$	1.3	1.70 ± 0.26

(iii) The other possibility is that $\lambda_{pd\mu-p} \neq \lambda_{pd\mu-d}$ or that $pd\mu-p$ preferentially populates $pd\mu$ states J = 0,1,1' and $pd\mu-d$ preferentially populates $pd\mu$ states J = 1,1',2.

This possibility (iii) provides a potential explanation of an earlier result by Bleser, et al. (6). They found that in liquid 75% $H_2 + 25\%D_2$ the yield Y_γ was enhanced over saturation by a factor of 1.18. This was taken as evidence for the Wolfenstein-Gershtein effect. However, they used $\lambda_d = 7 \times 10^6/s$! With the modern value $\lambda_d = 42 \times 10^6/s$. they should have seen an enhancement of Y_γ (25% Cd) = 1.56 *Y_γ (*saturation*).

Bleser, et al's. results were confirmed in a more recent measurement (7). What causes the discrepancy between the factor measured (1.18) and that predicted (1.56)?

From the description (6) of the gas filling technique it appears likely that D_2 and H_2 were simultaneously present in their palladium purifier. This procedure would tend to yield HD as an output along with H_2 and D_2. Hence, based on our results we might conclude that the results of ref. 6 show a smaller than expected fusion gamma ray yield because they had HD in their target and we now know that

$Y_\gamma(HD) < Y_\gamma(H_2 + D_2)$

Temperature Dependence of $pd\mu$ Formation

Still using the conventional model for $pd\mu$ formation we can predict what the 300°K gamma yield should be at liquid hydrogen density ($\phi = 1$). Extrapolating from our value ($\phi = .058\phi_o$) and using $\lambda_d = 42\ x\ 10^6/s$ we obtain $Y_\gamma(C_d = 0.17, T = 300K, \phi = 1) = 0.051 \pm 0.015$. But from Bleser, et al's. value at liquid temperature:

$$Y_\gamma(Sat,\ T = 22°K) = 0.14 \pm 0.02$$

and including the enhancement due to Wolfenstein-Gershtein effect $Y_\gamma(Cd = 0.17,\ T = 22°K) = 0.17 \pm 0.02$. There are other corraborating data that point to a temperature dependence listed in Table 4.

Table 4

Comparison of Room Temperature Yields to Liquid Temperature Yields of Fusion Gammarays.

Source	$\lambda_{pd\mu}(10^6S^{-1})$	T	$Y_\gamma(22°K)/Y_\gamma(300°K)$
ref. 6	5.8 ± 0.3	22°K	
ref. 8	1.8 ± 0.6	300°K	3.2 ± 1.0
ref. 9	2.0 ± 0.5	300°K	2.9 ± 0.7
current data		300°K	3.3 ± 1.0
average			3.1 ± 0.5

The results from Table 3 and ref (1) point to a temperature dependence for the fusion gamma ray from the $pd\mu$ molecule.

Conclusions:

The results reported here are based on about 50% of our new data. We are in the process of analyzing the rest. We draw two surprising conclusions:

$$Y_\gamma(H_2 + D_2) \quad \neq \quad Y_\gamma(HD) \tag{1}$$

$$\frac{Y_\gamma(H_2 + D_2,\ 22°K)}{Y_\gamma(H_2 + D_2,300°K)} \quad = \quad 3.1 \pm 0.5 \tag{2}$$

REFERENCES

1. K. Aniol, et.al., Muon Catalyzed Fusion -87, Gatchina, U.S.S.R. 26-29, May 1987.

2. L.I. Ponomarev and M.P. Faifman, Sov.Phys. JETP **44** 886 (1976).

3. V.M. Bystritskii, et.al., Sov.Phys. JETP **44** 881 (1976).

4. M.P. Faifman, et.al. Dubna preprint E4-86-541 (1986).

5. P. Kammel, et.al., Phys.Rev. **A28** 2611 (1983).

6. E. Bleser, et.a., Phys. Rev. **132** 2679 (1963).

7. W. Bertl et.al., Atomkernenergie/Kerntechnik **43** 184 (1983).

8. V.P. Dzhelepov, et.al., Sov.Phys. JETP **23** 820 (1966).

9. G.G. Semenchuk, et al., Muon Catalyzed Fusion-87, Gatchina, USSR 26-29 May 1987.

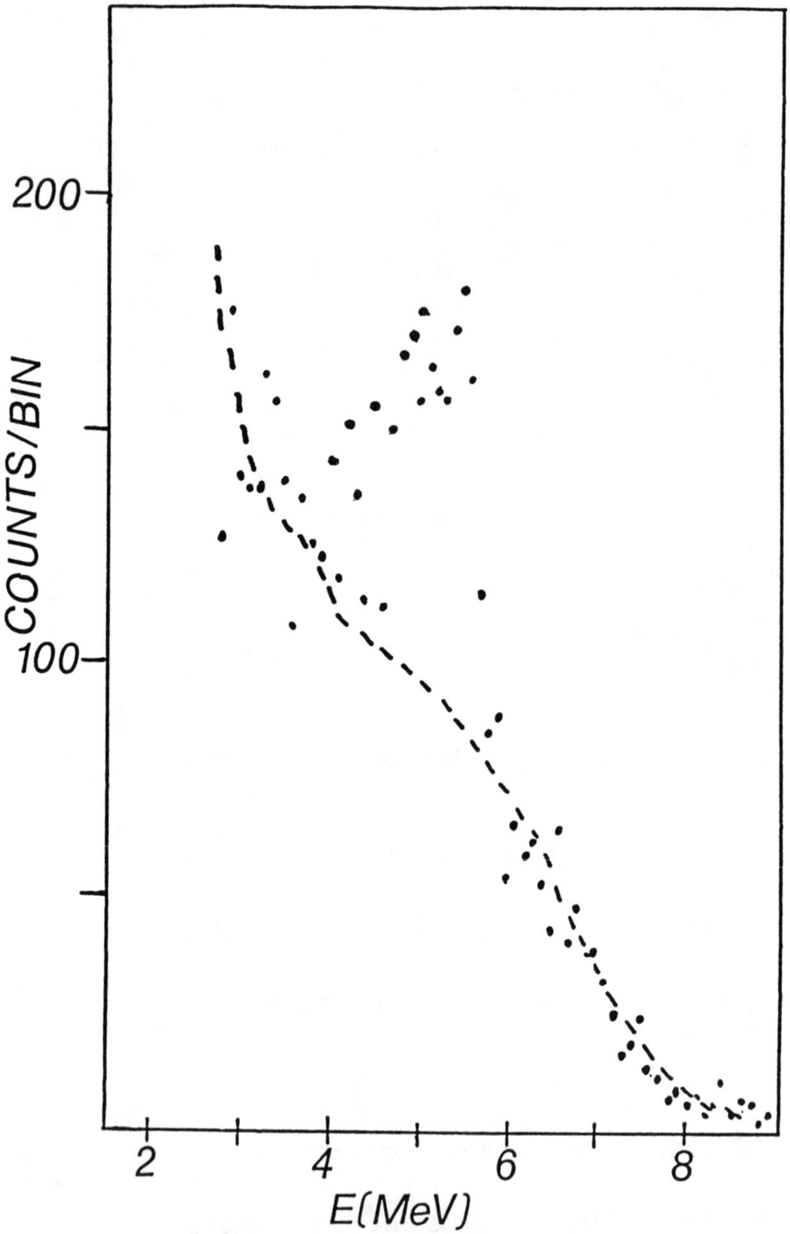

Figure 1: Gamma ray spectrum (dotted) from the $H_2 + D_2 = C(D_2) = 0.17$ target. The background spectrum is shown as the dashed line. The bin size is 0.1 MeV. The gamma ray from the $p + d \rightarrow {}^3He + \gamma$ reaction is visible.

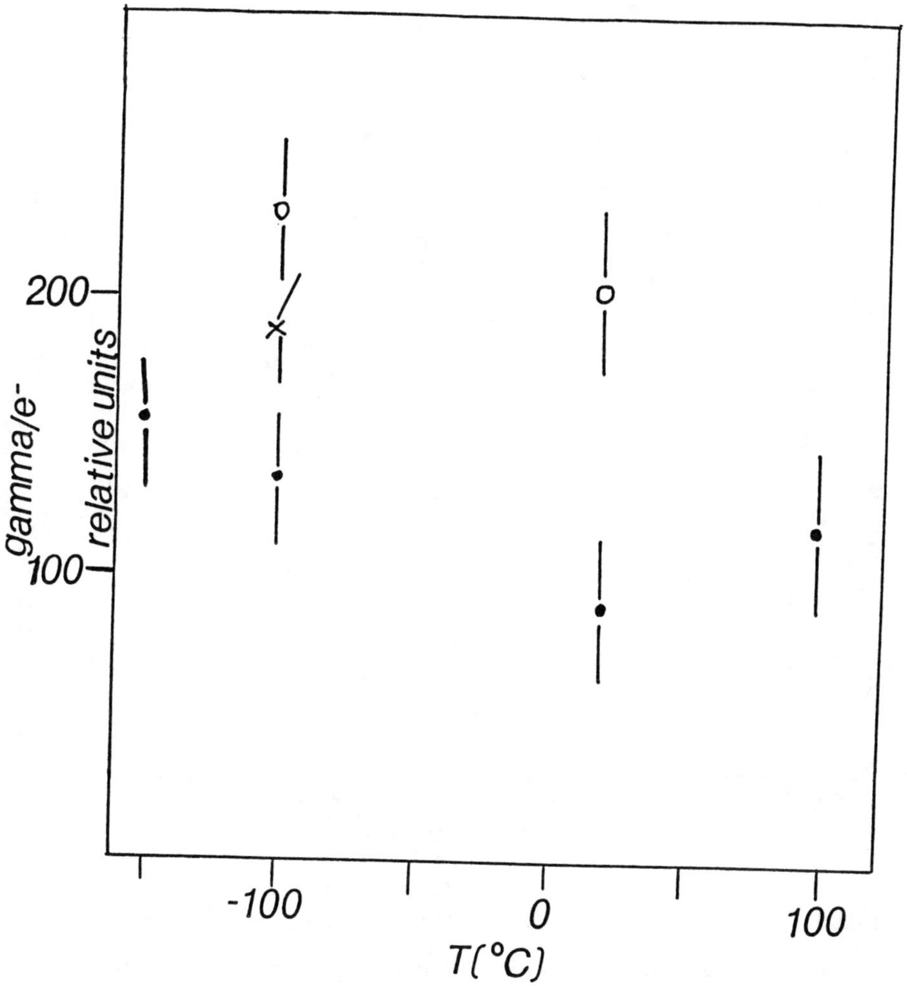

Figure 2: Relative yields of fusion gamma ray/muon decay electrons (arbitrary units). Open circles 0, 83% H_2 + 17%D_2; X , equilibrated mix from above, 56% H_2 + 38%HD + 6%D_2; closed circles 0, 48% H_2 + 52%HD

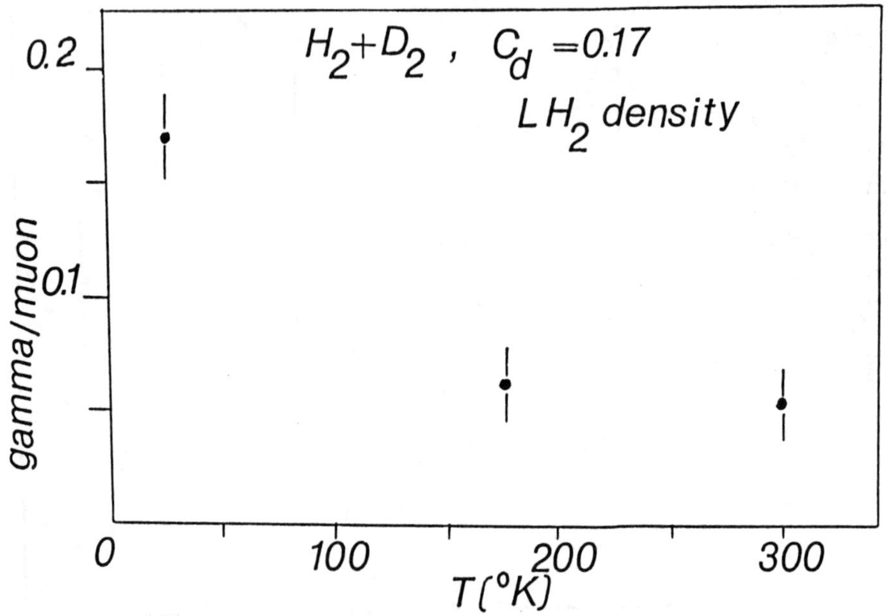

Figure 3: Absolute yield of fusion gamma rays per muon. The points at 173°K and 300° are our new data normalized to liquid hydrogen density using $\lambda_d(\frac{3}{2} \rightarrow \frac{1}{2}) = 42 \times 10^6\ s^{-1}$. The point at 22°$K$ is scaled value of ref. (6) for a 17% deuterium concentration.

PREDICTED METHODS OF CHANGING THE MUON CATALIZED FUSION CYCLING RATE

E.V. Sheely, S.E. Jones, L.M. Rees, J.K. Shurtleff,
S.F. Taylor, J.M. Thorne

Brigham Young University, Provo, Utah,

ABSTRACT

The following paper uses an easily modifiable computerized simulation procedure to predict reaction yields of muo-fusion. Reaction yields are predicted under varying conditions without the explicit solution of rate equations. Condititions and methods of optimizing reaction conditions by controlling the concentrations of hydrogen isotopes are discussed. The effects of non–equilibrated manipulation of reaction components and intermediates are predicted.

I. INTRODUCTION

There are three steps to solving muo-fusion rate equations: the determination of the Gibb's free energy, the solution of equilibrium equations, and the solution of differential kinetic equations. An easily modifiable method of predicting reaction yields under varying conditions without the explicit solution of rate equations will be presented. The method enables the effects of non-equilibrated reaction components and intermediates to be quickly and accurately predicted.

The number of fusions observed during muon–catalyzed fusion is dependant upon many factors. Among the most important of these is the concentration of the various isotopes. For temperatures ≤ 500 K optimum yields during muon catalyzed D-T fusion are obtained when the concentration of DT is small in comparison to the concentration of D_2 and T_2. Recent analyses indicate that at high temperatures the cycling rate increases as the concentration of DT increases. Recent predictions also indicate that the presence of free T atoms in a reaction chamber will increase the cycling rate. Methods and effects of manipulating these reaction concentrations are disscussed.

II. DERIVATION OF EQUATIONS

In order to derive muon-catalyzed fusion kinetic equations we make several simplifing assumptions. There is experimental evidence to support some, but not all of these assumptions.

Assumptions:

1) All reactions are either first order or pseudo-first order (they may be accurately considered first order in kinetic equations).

This assumption has been shown to be accurate in all cases in which it has been tested, and is in agreement with current theory on muon reaction mechanics in the rest of the cases.[1] There is no proof that this assumption will hold for all conditions.

2) The transfer reaction $t\mu \longrightarrow d\mu$ may be considered negligible.

3) The $tt\mu$ muo–molecules do not form via resonance.

4) There are no significant reactions taking place other than those represented in the following diagrams. (Epithermal and hyperfine effects have not been included.)

Reaction Diagrams and Equations

The following diagrams and equations represent reactions which are predicted to occur during muon-catalyzed-fusion:[2] parallel arrows (↑↑) represent the reaction path of molecules with parallel nuclear spin, anti-parallel arrows (↑↓) represent the reaction path of molecules with anti-parallel nuclear spin, and Lambda (λ) represent rate constants. Not included in the diagrams or equations is the effect of muon scavenging by impurities in reaction chambers.

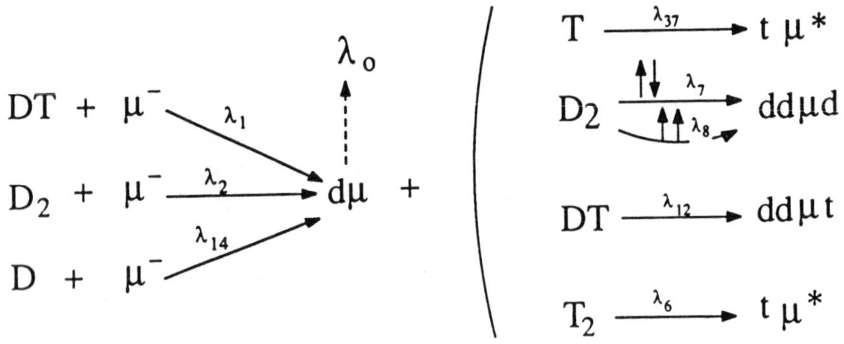

[1] Gershtein, **Sov. Phys. JETP**, 51(6), pp. 1053–1058.
[2] Jones, **Nature**, Vol 321, pp. 127–133.

$$\frac{d[d\mu]}{dt} = \lambda_1[DT][\mu^-] + [D_2][\mu^-] + \lambda_{14}[D][\mu^-] - \lambda_0[d\mu] - \lambda_{37}[d\mu][T] - (\lambda_7 + \lambda_8)[d\mu][D_2]$$
$$-\lambda_{12}[d\mu][DT] - \lambda_6[d\mu][T_2]$$

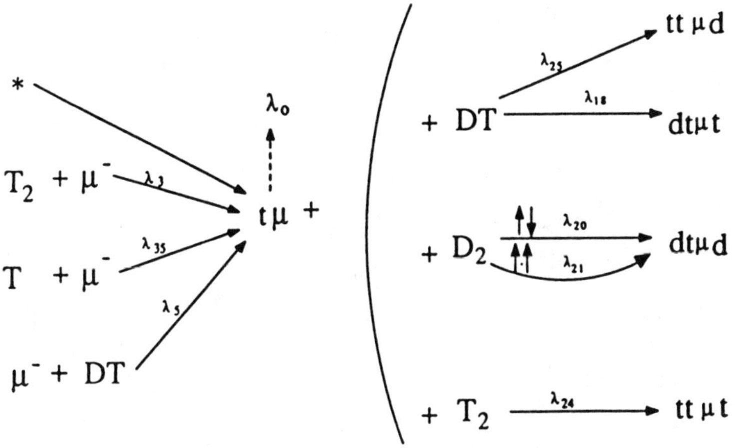

$$\frac{d[t\mu]}{dt} = \lambda_6[d\mu][T_2] + \lambda_{37}[d\mu][T] + \lambda_3[\mu^-][T_2] + \lambda_{35}[\mu^-][T] + \lambda_5[\mu^-][DT] - (\lambda_{25} + \lambda_{18})[t\mu][DT]$$
$$-(\lambda_{20} + \lambda_{21})[t\mu][D_2] - \lambda_{24}[t\mu][T_2]$$

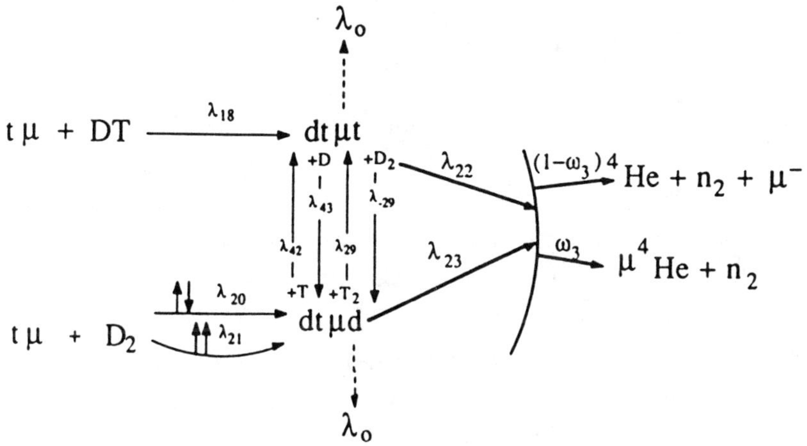

$$\frac{d[dt\mu t]}{dt} = \lambda_{18}[t\mu][DT] + \lambda_{42}[T][dt\mu d] + \lambda_{29}[T_2][dt\mu d] - \lambda_{43}[D][dt\mu t]$$
$$-\lambda_{-29}[D_2][dt\mu t] - (\lambda_0 + \lambda_{22})[dt\mu t]$$

$$\frac{d[dt\mu d]}{dt} = (\lambda_{20} + \lambda_{21})[D_2][t\mu] + \lambda_{43}[D][dt\mu t] + \lambda_{-29}[D_2][dt\mu t] - (\lambda_0 + \lambda_{23})[dt\mu d]$$
$$-\lambda_{42}[T][dt\mu d] - \lambda_{29}[T_2][dt\mu d]$$

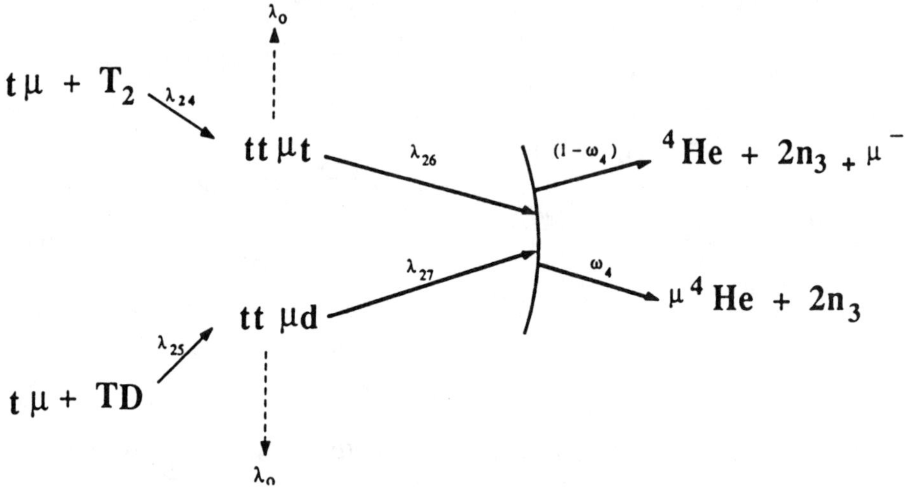

$$\frac{d[tt\mu t]}{dt} = \lambda_{24}[T_2][t\mu] - (\lambda_0 + \lambda_{26})[tt\mu t]$$

$$\frac{d[tt\mu d]}{dt} = \lambda_{25}[DT][t\mu] - (\lambda_0 + \lambda_{27})[tt\mu d]$$

$$\frac{d[dd\mu d]}{dt} = (\lambda_7 + \lambda_9)[D_2][d\mu] + \lambda_{38}[D][dd\mu t] + \lambda_{11}[D_2][dd\mu t] - (\lambda_0 + \lambda_8)[dd\mu d]$$
$$-\lambda_{34}[T][dd\mu d] - \lambda_{-11}[T_2][dd\mu d]$$

$$\frac{d[dd\mu t]}{dt} = \lambda_{12}[DT][d\mu] + \lambda_{34}[T][dd\mu d] + \lambda_{-11}[T_2][dd\mu d] - (\lambda_0 + \lambda_{10})[dd\mu t]-$$
$$\lambda_{38}[D][dd\mu t] - \lambda_{11}[D_2][dd\mu t]$$

$$\frac{d[\mu^-]}{dt} = \frac{1}{2}\{\lambda_8[dd\mu d] + \lambda_{10}[dd\mu t]\}(2 - \omega_1 - \omega_2) + \{\lambda_{22}[dt\mu t] + \lambda_{23}[dt\mu d]\}(1 - \omega_3)$$
$$+\{\lambda_{26}[tt\mu t] + \lambda_{27}[tt\mu d]\}(1 - \omega_4) - [\mu^-]\{\lambda_0 + \lambda_1[DT] + \lambda_2[D_2] + \lambda_{14}[D]$$
$$+\lambda_3[T_2] + \lambda_{35}[T] + \lambda_5[DT]\}$$

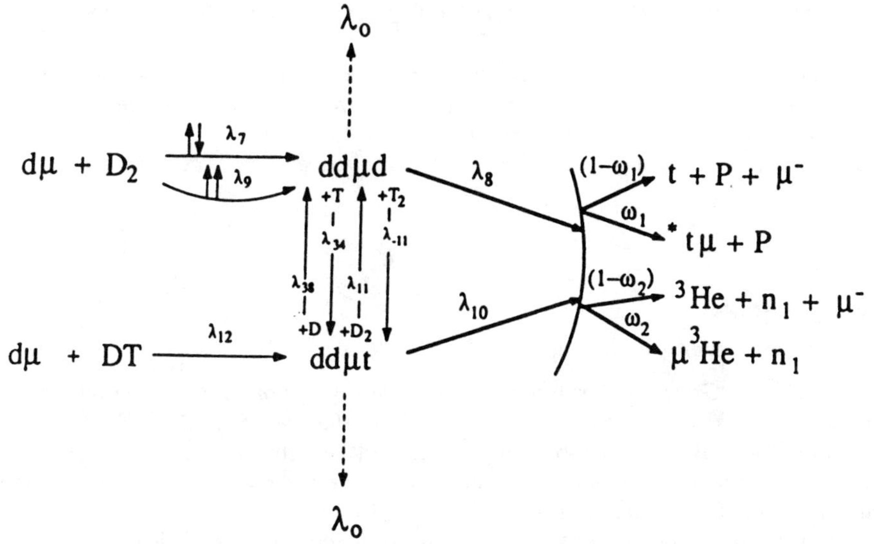

$$\frac{d[^3He]}{dt} = \frac{1}{2}\{\lambda_8[dd\mu d] + \lambda_{10}[dd\mu t]\}(1-\omega_2)$$

$$\frac{d[^3_\mu He]}{dt} = \frac{1}{2}\{\lambda_8[dd\mu d] + \lambda_{10}[dd\mu t]\}\omega_2$$

$$\frac{d[^4He]}{dt} = \{\lambda_{22}[dt\mu t] + \lambda_{23}[dt\mu d]\}(1-\omega_3) + \{\lambda_{26}[tt\mu t] + \lambda_{27}[tt\mu d]\}(1-\omega_4)$$

$$\frac{d[\mu^4He]}{dt} = \{\lambda_{22}[dt\mu t] + \lambda_{23}[dt\mu d]\}\omega_3 + \{\lambda_{26}[tt\mu t] + \lambda_{27}[tt\mu d]\}\omega_4$$

$$\frac{d[n_1]}{dt} = \frac{1}{2}(\lambda_8[dd\mu d] + \lambda_{10}[dd\mu t])$$

$$\frac{d[n_2]}{dt} = \lambda_{22}[dt\mu t] + \lambda_{23}[dt\mu d]$$

$$\frac{d[n_3]}{dt} = 2(\lambda_{26}[tt\mu t] + \lambda_{27}[tt\mu d])$$

$$\frac{d[n]}{dt} = \frac{d[n_1]}{dt} + \frac{d[n_2]}{dt} + \frac{d[n_3]}{dt}$$

III. DETERMINATION OF EQUILIBRIUM CONSTANTS

In order to determine the relative amounts of various hydrogen molecules which will be present in a reaction chamber at a given temperature and pressure, it is necessary to know the equilibrium constants (K_{eq}) of the gases involved. Equilibrium constants are most often calculated from the Gibb's free energy (ΔG) by the equation:[3]

$$\Delta G = -RT \ln K_{eq}$$

R = 8.3144
T is the temperature in degrees Kelvin

Values of ΔG for non-tritium containing hydrogen molecules can be obtained from *J. phys. chem Ref. Data*, Vol 14, suppl. 1, 1985 pg. 991,1211. For tritium containing molecules, the Gibb's free energy and equilibrium constants must be determined from the partition function. Due to the radioactivity of tritium, the standard form of the partition can not be used. Corrections need to be made.[4]

In order to expand known values to all possible temperatures, it is necessary to determine an equation that fits the data or to perform interpolation and extrapolation. Iterative non-linear interpolation can give greater accuracy, however, it has the disadvantage of taking longer to calculate. The iterative technique which I have used is Nevil's non-linear interpolation method.[5]

IV. EQUILIBRIUM EQUATIONS

In order to determine the equilibrium concentration of hydrogen isotopes, it is necessary to consider the reactions:

$$HD \longrightarrow \frac{1}{2}H_2 + \frac{1}{2}D_2$$

$$HT \longrightarrow \frac{1}{2}H_2 + \frac{1}{2}T_2$$

$$DT \longrightarrow \frac{1}{2}D_2 + \frac{1}{2}T_2$$

$$\frac{1}{2}H_2 \longrightarrow H$$

$$\frac{1}{2}D_2 \longrightarrow D$$

---

[3] Levine, **Physical Chemistry** P. 166.

[4] Souers,**Hydrogen Properties for Fusion Energy**, Pg 20-24,160,295

[5] Burden, **Numerical Analysis**, P. 97

$$\frac{1}{2}T_2 \longrightarrow T$$

Using the above reaction equations it is possible to derive the following equilibrium equations

$$K_{X,H} \equiv \frac{X_H}{\sqrt{X_{H_2}}} \qquad K_{X,HD} \equiv \frac{\sqrt{X_H X_D}}{X_{HD}}$$

$$K_{X,D} \equiv \frac{X_D}{\sqrt{X_{D_2}}} \qquad K_{X,HT} \equiv \frac{\sqrt{X_H X_T}}{X_{HT}}$$

$$K_{X,T} \equiv \frac{X_T}{\sqrt{X_{T_2}}} \qquad K_{X,DT} \equiv \frac{\sqrt{X_D X_T}}{X_{DT}}$$

These reaction diagrams and equations do not include all possible components, however, they do include all components which occur in quantities which can effect normal muo-fusion reaction conditions.[6]

It is necessary to evaluate the initial concentrations of the isotopes.

$$X_{H_2} = X_{i,H} - \frac{1}{2}X_H - \frac{1}{2}X_{HD} - \frac{1}{2}X_{HT}$$

$$X_{D_2} = X_{i,D} - \frac{1}{2}X_D - \frac{1}{2}X_{HD} - \frac{1}{2}X_{DT}$$

$$X_{T_2} = X_{i,T} - \frac{1}{2}X_T - \frac{1}{2}X_{HT} - \frac{1}{2}X_{DT}$$

where $X_H, X_D, X_T, X_H, X_D, X_T, X_{HT}, X_{DT}$ and X_{HD} are the equilibrium concentrations, and $X_{i,H}, X_{i,D}$ and $X_{i,T}$ are the initial molecular concentrations of H_2, D_2 and T_2 respectively.

K_x is the molar equilibrium constant and can be calculated from the equilibrium pressure constant K_p.[7] Assuming Ideal gas behavior $K_p = K_{eq}$.

$$K_x = P^{-\Delta n} K_{eq}$$

[6] Gill, **Tritium Handling in Vacuum Systems,** Los Alamos National Lab (1983), P. 20.

[7] Lewis, **Thermodynamics**, pp. 169–175.

At temperatures less than 2000 K, when only D_2,DT and T_2 are present many of the terms in the above equations approach zero. It is then possible to obtain a quadratic equation which can be used to solve the above equilibrium equations. **Care must be taken**, however, since when K_x becomes large, loss of significant figures due to rounding error makes it impossible to get accurate results using the normal form of the quadratic formula. Accurate approximations can be obtained using numerical methods of solution.

When protium, deuterium and tritium are present, the solution of the equilibrium equations becomes more complicated, especially at high temperatures. In order to analyze the full effects of manipulating reaction conditions, it is necessary to have a method of solution which will work when all reaction components are present. Newton's method[8] of solving systems of non-linear equations is a quadratically converging iterative technique which works well with these equations. Convergence can be speeded by decreasing the number of equations and unknowns. This can be done by combining the above equilibrium equations to form the following three equations:

$$X_{H_2} = X_{i,H} + \frac{K_H\sqrt{X_{H_2}}}{2} + \frac{K_{HD}\sqrt{X_{H_2}X_{D_2}}}{2} + \frac{K_{HT}\sqrt{X_{H_2}X_{T_2}}}{2}$$

$$X_{D_2} = X_{i,D} + \frac{K_D\sqrt{X_{D_2}}}{2} + \frac{K_{HD}\sqrt{X_{H_2}X_{D_2}}}{2} + \frac{K_{DT}\sqrt{X_{D_2}X_{T_2}}}{2}$$

$$X_{T_2} = X_{i,T} + \frac{K_T\sqrt{X_{T_2}}}{2} + \frac{K_{HT}\sqrt{X_{H_2}X_{T_2}}}{2} + \frac{K_{HT}\sqrt{X_{D_2}X_{T_2}}}{2}$$

When using Newton's method to solve the equilibrium equations some problems could result if special care is not taken. The equations should be solved in terms of X_{H_2}, X_{D_2} and X_{T_2} or in terms of X_{HD}, X_{HT} and X_{DT}, not in terms of X_H, X_D and X_T. When solved in terms of free H, D and T atoms, convergence is slow, and in some cases does not occur.

[8] Burden, pp. 498–499.

V. SOLUTION OF DIFFERENTIAL KINETIC EQUATIONS

The differential kinetic equations presented in this paper are all linear and can therefore be solved using eigenvectors. The exact solution is, however, difficult to obtain, and is difficult to alter as new, more accurate factors are determined. It is therefore desirable to have a numerical method of solution which can be altered easily. The method I have used is the forth order Runge-Kuta method for systems of equations. [9]

The results obtained are equivalent to those published earlier with one marked exception. The reaction rate of muon catalyzed fusion can be expected to make an increase near 1500 K due to the presence of the free T and D radicals. The reactions:

$$1)\quad D + \mu \longrightarrow d\mu + e^-$$

$$2)\quad T + \mu \longrightarrow t\mu + e^-$$

$$3)\quad d\mu + T \longrightarrow t\mu + D$$

are expected to proceed much more rapidly than do the competing reactions:

$$D_2 + \mu \longrightarrow d\mu + D + e^-$$

$$T_2 + \mu \longrightarrow t\mu + T + e^-$$

$$DT + \mu \longrightarrow d\mu + T + e^-$$

$$DT + \mu \longrightarrow t\mu + D + e^-$$

$$d\mu + T_2 \longrightarrow t\mu + D + T$$

$$d\mu + DT \longrightarrow t\mu + 2D$$

These predictions are made acording to electron shell stability and calculations preformed by Johann Rafelski.[10] At low temperatures the slowest initial reactions which normally occur are those which contain DT. DT reacts more slowly than D_2 and T_2 in rate determining steps.[11] Recent predictions of Mel Leon and data analyses performed by Al Anderson indicate that at high temperatures (above ≤ 500 K) DT reacts faster than D_2 or T_2. In order to optimize reaction conditions it is necessary to vary DT concentration and to have free D and T atoms present in the reaction chamber.

---

[9] Burden, pp. 225, 264.

[10] Rafelski, personal conversations

[11] Jones, pp. 127–133.

In most muon–catalyzed fusion cycles the transfer rate of $d\mu$ to $t\mu$ is a rate determining step (one of the slowest steps in the reaction chain), and is therefore important to the overall fusion rate.

Rate limiting reactions:

$$d\mu + D_2 \longrightarrow dd\mu d$$

$$d\mu + DT \longrightarrow dd\mu t$$

$$d\mu + DT \longrightarrow t\mu + 2D$$

$$d\mu + T_2 \longrightarrow t\mu + T + D$$

$$t\mu + D_2 \longrightarrow dt\mu d$$

$$t\mu + T_2 \longrightarrow tt\mu t$$

$$t\mu + DT \longrightarrow tt\mu d$$

$$t\mu + DT \longrightarrow dt\mu t$$

$$d\mu + T \longrightarrow t\mu + D$$

As can be seen from the above equations, most of the rate limiting steps are independent of the concentration of T. As the concentration of D and T increases, the concentration of the reactants of many of the rate limiting step decrease. Many of the rate determining steps will therefore be hindered by a large excess of free D and T atoms. In order to determine the exact concentrations necessary to optimize reaction conditions it will be necessary to have more accurate determinations of rate constants than now exist*.

VI. METHODS OF ALTERING KINETICS

In order to increase the reaction rate, it is necessary to alter the concentration of reaction components. This can be done in several ways with varying degrees of efficiency. Through the use of electromagnetic radiation, it may be possible to destroy undesirable molecules, thus freeing muons and reaction components to react in desired manners. It may also be possible to speed reaction rates through the use of radiation. A thorough investigation of the use of radiation in the above–mentioned manner has not yet been performed.

The most straightforward and perhaps optimual method of causing reactions to go in desired manners is to control the concentration of the components in a reaction chamber. In order to optimize reaction conditions, it is necessary to control the concentration of DT in reaction chambers. This can be done in several ways.

* These rate constants are constant with respect to concentration. They are not constant with respect to temperature or pressure.

At low temperatures it is possible to run muon–catalyzed fusion reactions with non-equilibrated mixtures of D_2 and T_2 there–by retarding the interference due to DT molecules. As the temperature is raised, however, equilibrium is quickly reached. The equilibrium concentration of DT is temperature dependant and reaches a maximum $\sim$ 50% at temperatures $\sim$ 1500 K (considering equal consentrations of deuterium and tritium). The concentration of DT remains near constant at $\sim$ 50% between 1500 K and 3000 K, then the concentration drops. This drop in DT concentration is due to the competing reactions:

$$D_2 \longrightarrow 2D$$

$$T_2 \longrightarrow 2T$$

$$DT \longrightarrow D + T$$

Another method of controlling DT within a reaction chamber involves selective laser-induced breakdown.[12] Lasers tuned to the proper vibrational frequency can selectively dissociate molecules while leaving other molecules unaffected. The principle disadvantage to eliminating DT in this manner appears to be the need of having an optical window in the reaction chamber. Although a need for an optical window is a disadvantage, it will not prevent this method from being used effectively. Further research needs to be performed in order to explore all the possibilities of this process.

A method of separating isotopes within a reaction chamber which could be used is absorption of gases on synthetic zeolites.[13] The hydrogen isotopes are absorbed by the zeolite, which is then heated slowly. The various molecular species come off at different temperatures, thus allowing the gases to be separated. This process has three main advantages over most other methods which could be used to separate gases within a reaction chamber. One advantage is that it can be used to remove helium from the chamber. This will be necessary when using reactors for long periods of time. Another advantage is that the method can be designed to operate without the use of a flow system. The use of flow systems increases the possibility of gas leakage. A third advantage of using zeolites to separate gases in a chamber is that it is relatively inexpensive to operate and maintain. The principle disadvantage of using zeolite absorption to seperate gases is that the process cannot be performed while reactions are taking place.

Two methods of isotope separation which are often coupled together are thermal diffusion and gas chromatography. [14] The coupling of these processes have the disadvantage that they require a flow system. These methods have many advantages, however, which far outweigh this one disadvantage. One advantage is

---

[12] Blazejowski, **Applied Physics B.** 41, pp. 109–117.

[13] Aledseev, **Atomnaia Energia** vol. 54, No. 6, pp. 409–411.

[14] Takayasu, **Kenkyu Hokoku — Toyama Daigaku Torichumu Kagaku Senta**, vol. 2, pp. 45-52.

that separation can be obtained in a relatively short period of time. Gas chromatographs have been designed which can separate large quantities of hydrogen isotopes in as little as one minute. By coupling a thermal diffusion column to the gas chromatograph a yield of near 100% separation of the desired isotopes can be obtained. Another advantage of this method is that ortho and para isotopes can be separated. Ortho and para isotopes have been postulated to have different fusing rates. If this proves to be true, they will need to be separated in-order to optimize reaction conditions. A third advantage of this process is that it can be used to eliminate helium from a reaction chamber as can the method of absorption on synthetic zeolites.

Another process which can be successfully coupled with a gas chromatograph is isotopic thermal decomposition.[15] This process occurs when gases are run over hot metal at temperatures near 1,000 K. When DT, which can be separated using a gas chromatograph, is run through a column containing hot metals it decomposes into D and T atoms, then recombines in an equilibrated mixture of DT, D_2 and T_2. This mixture can be run through the gas chromatograph and the desired components removed. The metals which appear to work best for this process are uranium, thallium and palladium.

Other methods of isotope separation which could be applied to systems of nuclear fusion are pulse-transient gas diffusion, D.C. gas discharge, and low temperature distillation. None of these processes offer advantages over those which have been discussed above.

VII. CONCLUSIONS

In order to accurately determine the optimum reaction conditions for muon catalyzed fusion it is necessary to consider the concentration and the reaction rates of D, T, D_2, T_2 and DT as well as the concentration of any impurities in the reaction chamber.

The use of numerical methods to solve equilibrium and kinetic equations predicted to occur during muon catalyzed fusion makes it possible to easily test the effects of altering reaction conditions and to alter reaction equations as more accurate data is determined. Newton's method of solving non-linear systems of equations is a quadratically converging method of solution which works well for solving equilibrium equations encountered while working with muo-fusion. The Runge-Kuta method of solving systems of first order differential equations is an easily modifiable method of solution which can be applied to muo-fusion rate equations.

Solution of muon catalyzed fusion rate equations indicates that an increase in the fusing rate should occur above 1500 K due to the presence of free T atoms in the reaction chamber.

In order to optimize reaction conditions it is necessary to control the concentration of DT in the reaction chamber. The most promising methods of controlling DT

[15] Masakazu, **J. Appl. Radiat. Isot.** vol. 34, No. 4, pp. 687–691.

concentration appear to be laser-induced breakdown of the DT molecule, absorption and removal using synthetic zeolites, or gas chromatographic separation coupled with a thermal diffusion column or an isotopic thermal decomposition chamber.

Chapter 2
Theoretical Muochemistry

THEORETICAL SURVEY OF μCF

M. Leon
Los Alamos National Laboratory

INTRODUCTION

Since not everyone attending this Workshop is a μCF expert, I will try to give a helpful introduction to the theory of this fascinating subject, concentrating on the d-t fusion cycle. As many of you know, the largest part of this body of theory has been developed by Leonid Ponomarev and his colleagues and collaborators over the last two decades. We are fortunate in having Professor Ponomarev with us at this Workshop, and look forward to hearing from him later this morning.

The main steps in the muon-catalyzed d-t fusion cycle are shown in Fig. 1. Most of the stages are very fast, and therefore do not contribute significantly to the cycling time. Thus at liquid H_2 densities (ϕ=1 in the standard convention) the time for stopping the negative muon, its subsequent capture and deexcitation to the ground state is estimated to be $\sim 10^{-11}$ sec.[1] The muon spends essentially all of its time in either the (dμ) ground state, waiting for transfer to a (tμ) ground state to occur, or in the (tμ) ground state, writing for molecular formation to occur. Following the formation of this "mesomolecule" (actually a muonic molecular ion), deexcitation and fusion are again fast. Then the muon is (usually) liberated to go around again. We will now discuss these steps in some detail.

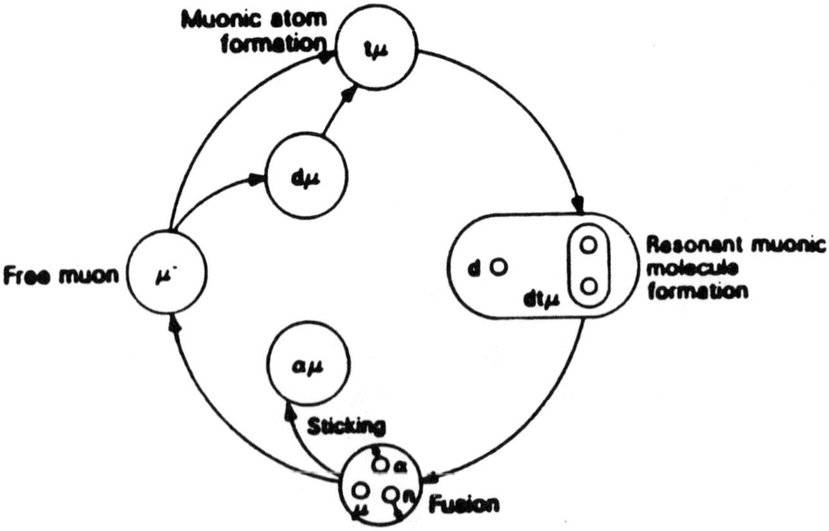

Fig. 1. The muon-catalyzed d-t fusion cycle.

MOLECULAR FORMATION

The *Auger* mechanism, where the energy released in forming the mesomolecule is carried off by an ejected electron, has been calculated by Ponomarev and Faifman,[2] using a two-level (or Born-Oppenheimer) approximation. The resulting molecular formation rate is quite slow for $dt\mu$: $\lambda^{e}_{dt\mu} \sim 3x10^{-4}s^{-1}$ at room temperature. (These rates are conventionally normalized to liquid H_2 target density, 4.25 x 10^{22} atoms/cm^3.)

This Auger rate is completely dwarfed by the rate for *resonant molecular formation*, the mechanism first suggested by Vesman[3] in 1967 to explain the striking observed temperature dependence for $dd\mu$ formation. This mechanism requires the existence of a loosely bound state, with binding energy $\lesssim 4$ eV, so that the energy released upon mesomolecular formation can go into the vibration and rotation of the resulting compound molecule:

$$t\mu + (D_2)_{0,0} \rightarrow [(dt\mu)^{*}dee]^{*}_{\nu,J} \tag{1}$$

This one-body final state means that the cross section for this process contains δ-functions at the resonance energies; temperature dependence of $\lambda_{dt\mu}$ from this mechanism comes from the overlap of these δ-functions with the Maxwell distribution of $t\mu + D_2$ kinetic energy.

The first requirement in proving the validity of this resonance idea is an accurate calculation of the binding energies of the postulated bound states. This task was embarked upon by Ponomarev and his collaborators following Vesman's suggestion, and required their developing the adiabatic representation (method of perturbed stationary states) into a powerful tool for calculating mesomolecular wave functions (solving the non-relativistic Coulomb three-body problem). By 1977 Vinitsky and Ponomarev[4] established that for the (J,v)=(1,1) (J $\equiv$ angular momentum, v $\equiv$ vibration quantum number) state of $dd\mu$ and $dt\mu$, $\epsilon_{11} \simeq -2.2$eV and -1.1eV respectively. This enabled Gershtein and Ponomarev[5] to point out that the resonant molecular formation rate $\lambda_{dt\mu} \simeq 10^8 s^{-1}$, so that $\simeq$ 100 fusions per muon in a dense D-T target is possible. This prediction of a hundred fusions per muon did much to trigger the present blossoming of interest in μCF.

The adiabatic representation has continued to develop,[6] but was overtaken (at least as far as claimed precision in the binding energy is concerned) about two years ago by variational calculations. The recent values $\epsilon_{11} = -1.975$ eV (-0.660 eV) for $dd\mu(dt\mu)$ should be accurate to a fraction of an meV, so that it is the "relativistic" corrections to these energies that are of greatest current interest. For $dd\mu$, the vacuum polarization term is completely dominant, providing + 10 meV out of a total correction to ϵ_{11} of +11 meV.[7] In contrast, for $dt\mu$ the nuclear charge distribution is dominant, contributing 15 meV of a total + 23 meV correction. Higher precision calculations are now under way, and we can hope to hear about them soon.

There are two channels of molecular formation that are especially sensitive to these energies, because the resonance energies turn out to be small. First, in the $dd\mu$ reaction

$$(d\mu)^F + (D_2)_{0,0} \rightarrow [(dd\mu)^S_{11}dee]_{7,1} \tag{2}$$

with dμ hyperfine state F=$\frac{3}{2}$ and ddμ total spin S=$\frac{1}{2}$, the resonance energy according to Zmeskal et al.[8] is only about 4 meV; this gives rise to the spectacular (and unexpected) hyperfine effect observed at SIN.[9] Since the temperature dependence of $\lambda_{dd\mu}^{\frac{3}{2}\rightarrow\frac{1}{2}}$ should eventually allow the determination of this resonance energy to within about 0.1 meV, this case will provide an extremely stringent test of the calculated ddμ J=1, v=1 binding energy, including all the corrections.

The other sensitive reaction is in dtμ formation:

$$(t\mu)^F + (D_2)_{0,J} \rightarrow [(dt\mu)_{11}^S dee]_{2,J+L} \qquad (1')$$

with F=0, S=1. Here the strongest resonances (those with orbital angular momentum L=0,1) actually lie *below* threshold! This circumstance is believed to lead to the three-body contribution to $\lambda_{dt\mu}$ discovered at LAMPF[10] (see below).

The pioneering calculation of the rates for resonant molecular formation (the Vesman mechanism) was made by Vinitsky et al.[11] in 1977; the importance of L,J and hyperfine effects,[12] and electron screening[13] was pointed out somewhat later. This left the computed molecular formation rates significantly smaller than the experimental value.[14,15,8] The situation was remedied when Menshikov and Faifman[16] pointed out the importance of using *undistorted* dμ + d and tμ + d wave functions in calculating the transition matrix element. Finally Menshikov et al.[17] for the ddμ case pulled all the many strands together – including "back decay" (the reverse of reaction 2)[18] and transitions of the [(ddμ)dee]* complex – in a beautiful *ab initio* calculation, which fits the experimental results[8,14,15] very well indeed.

The extension of this calculation to $\lambda_{dt\mu}$ is straight forward for all channels except the crucially important F=0, D_2 reaction, with its strong, below-threshold resonances (presumably J $\rightarrow$ J': 0 $\rightarrow$ 1, 0 $\rightarrow$ 2, 1 $\rightarrow$ 2, etc.). The precise location of these resonances will soon be known, when the precise calculations of the relativistic corrections are completed. At low enough density, treatment of the F=0, D_2 channel is also straightforward: the Auger deexcitation of the [(dtμ_{11}dee]* complex broadens the resonance δ-functions into Lorentzians,[19] with width $\Gamma_e \simeq 0.8$ meV.[20] As a result the contributions of the below-threshold resonances are dominant, the higher-L resonances being suppressed by a centrifugal barrier factor.[12] (This assumes that the tμ atoms are thermalized at the target temperature.)

At higher densities ($\phi >$ few %), three-body contributions to molecular formation become signification.[21] The *impact approximation*, so important for the theory of collisional line broadening in spectroscopy, has been applied extensively to this problem.[19,22–25] This entails including in the total width of the Lorentzians a *collisional contribution*, expressing the effect of collisions with neighboring molecules:

$$\Gamma = \Gamma_e + \phi\Gamma_c \qquad (3)$$

with

$$\Gamma_c = n_o < v\sigma > . \qquad (4)$$

But what collision cross section σ should be used here? Petrov[19] and others[22,23,25] have used only the inelastic cross section on the final state, while in optics the initial state inelastic scattering plus the integral of the square of the difference of the elastic scattering amplitudes also enters:[27]

$$\sigma = \sigma_{inel}^{i} + \sigma_{inel}^{f} + \int d\Omega \mid f_{el}^{i}(\Omega) - f_{el}^{f}(\Omega) \mid^2 . \tag{5}$$

In contrast, Menshikov[24] claims that, because of the momentum carried by the $t\mu$, the appropriate cross section is the final state *total* cross section:

$$\sigma = \sigma_{el}^{f} + \sigma_{inel}^{f}. \tag{6}$$

However, in spite of its seductiveness, it is evident that the basic conditions for the impact approximation are *not* met for this problem of molecular formation.[26] There are two conditions: First, the "detuning" (displacement from the imperturbed energy) ΔE must be related to the collision duration τ_c by[27]

$$\mid \Delta E \mid << \frac{\hbar}{\tau_c}. \tag{7}$$

For a realistic inter-molecular interaction and the temperature range of interest, the RHS of this relation is ~1 meV; thus the impact approximation is confined to a completely uninteresting energy range. This limitation arises because, for $\Delta E \gtrsim \hbar/\tau_c$, information about the scattering wave function is needed for times $t < \tau_c$, *not* just the asymptotic properties of this wave function (phase shifts, etc.). A second condition is the requirement of purely binary collisions; this leads to[26]

$$\phi < \begin{cases} 0.14 & @300K \\ 0.02 & @\ 30K \end{cases} , \tag{8}$$

which by itself removes most of the experimental data points!

It appears that what is needed is the generalization (to the massive incoming $t\mu$ as opposed to the photon) of the (many-body) *quasistatic approximation* of collisional line-broadening theory[28] (which may be related to the "Quasi-resonant molecular formation" of Menshikov and Ponomarev[21]). In my opinion, the theory of three-body molecular formation barely exists at present.

A possibility mentioned several times[29,30] for solid and liquid targets is that one or more *phonons* carry off enough energy to make the strongest resonance ($0 \rightarrow 1$, at ~ - 12 meV) accessible to the physical region. So far, only a calculation for a *metallic*, rather than molecular, hydrogen target has appeared.[31]

DEEXCITATION AND FUSION

Once the $[(dt\mu)_{11}^{*}dee]^{*}$ complex is formed, Auger transitions will carry it quickly to a $J=0$ state of $dt\mu$ where fusion is very rapid. The chain of transitions has been studied by Bogdanova et al.[20]

The proper method of calculating the fusion rate λ_f (and the sticking, discussed below) has generated some controversy. What we shall refer to as the *orthodox* view, expounded by Bogdanova,[32] Markushin[33] and others, holds that a good approximation is provided by the simple formulae

$$\lambda_f = \kappa \cdot p \tag{9}$$

$$p \equiv \int d^3\mathbf{r} \mid \Psi^{o,v}(\mathbf{r},\mathbf{o}) \mid^2, \tag{10}$$

$$\kappa \equiv \lim_{v \to 0} [v\sigma C^{-2}], \tag{11}$$

where C is the Gamow factor for the d-t system, σ is the reaction cross section (d + t $\rightarrow ^4$He + n), and $\Psi^{J,v}$ ($\mathbf{r}$,$\mathbf{R}$) is the three-body wave function for the $(dt\mu)_{Jv}$ mesomolecule ($\mathbf{R}$ being the d-t seperation and $\mathbf{r}$ the muon position relative to the d-t CM). This approximation could fail (in the orthodox view) only if significant *rearrangement* of the dtμ spectrum by the strong interaction occurs. This would happen if the ^{5}He$^*(\frac{3}{2}^+)$ resonance, which dominates the fusion reaction, were to be closely degenerate with a dtμ bound state. However, both the resonance energy and its width serve to prevent this rearrangement.

This point of view is supported., e.g., by the recent R-matrix calculation of Struensee et al.,[34] which finds λ_f values close to the orthodox results. Obviously, the *dissenters* don't agree, and I am sure we will be told why during this Workshop.

STICKING AND REACTIVATION

Sticking of the negative muon to the daughter He nucleus limits the number of fusions per muon that can be attained. Two factors are involved: *initial sticking* $\omega_s^0(n,\ell)$ in the (n,ℓ) state of the $(\mu\alpha)$ system, and the probability $[R(n,\ell;\phi)]$ for *reactivation* of the muon during the slowing-down of the $(\mu\alpha)^+$:

$$\omega_s(\phi) = \sum_{n,\ell} [1 - R(n,\ell;\phi)]\omega_s^0(n,\ell). \tag{12}$$

The simplest approximation for the initial sticking uses the *adiabatic* (Born-Oppenheimer) approximation for the muon wave function as the d and t approach one another, and takes the overlap (*sudden* approximation) with the final state wavefunction of a muon traveling in a bound state around the retreating α-particle:

$$\omega_s^0(n,\ell) = \mid < \Psi_{1s}(\mu -^5 He) \mid e^{iM\mathbf{V}\cdot\mathbf{r}} \Psi_{n\ell 0}(\mu - \alpha) > \mid^2 . \tag{13}$$

Calculated in this way the total initial sticking comes out to be[35]

$$\sum_{n,e} \omega_s^0(n,e) = 1.16\%. \tag{14}$$

The largest correction to this approximation comes from the fact that the true wave function does not adiabatically adjust as the d-t separation goes to zero, but rather "lags-behind". Thus instead of $\Psi_{1s}(\mu -^5 He)$ in Eq.

13, the true three-body wave function, with $\mathbf{R} \rightarrow o$, is needed. The results with the wave function calculated by several different methods (quantum Monte Carlo,[36] adiabatic representation,[35] variational[37]) agree quite well and reduce the sticking by about 25%:[38]

$$\Sigma\omega_s^0 = 0.85\%. \tag{15}$$

Next, we consider strong interaction effects on ω_s°, in particular the effect of the energy dependence of the $\frac{3}{2}^+$ $^5He^*$ t-matrix element. According to the orthodox view, this affects only the non-adiabatic corrections to ω_s°, and merely increases $\Sigma\ \omega_s^\circ$ by $< 3\%$.[32,33] The dissenting view is quite different; e.g., Rafelski et al.[39] claim the strong interaction effects reduce $\Sigma\omega_s^\circ$ by a factor of two! We will certainly hear from both sides during the Workshop.

The reactivation of muons in stripping or transfer collisions is very important, especially for dtμ because here the recoil velocity of the α is so large ($v \simeq 6$ a.u.). The reactivation depends on the competition among all the excitation, deexcitation, Stark mixing and transfer processes, and the slowing of the $(\alpha\mu)^+$. Since the R(n,ℓ;ϕ) obviously depends drastically on the (n,ℓ) values, there is an important dependence on the initial populations ω_s°(n,ℓ). The most complete calculations are those of Cohen[40] and Markushin,[41] which in fact agree very well for R(ϕ) and the x-ray intensities $K_\alpha(\phi)$, $K_\beta(\phi)$, ... The x-ray intensities (ddμ as well as dtμ) are in fair agreement with SIN results.[42] The theoretical R(ϕ) variation with density ϕ is too small to account for the variation in $\omega_s(\phi)$ reported from the LAMPF neutron data.[10]

SCAVENGING BY HELIUM

^{3}He from tritium decay appears in any d-t experiment, so He scavenging will always be present. There are obviously six possible reactions:

$$x\mu + ^A\text{He} \rightarrow ^A\text{He}\mu + x \tag{16}$$

(x = p,d,t; A = 3,4).
The reactions are believed to proceed via the (Auger) formation of the (xHeμ) mesomolecular state 2pσ, followed by radiative dissociation:[43]

$$\begin{aligned} x\mu + He &\rightarrow (xHe\mu)_{2p\sigma} + e^- \\ &\hookrightarrow (He\mu)_{1s} + x + \gamma \end{aligned} \tag{17}$$

The rates $\lambda_{x\,He}^{\;A}$ as functions of temperature have been calculated by Fomichev et al.[44] The predictions that (1) $\lambda_{x\,He}^{\;A}$ (T) increases as T decreases, and (2) $\lambda_{t\,He}^{\;3}$ (T) is an order of magnitude larger than $\lambda_{d\,He}^{\;3}$ (T), appear to be experimentally verified.[43,45]

In addition to this ground-state transfer, for d-t targets (with their large cycling rates) the initial capture by, and excited-state transfer to, He is significant; the total scavenging rate will have both contributions:

$$\lambda_{He} = \lambda_{He}(g.s.) + \lambda_c\omega_{He}. \tag{18}$$

The quantity ω_{He} is analogous to $(1-q_{1s})$ (see below). So far, there is no calculation of this quantity.

ELASTIC SCATTERING

Elastic scattering of muonic atoms from target molecules plays an extremely important role in the catalysis cycle, since it determines the rate of thermalization of the dμ and tμ atoms. The early calculations of Matveenko and Ponomarev[46] used a simple two-level (Born-Oppenheimer) approximation. More recent work (Melezhik and collaborators[47,48]) uses the adiabatic representation (PSS method) for calculating the nuclear scattering (dμ + d, etc.). To this is added the effect of electron screening (in Born approximation), and of molecular structure (using the Fermi pseudo-potential method). An example of the resulting cross sections, for dμ (F=$\frac{1}{2}$) scattering from d,D, and D_2 is shown in Fig. 2.

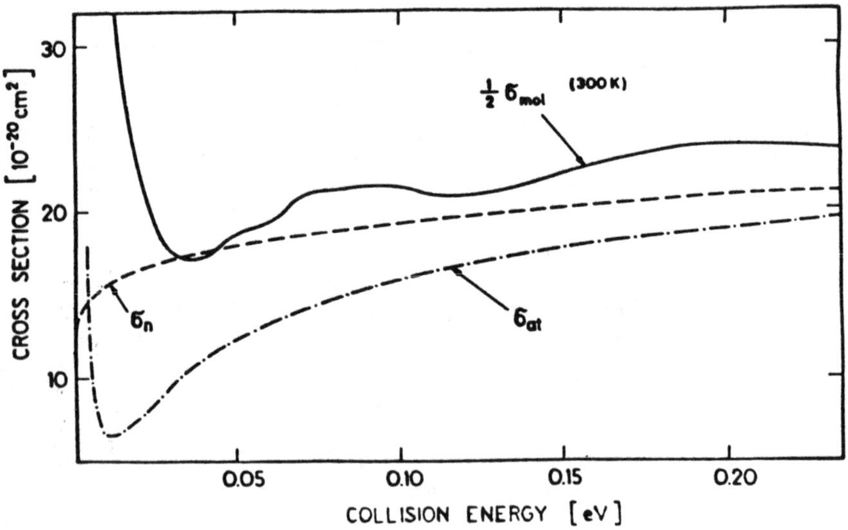

Fig. 2. Elastic scattering of F=$\frac{1}{2}$ dμ atoms from d (σ_n), D atoms (σ_{at}) and D_2 molecules (σ_{mol}) (from ref. 48).

HYPERFINE TRANSITIONS

The hyperfine transitions

$$(d\mu)^{F=\frac{3}{2}} + d \rightleftarrows (d\mu)^{F=\frac{1}{2}} + d \tag{19}$$

$$(t\mu)^{F=1} + t \rightarrow (t\mu)^{F=0} + t \tag{20}$$

come about because of *exchange scattering*. Matveenko and Ponomarev[46] calculated these rates using the two level approximation. Recent calculations of Melezhik and collaborators[47] use the adiabatic representation. Results for the rates for $d\mu + d$ are shown in Fig. 3, along with the low temperature SIN point.[9] While this looks like good agreement, the theoretical value will be significantly raised when the contribution from *resonant hyperfine quenching*,[17,49], i.e., the sequence

$$(d\mu)^{\frac{3}{2}} + D_2 \rightarrow [(dd\mu)^{*}dee]^{*} \rightarrow (d\mu)^{\frac{1}{2}} + D_2, \tag{21}$$

is included, thus leaving a discrepancy.

For quenching of the triplet $t\mu$ state the older two-level calculations[46] giving $\lambda_t = 9 \times 10^8 s^{-1}$ agree quite well with adiabatic representation result $9.1 \times 10^8 s^{-1}$ of Melezhik.[47] Kammel et al.[50] at SIN looked for the 'build-up' in time of the neutron signal at very low C_t ($\equiv$ tritium fraction) but saw only a very much faster build-up; this *may* indicate an experimental λ_t significantly larger than the predicted value. More experimental information is sorely needed.

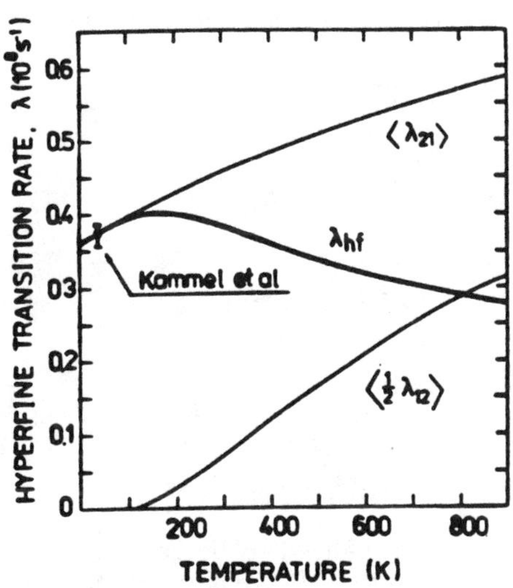

Fig. 3. Hyperfine transition rates for $d\mu$ atoms colliding with d's (from ref. 47).

d → t TRANSFER

The two-level calculation of Matveenko and Ponomarev[46] gave $1.9 \times 10^8 s^{-1}$ for the (ground state) d → t transfer, while the recent calculations of Melezhik[51] and Kobayashi et al.[52] give 2.7 and $2.6 \times 10^8 s^{-1}$ respectively. In discussing the measured values of λ_{dt}, we must keep in mind that these depend on what is ***assumed*** about the triplet quenching rate λ_t. The experimental numbers from several groups, $\simeq 2.8 \times 10^8 s^{-1}$,[14,15,53,54] correspond to neglect of the tμ triplet state, i.e., $\lambda_t = \infty$; smaller values of λ_t give larger λ_{dt} values.

KINETICS OF THE d-t CYCLE

In general, in order to calculate the rate of production of fusion neutrons as a function of time in terms of the underlying physical rates, it is necessary to solve the kinetic equations describing the system.[55] However, for the steady-state cycling rate λ_c (i.e., after any transients have become negligible), it is sufficiently simply to add the times the muon spends in each state, to get the cycle time[12]:

$$\lambda_c^{-1} = T_d + T_t, \tag{22}$$

$$T_d = \frac{q_{1s} C_d}{\lambda_{dt} C_t} \qquad (d\mu \text{ ground state}) \tag{23}$$

$$T_t = T_t^1 + T_t^\circ \qquad (\text{triplet and singlet } t\mu \text{ ground state}); \tag{24}$$

here,

$$T_t^1 = \frac{\frac{3}{4}}{\lambda_t C_t + \lambda^1_{dt\mu} C_d}, \tag{25}$$

$$T_t^\circ = \frac{\frac{1}{4} + \frac{3}{4}\chi}{\lambda^\circ_{dt\mu} C_d}, \tag{26}$$

and the branching ratio χ is given by

$$\chi = \frac{\lambda_t C_t}{\lambda_t C_t + \lambda^1_{dt\mu} C_d}. \tag{27}$$

These times and rates are normalized to liquid-hydrogen density, and the ddμ and ttμ channels are neglected. If the high temperature ratio of D_2 and DT molecules holds, we can write for the molecular formation rates

$$\lambda^F_{dt\mu} = C_d \lambda^F_{dt\mu - d} + C_t \lambda^F_{dt\mu - t}. \tag{28}$$

The factor $q_{1s}(\phi, C_t)$ is the probability of a muon, initially captured into a highly excited dμ atom, *not* transferring to a t during the (mainly collisional) deexcitation cascade. While some calculations of q_{1s} have been carried out,[56,57] the experiments seem to favor less drastic ϕ and C_t dependence than predicted. It seems likely that the deexcitation cascade is much more complex than the present models allow.

CONCLUSIONS

While much has been accomplished toward a complete quantitative description of the d-t catalysis cycle, the job is by no means finished. In particular, a really quantitative theory of three-body (many-body) molecular formation is sorely lacking. A more detailed treatment of the deexcitation cascade, from the initial atomic capture of the μ^- to the ground state dμ or tμ, is needed for calculating ω_{He} and more realistic $q_{1s}(\phi, C_t)$ values. And according to the dissenting view mentioned above, more careful treatment of the strong-interaction effects on λ_f and ω_s° is called for.

More precise calculations of the relativistic corrections to the dtμ and ddμ binding energies will soon become available, so that the two-body molecular formation rates $\lambda_{dd\mu}$ and $\lambda_{dt\mu}$ can be computed more precisely. Presumably, more complete results for elastic scattering, hyperfine quenching and d $\rightarrow$ t transfer will also soon be completed, allowing more definitive conclusions about thermalization, kinetics, etc. Then comparison with experiment will show where problems remain.

ACKNOWLEDGEMENT

This survey has benefitted from many helpful and instructive conversations with various μCF colleagues, most especially with James S. Cohen.

REFERENCES

1. V. E. Markushin, Zh. Eksp. Teor, Fiz. 80, 35 (1981) [Sov. Phys. JETP 53, 16 (1981)].
2. L. I. Ponomarev and M. P. Faifman, Zh. Eksp. Teor. Fiz. 71, 1689 (1976) [Sov. Phys. JETP 44, 886 (1976)].
3. E. A. Vesman, Zh. Eksp. Teor. Fiz. Pisma 5, 113 (1967).
4. S. I. Vinitsky and L. I. Ponomarev, Zh. Eksp. Teor. Fiz. 72, 1670 (1977) [Sov. Phys. JETP 45, 876 (1977)].
5. S. S. Gershtein and L. I. Ponomarev, Phys. Lett. 72B, 80 (1977).
6. E.g., A. D. Gocheva et al., Phys. Lett. 153B, 349 (1985).
7. L. I. Ponomarev and G. Fiorentini, Muon Catalyzed Fusion 1, 3 (1987).
8. J. Zmeskal et al., Muon Catalyzed Fusion 1, 109 (1987).
9. P. Kammel et al., Phys. Lett. 112B, 319 (1982); Phys. Rev. A 28, 2611 (1983).
10. S. E. Jones et al., Phys. Rev. Lett. 56, 588 (1986).
11. S. I. Vinitsky et al., Zh. Eksp. Teor. Fiz. 74, 849 (1978) [Sov. Phys. JETP 47, 444 (1978)].
12. M. Leon, Phys. Rev. Lett. 52, 605 (1984).
13. J. S. Cohen and R. L. Martin, Phys. Rev. Lett. 53, 738 (1984).
14. S. E. Jones et al., Phys. Rev. Lett. 51, 1757 (1983).
15. W. H. Breunlich et al., Muon Catalyzed Fusion 1, 67 (1987).
16. L. I. Menshikov and M. P. Faifman, Yad. Fiz. 43, 650 (1986) [Sov. J. Nucl. Phys. 43, 414 (1986)].
17. L. I. Menshikov et al., Zh. Eksp. Teor. Fiz. 92, 1173 (1987) [Sov. Phys. JETP 65, 656 (1987)].
18. A. M. Lane, Phys. Lett. 98A, 337 (1983).
19. Yu. V. Petrov, Phys. Lett. 163B, 28 (1985).

20. Bogdanova et al., Zh. Eksp. Teor. Fiz. 83, 1615 (1982) [Sov. Phys. JETP 56, 931 (1982)].
21. L. I. Menshikov and L. I. Ponomarev, Phys. Lett. 167B, 141 (1986).
22. M. Leon, Muon Catalyzed Fusion 1, 163 (1987).
23. Yu. V. Petrov et al., Muon Catalyzed Fusion 2, 261 (1988).
24. L. I. Menshikov, ibid, p. 273.
25. A. M. Lane, to be published.
26. J. S. Cohen and M. Leon, preprint LA-UR-88-1073.
27. M. Baranger, in Atomic and Molecular Processes, ed. D. R. Bates (Academic, NY, 1962), ch. 13.
28. E.g., I. I. Sobelman et al., Excitation of Atoms and Broadening of Spectral Lines (Springer-Verlag, Berlin, 1981), ch. 7.
29. M. Leon, Proc. Workshop on Fundamental Muon Physics, Los Alamos (1986), LA-10714-C, p. 151.
30. L. I. Ponomarev, Muon Catalyzed Fusion 3, 629 (1988).
31. K. Fukushima and F. Iseki, Muon Catalyzed Fusion 1, 225 (1987).
32. L. N. Bogdanova, Muon Catalyzed Fusion 3, 359 (1988).
33. V. E. Markushin, Muon Catalyzed Fusion 1, 297 (1987).
34. M. C. Struensee et al., Phys. Rev A 37, 340 (1988).
35. L. N. Bogdanova et al., Nucl Phys. A454, 653 (1986); see also L. Bracci and G. Fiorentini, Nucl. Phys. A364, 383 (1981).
36. D. Ceperley and B. J. Alder, Phys. Rev. A 31, 1999 (1985).
37. C.-Y. Hu, Phys. Rev. A34, 2536 (1986).
38. See also N. Takigawa and B. Müller, Muon Catalyzed Fusion 1, 341 (1987).
39. H. Rafelski et al., Muon Catalyzed Fusion 1, 315 (1987).
40. J. S. Cohen, Phys. Rev. Lett. 58, 1407 (1987).
41. V. E. Markushin, Muon Catalyzed Fusion 3, 395 (1988).
42. H. Bossy et al., Phys. Rev. Lett. 59, 2864 (1987).
43. Yu. A. Aristov et al., Yad. Fiz. 33, 1066 (1981) [Sov. J. Nucl. Phys. 33, 564 (1981)].
43. S. E. Jones et al., Phys. Rev. Lett. 51, 1757 (1983).
44. V. I. Fomichev et al., L.N.P.I. preprint 1177 (1986).
45. M. Leon et al., Muon Catalyzed Fusion 2, 231 (1988).
46. A. V. Matveenko and L. I. Ponomarev, Zh. Eksp. Teor. Fiz. 59, 1593 (1970) [Sov. Phys. JETP 32, 871 (1971)].
47. V. S. Melezhik, Muon Catalyzed Fusion 1, 205 (1987).
48. A. Adamczak and V. S. Melezhik, Muon Catalyzed Fusion 2, (1988).
49. M. Leon, Phys. Rev. A 33, 4434 (1986).
50. P. Kammel et al., Muon Catalyzed Fusion 3, 483 (1988).
51. V. S. Melezhik, to be published.
52. K. Kobayashi et al., Muon Catalyzed Fusion 2, 191 (1988).
53. V. M. Bystritsky et al., Zh. Eksp. Teor. Fiz. 80, 1700 (1981) [Sov. Phys. JETP 53, 877 (1981)].
54. D. V. Balin et al., Zh. Eksp. Teor. Fiz. 92, 1543 [Sov. Phys. JETP 65, 866 (1987)].
55. S. S. Gershtein et al., Zh. Eksp. Teor. Fiz. 78, 2099 (1980) [Sov. Phys. JETP 51, 1053 (1980)].
56. L. I. Menshikov and L. I. Ponomarev, Pisma Zh. Eksp. Teor. Fiz. 39, 542 (1984) [JETP Lett. 39, 663 (1984)]; –Pisma Zh. Eksp. Teor. Fiz. 42, (1985) [JETP Lett. 42, 13 (1985)].
57. A. V. Kravtsov et al., Phys. Lett. A, to be published.

POSSIBLE INFLUENCE OF VACUUM POLARIZATION ON Q_{1s} IN MUON CATALYZED D-T FUSION

B. Müller[1,2] and J. Rafelski[1]
[1]Dept. of Physics, University of Arizona, Tucson, AZ 85721
[2]Inst. f. Theor. Physik, Universität, D-6000 Frankfurt, West Germany

M. Jändel
AFI, Roslagsvägen 100, S-10405 Stockholm, Sweden

S.E. Jones
Dept. of Physics, Brigham Young University, Provo, UT 84602

ABSTRACT

The vacuum polarization splitting of the M-shell states in muonic hydrogen can have a profound influence on the muonic de-excitation cascade in deuterium and tritium targets. The cascade also shows sensitive dependence on the precise rate of transfer processes between certain excited muonic deuterium and tritium atoms. Recent experimental data, where a much greater population of the ($d\mu$) 1s state (q_{1s}) was found than previously predicted, can be explained if the transfer rates from the ($d\mu$) M-shell are assumed to be strongly suppressed.

INTRODUCTION

Understanding of the atomic capture and de-excitation cascade of muons is of profound importance in the study of muon catalyzed fusion [1], because the fraction of muons reaching the deuterium ground state has a large influence on the overall fusion rate. In particular, the fusion cycle is inhibited by the slow transfer of muons between the ground states of deuterium and tritium. The muonic cascade is determined by a competition between radiative transitions, density dependent external Auger transitions, density dependent quenching of the muonic levels, and transfer processes which depend both on density ϕ and tritium concentration c_t. As the rates for these processes differ widely between atomic shells, and also within each shell, the cascade can take very different routes depending on the actual population of these states. The original prediction of this cascade by Menshikov and Ponomarev [2] has been observed to differ significantly from the experimental results [3-5], in particular, q_{1s} was found to fall much less rapidly with c_t than predicted.

THE MECHANISM: M-SHELL SPLITTING AND MOLECULAR TRANSFER SUPPRESSION

Our present conjecture is based on the observation that the splitting due to vacuum polarization between states in the M-shell of muonic hydrogen is about 70 meV, of the order of thermal energies in the experiments. This splitting is larger than the rates of Stark mixing between these states, so that they should retain approximately good angular momentum. On the other hand, as Stark mixing is believed [6] to be sufficiently fast so that the substates of the M-shell are populated statistically, for densities not less than ϕ=0.01, the population of the 3s, 3p, and 3d states will be strongly temperature dependent below 500 K.

Of course, this observation relies on the fact that thermal equilibrium among the substates of the M-shell is established. We rely here on a detailed study of (μd) elastic and charge-exchange cross sections by Menshikov and Ponomarev [7], which shows that the thermalization rate of excited muonic atoms is $10^{12}s^{-1}$ at ϕ=1, i.e. one order of magnitude larger than the Auger decay rates of M-shell states. Radiative transition rates are about $10^{10}s^{-1}$ for n>2 and hence will not cause any significant deviation from a thermal distribution for densities larger than $\phi = 10^{-2}$. The thermal equilibrium will not prevail in the more strongly split L-shell, since the rate of quenching at $T < 500$K is smaller than the rate of de-excitation of the 2p-level.

The vacuum polarization splitting of the M-shell would not influence the muon cascade, if the transfer rates had the strengths computed in refs. [2,7]. However, if the observation is combined with the conjecture of strong suppression of the M-shell transfer rate, our analysis shows that a major change in the cascade occurs. At low temperatures, when the 3s state is dominantly populated, the muon falls into the 2p state which rapidly decays to the K-shell by radiation emission. At higher temperatures, a significant fraction of muons is in the 3p state which, after decaying into the metastable 2s state, mostly leads to transfer of the muon to a tritium atom. Thus the combination of M-shell splitting and transfer suppression has the effect of (a) enhancing the population, q_{1s}, of the muonic ground state in deuterium - in agreement with experimental observations, and (b) yielding a functional dependence of q_{1s} that is falling with temperature in accordance with experimental results [3].

It must be noted that, at present, such a reduction in the transfer rate with respect to previous calculations is to a large degree hypothetical but not implausible, given the peculiar properties of the M-shell states: The energy gain in the transfer to the tritium M-shell (5.3eV = 48eV/3^2) is very close to the dissociation energy of the target molecule (4.6eV), and the subshell splittings closely match the rotational energies in the hydrogen molecule. As we shall see, we require a suppression of the M-shell transfer by about two orders of magnitude. At this moment, we do not have a satisfactory, quantitative explanation for such a large suppression factor. (A calculation of the influence of molecular binding along the lines of neutron scattering theory [8] exhibits both enhancing and suppressing effects.) In view of the very delicate molecular structure effects involved in the computation of the M-shell transfer rate we will, therefore, use this suppression as a free parameter of our calculation, and concentrate on analyzing the cycle dynamics in terms of experimentally observed effects.

1S-POPULATION OF MUONIC DEUTERIUM IN A D-T MIXTURE

The relevant level structure and decay scheme of a μd atom is shown in Fig. 1. Level splittings in the L- and M-shell are determined mainly by the vacuum polarization corrections, which amount to a relative shift of 220 meV between the 2s and 2p states, and 66.5 meV and 72.1 meV between the 3s and

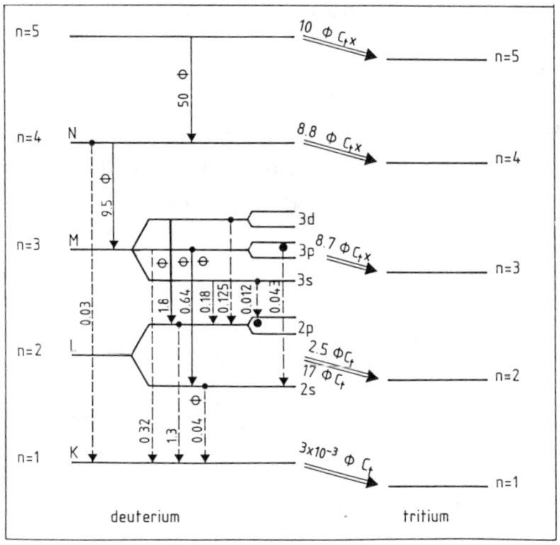

Fig. 1: Level scheme of muonic deuterium with transition and transfer rates. Dashed lines indicate radiative, solid lines Auger transitions.

3p and the 3s and 3d states, respectively. (The values used here also include the much smaller fine structure and hyperfine structure effects.) The strengths of the radiative and external Auger transitions, and of the transfer rates of Menshikov and Ponomarev [2] are shown in Fig. 1, in units of 10^{11} s^{-1}. The Auger and transfer rates depend linearly on the target density ϕ. Note the competition between Auger and radiative transitions of the 3p state, with the radiative decay into the 1s ground state dominating at lower densities (ϕ <0.5). The populations n_{3s}, n_{3p} and n_{3d} are related by,

$$n_{3p} = 3n_{3s}\exp(-\Delta_{3p}/T), \qquad n_{3d} = 5n_{3s}\exp(-\Delta_{3d}/T), \tag{1}$$

where Δ_{3p} and Δ_{3d} are the energy differences to the 3s state, respectively and T is the temperature. Since this splitting influences the statistical populations, there will be a significant temperature dependence of the muonic cascade in deuterium, in particular for $T < 500K$, which in turn influences the kinetics of the muon catalysis cycle.

The rates for the transfer process

$$(d\mu)_n + t \quad \rightarrow \quad d + (t\mu)_n \tag{2}$$

have been calculated [2] for collisions of d atoms with tritium atoms, not molecules. For atom-molecule collisions the transfer process can be strongly influenced by a substantial change of the final state density, as remarked above. For the K- and L-shells the transfer can easily be accompanied by dissociation of the target molecule, whereas those for n=3 and higher shells cannot. Thus for K- and L-shells the values calculated by Menshikov and Ponomarev probably apply, but for the M-shell a molecular suppression mechanism may well be active.

We now turn to consider the effects of M-shell splitting and transfer suppression in muon catalyzed fusion. If the muon reaches the 1s state in deuterium the (dtμ) fusion cycle is significantly delayed due to the very low transfer rate (3×10^8s$^{-1}\phi C_t$) from this state. The (dμ) K-shell population probability is, therefore, a quantity of considerable practical interest. This probability can be written $C_d q_{1s}$ where q_{1s} is the probability for a muon, initially captured by a deuteron, to reach the (dμ) groundstate. C_d is the fraction of deuterium in the target. Calculations by Menshikov and Ponomarev [2], which are based on uninhibited transfer from all excited states and ignore the splitting of the M-shell, predict very small values for q_{1s}. Aside from the quoted muon catalyzed fusion experiments [3-5], other work, specifically designed to measure q_{1s} and recently carried out [9], has also not been consistent with a small value of q_{1s}.

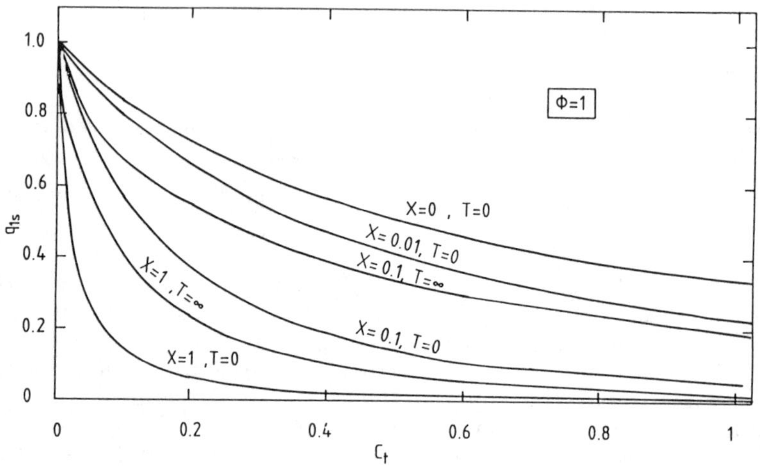

Fig. 2: Fraction of muons reaching the deuterium 1s state for various values of the suppression factor x, as function of tritium concentration for $\phi = 1$.

As discussed above, a molecular suppression factor, denoted by x, is conjectured to strongly inhibit the transfer from states in the (dμ) M-shell. The dependence of q_{1s} on the reduced strength of the transfer from higher states is shown in Fig. 2, which gives q_{1s} as function of C_t for several values of the parameter x. Our calculation with the value x=1 corresponds to Ponomarev's model, but includes the effects of vacuum polarization on the populations of the L- and M-shell substates. Assuming total transfer suppression (x=0) and a thermal distribution in the M-shell, we find much larger values, e.g. q_{1s}=0.5 at ϕ=1 and C_t =0.5 as compared with q_{1s}=0.08 predicted in the absence of a molecular suppression effect. The dependence on tritium concentration was experimentally measured by Jones et al. [3] for ϕ=0.72 and T=300K, who found $q_{1s}(C_t=0.5)/q_{1s}(C_t=0.04) = 0.72\pm0.15$. Our result for this ratio is 0.67, whereas Menshikov and Ponomarev find the much smaller value of 0.14. The measured ϕ-dependence [3] is also in better agreement with our calculations.

The temperature dependence of q_{1s} is a sensitive probe for the value of x. We find that q_{1s} *decreases* with increasing temperature for x=0, whereas this trend is reversed even for values as small as x = 0.01. In the Menshikov-Ponomarev model (x = 1), modified to include a thermal distribution of the M-shell sublevels, q_{1s} increases by more than a factor 3 within the interval T = 0 - 500 K. (The original Menshikov-Ponomarev model has no temperature dependence.) Using data of Jones et al. [3] for λ_{dt} and q_{1s}, and assuming that λ_{dt} is independent of the temperature instead of q_{1s}, we find that $(q_{1s}^{-1}\text{-}1)$ is proportional to $(6\pm1)\times10^{-4}$T. This result is displayed in Fig. 3 where it is seen to be in reasonable agreement with the case x = 0, i.e. total suppression of the transfer from $(d\mu)_{n>2}$ atoms to (molecular) tritium.

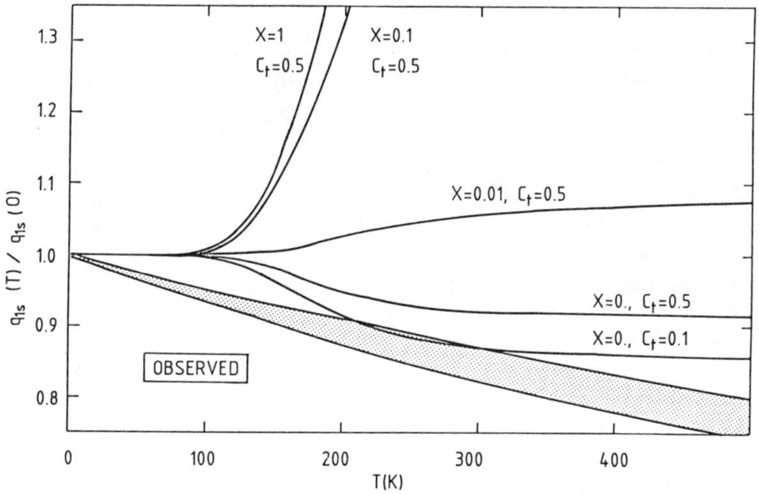

Fig. 3: Temperature dependence of q_{1s} for various values of c_t and x. The observed temperature dependence, indicated by the shaded area, is that of Jones et al. (ref.[3]).

MUONIC X-RAYS IN PURE DEUTERIUM

The muonic cascade in pure deuterium can be employed to verify the T-dependence in the M-shell population. As before, we assume here that the population of the substates of the M-shell is thermally equilibrated for not too low densities ($\phi > 0.1$). Due to continuous fast repopulation of all states within the shell the independent populations n_{3s}, n_{3p}, and n_{3d} decay with the same average rate. However, only muons populating the 3p state contribute to the yield of K_β (3p-1s) radiation. This yield is hence directly proportional to the 3p partial population and the branching ratio into this radiative channel. Decays from both the 3s and 3d states populate the 2p level which also decays into the 1s level. Taking the ratio of the K_β and prompt K_α radiation we eliminate to a large extent our ignorance about other details in the cascade.

Calculations of the ratio K_β/K_α as a function of T for different ϕ show a strong dependence on the precise splitting between the 3p and 3s states indicating the sensitivity of this quantity to the hypothesis of thermal equilibrium in the M-shell. This temperature dependence provides also for a measure of the energy difference between the 3s and 3p states and is hence an indirect measurement of the vacuum polarization effect.

CONCLUSIONS

We have shown that a significant temperature dependence of q_{1s} can result from the vacuum polarization splitting in the M-shell of the (μd) atom, if the M-shell transfer rates of muons from deuterium to tritium atoms bound in molecules are strongly suppressed. Our conjecture draws heuristic support mainly from the fact that it allows for a much better description of recent experimental results [3-5].

We note that, if our line of argument for molecular transfer suppression is correct, the suppressing mechanism is probably not active for *atomic* tritium. Calling the fraction of atomic tritium x' and using Fig. 2 we find that $q_{1s} = 0.5 - 10x'$ at $\phi=1$. Increasing x' to 1% hence would yield a 20% increase in the cycling rate, for large $\lambda_{dt\mu}$. Obviously, this effect could be used to test the presence of our conjectured molecular M-shell transfer suppression experimentally. For this it would suffice to add a small fraction of atomic tritium to the target. The temperature dependence of the cascade due to vacuum polarization splitting, on the other hand, can be studied by measuring the K_β/K_α ratio in pure deuterium.

Acknowledgement: We would like to thank E.Zavattini for the communication of unpublished results. This work was supported in part by the U.S. Department of Energy.

REFERENCES

1. L.Bracci and G.Fiorentini, Phys. Rep. **86**, 169 (1982).
2. L.I.Menshikov and L.I.Ponomarev, Pis'ma ZhETF **39**, 542 (1984) [JETP Lett. **39**, 663 (1984)].
3. S.E.Jones *et al.*, Phys. Rev. Lett. **56**, 588 (1986).
4. A.N.Anderson, Report at this conference.
5. W.H.Breunlich, Report at this conference.
6. L.Bracci and G.Fiorentini, Nucl. Phys. **A364**, 383 (1981).
7. L.I.Menshikov and L.I.Ponomarev, Z. Phys. **D2**, 1 (1986).
8. S.W.Lovesey, *Theory of Neutron Scattering*, vol.1, Clarendon Press, Oxford 1984.
9. M.Leon, Proc. Int. Symp. on Muon Catalyzed Fusion, Gatchina (1987).

PRESSURE BROADENING OF THE $[(dt\mu)dee]^*$ FORMATION RESONANCES

James S. Cohen, M. Leon, and N. T. Padial
Los Alamos National Laboratory, Los Alamos, New Mexico 87545

ABSTRACT

The treatment of $[(dt\mu)dee]^*$ formation at high densities as a pressure broadening process is discussed. Cross sections for collisions of the complex $(dt\mu)dee$, and of the D_2 molecule from which it is formed, with the bath molecules have been accurately calculated. These cross sections are used to calculate the collisional width in three variations of the impact approximation that have been proposed for this problem. In general, the quasistatic approximation is shown to satisfy the usual conditions of muon-catalyzed fusion better than does the impact approximation. A preliminary rough treatment is presented to illustrate the quasistatic approximation.

I. INTRODUCTION

The diagram in Fig. 1 shows the currently predicted positions[1] of the resonances that may contribute to $dt\mu$ molecular formation (mf) in a collision of $t\mu$ with D_2 at low temperature,

$$t\mu + D_2\,[\nu_i{=}0,\,J_i] \longrightarrow (dt\mu)dee\,[\nu_f{=}2,\,J_f]\ . \tag{1}$$

In all cases considered here the target D_2 is in its ground vibrational state and the complex $(dt\mu)dee$ is formed with electronic molecular vibrational quantum number of 2, so the various possible transitions will be designated by the initial and final rotational quantum numbers, $J_i \to J_f$ The amplitude shown for each resonance in Fig. 1 is *roughly* proportional to its mf matrix element *and* to the abundance of the initial state in a low-temperature target. Until Petrov[2] published his germinative paper in 1985, each of these resonances was viewed as a δ function.[3] The δ functions above threshold can be reached by energetic $t\mu$ atoms in the Maxwellian distribution, but those below threshold are completely inaccessible in this picture. Petrov[2] pointed out that each resonance actually has a finite width due both to intramolecular (electronic Auger) contributions as well as intermolecular (collisional) effects. Menshikov and Ponomarev[4] called attention to the possibility that three-body effects,

$$t\mu + D_2 + X \longrightarrow (dt\mu)dee + X \tag{2}$$

where X is D_2, DT, or T_2, could be responsible for the observation of Jones *et al.*[5] of a nonlinear dependence of the mf rate on density. We believe that these three (*or more*) body effects are very usefully interpreted as pressure broadening of the resonances.

II. PRESSURE BROADENING APPROXIMATIONS

In the diagram of Fig. 2, we have picked out the two most promising below-threshold resonances and attached a *hypothetical* line-shape function to

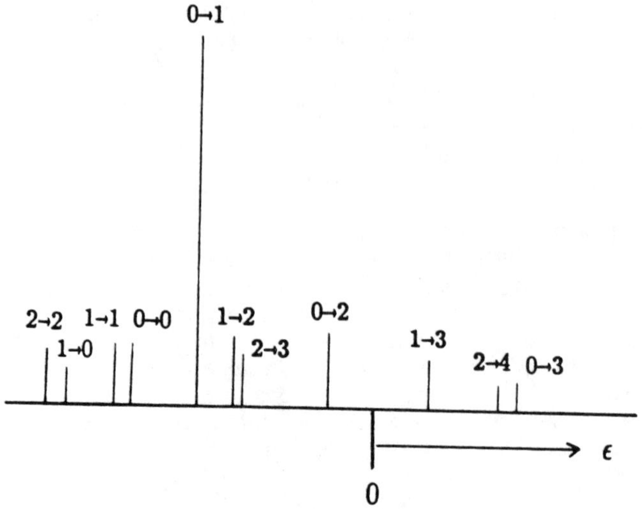

Fig. 1. Schematic of $[(dt\mu)dee]^*$ formation resonances, $J_i \rightarrow J_f$, with heights roughly indicative of the size of the matrix element and abundance of the initial state in a low-temperature target.

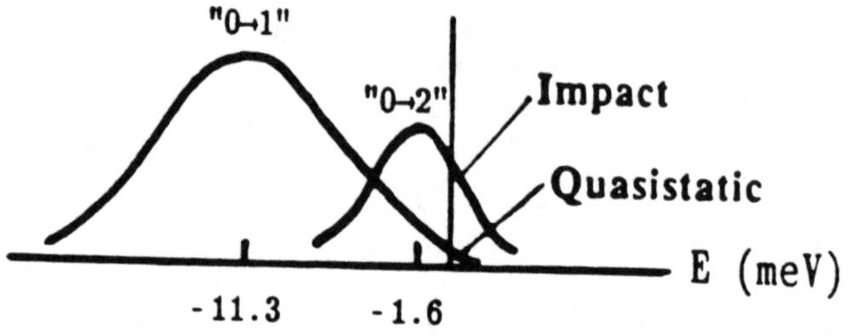

Fig. 2. The two most important below-threshold resonances. The line shapes here are drawn to exhibit the impact and quasistatic approximations and have no quantitative significance.

them. The contribution of a negative-energy resonance to the mf rate is given by the overlap of the line profile with the thermal (e.g., Maxwellian) distribution — this occurs, of course, only at positive energies. Here we have drawn these two profiles as if one, the 0→2 resonance, lies very close to threshold, whereas the other, the 0→1 resonance, reaches threshold only with its far wing. This situation illustrates, though not completely as we shall see, the two main line-broadening approximations: the impact approximation valid for small $|\Delta E|$ and the quasistatic approximation valid for large $|\Delta E|$.[6] We will say more in Sec. III about the validity of these two mutually exclusive approximations, but for now we just want to show what is needed to carry them out. The most used approximation in muon-catalyzed fusion (μCF) in the past has been the impact approximation, sometimes called the Lorentzian approximation because it always yields a line shape with Lorentzian functional form,

$$I_{\rm imp}(\omega) = \frac{\Gamma_{\rm imp}/\pi}{(\omega-\omega_0)^2 + (\Gamma_{\rm imp}/2)^2} \; . \tag{3}$$

Three formulas for the width $\Gamma_{\rm imp}$ of the Lorentzian have been proposed: the usual one from optics[6-7] and one recently proposed by Menshikov[8] as well as the formula originally proposed by Petrov.[2] In all three, the width is simply proportional to the density. In optics where the photon carries negligible momentum, $\Gamma_{\rm imp}$ depends on the incoherent sum of the inelastic cross sections for the initial *and* final states plus the coherent difference of the elastic scattering amplitudes,[7]

$$\Gamma^{\rm op}_{\rm imp} = \hbar \left\langle\, nv \,[\, \sigma_{\rm i}^{\rm inel} + \sigma_{\rm f}^{\rm inel} + \int\!\!\int |f_{\rm i}^{\rm el}(\theta,\varphi) - f_{\rm f}^{\rm el}(\theta,\varphi)|^2 \sin\theta \; d\theta\, d\varphi] \right\rangle_{\rm T} , \tag{4}$$

where the indicated average is over the velocity distribution and n is the density.

However, when the photon is replaced by the massive $t\mu$ atom, Menshikov[8] has asserted that the impact of $t\mu$ completely disrupts the effect of the initial state and the result is the same as Eq. (4) *except* with the initial state deleted; i.e., it is given by the total cross section for the final state only,

$$\Gamma^{\rm sr}_{\rm imp} = \hbar \left\langle\, nv \,[\, \sigma_{\rm f}^{\rm inel} + \sigma_{\rm f}^{\rm el} \,] \right\rangle_{\rm T} \tag{5}$$

where "sr" stands for strong recoil. In both of the above formulas elastic scattering makes a large contribution, in contrast to the formula of Petrov which is like Eq. (5) *except* that it contains only $\sigma_{\rm f}^{\rm inel}$,

$$\Gamma^{\rm inel}_{\rm imp} = \hbar \left\langle\, nv\, \sigma_{\rm f}^{\rm inel} \right\rangle_{\rm T} \; . \tag{5$'$}$$

No rigorous derivation of this last formula exists, but it is at least interesting in a heuristic sense and *appears* to have some features in common with the observed mf rate. We have accurately calculated all these cross sections; the results are given in Sec. IV.

The treatment *opposite* to the impact approximation is known as the quasistatic approximation.[6] It depends on the wave function *during* the collisions, not just on asymptotic properties like cross sections. In the usual quasistatic formula,

$$I_{qs}(\omega) = \int \rho(\vec{R})\ \delta[V_f(\vec{R}) - V_i(\vec{R}) - \hbar\omega]\ d^{3N}R \ , \qquad (6)$$

where the integration is over the position vectors of all the neighbors. Here $\rho(\vec{R})$ is the spatial distribution of all molecules, which depends on the system potential energy and the temperature, and serves in Eq. (6) as the weighting function of the energy shift of the target molecule due to the different potential it sees before and after the transition. Equation (6) does not yield a universal line-shape function analogous to Eq. (3); however, the shape is generally exponential in the line wings. Also, the quasistatic width is not simply proportional to the density like the impact width.

The pressure-broadened mf rate is usually written

$$\lambda^{i}_{mf}(T) = \int_0^\infty \lambda^{(2)i}_{mf}(\epsilon,T)\ I(\epsilon-\epsilon_i,T)\ d\epsilon \qquad (7)$$

where ϵ_i is the unperturbed energy of the ith resonance, $\lambda^{(2)i}_{mf}(\epsilon,T)$ is the two-body rate at the perturbed energy ϵ, and $I(\epsilon-\epsilon_i,T)$ is the line-broadening function. It is usually assumed that the broadening function does not depend in any essential way on the particular resonance being considered. Though this assumption is intuitively appealing and operationally convenient, its validity is not really obvious. First of all, the cross sections for collisions with the bath molecules depend on the quantum numbers of the molecule, but this dependence has been shown to be relatively weak.[9-10] Possibly more important is the dependence that comes about because the center of mass of $(dt\mu)dee$ does not coincide with that of the D_2 molecule from which it is formed.[11] This shift causes the broadening to depend on the angular momentum of the transition; i.e., the broadening is different for the 0→1 and 0→2 transitions. Numerically the difference could be as much as a factor of 2 for a given $\Delta E\ (=\epsilon-\epsilon_i)$. Of course, as a center-of-mass effect it can change things only to the extent that the bath molecules alter the recoil energy. The actual importance of this observation is not yet known.

III. CRITERIA FOR LINE-BROADENING APPROXIMATIONS

The fundamental conditions for the impact approximation come from the requirement that the Fourier integral (note $\Delta E = \hbar\omega$)

$$I(\omega) = \frac{1}{\pi}\,\mathcal{R}e\left[\int_0^\infty \Phi(t)\ e^{-i\omega t}\ dt\right] \qquad (8)$$

of the correlation function

$$\Phi(t) = \langle\, f^*(0)\ f(t)\,\rangle_T \qquad (9)$$

yield a Lorentzian.[6] $\Phi(t)$ is the ensemble average of the overlap of the functional f describing the oscillation of the system at different times, subject to the interactions with the neighboring bath molecules. For Eq. (8) to yield a Lorentzian, $\Phi(t)$ must be an exponential at times that contribute most to the integral; i.e., the correlation function must be exponentially decaying by times $t \sim |\omega|^{-1}$. This condition requires, first of all, binary collisions since otherwise the interaction is maintained by additional collisions and can never exponentially decay. Secondly, the detuning $|\omega|$ must not be too large since otherwise the interaction will still be dynamically developing during the transition. The first condition is given by the inequality

$$\Delta t_c >> \tau_c \tag{10}$$

where

$$\Delta t_c = \frac{1}{nv\sigma} \tag{11}$$

is the time *between* collisions and

$$\tau_c = \frac{\rho_c}{v} \simeq \frac{1}{v}\left[\frac{\sigma}{\pi}\right]^{\frac{1}{2}} \tag{12}$$

is the *duration* of a collision. The second condition is given by the inequality

$$|\omega| << \tau_c^{-1} \ . \tag{13}$$

Note that a different criterion has often been stated to justify the application of the impact approximation to $dt\mu$ formation, namely,

$$|\Delta E| \lesssim \Gamma \ . \tag{14}$$

To see how this inequality is related to the *fundamental* conditions above, Eqs. (10) and (13), we can use the relation

$$\Gamma \simeq \frac{\hbar}{\Delta t_c} \ . \tag{15}$$

Hence we can rewrite Eq. (10) as

$$\Gamma << \hbar/\tau_c \tag{16}$$

and Eq. (13) as

$$|\Delta E| << \hbar/\tau_c \ ; \tag{17}$$

however, these two inequalities imply nothing about the relation of Γ to $|\Delta E|$, and, in fact, Eq. (14) is neither a necessary nor a sufficient condition for the impact approximation.

It is convenient to rewrite the criteria of Eqs. (10) and (13) as

$$\phi << \frac{\sqrt{\pi}}{n_0 \sigma^{3/2}} = \frac{8.3\times10^{-23}}{\sigma^{3/2}} \, , \tag{10'}$$

where n_0 is the density of liquid hydrogen (LHD) and ϕ is the target density in LHD units, and

$$|\Delta E| << \hbar v \sqrt{\pi/\sigma} = 1.17\times10^{-12} \, v/\sqrt{\sigma} \text{ meV} \, . \tag{13'}$$

These criteria are written in terms of a cross section σ that depends on what variation of the impact approximation is being used. The most important question is what contribution is made by elastic scattering. As it turns out, the answer to this question may determine whether or not the impact approximation is useful for calculations of $(dt\mu)dee$ formation. In available derivations, elastic scattering makes the dominant contribution to Γ_{imp}.

The effective cross sections in Eqs. (4) and (5) happen to be almost the same so the criteria can be evaluated together for these two formulas. At 300° K, $v \simeq 1.6\times10^5$ cm/s and $\sigma \simeq 7\times10^{-15}$ cm^2, yielding conditions

$$\phi << 0.14$$

and

$$|\Delta E| << 2.2 \text{ meV}$$

for validity of the impact approximation. At 20° K, the velocity, $v \simeq 4.1\times10^4$ cm/s, is lower and the cross section, $\sigma \simeq 2.6\times10^{-14}$ cm^2, is larger, so the situation is even worse for the impact approximation,

$$\phi << 0.02$$

and

$$|\Delta E| << 0.3 \text{ meV}$$

being required. These inequalities indicate that the standard impact approximation is not valid for the usual experimental target condition of near-liquid density, and that it will *never* be valid for the 0→1 resonance *if* that resonance lies ~11 meV below threshold as predicted.

On the other hand, if only the inelastic cross section is used, then σ is reduced by a factor of 10 at 300° K and a factor of 45 at 20° K. In this case, the density criterion would be satisfied, and the $|\Delta E|$ criterion would be satisfied for the 0→2 resonance though still not for the 0→1 resonance. We emphasize, however, that no current theory justifies ignoring the elastic contribution.

The quasistatic approximation is valid if the inequality of *either* Eq. (10) or (13) is reversed. Generally a many-body, rather than a binary, quasistatic calculation, will be required at high densities; however, a binary approximation may still be valid in the far wing of a line. To the extent that the bath molecules alter the target recoil from the $t\mu$ impact there may be an additional kinetic-energy effect. Hence the criteria for validity of Eq. (6) may be somewhat more stringent than in optics.

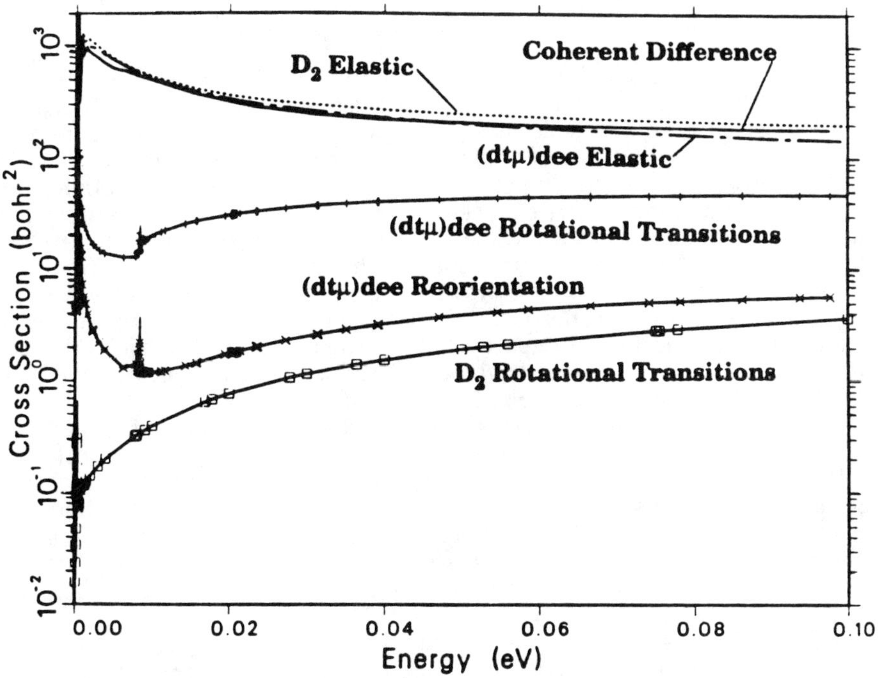

Fig. 3. Cross sections for calculation of the resonance width in the impact approximation for $t\mu + D_2 + D_2 \longrightarrow (dt\mu)dee + D_2$.

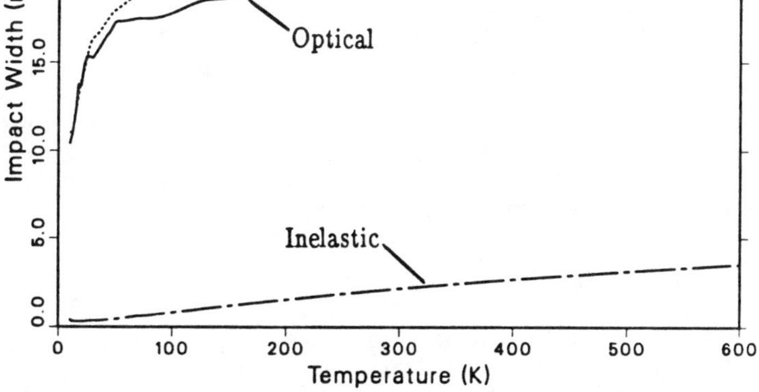

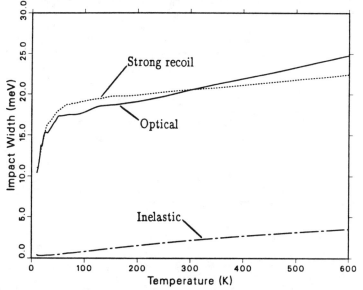

Fig. 4. Impact-approximation widths at LHD (ϕ=1). The solid curve is calculated with Eq. (4), the dashed curve with Eq. (5), and the chained curve with Eq. (5′).

IV. RESULTS IN THE IMPACT APPROXIMATION

We have calculated elastic[10] and rotationally inelastic[9] cross sections using the quantum-mechanical close-coupling method for collisions of $(dt\mu)dee$ and $(dd\mu)dee$ with normal molecules in the μCF targets. Some of the inelastic cross sections contribute to stabilization of the resonant complex initially formed; they are substantially smaller than the cross sections obtained by Ostrovsky and Ustimov[12] using a scaling law derived in the Born approximation. For example, the rate for the $J=0 \rightarrow J=1$ transition in $(dt\mu)dee + D_2$ collisions at 100° K is about 6 times smaller than calculated by Ostrovsky and Ustimov. However, the present interest is in the total broadening of the resonance due to these collisions. Both the elastic and the inelastic cross sections are required in the impact-approximation formulas (4) and (5). The calculations of these quantities have been described in detail previously,[9-10] so the important results will just be summarized here.

The cross sections contributing to the impact broadening of reaction (1) are shown in Fig. 3. The broadening is clearly dominated by elastic scattering whether the coherence difference, required for Eq. (4), or just the final-state $[(dt\mu)dee]$ cross section, required for Eq. (5), is used. Next in importance are the rotationally inelastic cross sections of $(dt\mu)dee$. These cross sections are dominated by $\Delta J=1$ transitions that are made possible by the unequal masses of $dt\mu$ and d. In D_2, only transitions with ΔJ *even* are allowed so the corresponding inelastic cross section is much smaller. Although the inelastic cross sections are increasing while the elastic cross sections are decreasing as the collision energy increases, elastic scattering is still the dominant contributor to the line broadening at 100 meV. The reorientation (m changing) cross sections are unimportant at all energies.

The above cross sections were actually calculated for $J_i=0$ and $J_f=1$, but they are rather insensitive to these choices. Hence it is reasonable to use the same cross sections for other J_i and J_f. The results of using these cross sections in Eqs. (4) and (5) are shown in Fig. 4. Remarkably, the calculated widths are about the same in the optical and strong-recoil formulations, and hence the impact analysis of line broadening will be insensitive to this choice. The width is more than three times that previously calculated[8] for a temperature of 23° K. The width calculated using only the final-state inelastic cross section [Eq. (5′)] is also shown in Fig. 4.

The very restrictive conditions for applicability of the impact approximation to pressure broadening of the $(dt\mu)dee$ formation resonances should be kept in mind. Disregarding this caution for the moment, we show in Fig. 5 the resulting mf rates as a function of density for the 0→1 and 0→2 resonances at 20° K. At this temperature the impact width is 14ϕ meV including the elastic contribution and 0.31ϕ meV taking into account only inelastic scattering. The total width includes the vacuum contribution; i.e.

$$\Gamma = \Gamma_{vac} + \phi\Gamma_{imp}(\phi=1) \tag{18}$$

where $\Gamma_{vac} = 0.84$ meV.[13] The value of the $dt\mu$ dipole matrix element (which does not affect the density dependence) is taken from Petrov *et al.*[14]

In the impact approximation the mf matrix element is clearly more important than the detuning; i.e., the 0→1 resonance at 11 meV below threshold still contributes much more than the 0→2 resonance at only 1.6 meV below threshold. The density dependence of the mf rate calculated using the width

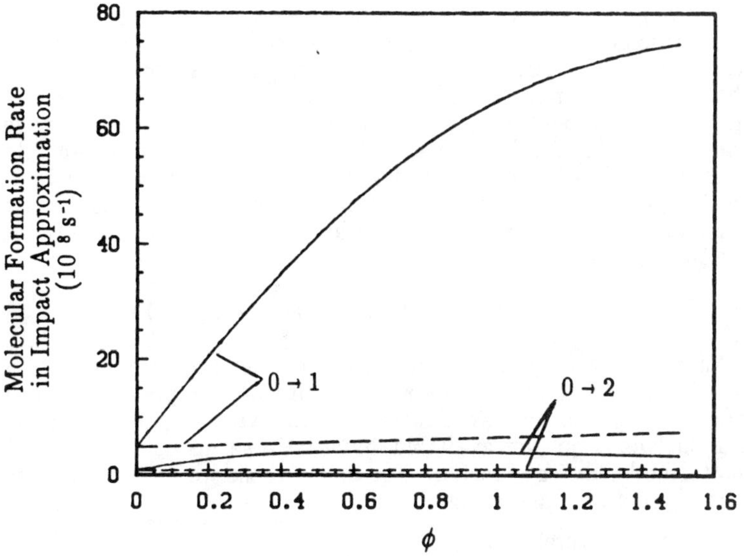

Fig. 5. Components of the molecular-formation rate at 20° K due to the 0→1 and 0→2 below-threshold resonances, calculated using *impact-approximation* widths. The solid curves use the width including both elastic and inelastic scattering; the dashed curves use the width taking only inelastic scattering into account.

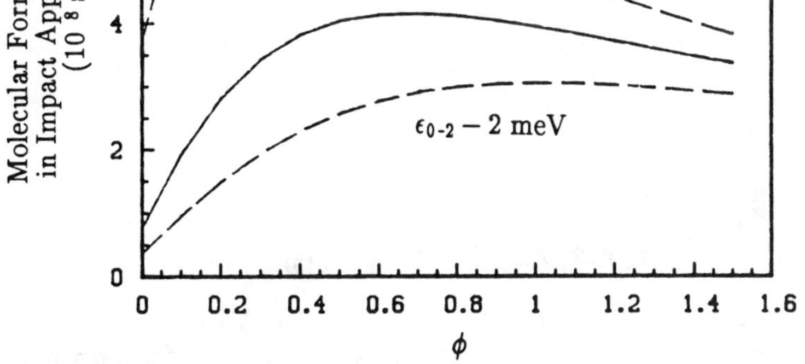

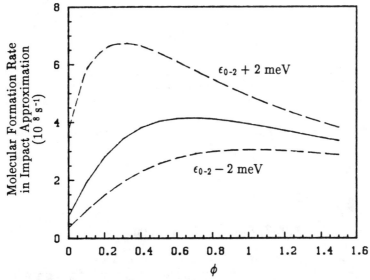

Fig. 6. Effect of raising or lowering the 0→2 resonance energy by 2 meV.

determined only by inelastic scattering actually seems to be in closer agreement with experiments.[5,15] However, it remains to be seen if this resemblance is significant since the criteria for impact broadening are not really satisfied even with the much smaller inelastic cross section. In Fig. 6 the effect of raising or lowering the 0→2 resonance energy by 2 meV is shown. The former moves the resonance above threshold. The actual resonance energies are uncertain by at least this amount. These curves demonstrate that an increase in the width (e.g., with ϕ) can actually decrease the mf rate when the resonance is very close to threshold; i.e., there exists an optimum width that maximizes the overlap with the thermal distribution.

V. QUASISTATIC APPROXIMATION

An accurate quasistatic calculation of the pressure broadening of the $(dt\mu)dee$ formation resonance would require a complete potential surface including the dependence on the intramolecular vibrational coordinate. Such a potential surface is not yet available, but it still seems desirable to exhibit now the qualitative behavior expected. For this purpose we have carried out a *rough* quasistatic calculation assuming:

1. Binary interactions
2. Exponential intermolecular potential
3. Interaction primarily due to nearest atoms (treating $dt\mu$ as a nucleus)
4. Rapid vibration and slow rotation compared with thermal motions.

The distribution of distances between nearest particles is given by

$$\rho(R)\; dR \propto 4\pi\, R^2\, n \exp\left[-\frac{4\pi}{3}\, nR^3 - \frac{V(R)}{kT}\right]\, dR \tag{19}$$

using the potential[16]

$$V(R) = V_0\, e^{-aR} \tag{20}$$

with $a = 1.7\; a_0^{-1}$ and $V_0 = 250$ eV. The quasistatic detuning is then given approximately by (for simplicity of notation, $\hbar$ is set to 1 in this section so $\omega = \Delta E$)

$$\omega = V_f - V_i \simeq \frac{dV}{dR}\,\Delta R = -\, a\; V \Delta R\;; \tag{21}$$

furthermore

$$\frac{d\omega}{dR} \simeq a^2\; V \Delta R\;. \tag{22}$$

Using Eqs. (19)–(22) in Eq. (6), we obtain

$$I_{qs}(\omega) = \frac{C}{\omega}\int_0^{\pi} \ln^2\left[\frac{V_0 a\,\Delta R}{\omega}\right] \exp\left[-\frac{4\pi n}{3a^3}\ln^3\left[\frac{V_0 a\,\Delta R}{\omega}\right] - \frac{\omega}{a\,\Delta R\, kT}\right]\sin\theta\; d\theta \tag{23}$$

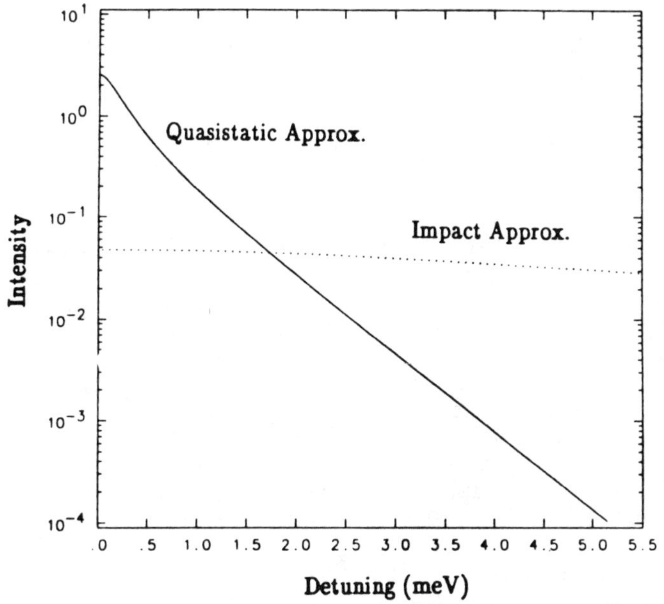

Fig. 7. Comparison of a crude quasistatic-approximation line shape (solid curve) with the impact-approximation line shape (dotted curve) at 20°K and LHD.

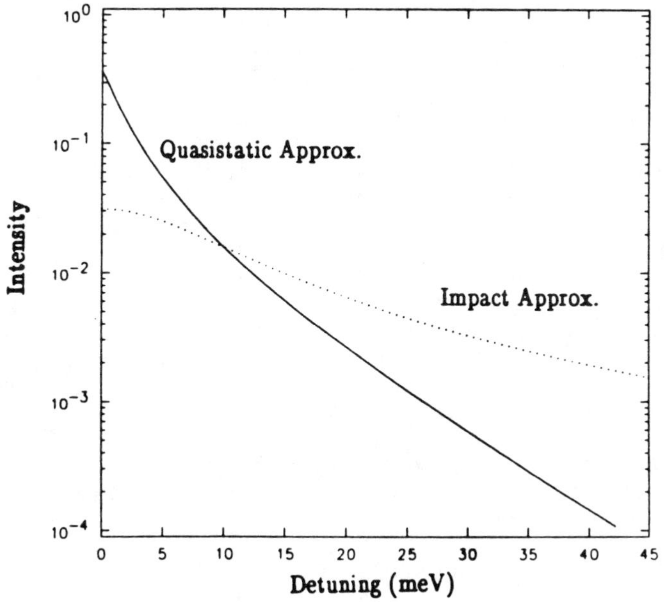

Fig. 8. Comparison of a crude quasistatic-approximation line shape (solid curve) with the impact-approximation line shape (dotted curve) at 300°K and LHD.

where ΔR is a function of θ and C is a density-dependent normalization constant. If ΔR is approximated by its average value, ~0.09 a_0, I_{qs} can be written as a simple analytic formula,

$$I_{qs}(\omega) \simeq \frac{C}{\omega} \ln^2\left[\frac{4\times 10^4}{\omega}\right] \exp\left[-0.0054\, \phi \ln^3\left[\frac{4\times 10^4}{\omega}\right] - 74\frac{\omega}{T}\right] \tag{24}$$

for ω in meV and T in °K.

Figures 7 and 8 show the quasistatic line shape, evaluated by Eq. (23), for temperatures of 20 and 300°K at liquid-hydrogen density. For comparison, the Lorentzian from the impact approximation is also shown. Obviously the Lorentzian has a much broader wing. In contrast to the impact approximation, which predicts dominance of the 0→1 resonance over the 0→2 resonance at low temperatures, the quasistatic approximation predicts just the opposite; i.e., in the quasistatic case, the effect of the large detuning of the 0→1 resonance outweighs its larger matrix element. The profile in the quasistatic approximation also has a much stronger temperature dependence than in the impact approximation. For example, the impact width at 300°K is only about 50% greater than at 20°K, whereas the quasistatic width increases by an order of magnitude.

It should be emphasized that the present treatment can be expected to bear only a qualitative resemblance to an accurate calculation for $(dt\mu)dee$ in a D_2 bath. Important interactions, at least in the near wing, are certainly not binary at liquid-hydrogen density. The exponential interaction potential may also be misleading; e.g., it leads to a one-sided line shape having a blue wing only.

VI. CONCLUSIONS

Our understanding of the pressure broadening of the $(dt\mu)dee$ formation resonances is still incomplete. We are not yet able to make quantitative predictions, in part because of lack of a complete theory and in part because of inadequate knowledge of the binding energies. The impact approximation utilizing only the inelastic cross section seems to reproduce qualitatively the observed density dependence, but the exclusion of the effect of elastic scattering does not appear to be justified. With elastic scattering included, the fundamental line-broadening criteria suggest that a quasistatic approximation will generally be more applicable to $(dt\mu)dee$ formation than will the impact approximation. This judgment is true in spite of the fact that $|\Delta E|$ even for the 0→1 resonance is smaller than the impact width Γ_{imp}. However, a rough quasistatic approximation seems to predict a stronger temperature dependence than has been experimentally observed. One possible explanation is that the resonance actually lies closer to threshold than predicted. Another possible explanation is that the quasistatic treatment needs to be generalized to take target recoil into account.

ACKNOWLEDGMENT

This work was supported by the U.S. Department of Energy, in large part by the Division of Advanced Energy Projects.

REFERENCES

1. M. P. Faifman, L. I. Menshikov, and L. I. Ponomarev, Muon Catalyzed Fusion 2, 285 (1988).

2. Yu. V. Petrov, Phys. Lett. 163B, 28 (1985).

3. S. I. Vinitsky *et al.*, Zh. Eksp. Teor. Fiz. 74, 849 (1978) [Sov. Phys. JETP 47, 444 (1978)].

4. L. I. Menshikov and L. I. Ponomarev, Pis'ma Zh. Eksp. Teor. Fiz. 45, 471 (1987) [JETP Lett. 45, 602 (1987)].

5. S. E. Jones *et al.*, Phys. Rev. Lett. 56, 588 (1986).

6. See, e.g., I. I. Sobelman, L. A. Vainshtein, and E. A. Yukov, *Excitation of Atoms and Broadening of Spectral Lines* (Springer-Verlag, Berlin, 1981), Chap 7.

7. M. Baranger, in *Atomic and Molecular Processes*, edited by D. R. Bates (Academic, New York, 1962), Chap 13.

8. L. I. Menshikov, Muon Catalyzed Fusion 2, 273 (1988).

9. N. T. Padial, J. S. Cohen, and R. B. Walker, Phys. Rev. A 37, 329 (1988).

10. N. T. Padial, J. S. Cohen, and M. Leon, Phys. Rev. A (to be published).

11. J. S. Cohen and M. Leon, Phys. Rev. A (submitted).

12. V. N. Ostrovsky and V. I. Ustimov, Zh. Eksp. Teor. Fiz. 79, 1228 (1980) [Sov. Phys. JETP 52, 620 (1980)].

13. L. N. Bogdanova, V. E. Markushin, V. S. Melezhik, and L. I. Ponomarev, Zh. Eksp. Teor. Fiz. 83, 1615 (1982) [Sov. Phys. JETP 56, 931 (1982)].

14. Yu. V. Petrov, V. Yu. Petrov, and A. I. Shlyakhter, Muon Catalyzed Fusion 2, 261 (1988).

15. C. Petitjean *et al.*, Muon Catalyzed Fusion 2, 37 (1988).

16. A. A. Radzig and B. M. Smirnov, *Reference Book on Atomic and Molecular Physics* (Atomizdat, Moscow, 1980), quoted in Ref. 8.

FORMATION OF HYDROGEN MESIC MOLECULES AT MODERATE GAS DENSITIES

V.Yu.Petrov and Yu.V.Petrov
Leningrad Nuclear Physics Institute
Gatchina, Leningrad 188350, USSR

Abstract

Formation probability is calculated for mesic molecular complexes of heavy hydrogen isotopes. Account is taken of the contribution by binary elastic collisions with surrounding molecules. Equations are obtained that describe deformation of the resonance contour due to collisions. The resonance shift and width are calculated as function of density. Energy behaviour of the complex formation cross-section far from resonance centre is discussed. The limit low densities is considered.

A high formation rate of the $dt\mu$ mesic molecules, or rather mesic molecular ions (MMI), predicted theoretically by L. Ponomarev et al. [1)], was soon demonstrated experimentally [2,3,4)]. According to the resonance mechanism, first suggested by E. Vesman [5)], a collision of a $t\mu$ mesic atom (MA) with a D_2 molecule results in formation of a large mesic molecular complex (MMC) :

$$(t\mu)_F + (D_2)_{\nu_i K_i} \longrightarrow \left[(dt\mu)_S^{Jv}\, dee\right]_{\nu_f K_f} \qquad (I)$$

which leads to a large cross-section of the reaction. Here F and S are the total spins of MA and MMI, respectively, J=v=I are the quantum numbers of the weakly bound excited state of MMI; ν_i and ν_f are the vibrational, and K_i and K_f - the rotational quantum numbers of D_2 and MMC, respectively (for details see [6,7]). The high rate of the reaction (I) is one of main reasons why a muon catalyzes a great number of dt-fusions [8] during its life $\tau_0=2.2\cdot 10^{-6}$ s. Over 100 fusions per μ^- were measured experimentally in a dense mixture[9,10] which makes it possible to rise the question of practical use of muon catalysis [11,12].

In this paper we discuss the effects of the environment on the reaction (I). Section 1 deals with the case when this effect can be neglected, Section 2 is devoted to derivation of general formulae of MMC formation probability at a moderate gas densities. In Section 3 a specific potential for scattering MMC on D_2 is considered as an application example. Final conclusions are formulated in Section 4.

1. LOW GAS DENSITY

The MMC formation probability normalized to the liquid hydrogen density $N_0 = 4.25\cdot 10^{22}$ cm^{-3} is

$$\lambda_{dt\mu} = N_0 \int \sigma_{dt\mu}(E)\, f(E) V_{CM} dE \qquad (1)$$
$$\sigma_{dt\mu} V_{CM} = \sum_{K_i K_f} W(K_i) |V|^2_{K_i \to K_f}\, 2\pi I(E) \frac{\Gamma_a}{\Gamma_v}$$

where $f(E)$ is the distribution in the $D_2 + t\mu$ total energy in their centre of mass system; $W(K_i)$ are the populations of states with given K_i (at given temperature); $|V|^2_{K_i \to K_f}$ are the matrix elements of transition from the state with K_i to the MMC state with the K_f. Finally, $I(E)$ is a form of the resonance cross-section contour normalized to unity. Within the low density limit, $I(E)$ is described by the Breit-Wigner formula (the resonance quantum numbers are omitted)

$$I(E) = \frac{1}{2\pi} \frac{\Gamma_v}{(E - E_{res})^2 + \frac{\Gamma_v^2}{4}} \qquad (2)$$

with a finite width [13,14,15] $\Gamma_v = \Gamma_{el} + \Gamma_a$ where Γ_{el} is the back decay width and $\Gamma_a = \Gamma_{em} + \Gamma_f$, the sum of the Auger width of transition of MMI to lower levels Γ_{em} and the fusion width Γ_f from the state J=v=I.

The width Γ_v for the $dt\mu$ and $dd\mu$ mesic molecules is determined mainly by Γ_a. For the $dd\mu$ mesic molecules, $\Gamma_a = 3\ 10^{-4}$ meV is small compared to all characteristic energies, and for positive resonances the contour $I(E)$ can be substituted by $\delta(E - E_{res})$ [16]. It means that at the resonance the sum of the kinetic energy of the incident mesic atom and the energy released during the formation of a mesic molecule is precisely equal to the MMC excitation energy, Hence the Vesman resonance formula [5] is valid.

For $dt\mu$ the width Γ_a=0,84 meV is not small, and the contribution of negative resonances cannot be neglected. Moreover, as was demonstrated in refs.[13,14,15,18], the main contribution to probability of dt mesic molecula formation comes from the transition from $K_r=0$ to $K_f=1$ which has negative resonance energy. Fig. 1 shows the probability of $dt\mu$ formation calculated with and without contributions of negative levels. At low temperatures the values differ by an order of magnitude.

The spectrum f(E) in Eq. (1) can be considered Maxwellian if the time T_{eq} needed for the equilibrium spectrum to establish, is considerably smaller than the muon lifetime τ_o. The value of T_{eq} is equal to the collision time of $t\mu$ with D_2 by the order of magnitude: $T_{eq}=(N_o \sigma_s V_{cm} \varphi)^{-1}$. Then at $\sigma_s=2\ 10^{-19}$ cm^2 [19] the density φ must be larger than $3\ 10^{-4}$. At large densities the shape of the contour I (E) is distorted because of collisions of the complex with surrounding molecules. The condition for applicability of Eq. (2) for I(E) will be considered below.

Note that eq. (2) corresponds to the diagram of Fig. 2, with the Breit-Wigner propagator

$$G(\omega, P) = \frac{1}{\omega - \varepsilon(\vec{P}) + i\Gamma_0/2} \qquad \begin{aligned} &\omega = E_{t\mu} + E_{D_2} - E_{res} \\ &\varepsilon(\vec{P}) = P^2/2\,MC, \\ &\vec{P} = \vec{P}_{D_2} + \vec{P}_{t\mu} \end{aligned} \tag{3}$$

The elastic amplitude A_{el}(diagram 2a) is obtained by multiplying eq.(3) by the vertices $\sqrt{\Gamma_{el}}$. The reaction amplitude A_r (diagram b) is obtained from A_{el} by substituting Γ_{el} by Γ_a in one of the vertices. Thanks to unitarity, the total cross-section σ_{tot} is related to the imaginary part of the elastic amplitude. The reaction cross-section differs from the total one only by a factor :

$$\sigma_{reac} = \frac{\Gamma_a}{\Gamma_a + \Gamma_{el}} \sigma_{tot} \tag{4}$$

However, this factor is close to unity ($\Gamma_{el} \ll \Gamma_a$), and further on we shall ignore the difference between σ_{reac} σ_{tot}. Then the contour shape is directly related to the imaginary part of the propagator

$$I(E) = -\frac{1}{2\pi} \operatorname{Im} G(\omega, \vec{P}) \tag{5}$$

In the following section we shall demonstrate that eq.(5) is valid even if one takes into account the effects of the surrounding gas D_2 on the mesic molecula formation.

2. MODERATE GAS DENSITY

Now let us take into account the possibility for the

complex to rescatter after formation on molecules of the ambient gas X. The elastic amplitude for these processes is described by the diagrams of Fig.3. We consider MMC and a molecule of the gas X as pointlike objects neglecting the possibility of their transition to other states. The dot line corresponds to the interaction between MMC and X. Let us calculate the imaginary part(at zero angle) of the sum of diagrams in Fig.3. Owing to the optical theorem we actually obtain a sum of cross sections of elementary processes with one, two, ... rescatterings. In addition, the result should be averaged over the Maxwellian spectrum of the gas X. This averaging is denoted by closing gas particle lines. The X lines correspond to the Keldysh Green functions [20]:

$$D(\omega,\vec{K}) = \frac{1}{\mu+\omega-\varepsilon(\vec{K})+i0} + 2\pi i n_x(\vec{K})\delta(\mu+\omega-\varepsilon(\vec{K})) \qquad (6)$$

$$\varepsilon(\vec{K}) = \frac{\vec{K}^2}{2M_c}$$

Here μ is the chemical potential of the gas, and $n_x(\bar{K})$ is the momenta distribution of gas particles:

$$\int n_x(\vec{K})\frac{d^3K}{(2\pi)^3} = N_x = \varphi_x N_0/2 \qquad (7)$$

(N_x is the density of the gas).

Typical diagrams for the gas-averaged scattering amplitudes are shown in Figs. 4 and 5. The diagram 4a describes interaction of the incident molecule D_2 with gas particles; the diagram 4b corresponds to renormalization of the entrance width Γ_{el} by the medium. Such diagrams change neither the position nor the width of a resonance, therefore we shall not take them into account in this paper. (Note that it was these diagrams, as well as the first diagram of Fig.5, that were taken into consideration in refs. [21-24].).

Summing up the diagrams of Fig.5 we obtain the probability of the reaction in the form of eq.(1). With contour shape given by eq.(5) where $G(\omega,\vec{p})$ is now the exact Green function of the complex in the medium:

$$G(\omega,\vec{P}) = \frac{1}{\omega - \varepsilon(\vec{P}) - \Sigma(\omega,\vec{P}) + i\Gamma_{\delta}/2} \tag{8}$$

Here $\Sigma(\omega,p)$ is the self-energy part which contains irreducible diagrams only. The imaginary part of $\Sigma(\omega,\vec{p})$ determines the width, and the real part corresponds to the energy shift of a resonance.

The value $\Sigma(\omega,p)$ can be expanded in series of the gas density, N_x. In a case of moderate gas density, we can use linear approximation in density. The only irreducible diagram is that shown in Fig.6 which corresponds to

$$\Sigma(\omega,\vec{P}) = -\frac{2\pi}{m}\int\frac{d^3K}{(2\pi)^3}n_x(\vec{K})f_{\omega}(\vec{P},\vec{K};\vec{P},\vec{K}) \tag{9}$$

Here f_{ω} is the scattering amplitude of the gas particle on the complex.

The physical meaning of the diagrams in Figs. 5, 6 is

clear. They describe the interaction of the gas X with the complex through successive independent scatterings. Further expansion of $\Sigma(\omega,\vec{p})$ in powers of density takes into account quantum interference of various collisions. They can be neglected if the wavelength of all particles is much less than the mean free path of the complex in the gas. It will be assumed below that the density is sufficiently small to comply with this condition.

The scattering amplitude f_ω in eq. (9) coincides with the physical scattering amplitude only exactly in the resonance, $\omega=\varepsilon(\vec{P})$. Otherwise, the complex is not on the mass shell, the role of its energy is played by ω rather than by $p^2/2M_c$, and we deal in fact with a virtual amplitude. The equation for it can be derived from the common integral equation for scattering amplitudes [25] by substituting ω by the energy $\varepsilon(\vec{p})$. It will be convenient to introduce here the virtuality

$$\lambda=\omega-\varepsilon(\vec{P})=E_{t\mu}+E_{D_2}-E_{res}-\varepsilon(\vec{P}) \tag{10}$$

and to consider the scattering in the centre of mass system. Let us introduce also the quantities

$$Q=\frac{M_x\vec{P}-M_c\vec{K}}{M_x+M_c}\quad;\quad\vec{Q}'=\frac{M_x\vec{P}'-M_c\vec{K}'}{M_x+M_c} \tag{11}$$

Then the integral equation for the virtual amplitude has the form

$$f_\lambda(\vec{Q},\vec{Q}') =$$

$$= -\frac{m}{2\pi}U(\vec{Q},\vec{Q}') + \int \frac{d^3Q''}{(2\pi)^3} \frac{U(\vec{Q}'-\vec{Q}'')f_\lambda(\vec{Q},\vec{Q}'')}{\lambda + \frac{Q^2}{2m} - \frac{Q''^2}{2m} + i\,\Gamma_v/2} \tag{12}$$

m being the reduced mass of X and MMC. Here U $(\vec{q})$ is the Fourier transformation of the potential between X and MMC. The above integral equation is inconvenient for practical applications. However, there is a standard way to pass to a differential equation [25)]. Introducing a new unknown function $\Phi_Q(\vec{Q}')$

$$f_\lambda(\vec{Q},\vec{Q}') =$$

$$= \frac{m}{2\pi}\left(\lambda + i\frac{\Gamma_v}{2} + \frac{Q^2}{2m} - \frac{Q'^2}{2m}\right)\left[\Phi_Q(\vec{Q}') + (2\pi)^3\delta^{(3)}(\vec{Q}-\vec{Q}')\right] \tag{13}$$

we obtain an inhomogeneous Schroedinger equation in the coordinate representation $(\hbar = 1,\ c = 1)$

$$\left\{-\frac{\nabla^2}{2m} + \left[\lambda + i\frac{\Gamma_v}{2} + \frac{Q'^2}{2m} - U(\vec{x})\right]\right\}\Phi_Q(\vec{x}) = -e^{i\vec{Q}\vec{x}}\left(\lambda + i\frac{\Gamma_v}{2}\right) \tag{14}$$

Now, using eqs. (5),(8),(9) let us rewrite the function I(E) that determines the resonance contour (in the MMC rest frame: $\vec{p} = 0$) in the following form

$$I(E) = \frac{1}{2\pi}\,\frac{\Gamma_{coll}(\lambda) + \Gamma_v}{(E - E_{res} + \Delta)^2 + \frac{1}{4}(\Gamma_{coll} + \Gamma_v)^2};\quad \lambda = E - E_{res} \tag{15}$$

Here $\Delta(\lambda)$ and $\Gamma_{coll}(\lambda)$ are the resonance energy shift and the collision width, respectively. Both quantities are functions of the distance to the resonance:

$$\Gamma_{coll}(\lambda) = \frac{4\pi}{m}\int \frac{d^3K}{(2\pi)^3}\, \mathrm{Im}\, f_\lambda(\vec{Q},\vec{Q})\, n_x(\vec{K})$$
$$\Delta_{coll}(\lambda) = \frac{2\pi}{m}\int \frac{d^3K}{(2\pi)^3}\, \mathrm{Re}\, f_\lambda(\vec{Q},\vec{Q})\, n_x(\vec{K}) \qquad (16)$$

Eqs. (16) together with eqs. (13),(14) solve completely the problem of the distortion of the Breit-Wigner contour by collisions in the linear approximations in the gas density.

3. ESTIMATE OF COLLISION WIDTH AND RESONANCE SHIFT

With a 20-30% accuracy the interaction potential between MMC and X can be presented in the form [21]:

$$U(r) = U_0 \exp(-r/a). \qquad (17)$$

Here u_0= 246 eV and a = $0.31\cdot10^{-8}$ cm. At low temperatures (e.g., T = 23K) one has $Q_a \ll 1$. In addition, $U_0 \gg Q^2/2\,m$. Therefore instead of the potential (17) one can use the hard core potential,

$$U(r) = \begin{cases} \infty & r < R \\ 0 & r > R \end{cases} \qquad (18)$$

The value of R is to be found from the turning point:

$$U(R) = \lambda + Q^2/2m. \qquad (19)$$

The solution of eq.(14) can be easily obtained in terms of radial functions with given orbital momentum l:

$$\Phi_l(r) = \begin{cases} -j_l(Qr) + j_l(QR)\,\dfrac{h_l^{(1)}(\xi r)}{h_l^{(1)}(\xi R)}, & r > R \\ 0, & r < R \end{cases} \tag{20}$$

$$\xi^2 = 2m\lambda + Q^2$$

Here j_l and $h_l^{(1)}$ are the spherical Bessel and Hankel functions, respectively. In this calculation we have neglected the vacuum width Γ_γ. Substituting eq. (20) into eq. (13) we obtain the expression for the virtual scattering amplitude at zero angle. In the limit $QR \gg 1$ and $\xi R \gg 1$ we deal actually with the semiclassical scattering. Therefore, we may change the summation over l by the integration over the impact parameter ρ. We have

$$f_\lambda = \sum_{l=0}^{\infty} (2l+1) f_l = 2Q^2 \int f(l = Q\rho)\, \rho\, d\rho$$

$$f_l = 2m\lambda \int_0^{\infty} [\Phi_l(r) + j_l(Qr)]\, j_l(Qr)\, r^2 dr \tag{21}$$

$$f_\lambda = \frac{2m\lambda}{3} R^3 + \xi \int_0^R \frac{\rho\, d\rho}{\sqrt{1-\rho^2/R^2}} \left[i\sqrt{1 - \frac{Q^2}{\xi^2}\frac{\rho^2}{R^2}} \right]$$

Finally, substituting the amplitude (21) in eqs.(16) we find for Γ_{coll} and for the shift of negative-energy resonance:

$$\Gamma_{coll}(\lambda) = 2\pi R^2 \left\langle \frac{\xi}{m} + \frac{\lambda}{Q} \ln \frac{\xi+Q}{\xi-Q} \right\rangle_M \frac{\varphi_x N_0}{2}$$

$$\Delta(\lambda) = \left[\frac{4\pi}{3} R^3 \lambda\right] \frac{\varphi_x N_0}{2} \tag{22}$$

$$\langle v \rangle = \left\langle \frac{\xi}{m} \right\rangle_M$$

Eqs. (22) describe the distortion of the shape of the Breit-Wigner resonance caused by collisions. At the resonance centre ($\lambda \ll Q^2/2m$) the collision width is expressed via the collision cross section averaged over the Maxwell gas spectrum,

$$\Gamma_{coll} = \langle \sigma_{tot} \cdot v \rangle \frac{\varphi_x N_0}{2} = \tag{23}$$

$$= 2\pi R^2 \langle v \rangle_M \frac{\varphi_x N_0}{2} = 2R^2 \sqrt{\frac{2\pi T}{M_x}} \cdot N_0 \varphi_x$$

$$\Delta = 0 .$$

This is a general formula following from unitarity:at $\lambda \ll Q^2/2m$ the virtual amplitude is identical to the real one whose imaginary part is equal to the total cross section. In case of a hard core potential (18) $\sigma_s = 2\pi R^2$. Actually, the main contribution to the probability of a mesic molecule formation comes from the transition from $K_i = 0$ to $K_f = 1$ lying far in the negative energy region. It follows then from eq.(22) that in the range $\lambda \gg T$ one has

$$\Gamma_{coll}(\lambda) = 4\pi R^2 \sqrt{\frac{2\lambda}{m}} \cdot \frac{\varphi_x N_0}{2}, \quad \lambda > 0 \tag{24}$$

$$\Delta(\lambda) = \frac{4\pi}{3} R^3 \lambda \cdot \frac{\varphi_x N_0}{2} .$$

Note that eqs.(22) are valid over the entire range of resonance energies and in contrast to refs.[22-24,26] yield a smooth, gap-free transition from the resonance centre to the resonance wing.

For positive-energy resonances ($\lambda < 0$) and $2m\lambda + Q^2 > 0$ we have $\xi < Q$ and the integral for Γ_{coll} in the last of eqs.(21) is cut by $\rho < \xi R/Q$. For this reason Γ_{coll} decreases with the distance to the resonance centre. For far positive resonances ($|\lambda| \gg T$) it is exponentially small. For the resonance shift in this case we have

$$\Delta(\lambda) = \frac{4\pi}{3} R^3 \lambda \frac{\varphi_x N_0}{2} , \quad \lambda < 0. \qquad (25)$$

i.e. collisions shift the resonance further to the positive region. Thus, in contrast to the negative-energy resonances, far positive-energy resonances cannot play an essential role in the MMC formation.

Substituting numerical values of the constants ($R = a \ln U_0 / E_{res} = 2.9 \cdot 10^{-8}$ cm, $\sigma_s = 2\pi R^2 = 53 \cdot 10^{-16} cm^2$, $m = 2.40$ GeV) and using formulae (24) we obtain for a far negative-energy resonance

$$\Gamma_{coll} = 4.1 \varphi_x \sqrt{\frac{|E_{res}|}{1\ \text{meV}}}\ ,\ \text{meV}; \qquad \Delta = -2.2\ E_{Res} \varphi_x\ . \qquad (26)$$

Using the value $E_{res} = -21$ meV for $K_i = 0 \rightarrow K_f = 1$ resonance[18] we have $\Gamma_{coll} = 18.6\ \text{meV} \cdot \varphi_x$. Comparing this with $\Gamma_v = 0.84$ meV we see that one can neglect the collisions and use the Breit-Wigner formula (2) at densities $\varphi \ll 5 \cdot 10^{-2}$.

The density dependence of $\lambda_{dt\mu}(\varphi_x)$ is plotted in Fig.7. This plot has a characteristic crossover at $\varphi_x \sim 0.2$ owing to the resonance shift. Although the results are preliminary it is of interest to compare them with experimental data. Unfortunately, at present there are no direct measurements of the dependence of $\lambda_{dt\mu}$ on density at low temperatures. However, the dependence on φ_x a more complex value $\lambda_c(\varphi_x)$, the cycle time, as measured by the SIN group [27], reveals the same crossover at $\varphi_x \sim 0.2$.

4. CONCLUSION

The study of the influence of surrounding molecules on the resonance cross section of MMC formation demonstrates that at low densities the resonance contour has a classical Breit-Wigner character. At such densities and at a low temperatures it is most easy to find the resonance position.

At the density increase, the contour widens owing to elastic scatterings of the complex on surrounding molecules and the resonance shifts to the negative-energy region. It is this shift that leads to the saturation of cross section increase and thus to a characteristic crossover at $\varphi_x \sim 0.2$.

In this paper we confine ourselves to elastic processes because both theoretical and experimental evidence shows that the relative contribution of non-elastic processes is small [28,29]. Calculations with more realistic potential than (18) are now in progress.

This report had been completed when the authors got acquainted with the paper[30] (based on another approach) which also dealt with the influence of collisions on the mesomolecular formation rate.

The authors are grateful to V.N.Gribov for interest in this work and aknowledge fruitful discussions with Prof. L.I.Ponomarev and L.I.Menshikov. Thanks are also due to A.I.Shlyakhter for his cooperation and to G.R.Dyck and G.V.Samsonova for their help in preparing the manuscript.

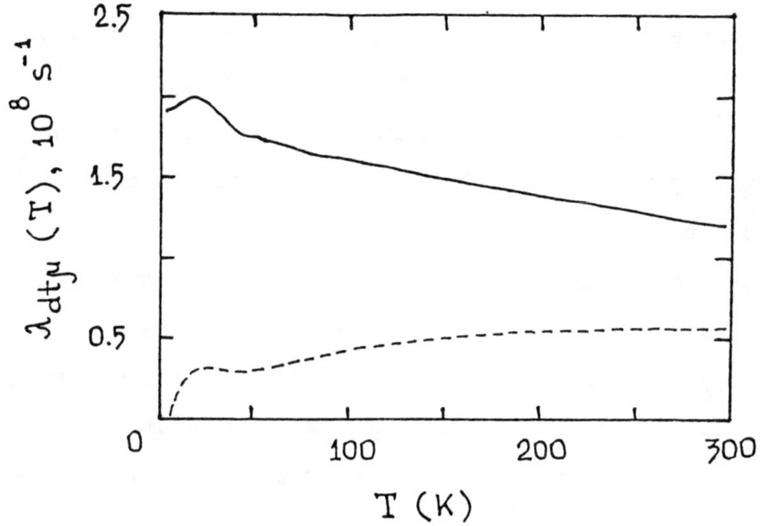

Fig.1. Rate of dt mesic molecule formation $\lambda_{dt\mu}$ versus temperature T at low densities ($E_{res}(0 \to 1)$=-21meV), For comparison, the dash line shown $\lambda_{dt\mu}$ at zero width when negative resonances do not contribute[18].

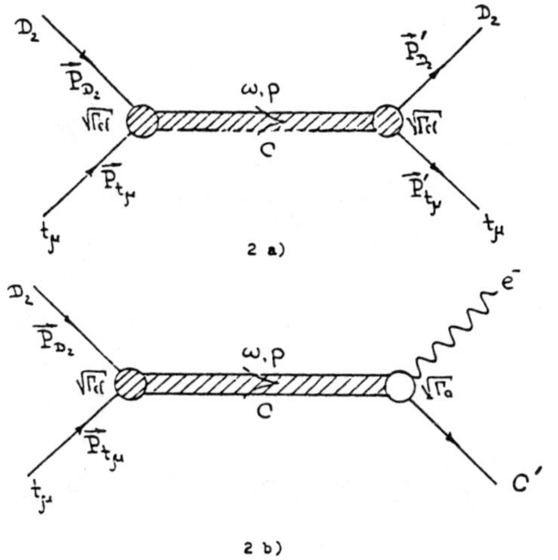

Fig.2. Resonance diagrams for amplitude of a) elastic scattering and b) reaction.

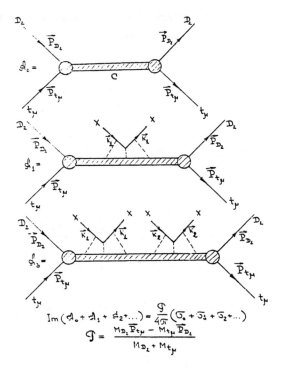

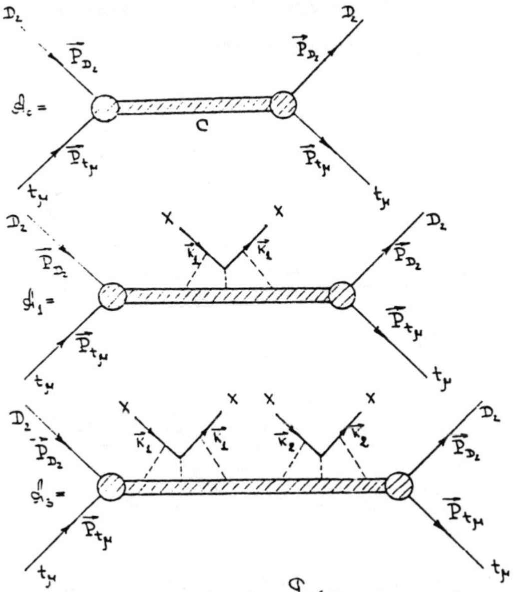

$$\mathrm{Im}(\mathcal{A}_0 + \mathcal{A}_1 + \mathcal{A}_2 + \ldots) = \frac{\mathcal{P}}{4\pi}(\sigma_0 + \sigma_1 + \sigma_2 + \ldots)$$

$$\mathcal{P} = \frac{M_{D_2}\vec{P}_{t_\mu} - M_{t_\mu}\vec{P}_{D_2}}{M_{D_2} + M_{t_\mu}}$$

Fig.3. Diagrams for elastic amplitude with multiple rescattering.

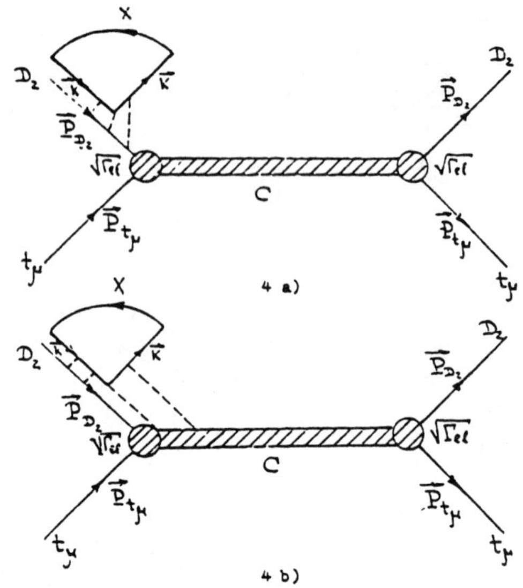

Fig.4. Diagrams for interaction in initial state renormalizing the entrance width.

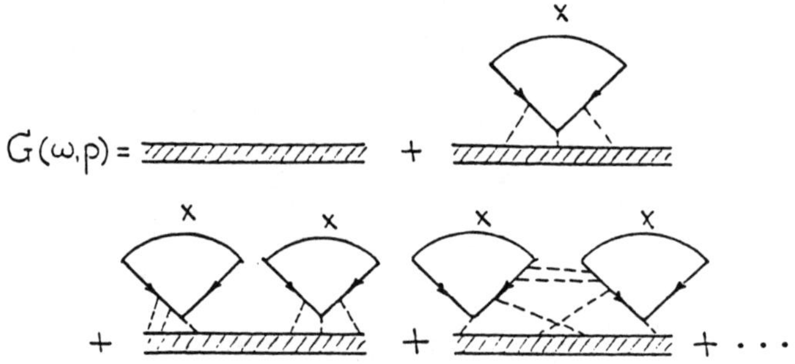

Fig.5. Green function of the complex in the medium.

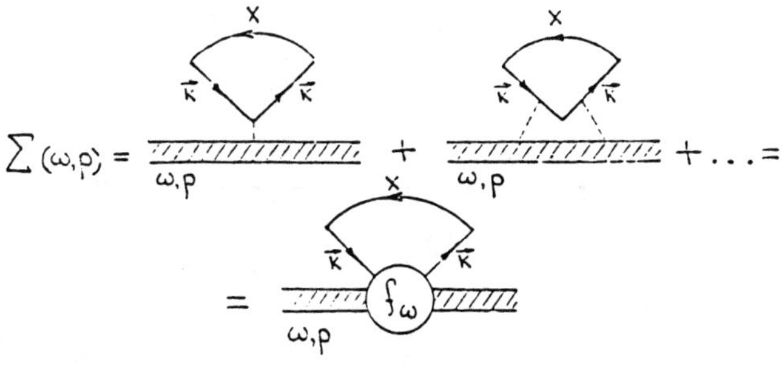

Fig.6. Irreducible diagrams for self energy in a linear approximation in N_x.

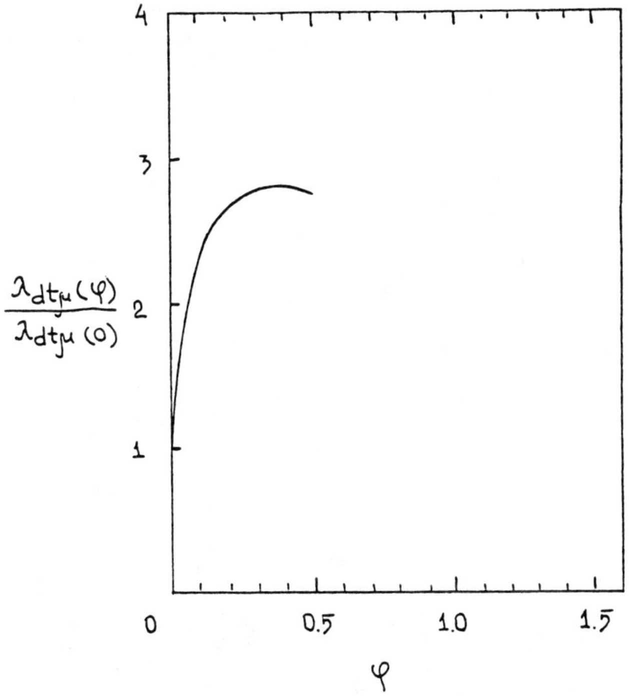

Fig.7. Value of $\lambda_{dt\mu}(\varphi_x)/\lambda_{dt\mu}(0)$ vs. density φ_x at $T \ll E_{res}$ ($E_{res}(0 \rightarrow 1) = -21$ meV).

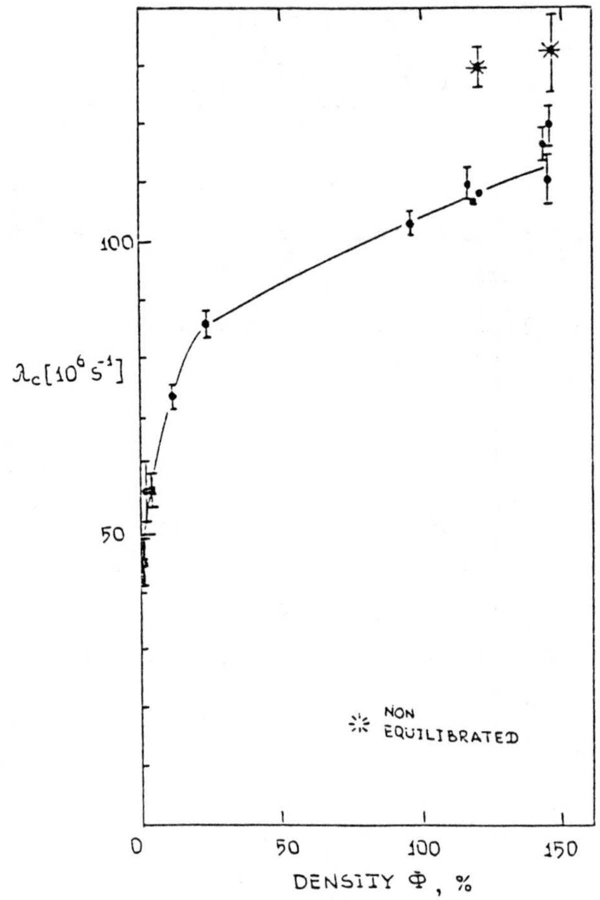

Fig.8. Cycle rate λ_c vs. gas density φ_x at tritium concentration C_t=40% and T< 50 K [27].

R E F E R E N C E S

1. Vinitsky S.I. et al., Zh.Eksp.Teor.Fiz., 74 (1978), 849 (Sov.Phys. JETP, 47 (1979), 444 .
2. Bystritsky V.M. et al., Phys.Rev.Lett., 94B (1980),476.
3. Jones S.E. et al., Phys.Rev.Lett., 51 (1983), 1757.
4. Breunlich W.H., Cargnelli M., Kammel P., Marton J. et al., Muon Catalyzed Fusion, 1 (1987), 67.
5. Vesman E.A., Zh.Eksp.Teor.Fiz. Pis'ma, 5 (1967), 113. (Sov. Phys. JETP Lett., 5 (1967) ,91); see also A.M.Lane, Phys.Lett., 98A (1983), 337.
6. Ponomarev L.I., Atomkernenergie-Kerntechnik, 43 (1983), 175.
7. Petitjean C. et al., Muon Catalyzed Fusion, 1 (1987),89.
8. Gershtein S.S. and Ponomarev L.I., Phys.Lett., 72B (1977), 80.
9. Jones S.E. et al., Phys.Rev.Lett., 56 (1986), 588.
10. Jones S.E., Nature, 321 (1986), 12 7.
11. Petrov Yu.V., Nature, 285 (1980), 466.
12. Petrov Yu.V., Muon Catalyzed Fusion, 1 (1987), 351.
13. Petrov Yu.V., Phys.Lett., 163B (1985), 28; Preprint LNPI-1058, Leningrad (1985); Muon Catalyzed Fusion, 1 (1987), 219.
14. Lane A.M., J.Phys.B, 20 (1987), 2911.
15. Leon M., Muon Catalyzed Fusion, 1 (1987), 163.

16. Menshikov L.I., Ponomarev L.I., Strizh T.A. and Faifman M.P., ZhETF, 92(1987), 1173 (Sov.phys.JETP).

17. Bogdanova L.N., Markushin V.E., Melezhik V.S and Ponomarev L.I., Zh.Eksp.Teor.Fiz., 83 (1982), 1615.

18. Petrov Yu.V., Petrov V.Yu. and Shlyakhter A.I., Preprint LNPI-1279, Leningrad (1987); Muon Catalyzed Fusion, 2 (1988), ch.5.

19. Adamczak A., and Melezhik V.S., Muon Catalyzed Fusion, 2 (1988), Ch.3.

20. Landau L.D. and Lifshitz E.M., "Physical Kinetics", Vol.10, Ch.10, Moscow, Nauka, 1979.

21. Menshikov L.I. and Ponomarev L.I., Phys.Lett., 167B (1986), 141; Preprint IAE-4202/12, Moscow (1985).

22. Menshikov L.I., Muon Catalyzed Fusion, 2 (1988),Ch.5.

23. Faifman M.P., Menshikov L.I. and Ponomarev L.I., Muon Catalyzed Fusion, 2 (1988), Ch. 5.

24. Ponomarev L.I., Muon Catalyzed Fusion, 2 (1988),Ch.5.

25. Landau L.D. and Lifshitz E.M., "Quantum Mechanics" (Non-Relativistic Theory", Moscow, Nauka (1979).

26. Faifman M.P., Muon Catalyzed Fusion, 2 (1988), Ch.5.

27. Petitjean C. et al., Muon Catalyzed Fusion, 2(1988), Ch.2.

28. Padial N.T., Cohen J.S. and Walker R.B., Phys.Rev., A37 (1988), 329.

29. Chandler D.W. and Farrow R.L., J.Chem.Phys., 85(1986), 810.

30. Men'shikov L.I. Mechanisms of mesic molecules dtu and ddu formation in gas. Preprint IAE-4606/2 Moscow, 1988.

SLOWING-DOWN AND COULOMB CAPTURE OF NEGATIVE MUONS IN MOLECULAR HYDROGEN

G. Ya. Korenman and V.P. Popov
Institute of Nuclear Physics, Moscow State University, Moscow 119899, USSR

ABSTRACT

The processes of slowing-down and Coulomb capture of negative muons in molecular hydrogen are considered using the microscopic model, which is based on the idea of the "black" surface of inelastic interaction. The model parameters are calculated. The kinetic characteristics of Coulomb capture are obtained. The decay of a high-excited neutral $(ab\mu e)^*$-molecule is discussed.

INTRODUCTION

The slowing-down and Coulomb capture of negative muons in molecular hydrogen form the first stage of the processes occuring in each cycle of the muon-catalyzed fusion. They take place after the next nuclear reaction in the $dt\mu$-molecule with the release of a muon with energy $E \lesssim 10$ keV. The slowing-down time of muon from this energy to Coulomb capture in liquid hydrogen is very small ($\sim 10^{-12}$s) and, therefore, the slowing-down process does not give any appreciable restrictions on the number of the μCF-cycles per muon. However, it is essential for development of the subsequent process which states the muons gets in after Coulomb capture, which the primary kinetic energy distribution of muonic atoms is like, etc. In this connection, a theoretical study of inelastic muon- molecular collisions seems to be necessary for the physics of muon catalysis.

Different theoretical approaches have been proposed in recent years for describing inelastic collisions (ionization and Coulomb capture) of muons with the atoms of hydrogen [1-4] and helium [5-7]. An analogous treatment of the collisions with molecules, however, has been lacking up until recently. Only in refs. [6,7] a quantitative model of the ionization and Coulomb capture of muons by the hydrogen molecule was formulated. The present paper is the further development and refinement of this model and, besides, a discussion of the decay of highly-excited muonic molecule and the formation of muonic atoms.

THE MODEL OF INELASTIC COLLISIONS OF NEGATIVE MUONS WITH A HYDROGEN MOLECULE

At energies of several eV to several hundreds of eV the muon velocity U_μ lies between the characteristic velocities U_n and U_e of nuclei and electrons in molecule, $U_n < U_\mu < U_e$. Under these conditions the muon-induced transitions of the electron into the continuum can be considered first at a fixed position of nuclei. Let us introduce the potentials $V_o(\vec{r},\vec{R})$ and $V_1(\vec{r},\vec{R})$ describing the interaction of the muon with the neutral molecule AB in the entrance channel and with the molecular ion AB^+ in the ionization channel. These potentials depend on the muon coordinates $\vec{r}$ and the internuclear coordinates $\vec{R} = \vec{R}_b - \vec{R}_a$ (a and b are nuclei of the hydrogen isotopes). They include the Coulomb muon-nucleus interaction and the averaged muon-electron unteraction,

$$V_q^{\mu e}(\vec{r},\vec{R}) = \int d^3r_e \; \rho_q(\vec{r},\vec{R})/|\vec{r} - \vec{r}_e|, \qquad (1)$$

where $\rho_q(\vec{r},\vec{R})$ is the electron density in the molecule ($q = 0$) or in the ion ($q = 1$) at a given position of nuclei. The system of units used is $\hbar = e = m_e = 1$. We introduce also the difference between the entrance channel term and the electron continuum boundary,

$$\omega(r, R, \cos\theta) = V_o(\vec{r}, \vec{R}) + E_o(R) - \left[V_1(\vec{r}, \vec{R}) + E_1(r)\right], \quad (2)$$

where $E_o(R)$ and $E_1(R)$ are the electron terms of the molecule AB and of the ion AB^+, respectively; θ is the angle between the vectors $\vec{r}$ and $\vec{R}$.

In the classical description of heavy particles (nuclei and muon) the ionization is possible only in the region of the configurational space ($\vec{r}$, $\vec{R}$) where $\omega(r, R, \cos\theta) \geqslant 0$. To each point of this region there corresponds the electron yield with a definite energy $\mathcal{E} = \omega(r, R, \cos\theta)$. The condition $\omega(r, R, \cos\theta) = 0$ determines the surface of the of the inelastic muon interaction $r = r_c(\cos\theta, R)$ that depend parametrically on R . As it follow from the calculations [3,6], the surfaces of the inelastic interaction of the muon with hydrogen and helium atoms are almost "black" at velocities $V < V_a$ and, so, the ionization with the probability close to unity occurs when the muon reaches this surface. The analogous calculations of the ionization probability of a molecule are rather difficult and have not been carried out as yet. The analysis of the calculations for the H and He atoms suggests that the surface of the muon-molecule inelastic interaction should also be almost "black". However , the calculation of the absorption by the molecular "black" surface is more difficult because of the absence of the spherical symmetry of the surface itself and of the potential in which the muon moves before the reaction.

Let the muon move initially along the Z-axis with energy E and the two-dimensional vector of the impact parameter $\vec{\rho}$ forming the angle φ with the X-axis. The direction of the axis of symmetry of a molecule $\vec{R}$ is given by the angles θ_R, φ_R. Then the reaction cross section can be represented as

$$\sigma_r (E, R, \theta_R) = \frac{1}{2} \int_0^{2\pi} d\varphi \, \rho_o^2 (E, R, \theta_R, \varphi - \varphi_R) \quad (3)$$

where $\rho_o (E, R, \theta_R, \chi)$ is the largest impact parameter at which the muon approaches the surface of the inelastic interactions. To determine ρ_o we use the "infinite order sudden approximation" (IOSA) which is frequently used in the theory of ion-molecular reactions [8]. In our case it reduces to the neglect of the change in the angular momentum of a molecule before the reaction. Then

$$\rho_o = r_c^2 (\cos\theta, R) \left[1 + W(\cos\theta, R) / E\right], \quad (4)$$

where θ depends on θ_R and $(\varphi_R - \varphi)$ and, also, on E and R ,

$$W(\cos\theta, R) = - V_o (r_c (\cos\theta), R ; \cos\theta). \quad (5)$$

Averaging the expression (3) over the orientations of the molecule, one can write

$$\sigma_r (E, R) = \pi \left[s(R) + W(R) / E\right] , \quad (6)$$

where

$$s(R) = \int r_c^2 (\cos\theta, R) \, d\Omega_R / 4\pi , \quad (7)$$

$$w(R) = \int r_c^2 (\cos\theta, R) \, W(\cos\theta, R) \, d\Omega_R / 4\pi. \quad (8)$$

To obtain the physical cross section, the quantities $s(R)$ and $w(R)$ should be averaged over R with the weight $u_n^2 (R)$, where $u_n(R)$ is the vibrational wave function of a molecule. Then

$$\sigma_r (E) = \pi (\bar{s} + \bar{w} / E) \quad (9)$$

where

$$\bar{s} = \int_0^\infty s\ (R)\ u_n^2\ (R)\ dR \qquad (10)$$

and analogously, for $\bar{w}$.

CALCULATIONS OF THE MODEL PARAMETERS

To calculate the potentials of the interaction of a muon with a molecule and an ion, we used the electron wave functions obtained in the Heitler - London approximation and in the LCAO - approximation, respectively. Figure 1 shows the typical results for the potentials

$$\tilde{U}_q\ (r\ ,\ R,\ \cos\theta\ ) = V_q^{\mu e}\ (r\ ,R\ ,\ \cos\theta\ ) + E_q(R) \qquad (11)$$

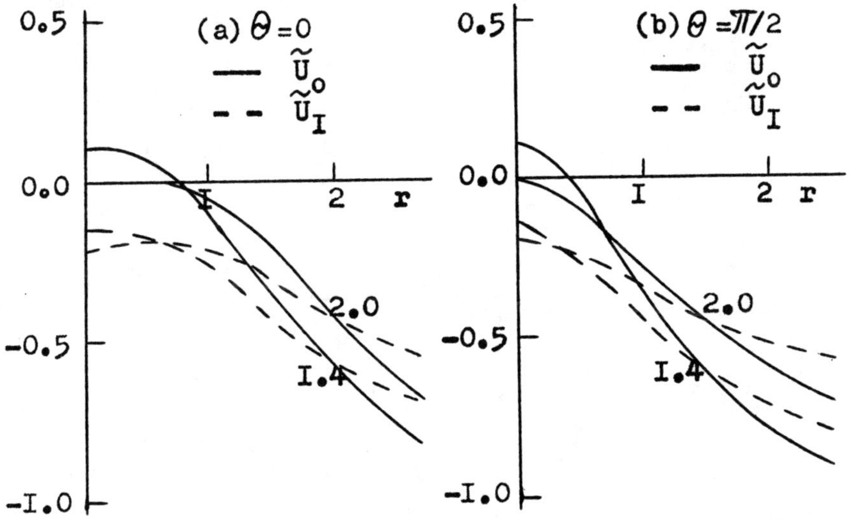

Fig. 1. Potentials $\tilde{U}_o$, $\tilde{U}_1$ for R = 1,4 and R = 2.0 a.u.; (a) $\theta = 0$, (b) $\theta = \pi/2$.

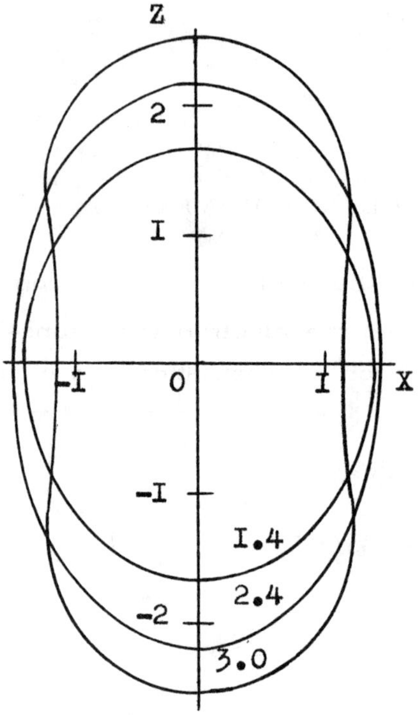

Fig. 2. **Polar diagram** $r = r_c(\theta, R)$ at R= 1.4, 2.4, and 3.4 a.u.

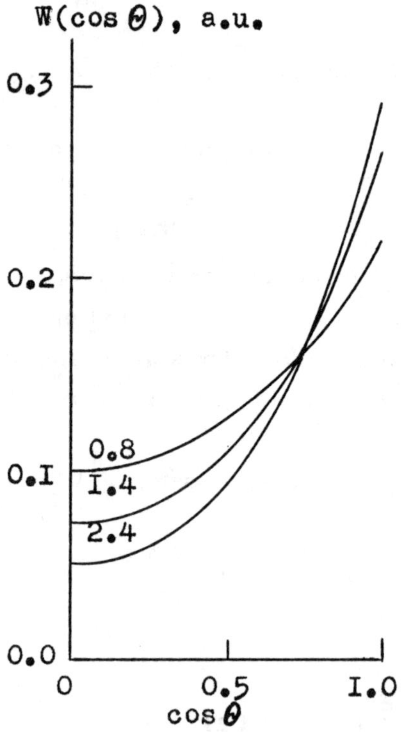

Fig.3 .Angular dependence of W $(\cos\theta, R)$ at R= 0.8, 1.4, and 2.4 a.u.

which do not include the Coulomb interaction of heavy particles. The potentials $\tilde{U}_0$, $\tilde{U}_1$ and their difference $\omega(r, R, \cos\theta)$ were obtained in a wide range of r, R and cos θ $(0 \leq r \leq 9$ a.u., $0 < R \leq 6$ a.u., $0 \leq \cos\theta \leq 1)$.

The equation $\omega(r, R, \cos\theta) = 0$ determines the axially symmetric surface in the space of muon coordinates. The results of the numerical solution of this equation are shown in fig. 2 as the polar diagram $r = r_c(\theta, R)$ at several R - values. These curves show the inelastic interaction surfaces crossed by the plane passing through the axis of symmetry of a molecule. At $R = 0$ (the united-atoms limit) the surface becomes a sphere of radius $r_c = 1.0\ a_B$ which coincides with the radius of the interaction of a muon with a helium atom in the effective-charges approximation [7]. As R increases, the surface stretches along the axis of a molecule and at $R = 3.8\ a_B$ it changes into two separated "drops". As the internuclear distance further increases, the "drops" become spheres with the centres on nuclei a and b which correspond to the limit of two separated hydrogen atoms.

To calculate the reaction cross section it is also necessary to know the potential in the entrance channel on the inelastic interaction surface, $W(\cos\theta, R)$. The dependence of this potential on $\cos\theta$ at several R is shown in fig. 3. For further calculations the functions $r_c(\cos\theta, R)$ and $W(\cos\theta, R)$ were expanded in terms of (even) Legendre polynomials $P_{2n}(\cos\ )$. The analysis of the results shows that in the most significant region of R-values (0.8 to 2.5 a.u.) it is sufficient to allow for the terms with $n \leqslant 2$.

When s(R) and w(R) were calculated it was assumed that the averaging over the orientations of a molecule with respect to the beam is equivalent to the averaging over $\cos\tilde{\theta}$, where $\tilde{\theta}$ is the angle between the axis of a molecule and the vector drawn from the centre-of-mass of a molecule onto the point of the surface $r = r_c(\cos\theta, R)$. For the homonuclear molecules the centre of masses coincides with the geometrical centre of a molecule so that $\tilde{\theta} = \theta$. The results of the calculation of s(R) and w(R) for this case are shown in Fig. 4. When there values are overaged over the vibrational ground state of a molecule, the main contribution to the integral of the type (10) is provided by the R-values close to the equilibrium internuclear distance of a molecule $R_e = 1.4\ a_B$. In this region for s(R) and w(R) it is possible to use the parabolic interpolation

$$s(R) = S_o + (R - R_e)\, S_1 + (R - R_e)^2 S_2 /2 \qquad (12)$$

As $u_o(R)$ we use the wave function of the one-dimensional harmonic oscillator that is shifted into the point $R = R_e$ and is characterized by the dimensional parameter x_o depending on the reduced mass of a molecule,

$$x_o(AB) = x_o(H_2)\left[m(H_2)/m(AB)\right]^{1/4}, \qquad (13)$$

where $x_o(H_2) = 0.233$ a.u.

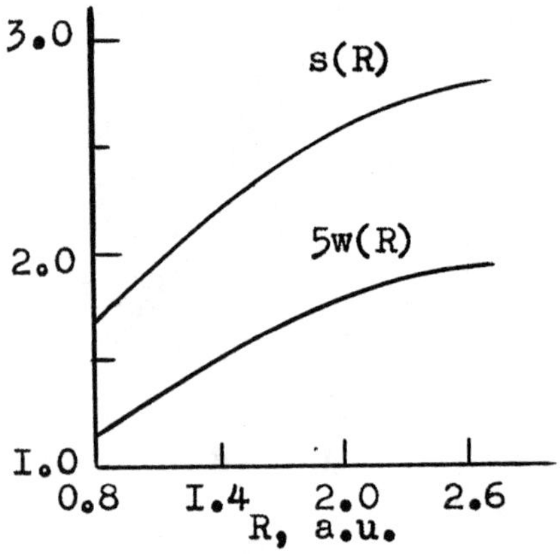

Fig. 4. The cross section parameters s(R) and w(R) versus internuclear distance R.

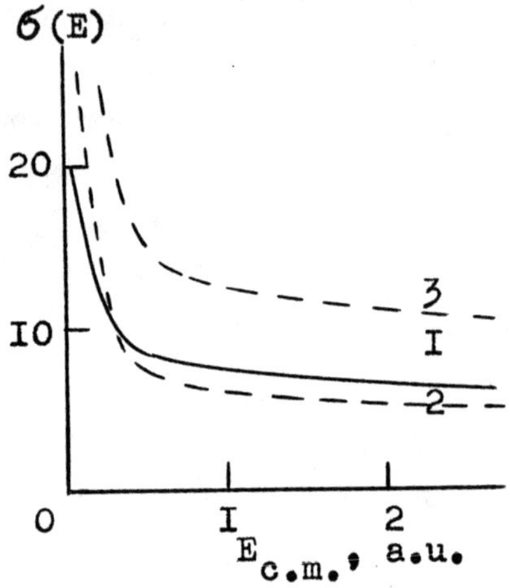

Fig. 5. Cross section of inelastic interaction of a muon with a H_2 molecule (curve 1). Curves 2 and 3 represent the atomic and double atomic cross sections.

In these approximations the parameters $\bar{s}$ and $\bar{w}$ entering into the inelastic interaction cross section (9) are of the form

$$s = s_o + s_2\, x_o^2/2, \qquad (14)$$

$$w = w_o + w_2\, x_o^2/2 .$$

For the H_2 molecule we obtained $\bar{s} = 2.18$ a.u., $\bar{w} = 0.293$ a.u. The dependence of these parameters on the reduced mass proves to be very weak. For D_2 and T_2 the parameters $\bar{s}$ and $\bar{w}$ increase only in the 4-th digit which seems to be outside the limits of the accuracy of the model.

For the heteronuclear molecules the centre of masses is shifted from the centre of symmetry by a distance $\varkappa R/2$, where $\varkappa = (M_b - M_a) / (M_b + M_a)$. Therefore the angular averaging is performed here with allowance for the difference between Θ and $\tilde{\Theta}$. This changes the parameters $\bar{s}$ and $\bar{w}$ by the magnitude

$$\delta \bar{s} = -\varkappa^2 R_e^2/12 \, , \quad (15)$$

$$\delta \bar{w} = 0.104 \, \delta \bar{s} \, . \quad (16)$$

The values of $|\delta \bar{s}|$ for the HD, HT and DT molecules are 0.018, 0.041 ad 0.0065, respectively. Thus, the asymmetry of a molecule affects the cross section parameters more strongly than does the reduced mass; yet, the asymmetry contribution to this model parameters does not exceed 2% . The calculated total cross section of the inelastic interaction (ionization and Coulomb capture) of a muon with a H_2 molecule are shown in Fig . 5 together with the atomic and doubled atomic cross sections from ref. [3] At all the energies, the molecular cross section is less than the doubled atomic cross section and at $E < 0.29$ a.u. it is even less than the cross section for one atom.

CROSS SECTIONS OF THE IONIZATION OF A MOLECULE AND THE COULOMB CAPTURE OF A MUON. THE EFFECTIVE SLOWING-DOWN OF MUONS .

In the above formulated model it is assumed that the inelastic interaction surface is "black" and, so, if the muon penetrates inside this surface, the system goes into the states of the electron continuum. But on the surface $\mathcal{E} = \omega' = 0$ and, so, the ejected electron possesses zero energy. Then the cross section (6) at a definite R value can be regarded as the ionization cross section if the energy E exceeds the "vertical" ionization potential $I(R) = E_1(R) - E_o(R)$ or as the Coulomb capture cross section if $E < I(R)$. In order to separate out the ionization cross section $\sigma_i(E)$ and the capture cross section $\sigma_c(E)$ from the total cross section $\sigma_r(E)$ it is necessary to average the quantities $\sigma_r(E,R)\,\theta(E-I(R))$ and $\sigma_r(E,R)\,\theta(I(R)-E)$ over R by analogy with (10). This procedure is, in fact, analogous to the division of the total cross section into the ionization cross section and the Coulomb capture cross section by the following formulae:

$$\sigma_i(E) = \sigma_r(E)\left[1 - \zeta(E)\right], \tag{17}$$

$$\sigma_c(E) = \sigma_r(E)\,\zeta(E), \tag{18}$$

where

$$\zeta(E) = \sum_n F_{no}\,\theta(I_o + \varepsilon_n - E), \tag{19}$$

F_{no} are the Frank-Condon factors for the transition from the vibrational ground state of the molecule AB into the excited state of the ion AB^+, I_o is the ionization potential of the ground-state molecule, ε_n is the excitation energy of the n-th vibrational level of the AB^+ ion. At $E < I_o$ $\zeta(E) = 1$. As E increases in the region of $E > I_o$ the function $\zeta(E)$ decreases in a stepwise manner. In this transitional region $\sigma_c(E)$ rapidly tends to zero and the ionization cross section $\sigma_i(E)$ increases from zero to $\sigma_r(E)$. Figure 6 illustrates the behaviour of the ratios $\sigma_c(E)/\sigma_r(E)$ and σ_i/σ_r in the transitional region. The width of the transitional region is determined by the Frank-Condon factors and by the vibrational energies ε_n

$$\Delta/2 \sim \bar{\varepsilon}_v = \sum_n \varepsilon_n F_{no}. \tag{20}$$

Using F_{no} from ref.[9] and ε_n from ref.[10], we obtain that

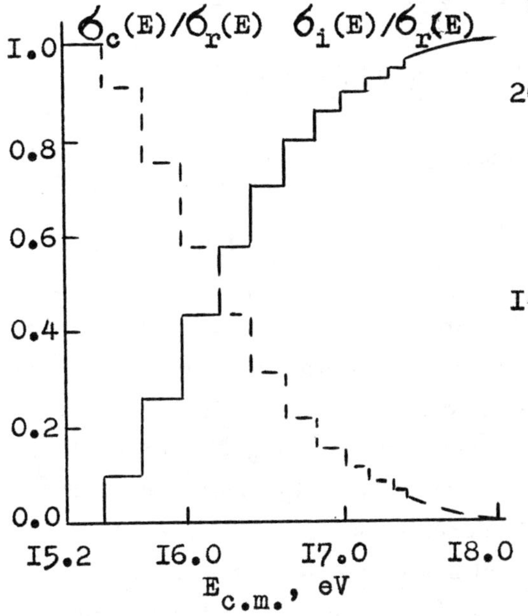

Fig. 6. The ratios of $\sigma_c(E)/\sigma_r(E)$ and $\sigma_i(E)/\sigma_r(E)$ in the transitional energy region.

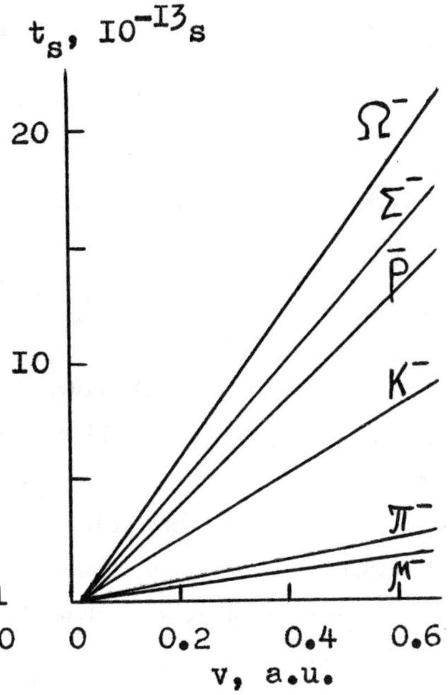

Fig. 7. The slowing-down time of different negative particle in liquid hydrogen versus the initial velocity.

for the H_2 molecule $\bar{\varepsilon}_v = 0.85$ eV. Taking into account (17)-(19) it is possible to introduce also the effective slowing-down (the losses of muon energy per unit path per one molecule) in the centre-of-mass system $\mathcal{H}(E) = \bar{I}\,\sigma_r(E)$, where $\bar{I} = I_o + \bar{\varepsilon}_v \simeq 0.6$ a.u. = 16.3 eV.

THE KINETIC CHARACTERISTICS OF COULOMB CAPTURE OF MUONS IN MOLECULAR HYDROGEN

The simple kinetic characteristic is the slowing-down time of muons from an initial energy E to Coulomb capture. Of greatest interest are the initial energies E corresponding to the velocities $v \lesssim v_a$. The calculations of $t_s(E)$ were performed in the continuous energy loss approximation

$$t_s^c(E) = \int_{E_c}^{E} dE' / \left[N \mathcal{H}_L(E') \cdot v(E') \right] , \qquad (21)$$

and in the discrete energy loss approximation

$$t_s^d(E) = \sum_k \Theta(E - I_L k) / \left[N \sigma_r (E - I_L k) v(E - I_L k) \right] . \qquad (22)$$

Here all the energies relate to the laboratory frame of reference; $I_L = \bar{I} \cdot (1 + m_\mu / M)$ is the mean energy loss in one inelastic collision in the L-system ; M is the mass of a molecule ; N is the number of molecules per unit volume. The formula (21) contains the parameter E_c which has the meaning of the mean energy at which the Coulomb capture occurs. At E_c- values between 0.3 I_L and I_L the difference in the results of the calculations by the formulae (21) and (22) is not higher than 2-3%. Fig. 7 is a plot of the slowing-down time of muons and other negative particles in liquid hydrogen as a function of the initial velocity v . Energy losses due to elastic collisions are not taken into account here and, therefore, for heavy particles the calculated values can give the upper boundary of t_s .

In the energy region under consideration the slowing-down time of muons at E = 2 keV is known from experiment[11]. In gaseous hydrogen at the pressure of 0.25 Torr it is equal to 670 ± 50 ns. Our calculation under the same conditions gives 750 ns. The agreement can be regarded as quite satisfactory.

Another significant kinetic characteristic is the reduced ratio of the probabilities of Coulomb capture of a muon by different components for the (H_2 + He) mixture

$$S(He,H) = (W(He)/W(H_2)) \cdot (n(H_2)/n(He)) , \qquad (23)$$

where $W(He)$ and $W(H_2)$ are the capture probabilities of a muon by the He atom and the H_2 molecule; $n(He)$ and $n(H_2)$ are the molar concentrations of the corresponding components. The experimental value of $S(He,H_2)$ is 0.88 ± 0.10 [12] at $n(He)/n(H_2) = 0.25$.

The detailed calculation of the kinetics of slowing-down and Coulomb capture for this mixture gives $S(He,H_2) = 0.80 \pm 0.02$ with a very weak dependence upon concentrations [7]. Within the experimental errors and the statistical error of calculation (by the Monte-Carlo method) the agreement with experiment is good .

From the calculations of the kinetics with the above-obtained cross sections it follows [7] that the distribution of the probability of Coulomb capture of muons is actually constant in interval from 0 to I_L . After the muon has been captured , there form the highly-excited states of the system $(ab\mu e)^*$ and the muon binding-energy distribution is homogeneous, $C(\epsilon) = \Theta(\bar{I} - \epsilon)/\bar{I}$. As a result of Coulomb capture the muon introduces into the molecule the angular momenta limited by the quantity $\rho_o(E, R, \theta_R)$. For the distribution of excited muon molecules over the total angular momentum J we obtain

$$C(\epsilon, J) = C(\epsilon)\,\Theta(J_o - J)(2J + I)/\left[J_o(\epsilon) + 0.5\right]^2 \qquad (24)$$

where $J_o(\epsilon) = \left\{2\mu\left[\bar{s}(\bar{I} - \epsilon) + \bar{w}\right]\right\}^{1/4}$, μ is the reduced mass of a muon. The expression (24) corresponds to the "cut-off" statistical distribution in J .

Finally, the distribution of the μ-molecules in kinetic energies T in the L-system is connected with the energy distribution of the capture probability of a muon since the momentum of the ejected electron is negligible. Then $W(T) = \Theta(T_m - T)/T_m$, where $T_m = \bar{I} \cdot m_\mu / M$. For the molecules H_2 , D_2, DT and T_2 we have, respectively , T_m = 0.92 , 0.46 ,

0.37 and 0.31 eV .

THE DECAY OF A HIGHLY-EXCITED MUONIC MOLECULE

As seen from the previous consideration, the Coulomb capture of a muon in hydrogen leads to formation of the excited muonic molecule $(ab\mu e)$ and the internuclear distance is initially $R_e(H_2)$ = 1.4 a.u. and the muon binding energy lies between zero and 0.6 a.u. These energies correspond to highly-excited states of a muon. The quasi-classical estimations show that the terms of most of these states are attracting and, therefore, the dissociation of the system can occur only as a result of the (g - u) electron transition. The Auger transitions of a muon $(ab\mu^- e) \rightarrow (ab\mu -)^+ + e$ are more probable. Let us consider in which states the mu - molecular ion can be formed as a result of the Auger transition .

If the binding energy of the muon is close to I , the muon and the electron are in the same region of space in the system $(ab\mu^- e)^*$ and, therefore , the Auger transition can occur before a substantial change of the internuclear distance. In this case the muon looses energy close to the binding energy of the electron in the H_2 molecule at $R = R_e(H_2)$, i.e. 0.6 a.u. As a result the mu-molecular ion is formed in the state with energy E_μ =-1.2 a.u. at the internuclear distance 1.4 a.u. If the binding energy of a muon in the neutral system is close to zero, the system is a little or no perturbed molecular H_2^+ ion holding the muon at a large distance . Therefore the Auger transition can occur after the nuclei have shifted to the equilibrium distance $R_e(H_2^+)$ = 2.0 a.u. After the transition the muon binding energy is $|E_\mu|$ = + 1.1 a.u.

Thus, after the Auger transition the muon energy $E_j(R)$ is from - 1.2 to - 1.1 a.u. at the internuclear distances of 1.4 to 2.0 a.u. In this energy region there are many terms of the system $(ab\mu^-)^+$. The nonadiabaticity can give rise to the transitions from the attractive to repulsive terms $W_k(R) = E_k(R) + 1/R$ which will lead to the dissociation of an ion into a muonic atom and one of nuclei . In this case $|E_k(\infty)| \sim |E_j(R)|$ so that the dissociation products will have the total kinetic energy $T \sim 1/R$. At equal masses of nuclei the muonic atom acquires the kinetic energy of the order of 10 eV. The possibility of the radiation decay $(a\,b\,\mu^-)^+$ of an ion and the outer Auger effect on electrons of the neighbouring molecules are not allowed for here. To elucidate the relating contribution of these processes and, also, to obtain the distribution of mu-atoms over quantum states, it is necessary to perform detailed calculations of the decay probabilities of an excited ion $(ab\,\mu^-)^+$ from the states with the above--noted binding energies.

In conclusion it is to be noted that the present model of inelastic interactions of a muon with hydrogen molecules gives satisfactory agreement with the available experimental data on the slowing-down time of muons in hydrogen and the relative probabilities of Coulomb capture of a muon in the H_2 + He mixture . This model enables one to consider the main kinetic characteristics of Coulomb capture of muons in molecular hydrogen and to proceed to consideration of the subsequent decay of an excited muon molecule. The model can also be applied to Coulomb captureof $\pi^-$and K^- mesons.

The authors wish to express their gratitude to V.V. Balashov and L.I. Ponomarev for helpful discussions and to V.K. Dolinov, S.V. Leonova and G.A. Fesenko for assistance with calculations.

1. J.S. Cohen , R.J. Martin and W.R. Wadt , Phys.Rev.A24, 33 (1981).

2. J.S. Cohen, Phys.Rev. A27, 167 (1983).

3. G.Ya. Korenman . Few body problems in physics. IX European Conf. (Tbilisi ,1984), p.44 .
4. J.D. Garcia , N.H. Kwong and J.S. Cohen, Phys.Rev.A35 , 4068(1987).
5. J.S. Cohen, R.J. Martin, ad W.R. Wadt ,Phys.Rev. A27 , 1821 (1983).
6. V.V. Balashov, V.K. Dolinov, G.Ya. Korenman et al. Intern. Symp. Muon Catalyzed Fusion -87 . Abstracts (Leningrad, 1987) , p.143 .
7. V.V. Balashov, V.K. Dolinov , G.Ya. Korenman et al. Preprint of the Institute of Nuclear Physics, Moscow State University - 87 -010, M., 1987 .
8. D.C. Clary , Molecular Physics, 53, 3 (1984).
9. S. Hara, H. Sato, S. Ogata and N. Tamba, J.Phys.B: At.Mol. Phys.,19 ,1177 (1986).
10. A.A. Radtzig and B.M. Smirnov . Handbook of atomic and molecular physics (M., Atomizdat , 1980).
11. H. Anderhub , J. Böcklin , Devereux M . et al. Phys. Lett., 101B , 151 (1981).
12. F. Kottmann , Contribution to the 11 Intern. Symp. on Muon and Pion Interaction with Matter (Dubna, 1987).

MUON LOSSES IN DEUTERIUM-TRITIUM MUON-CATALYZED FUSION DUE TO FAST TRANSFER REACTIONS TO HELIUM NUCLEI

A.Bertin(*), M.Bruschi(*), M.Capponi(*), J.D.Davies(**), S.De Castro(*), I.Massa(***), M.Piccinini(*), M.Poli(****), N.Semprini-Cesari(*), A.Trombini(*), A.Vitale(*) and A.Zoccoli(*)

(*) Dipartimento di Fisica dell'Università di Bologna, and Istituto Nazionale di Fisica Nucleare, Sezione di Bologna, Italy

(**) Department of Physics, University of Birmingham, UK

(***) Dipartimento di Fisica dell'Università di Cagliari, and Istituto Nazionale di Fisica Nucleare, Sezione di Cagliari, Italy

(****) Dipartimento di Energetica dell'Università di Firenze, Italy, and Istituto Nazionale di Fisica Nucleare, Sezione di Bologna, Italy

ABSTRACT

The results obtained at low density on the fast muon transfer from excited levels of μd muonic atoms to ^4He are discussed. The significant muon losses which fast transfer processes to ^4He nuclei may induce in a high-density deuterium-tritium target used for muon-catalyzed fusion are considered.

INTRODUCTION

It is well known that ^3He and ^4He nuclei are unavoidalby present in a deuterium-tritium (d-t) target[1] which is used for muon-catalyzed fusion (MCF).[2] The ^3He nuclei arise from tritium decay, while the ^4He ones are the product of the reaction

$$\mu^- + d + t \rightarrow {}^4He + n + \mu^-. \tag{1}$$

The ^3He contamination within the target increases at a rate which depends on the tritium concentration. In its turn, the number of ^4He nuclei increases at a rate which depends on the intensity of the incoming muon beam. This fact implies that, if MCF is used for industrial production of energy, the ^4He contamination within the d-t target after a given time ΔT will increase with the power of the MCF reactor.

The muon transfer reactions from deuterium muonic atoms (μd) to helium nuclei,

$$\mu d + {}^3He \rightarrow \mu{}^3He + d \tag{2}$$

$$\mu d + {}^4He \rightarrow \mu{}^4He + d \tag{3}$$

may then occur within the considered target, as well as the analogous processes

$$\mu t + {}^3He \rightarrow \mu{}^3He + t \tag{2'}$$

$$\mu t + {}^4He \rightarrow \mu{}^4He + t \tag{3'}$$

starting from tritium (μt) muonic atoms. All these reactions set the muon out of play for the purposes of MCF, in a way which may come to compete with the sticking effect[2] to the helium nucleus released in process (1). It is therefore important to clarify all the conditions under which these muon transfers may occur.

In the present report, we shall discuss some aspects of this problem, with particular regard to the occurrence of reaction (3) from excited levels of the μd system, and to the extension the problem assumes in a d-t target which is used for MCF.

Table I - Triple-gas mixture conditions and total number N of delayed electron events (after background subtraction) per stopped muon ([a])

RUN	Partial Pressure ([b]) (atm)			Initial time cut	Gate width	N
	Deuterium	Xenon	Helium	(μs)	(μs)	
1	6	4.5×10^{-4}	0.0	0.8	7.2	15366
2	6	4.5×10^{-4}	0.6	0.8	7.2	20660

[a] See reference 4.
[b] The temperature was 293 K.

PROMPT TRANSFER OF MUONS TO HELIUM FROM EXCITED LEVELS OF THE μd ATOM

In a deuterium-helium mixture where negative muons are stopped, reaction (3) may occur, starting both from $(\mu d)_{1S}$ atoms in their ground state, and from $(\mu d)^*$ systems in an excited level (which means any level higher than the 1S one).[3] In the latter case, the process takes place during the cascade time (t_C) of the deuterium muonic atom to its ground level: it is therefore a prompt reaction, which occurs at a high average rate $(\lambda_{\mu d,He})^*$.

Table II - Experimental results on the muon transfer rates from μd muonic atoms to ^{4}He nuclei

Authors	Target conditions: Pressure	Target conditions: Temp.	Observed events	Muon transfer rate (10^8 s^{-1}): Prompt transfer $(\lambda_{\mu d,He})^*$	Muon transfer rate (10^8 s^{-1}): Long-term transfer $(\lambda_{\mu d,He})$
Bertin et al.[a]	6 atm d_2 (+0.6 atm He +4.5x10^{-4} atm Xe)	Room temp.	Muon decay electrons	$\geqq 10^3$	$\leqq 10^{-1}$
Balin et al.[b]	85.9 atm d_2 +(1.89x10^{-2})He	Room temp.	First event of d-d fusion following muon stop		3.68±0.18
Matzusaki et al.[c]	liquid d_2 +430 ppm He	20 K	de-excitation x-rays from $(\mu He)^*$ atoms		13.1±1.2

[a] See reference 4
[b] See reference 5
[c] See reference 6

The renewed interest shed by MCF on the transfer processes ((2) to (3')) has suggested us to re-examine the consequences of the only experimental study[4] on process (3) referring to low-density conditions. The relevant results were provided by Bertin, Vitale and Zavattini, who analyzed a measurement carried out stopping negative muons in a low-density gaseous target (at the experimental conditions recalled in Table I); and looking at the yield of delayed muon decay electrons. Their analysis (see Table II) [5,6]provided the first experimental evidence of muon transfer to ^{4}He nuclei from excited levels of μd muonic atoms. Here we wish to emphasize that the following results of this pioneer work are still to be retained:

i) long-term reactions (3) (namely transfers starting from $(\mu d)_{1S}$ systems) occur at a rate $\lambda_{\mu d,He} \lesssim 10^7\ s^{-1}$ at the quoted conditions of measurement. (This information was obtained looking at the differential time distribution of the observed muon decay electrons; the rate is normalized to a density of ^{4}He atoms equal to $2.11 x 10^{22}\ cm^{-3}$).

ii)at the same low-density conditions, prompt reactions (3) occur at a high rate (for a fraction $F^* \cong 20\%$ of the stopped muons) from excited states of the μd systems. (Here "prompt" means at very early times ($\lesssim 1$ ns) following the formation of the μd systems themselves; this result was obtained from the total yield (N) of the detected muon decay electrons.)

The high rate at which the prompt transfer process occurs can be expressed, in terms of the helium density ρ_{He}, as $(\lambda_{\mu d,He})^* = \rho_{He}\, \sigma\, v$, where σ is the average cross-section for the relevant reaction, and v is the velocity of the excited $(\mu d)^*$ muonic atom. Correspondingly, the fraction F^* can be written

$$F^* = \rho_{He}\sigma v t_C \cong K\rho_{He}t_C \qquad (4)$$

where K is roughly constant. We can then extrapolate the $F^*= 20\%$ value observed in the low-density measurement to other density conditions, simply by normalizing to the value of the helium partial pressure p_{He}, namely

$$F^* = f(0.20/p_{He,ld})p_{He} \qquad (5)$$

where $p_{He,ld}$ is the helium partial pressure in the low-density measurement,[4] and f is a factor which is due to the variation of the cascade time t_C with the d-He mixture density.

Equation (5) contains the information that, even keeping into account the decrease in t_C at high target densities, muon transfer to helium from excited states of the μd atoms may play a significant role also at target densities which are suitable for the MCF purposes.

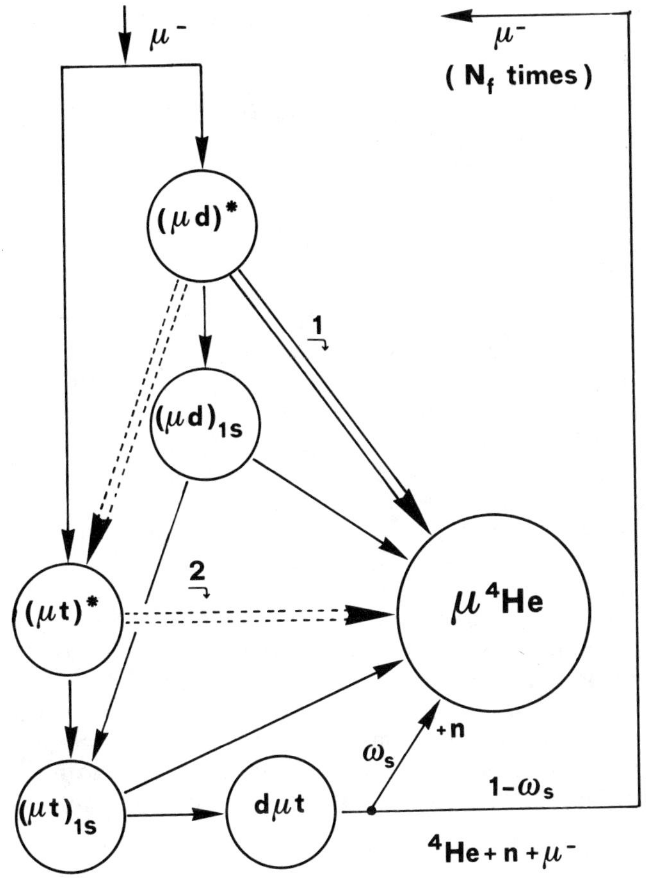

Fig. 1. Very simplified scheme of the muon-catalytic cycle in a deuterium-tritium target. The transfer processes from excited levels of the μd and μt atoms to ^{4}He nuclei are evidenced out (channels 1 and 2, respectively).

MUON LOSSES DUE TO PROMPT TRANSFER OF MUONS TO HELIUM IN A D-T TARGET

With regard to the more complex features of the negative muon cycles in a high-density d-t target (see Fig.1), the following considerations can be formulated:

i) only two experimental results on the rate of process (3) occurring at high-density conditions are available so far (see Table II).[5,6] Both of these data were obtained by analyzing delayed event decay rates, which essentially provide information on process (3) occurring - even if through an intermediate molecular state - from $(\mu d)_{1S}$ atoms which already have attained their 1S ground state.

ii) aside of the analogous process involving μd atoms in the 1S state, the isotopic exchange reaction

$$(\mu d)^* + t \rightarrow (\mu t)^* + d \tag{6}$$

where the muon is transferred between excited levels of the deuterium and tritium muonic atoms can occur. Its rate is expected to be quite high (of the order of 10^{12} s^{-1} at the tritium density of 4.2×10^{22} atoms cm^{-3}).[7]

iii)there are both experimental[8] and theoretical [3,9] indications on the fact that muons are transferred to helium at high rates, also starting from excited levels of the hydrogen (μp) and tritium muonic atoms.

Therefore, also for the more complicate case of a high-density d-t target sketched in Fig.1, fast muon transfers from excited states of muonic deuterium and tritium atoms can be expected to play an important role in determining muon losses in the MCF cycles.

If (as we expect) the rates for the transfer channels 1 and 2 of Fig.1 are similar, independently of the value of the rate for process (6), the fraction F^* of muons transferred to ^4He from excited states of μd and μt atoms can approximately be evaluated from Equation (5) also for the case of a high-density d-t target. If I is the stopped muon beam intensity (in particles s^{-1}), N_f the number of fusions induced by one single muon, ΔT the target irradiation time (s), V the target volume (cm^3), and $N_{He} = 2.68 \times 10^{19}$ cm^{-3} is the helium atomic density at 1 atm pressure, one has

$$(F^*)_{d-t} = f(0.20/0.6) \times (IN_f^2 \Delta T)/(VN_{He}) \tag{7}$$

For the ideal case of a negative muon beam having an intensity $I=10^{16}$ muons s^{-1},[10] which is totally stopping within a d-t high density target, having a volume V = 1 m^3, after one day of running one would get, assuming $N_f = 250$,[11] and a reduction f=0.25 of the cascade time due to the high-density condition,

$$(F^*)_{d-t} \cong 0.17 \tag{8}$$

which is quite of the same order (after the considered time interval) as for the low-density case.

One would note here that N_f enters as a square in Eq.(7) because the number of ^4He nuclei in the target is given by $IN_f\,\Delta T$, while each stopped muon restarts the cascade process within the $(\mu d)^*$ (or $(\mu t)^*$) atom N_f times due to the fusion cycles. Since in first approximation $N_f \cong 1/\omega_s$ (where ω_s is the sticking factor of the muon to the ^4He nucleus in the final state of process (1), see Fig.1), this means that $(F^*)_{d\text{-}t} \cong 1/(\omega_s)^2$. In other words, the smaller the sticking factor, the more important becomes the fast muon transfer from excited μd (μt) muonic atoms.

Let us consider, instead, the fraction $(F')_{d\text{-}t}$ of muons transferred to ^4He from $(\mu d)_{1S}$ atoms. One has in this case (see Fig.1)

$$(F')_{d-t} \cong (c_{He}\lambda_{\mu d,He})/(c_{He}\lambda_{\mu d,He} + c_t\lambda_{\mu d,t} + \lambda_o) \quad (9)$$

where c_{He} and c_t are the ^4He and tritium concentrations, respectively, $\lambda_{\mu d,t}$ is the rate at which process (6) proceeds when the μd atom is in its 1S ground state, and λ_o is the free muon decay rate. If one assumes for $\lambda_{\mu d,He}$ the value measured at liquid deuterium conditions by Matsuzaki et al.[6] (see Table II), and consider a d-t target with c_t=50%, one gets

$$(F')_{d-t} \cong 4x10^{-5} \quad (10)$$

for $c_{He} \cong 0.5 \times 10^{-5}$ (which corresponds - assuming N_f =250 - to the atomic concentration of ^4He nuclei in the 1 m^3 liquid d-t target after irradiating it with the muon beam previously considered for a time interval ΔT = 24 hours). Similar considerations can be formulated for the muons transferred to ^4He from $(\mu t)_{1S}$ atoms.

CONCLUSIONS

From the previous discussion, the following conclusions can be drawn:

i) Fast transfer processes from excited levels of the μd (μt) atom to ^4He nuclei may become quite significant when high-intensity muon beams are involved, after relatively short irradiation times. In estimating muon losses for MCF cycles, therefore, these processes should be carefully considered. Equation (8) gives a reference number, obtained from the extrapolation of an experimental result and in connection to a very high (still not available) muon beam intensity. Nevertheless, its value could even increase if targets of smaller volume could be used.

ii) The experimental results so far obtained on process (3) refer to significantly different experimental conditions, and are spreading over a wide range of values (see Table II). To provide systematical experimental results on this reaction, with particular regard to the rate dependences on the different experimental conditions and on the different time regimes of the process, is a necessary step to be undertaken.

iii)More generally, the problem of charge exchange (see Equation (6)) and transfer reactions to ^3He and ^4He nuclei starting from $(\mu d)^*$ or $(\mu t)^*$ excited levels shows up as a field which still demands experimental clarification for the purposes of the MCF community.

The authors are grateful to E.Zavattini for some useful discussions.

REFERENCES

1. S.S.Gershtein and L.I.Ponomarev, Phys.Lett. 72B, 80 (1977).
2. See e.g. S.E.Jones, Nature 321, 127 (1986); and references therein.
3. See e.g. A.V.Kravtsov and N.P.Popov, Z.Phys. 6, 61(1987); and references therein.
4. A.Bertin, A.Vitale, and E.Zavattini, Lettere al Nuovo Cimento 18, 381 (1977).
5. D.B.Balin, A.A.Vorob'ev, An.A.Vorob'ev, Yu.K.Zalite, A.A.Markov, V.I.Medvedev, E.M.Maev, G.G.Semenchuk and Yu.V.Smirenin, Pis'ma Zh. Eksp. Teor.Fiz. 42, 236 (1985); (English translation: JETP Lett: 42, 293 (1985)).
6. T.Matsuzaki, K.Ishida, K.Nagamine, Y.Hirata, and R.Kadono, (to be published on Muon Catalyzed Fusion).
7. See for instance A.V.Kravtsov, A.Yu.Mayorov, A.I.Mikhailov, S.Yu.Ovchimikov, N.P.Popov, V.M.Suvorov, A.I. Shchetkovsky, LNPI-1315, Leningrad (1987) (to be published on Muon Catalyzed Fusion); and references therein.
8. V.M.Bystritskii, V.P.Dzhelepov, V.I.Petrukhin, A.I.Rudenko, V.M.Suvorov, V.V.Filchenkov, N.N.Khovanskii, and B.A.Khomenko, Zh.Eksp.Teor.Fiz. 84, 1257 (1983); (English translation: Sov.Phys.JETP 57, 728 (1983)).
9. N.P.Popov, LNPI - 1359 (1988) (to be published on Proc.II Int. Symp. on Muon and Pion Interactions with Matter, Dubna, 30 June-4 July (1987)).
10. M.Weiss, CERN-PS/87-52(LI) (1987) (Proc. of the Course-Workshop on Muon-Catalyzed Fusion and Fusion with polarized Nuclei, Erice, Italy, 3-9 April 1987, to be published).
11. S.E.Jones, Proc. of the Course-Workshop on Muon-Catalyzed Fusion and Fusion with polarized Nuclei, Erice, Italy, 3-9 April 1987 (to be published).

Increase in Meso-Molecular Formation by Laser-induced Resonances

S. Barnett

A.M. Lane

Theoretical Physics Division, AERE Harwell,
Oxon OX11 ORA, U.K.

It is pointed out that the formation of meso-molecules, $(t\mu)+d\rightarrow(dt\mu)$, may be enhanced by the presence of a laser which implants a "pseudo-resonance" in the continuum, $(t\mu)+d$. Relative to the usual resonances in the collisions $(t\mu)+D_2$, the implanted resonance dominates for laser intensities $\gtrsim 10^{13}$W cm^{-2}.

In muon-catalysed fusion, the two key elements in determining the number of nuclear reactions catalysed by a muon are the meso-molecular formation rate and the sticking probability. (See e.g. Jones et al. 1986, Breunlich et al. 1987). The number increases with increasing formation rate, and decreasing sticking probability. In this work, we are concerned only with the former. It is accepted that the rate is dominated by resonances in the relevant cross-section, and we enquire whether the rate may be increased by adding laser-implanted resonances to those provided by Nature.

The idea of implanting resonances with a laser was put forward in an atomic physics context (Armstrong, Beers & Feneuille 1975). Such resonances have been seen experimentally (Heller et al. 1981, Hutchinson & Ness 1988). The essential aspects are readily conveyed. Consider an atom with a ground state, $|0\rangle$, an excited bound state close to threshold $|1\rangle$, and a continuum $|E\rangle$. Photoionisation from the excited state has the cross-section $\sigma_{\gamma 1} = \langle E|d|1\rangle^2$ (dropping irrelevant factors), where $\langle E|d|1\rangle$ is the dipole matrix element. Thus if the target atom were in its excited state, its rate of depletion under laser flux I (number of photons per unit area per unit time) will be $I\sigma_{\gamma 1} \equiv \Gamma_1$ (say). The laser frequency ω is related to the initial and final energies by the energy conserving condition $\omega=(E-E_1)$. (We set $\hbar=1$). We now consider the target atom in its ground state, and suppose that, in addition to the strong laser ω (which cannot ionise the ground state if $\omega < (E-E_o)$, the atom is exposed to a weak laser with frequency, ω', in the vicinity of $\omega'=\omega+(E_1-E_o)$. The photoionisation cross-section $\sigma_{\gamma o}=\langle\psi_E|d|o\rangle^2$ will now show a resonance arising from the fact that the continuum states $|\psi_E\rangle$ contain an induced structure from laser ω. Fano theory gives:

$$|\psi_E\rangle = (1+\varepsilon^2)^{-\frac{1}{2}} \left[\varepsilon|E\rangle + \frac{1}{\pi\langle E|d|1\rangle\mathcal{E}} \left(|1\rangle + \frac{\mathcal{E}}{E-H_o} d|1\rangle\right)\right] \quad (1)$$

where

$$\varepsilon \equiv \frac{(\omega'+E_o)-(\omega+E_1)}{\frac{1}{2}\Gamma_1} \quad (2)$$

H_o is the atomic Hamiltonian, and $\mathcal{E}$ is the electric field of the laser. Note that, since $|1\rangle$ and $|E\rangle$ have opposite parity, (1) is a mixed-parity state; also the relative sizes of the three terms in (1) are, as $\mathcal{E}\to o$, given by $1:\mathcal{E}:\mathcal{E}^2$. Taking the photoionisation matrix element with $\langle 0|d$ gives the Fano formula:

$$\sigma = \langle 0|d|\psi_E\rangle^2 = (1+\varepsilon^2)^{-1} \left[\varepsilon\langle 0|d|E\rangle + (\pi\langle E|d|1\rangle)^{-1}\langle 0|d(E-H_o)^{-1}d|1\rangle\right]^2 \quad (3)$$

where we have assumed $|0\rangle$, $|1\rangle$ have the same parity.

Our interest is not in photoionisation but in the integrated cross-section for exciting state $|1\rangle$ from the continuum channel. From (1), this is:

$$\int \sigma_1 dE = \frac{1}{\pi}\Gamma_1 \; . \quad (4)$$

Thus the integrated capture cross-section for forming low-lying states (those lying below E_1) via state $|1\rangle$ is:

$$\int \sigma_{cap}\, dE = \frac{1}{\pi}\frac{\Gamma_1\,\Gamma\!\downarrow}{\Gamma_1 + \Gamma\!\downarrow} \quad (5)$$

where $\Gamma\downarrow$ is the decay width of $|1\rangle$ for decay to such states. This is to be compared to the corresponding quantity from a true resonance r, say, in the absence of the laser. This is the same with Γ_r in place of Γ_1 where Γ_r is the decay width of the resonance into the continuum channel. (We may assume $\Gamma\downarrow$ to be the same for the resonance and for state 1). For the cross-section from the laser-induced resonance to be significantly larger, we need $\Gamma_1 >> \Gamma_r$ and $\Gamma\downarrow >> \Gamma_r$.

We now describe the essential features of resonant meso-molecular formation. The case of interest is:

$$(t\mu) + (D_2)_{K_i 0} \rightarrow (\text{"}D_2\text{"})_{k\nu} \equiv ((dt\mu)_{11}dee)_{k\nu}$$

This represents the collision of a triton-muon atom $(t\mu)$ with a deuterium molecule (in rotation-vibration state $K_i, \nu_i=0$) leading to the formation of the quasi-D_2 molecule, "D_2", in which a deuteron is replaced by the meso-molecule $(dt\mu)_{11}$. $K\nu$ signifies the rotation-vibration state of "D_2". The meso-molecule is in its fourth excited state (labelled 11), bound by 0.6 eV, and decays to lower states (mainly by the Auger process) with width, $\Gamma\downarrow$, estimated as $1.2\text{x}10^{12}s^{-1}$. (Bogdanova et al. 1982). This is the dominant decay of "D_2". The other main decay mode is "back-decay" into the entrance channel, for which the widths are $\Gamma_r < 10^{11}s^{-1}$. There are 17 expected resonances (all with $\nu=2$) in the range up to 100 meV above threshold. The four largest values of Γ_r in $10^{11}s^{-1}$, along with resonance parameters (K_i,K, energy) are: 1.01(4,7,49.4 meV), 0.94(4,6,15.8 meV), 0.48(2,4,14.8 meV) 0.46(3,5,16.8 meV). (Lane 1988). For all resonances, the above condition $\Gamma_r << \Gamma\downarrow$ is satisfied: $\Gamma_r < 0.08\ \Gamma\downarrow$. It follows that, if $\Gamma_1 > \Gamma_r$, the integrated capture cross-section is significantly increased by a laser-induced resonance (by factor $\Gamma_1 \; \Gamma{\downarrow}[(\Gamma_l+\Gamma{\downarrow})\Gamma_r]^{-1}$).

Estimate of Γ_1: The essential factor in Γ_1 is the square of the dipole matrix element

$$<E|d|1> = <E|\sum_i e_i z_i|\phi(\underline{r},\underline{r}_{t\mu})\Psi_{kv}(\underline{R})> \quad (6)$$

where $\underline{r}_i \equiv (x_i,y_i,z_i)$ is the co-ordinate of particle i (with charge e_i) relative to the overall centroid. $\phi(\underline{r},\underline{r}_{t\mu})$ represents the (tdμ) meso-molecule (in the (11) state), and $\Psi_{kv}(\underline{R})$ represents the overall motion of "D_2". $\underline{r}_{t\mu} \equiv (\underline{r}_\mu - \underline{r}_t)$ and $\underline{r}$ is the coordinate of the deuteron in (dtμ) relative to the centroid of (tμ). $\underline{R}$ is the coordinate of the centroid of (tdμ) relative to the second deuteron. ($\underline{R},\underline{r},\underline{\rho}$) are a complete set of independent normal coordinates.

$$\sum_i e_i z_i = e(-z_{t\mu} + 0.6\ z + 0.44\ Z) \quad (17)$$

(In keeping with previous work on the formation of "D_2" states, we suppress reference to electron motion; the presence of electrons is represented by screened potentials for the deuterons). We estimate <E|d|1> by assuming that the term in z is dominant on the following basis. The $z_{t\mu}$ coordinate is small, and we can expect that each side of <E|d|1> overlaps strongly on the (tμ) ground state, implying small dipole matrix element. The two sides of <E|d|1> necessarily differ in their dependence on $\underline{r}$, |E> containing a free state in $\underline{r}$, while ϕ is a bound state, these being orthogonal. Thus the matrix element of Z is strongly reduced by this near-orthogonality in $\underline{r}$. (Further, for E near threshold, there is a separate reduction due to the p-wave form of |E> in coordinate $\underline{r}$, implied by p-wave nature of ϕ). In contrast, on choosing $K=K_i$, $v=v_i(=0)$, the $\underline{R}$-overlap between the two sides of $\langle E|d|1\rangle$ can be large, ≈ 1. If the relative motion of $(t\mu)$ and D_2 in $|E\rangle$ is approximated as a plane wave, then, in terms of $\underline{R},\underline{r},\underline{r}_{t\mu}$:

$$\langle \underline{R},\underline{r},\underline{r}_{t\mu}|E\rangle \approx e^{i\underline{k}_i\cdot(f\underline{r}+g\underline{R})}\Psi_{K_i 0}(\underline{R}-h\underline{r})\eta(\underline{r}_{t\mu}) \tag{8}$$

where f,g,h are simple functions of masses involved (Lane 1987). The presence of ϕ in $\langle E|d|1\rangle$ means that the value (Lane 1987) of $h\underline{r}$ is appreciably smaller than $\underline{R}$, so we ignore it. The $\underline{r}$ dependence then arises solely from $\exp(if\underline{k}_i.r)$. To take account of the strong short-range interaction in $\underline{r}$, we replace this exponential (times $\eta(\underline{r}_{t\mu})$ by the scattering solution for $(t\mu)$+d for initial momentum $f\underline{k}_i$, say $\chi_{f\underline{k}_i}(\underline{r},\underline{r}_{t\mu})$ With the above approximations resulting in overlap ≈ 1 in $\underline{r}_{t\mu}$ and $\underline{R}$, the value of $\langle E|d|1\rangle$ is then given by the photo-ionisation matrix element for $(dt\mu)_{11} \rightarrow (t\mu)$+d at final momentum $f\underline{k}_i$

$$\langle E|d|1\rangle \approx 0.61\ e\ \langle\chi_{f\underline{k}_i}|z|\phi\rangle \tag{9}$$

This can be readily estimated on the basis that most of the norm of ϕ lies in the exponential tail in $\underline{r}$ (due to the light binding, (0.6 eV). If the scattering length for $(t\mu)$+d scattering is a, then, for r in the tail region and small fk_i, $\chi_{f\underline{k}_i}$ is $(1-ar^{-1})\eta$. Putting in all factors we find:

$$\Gamma_1 = \frac{4}{9}\left(\frac{me^2}{\hbar^2}\right)\left(\frac{1}{4}\,g^2\right)\left(f^2k_i\right)\left[\int_R^\infty (r-a)\langle\eta|\phi\rangle r^2 dr\right]^2 \tag{10}$$

where R is the point in r where $\langle\eta|\phi\rangle$ begins to take its asymptotic form (Menshikov 1985):

$$\langle\eta|\phi\rangle = 0.57\kappa^{3/2}\left(\frac{1}{x} + \frac{1}{x^2}\right)e^{-x} \tag{11}$$

with $x = \kappa r$, and κ is the binding wave-number and m the associated mass: $\hbar^2\kappa^2/2m = 0.6$ eV. The integral in Γ_1 is:

$$\frac{0.57}{\kappa^{5/2}}\left[X^2+2X+3 - X_o(2+X)\right]e^{-X}$$

where $X \equiv \kappa R$, $X_o \equiv \kappa a$. With $a \approx 3.8a_\mu$ (Melezhik and Ponomarev 1983), $R \approx 7a_\mu$, $\kappa^{-1} = 20.8a_\mu$, $X \approx 0.33$, $X_o \approx 0.20$ and the integral is $\approx 1.4\ \kappa^{-5/2}$. This leads to (with f=0.68, m=1.7x proton mass):

$$\Gamma_1 \approx 0.8\ 10^{12}\ s^{-1}\ (10^{-8}\ \mathcal{E}(\text{eV cm}^{-1}))^2(E_i(\text{ev}))^{\frac{1}{2}} \tag{12}$$

where E_i is the relative energy of $(t\mu)$ and D_2. Thus, for the induced resonance to have Γ_1 larger than the largest Γ_r listed above (at E_i=49.4 meV), we need:

$$\mathcal{E} \gtrsim 7.5 \times 10^7\ \text{eV cm}^{-1}$$

which corresponds to an energy flux $\gtrsim 7 \times 10^{12}$ W cm^{-2}. This is within the range of present laser technology. Note that, at such large power, the D_2 molecules will disintegrate, but (12) is still relevant since it also applies to the "bare" process $t\mu+d \rightarrow (dt\mu)_{11}$. The only effect is the omission of the factor f, and different mass factors; correcting for these means that E_1(ev) in (12) is replaced by 1.7 times the relative energy in the $(t\mu)+d$ system. (The $(t\mu)$ atom is not significantly excited for fields $\mathcal{E} < 10^{14}$ eV cm^{-1}). Note also that the $(dt\mu)$ yield does not continue to increase with intensity once Γ_1 is $> \Gamma\downarrow$, i.e. for intensity $> 10^{14}$W cm^{-2}. However, different induced resonances are additive in the yield, so the yield may still be increased by increasing the number of lasers (of different frequency).

In summary, we have seen that $(dt\mu)$ formation in $(t\mu)+D_2$ collisions will be enhanced by a laser of photon energy ~ 0.6 eV if the intensity is $\gtrsim 10^{13}$W cm^{-2}. At this power, the D_2 are rapidly ionised, but this does not matter since the basic process involved is $t\mu+d \rightarrow (dt\mu)_{11}$. The only drawback is heating due to the laser causing the spread of thermal relative velocities in $(t\mu)+d$ to increase, so the fraction of collisions at the resonance energy decreases.

Work described in this Letter was undertaken as part of the Underlying Research Programme of the UKAEA.

References

L. Armstrong, B.L. Beers and S. Feneuille 1975, Phys.Rev. A12, 1903.

L.N. Bogdanova, V.E. Markushin, V.S. Melezhik and L.I. Ponomanev 1982 Sov.J.Nucl.Phys. 56, 931.

W.H. Breunlich et al. 1987, Phys.Rev.Lett. 58, 329.

Yu.I. Heller, V.F. Lukinykh, A.K. Popov and V.V. Slabko 1981 Phys.Lett. 82A, 4.

M.H.R. Hutchinson and K.M.M. Ness 1988, Phys.Rev.Lett. 60, 105.

S.E. Jones et al. 1986, Phys.Rev.Lett. 56, 588.

A.M. Lane 1987, J.Phys.B. At.& Mol. 20, 2911.

A.M. Lane 1988, J.Phys.B, At.Mol. and Opt. in course of publication.

V.S. Melezhik and L.I. Ponomarev 1983, Sov.Phys. JETP 58, 254.

L.I. Menshikov 1985 Sov.J.Nucl.Phys. 42, 918.

LASER INDUCED μ-MOLECULAR FORMATION

S. ELIEZER AND Z. HENIS
Plasma Physics Department
Soreq Nuclear Research Center
Yavne 70600, Israel

ABSTRACT

It is suggested to enhance the μ-molecular formation by strong electromagnetic fields.

We consider a three level system for a μ-molecule, such as dtμ, dpμ etc. For the dtμ molecule these levels are : a state in the continuum of tμ+d (a), the bound state (J,v)=(1,1) of the molecule dtμ (c) and the bound state (J,v)=(0,1) of dtμ (b). Under the influence of an external field a Stokes transition from level (a) to level (c) takes place while the transition from (c) to (b) is occuring through an Auger process. The spontaneous decay from (a) to (c) and the decay of level (c) are included phenomenologically.

The probability amplitudes of the levels (a) and (c) (within the rotating wave approximation) are estimated assuming a dipole transition. The cross section for this two step process and the efficiency (defined as the ratio between Stokes induced transitions and spontaneous decay) are calculated.

The results show that for a resonant laser frequency and intensities of 6.10^{5} W/cm^{2} the Stokes effeciency is about 50.

INTRODUCTION

When an energetic negative muon (μ) enters a compressed hydrogen gas (e.g. hydrogen liquid density defined by n_o=4.25. 10^{22} atoms /cm^{3}) the following chain of reactions occurs [1,2,3]: a) slowing down of μ (b) capture of μ into atomic levels and the cascade to the atomic ground state. (c) μ-molecular resonant formation and de-exitation to the molecular ground state. (d) nuclear fusion of the nuclei which are kept close together by the negative muon. The μ is either released or captured by the nuclear fusion products. If the lifetime of the muon is long compared to the time scale of the other processes (a) to (c) then many fusion reactions can occur during the lifetime of a muon. This chain of reactions resulting in the nuclear fusion process is usually called muon catalysed fusion.

MOLECULAR FORMATION BY STOKES TRANSITIONS

In this paper it is suggested that μ-molecular formation may be enhanced by strong electomagnetic fields.

We consider a three energy-level system of a μ molecule, such as dtμ, ddμ, dpμ etc., as illustrated in fig. 1:

Fig. 1: Stokes (Resonance) Formation of dtμ

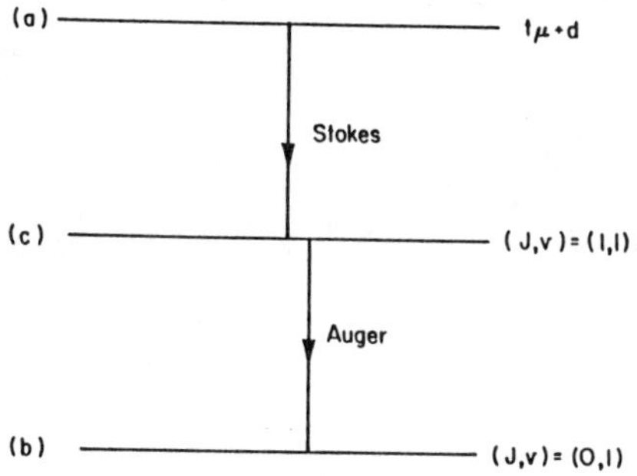

The quantum levels are denoted by (a),(b) and (c) where $E_a>E_c>E_b>$ are the appropriate energies. The lifetimes of these levels satisfy the condition $\gamma_a>>\gamma_c$.

For example for the dtμ molecule these levels are : a state in the continuum of the system tμ+d (a), the bound state (J,v)=(1,1) of the molecule dtμ (b) and the bound state (J,v)=(0,1) of dtμ (c).

The system is initially in the state (a). Under the influence of an external field with frequency ω we can induce a two-step decay of level (a). During the first step a virtual transition to level (c) is induced by absorbtion of a single photon while the second step is a transition to level (b) through a spontaneous decay or an Auger process. The electromagnetic interaction between the external field and the system is described by the Hamiltonian $H_{int} = V \cos \omega t$. V will be specified later. Our formal approch is analoguous to the treatment of Raman scattering. The Schrodinger equation yields a system of equations for the amplitudes ψ_a and ψ_c of the levels (a) and (c). The transitions from (a) to (c) without the laser field and the decay of level (c) are included phenomenolo - gically by the rates γ_a and γ_c. Within the rotating wave approximation [4] these equations are :

$$\dot{\psi}_a = -\frac{1}{2}\gamma_a\psi_a + \frac{1}{2}\frac{i}{h}V_{ca}\,e^{i\Delta t}\,\psi_c \qquad (1)$$

$$\dot{\psi}_c = -\frac{1}{2}\gamma_c\psi_c + \frac{1}{2}\frac{i}{h}V_{ac}\,e^{-i\Delta t}\,\psi_a \qquad (2)$$

$|V_{ac}|$ denotes the transition matrix element between the states (a) and (c) and Δ is the detuning given by

$$\Delta = \frac{E_a - E_c}{h} - \omega \tag{3}$$

The solution of the equation (1) and (2) for the initial conditions $\psi_a(0) = 1$ and $\psi_c(0) = 0$ is

$$|\psi_c(t)|^2 = \left(\frac{|V_{ca}|}{h}\right)^2 \frac{e^{-\gamma t}}{\mu\mu^*} \sin(\tfrac{1}{2}\mu t) \sin(\tfrac{1}{2}\mu^* t) \tag{4}$$

where γ is

$$\gamma = \frac{\gamma_a + \gamma_c}{2} \tag{5}$$

μ is the complex Rabi flopping frequency,

$$\mu = \left[(\Delta - i\delta)^2 + \left(\frac{|V_{ac}|}{h}\right)^2\right]^{\frac{1}{2}} \tag{6}$$

and δ is given by

$$\delta = \frac{\gamma_a - \gamma_c}{2} \tag{7}$$

The probability for the Stokes process $P_{st}(T)$ up to time T is given by:

$$P_{st}(T) = \gamma_c \int_0^T dt |\psi_c(t)|^2 \tag{8}$$

Since for our cases of interest $\tau_a >> \tau_c$ one can obtain the probability $P_o(T)$ for the transition from (a) to (b) switching off the laser during a time T by:

$$P_o = \gamma_a \int_0^T dt |\psi_a(t)|^2 \tag{9}$$

The efficiency η for Stokes transitions up to time T is given by:

$$\eta = \frac{P_{st}(T)}{P_o(T)} \tag{10}$$

By integrating (4) one obtains the following expression for $P_{st}(T)$:

$$P_{st}(T) = \frac{\gamma_c}{2\mu\mu^*}\left(\frac{|V_{ca}|}{h}\right)^2 \left[e^{-\gamma t}\left(\frac{-\gamma\cosh yT - y\sinh yT}{\gamma^2 - y^2} - \frac{\gamma\cos xT + x\sin xT}{\gamma^2 + x^2}\right) + \frac{2\gamma^2 + x^2 - y^2}{(\gamma^2 - y^2)(\gamma^2 + x^2)}\right] \tag{11}$$

x and y are the real and imaginary parts of μ (see eq. 6).

The first order one finds a simplified solution of equations (1) and (2):

$$|\psi_a(t)|^2 = e^{-\gamma_a t} \tag{12}$$

$$|\psi_c(t)|^2 = \left(\frac{|V_{ac}|}{2h}\right)^2 \frac{e^{-\gamma_a t} + e^{-\gamma_c t} - 2e^{-\gamma t}\cos\Delta t}{\Delta^2 + \delta^2} \tag{13}$$

We are interested in a system where (c) decays much faster than (a) i.e. $\gamma_c \gg \gamma_a$. Then and δ reduce to $\gamma = -\delta = \gamma_c/2$. According to (12) and (13) the population of the exited level $|\psi_c(t)|^2$ decays to first order for large times on the same slow time scale γ_a^{-1} as $|\psi_a(t)|^2$. We assume that the pulse of the external field is long compared to the lifetime of the level (c) and small compared to the lifetime of (a). ($\gamma_c^{-1} \ll T \ll \gamma_a^{-1}$). By integrating (12) and (13) one gets the following expression for $P_{st}(T)$ and η :

$$P_{st}(T) = \left(\frac{|V_{ac}|}{2h}\right) \frac{\gamma_c T}{\Delta^2 + \left(\frac{\gamma_c}{2}\right)^2} \tag{14}$$

$$\eta = \frac{|V_{ac}|^2 \gamma_c}{4h^2\gamma_a\left(\Delta^2 + \frac{\gamma_c^2}{4}\right)} \tag{15}$$

Consider the resonance case, $\Delta = 0$ one gets eqs. (14) and (15)

$$P_{st}(T) = \left(\frac{|V_{ac}|}{h}\right)^2 \frac{T}{\gamma_c} \tag{16}$$

and the induced Stokes efficiency:

$$\eta = \left(\frac{|V_{ac}|}{h}\right)^2 \frac{1}{\gamma_a \gamma_c} \tag{17}$$

DIPOLE TRANSITIONS

In order to calculate the interaction matrix element $|V_{ac}|$ we assume a dipole transition between quantum levels (a) and (c):

$$|V_{ac}|^2 = \frac{1}{3}(eE_L)^2 |d_{ac}|^2 \tag{18}$$

E_L is the electric field of the laser, related to the flux F by $E_L^2 = (4\pi)^2 h^2 F/\lambda_L$, where λ_L is the laser wavelength. In terms of the laser intensity defined by $I_L = cE_L^2/8\pi$, the matrix element $|V_{ac}|$ is given by:

$$|V_{ac}|^2 = \frac{8\Pi}{3}\frac{e^2}{c} I_L |d_{ac}|^2 \tag{19}$$

We use the matrix element $|d_{ac}|^2$ as calculated in Ref [5]:

$$|d_{ac}|^2 = \frac{1}{4\Pi} \sum_{m_J} \left| \int dR\, dr\, \psi_c(r,R)\, d(r,R)\, \psi_a(r,R) \right|^2 \tag{20}$$

where R is the radius vector joining the nuclei t and, r is the radius vector of the μ with origin at the geometric center of segment R and **d** is the dipole moment of the system comprising the mesic atom and the nucleus of the D_2 molecule, relative to the mass center of dtμ. Following [5]:

$$\mathbf{d} = -\frac{1}{2}\frac{M_d - M_t}{M_d + M_t}\left(1 - \frac{m_\mu}{M_{tot}}\right) R - \left(1 + \frac{m_\mu}{M_{tot}}\right) r \tag{21}$$

M_μ , M_d, M_t are the masses of μ,d,t and $M_{tot}= M_\mu+M_d+M_t$.

CONCLUSIONS

Table I shows the numerical values of the dipole matrix element $|d_{ac}|$ as calculated in 5, the lifetimes of the levels (a) and (c) [6,7] and the estimated efficiency for resonant Stokes transitions as a function of the laser intensity, for the molecules dtμ and ddμ . The results show that for moderate laser intensities of 10^6 W cm^{-2} the efficiency for resonance formation of dtμ by induced Stokes transitions is about 50, while for ddμ it has the huge value of 10^7.

Table I. Quantities that determine the efficiency of resonant Stokes induced molecular formation of ddμ and dtμ

Quantity	ddμ	dtμ
γ_a	3.10^6sec^{-1}	10^9sec^{-1}
γ_c	3.10^8sec^{-1}	10^{12}sec^{-1}
$\|d_{ac}\|$	$1.058.10^{-8}$cm	$1.005.10^{-8}$cm
η	70 I_L (W/cm^2)	$5.10^{-5}I_L$ (W/cm^2)

However it must be emphasized that the resonance condition is extremely important for the above estimation of η . The resonance condition $\Delta < \gamma_c /2$ is equivalent to $\alpha < 3.10^{-16} \gamma_c /(E_a-E_c$ (eV) where $\alpha=h\ \Delta/(E_a-E_c)$. In particular for the dtμ case the resonance condition implies $\alpha<5.10^{-4}$ while for ddμ, $\alpha<5.10^{-8}$. Moreover if $\Delta > \gamma_c/2$ then the efficiency of the Stokes induced transitions is reduced significantly. In this case one obtains $\eta \approx 10^{-11}$ I_L (W/cm^2)/α^2 . For example requiring a resonance detuning of 10% (α = 0.1) and a Stokes efficiency η=10 one needs a laser field of 10^{10}W/cm^2.

REFERENCES

1. S.S. Gershtein and L. I. Ponomarev, in : Muon Physics, Vol. III, eds V. W Hughes and C. S. Wu (Academic Press), N. Y.1975 p. 141

2. L. Bracci and G. Fiorentini, Phys. Rep. 86, 169(1982)
3. S. E. Jones, Nature , 321, 127 (1986)
4. M. Sergent III, M. O. Scully and W. E. Lamb. Jr., Laser Physics (Addison - Wesley, Reading

Mass 1974)
5. S. I. Vinitskii et al, Sov. Phys. JETP 47, 444(1978)
6. L. N. Bogdanova et al, Sov. Phys. JETP 56, 931(1982)
7. A. M. Lane, Phys. Lett. 98A,337(1983)

MUONIC MOLECULAR FORMATION UNDER LASER IRRADIATION AND IN THE CLUSTERED ION MOLECULE (The Effect of Protonium Additive on the Muon Catalyzed Fusion cycle)

Hiroshi Takahashi
Brookhaven National Laboratory, Upton, NY 11973

INTRODUCTION

The number of fusions catalyzed by one muon depends on the rate of formation of the muonic molecule. The fast resonance formation rate for the dtμ molecule due to the shallow binding energy of (J=1, v=1) state produces more than 140 fusions per one muon for a liquid hydrogen density DT target.[1,2,3]

The formation rate of the dtμ molecule is very sensitive to the differences in the vibrational and rotational level between D_2 and [(dtμ)-d-2e] molecules.[4,5,6,7] The density effect of the normalized reaction rate has been studied by the resonance broadening due to collisional quenching.[8,9,10] The surrounding molecules of the molecule forming dtμ act as the third body which takes out the excess energy forming dtμ from tμ, and the formation reaction occurs with the excitation of the vibrational state just below the threshold energy. By using the laser as the third body, the rate of resonance formation can be increased. In my last paper,[11] I calculated the formation rate under high-intensity laser irradiation, using Vinitsky's model assuming that the laser interacts directly with the deuteron and modulates the interaction between tμ and d_2 nuclei. However, the laser interacts more strongly with the electrons, because the interaction energy of the laser and the charged particle is proportional to the velocity of the particle's motion, and the velocity of the electron is a few thousand times greater than the velocity of the nuclei. This interaction with electron was neglected in my last paper. In the present paper, the enhancement of the dtμ formation rate by the strong laser irradiation is studied, taking into account the laser electron interaction; I showed that the enhancement can be achieved by an intensity lower than the one described previously.

Instead of using the surrounding molecule as the third body enhancing or changing the rate of dtμ formation, the clustered ion can[12] also change the dtμ formation rate drastically because it has many more vibrational modes than the diatomic molecule.

Many clustered hydrogen ions exist as stable ions, and they are created at very high temperatures. Recently Jändel et al. discussed the dtμ fusion reaction in the high temperature environment created by the heat spike,[13] where there is the possibility that D_2,T_2 molecules are transformed into the (D_3^+,e) (T_3^+e) states. Thus it is worthwhile to study the rate of dtμ formation from these clustered ions. In this paper, the most simple clustered ion (the D_3^+ ion) is studied,[14,15] as a first step. Due to the many modes which can be excited in the clustered ion, the temperature dependence of $dt\mu$ molecular formation is smoother than when the $dt\mu$ molecules are formed from diatomic molecules.

The addition of protonium would be beneficial, giving savings in fuel cost and also reducing the neutron wall loading: both factors have been criticized as regards their practical use in this fusion approach to energy production. Also, it will be beneficial to use this method as an energy source for future space propulsion.

Three effects of the addition of protonium to the D.T. mixture target can be considered. The first is an increase of the $dt\mu$ fusion cycle in a very dilute D.T. mixture. By adding the protonium to the low density D.T. gas mixture, the effective density becomes high, and $dt\mu$ molecular formation rate can be increased as the density increases by the third body effect.[8,9,10] The second effect is the population change of the $d\mu$ ground state[16] which may become a bottle neck in the fusion cycle. The third effect is a change in $dt\mu$ molecular formation when the protonium is in the form of a D-H molecule instead of a D_2 and H_2 molecule. These subjects will be discussed.

1. $dt\mu$ Formation under Laser Irradiation

When the laser irradiates the $t\mu$-D_2 system, the kinetic part of the Hamiltonian $P_i{}^2/2M_i$ ($i = t,\mu,d_2,d_2,e_1,e_2$) is replaced by the $(\vec{P} + ie\vec{A}/C)^2/2M_i$ where $\vec{A}$ is the vector potential corresponding to the laser field, which is treated as the classical field due to the strong coherency. The term, $(e_i{}^2A_i{}^2/2M_iC^2)$, expresses the interaction energy due to the ponderomotive force and changes only the phase of wave function. This term becomes large only when the laser intensity is much higher than the laser intensity considered here, therefore, this term can be neglected.

The linear term in A is $e\vec{P}_i\vec{A}/M_iC = e\vec{A}_i\vec{v}_i/C$ and it is proportional to the velocity v_i of the particle i. The velocities of the electron and muon are, respectively, of the order of C/137 and they are about a few thousand times the velocity of the nuclei motion, so that the interaction of the laser with electron cannot be neglected.

The interaction of the laser with the muon is comparable to the electron, but the muon is 200 times more strongly bound than the electron. Thus muon wave function is hardly changed by the laser field intensity which we are considering here. Therefore, we can neglect this muonic term.

The initial wave function of the $t\mu$-D_2 molecule system and the final wave function of ($dt\mu$-d-2e) molecular system under the laser with single frequency ω irradiation can be expressed as

$$\Psi_{\substack{i\\f}}(r_k,t) = \sum_m \Phi_{\substack{imA\\f}}(r_k)e^{i(E_{\substack{iA\\f}} + m\omega)t}, \tag{2.1}$$

where k stands for the muon (μ), triton (t), deuteron (d_1 and d_2) and electrons (e_1 and e_2). The wave functions are composed of many modes whose energies are separated with frequency ω.

The amplitude functions of $\Phi_{imA}(r_k)$, $\Phi_{fm'A}(r_k)$ and initial and final energies E_{iA} and E_{fA} are functions of the laser intensity A, and when the laser intensity increases, the amplitude of the high $|m|$ mode behaves like the function of $J_m\left(\frac{ePA}{\hbar c\,\omega}\right)$ as discussed in the previous paper.[11] The rate of dtμ molecular formation rate is expressed as

$$\lambda = \sigma v\, N_o\ [\mathrm{sec}^{-1}]\ , \qquad (2.2)$$

where v is the relative velocity of the colliding beam, N_o is the number density of the deuteron in the D,T mixture,

$$v\,d\sigma = 2\pi\hbar^{-1} \sum_{m,m'} \left|T_{fmim'}\right|^2 \delta(E_{fA} - E_{iA} + (m-m')\omega)\ \gamma(\varepsilon, \varepsilon_T)d\varepsilon. \qquad (2.3)$$

$\gamma(\varepsilon, \varepsilon_T)$ is the Maxwellian distribution function.

$$T_{fm,im'} = \int d\, r_k\ \left(\Phi_{fmA}(r_k)\ H_{int}\ \Phi_{im'A}(r_k)\right) \qquad (2.4)$$

H_{int} is the interaction Hamiltonian of the tμ atom and D_2 molecule and is expressed as

$$H_{int} = \sum_{k',k''} \frac{e_{k'}e_{k''}}{|r_{k'} - r_{k''}|} \qquad (2.5)$$

where $k' = t, \mu;\ k'' = d_1, d_2, e_1, e_2$.

This interaction energy becomes large only when the tμ atom comes close to the deuteron, because of the neutral charge on the tμ atom. Thus, this can be approximated by the interaction energy between the dtμ dipole moment and the deuteron (d_2) and the two electrons (e_1,e_2) as shown in Eq. (6):

$$H_{int} = \vec{d}\cdot[\vec{\rho}/\rho^3 - (\vec{r}_{1e}/r_{1e}^3 + \vec{r}_{2e}/r_{2e}^3)] \qquad (2.6)$$

In the paper by Vinitsky et al.[4] and my previous paper,[11] only the first term of the interaction with the deuteron (d_2) is considered, and the second term of the interaction with the electron[17] was neglected. When a strong laser field is applied to the D_2,T_2 (DT) mixture, the many modes with energies of $E_{Ai} + m'\omega h$ and $E_{fA} + m\omega h$ are excited, and the dtμ molecule is formed by satisfying the energy conservation requirement through energy transfer between these modes. When the laser field is weak, the amplitudes of the model with high m,m' value are small: the transition matrix becomes small and it results in a small rate of formation. The amplitude change of the electronic wave function is calculated as a function of the strength of the laser field.

Under the single-mode laser field which is treated as the classical field, the schrödinger equation can be written as:

$$[H + \sum_i e\, \vec{P}\cdot\vec{A}(t)/C\, M_i]\ \Psi(r_k,t) = -i\hbar\ \partial\Psi(r_k,t)/\partial t \qquad (2.7)$$

where H is the original Hamiltonian, and $\vec{A}(t)$ is the vector potential expressing the laser field which can be written as

$$\vec{A}(t) = \vec{A}\sin \omega t = \vec{A}\ [e^{i\omega t} - e^{-i\omega t}]/2i \qquad (2.8)$$

The wave function $\psi(r_k,t)$ is expanded as:

$$\Psi(r_{k,t}) = \sum_{m=-\infty}^{\infty} \Phi_m(r_k,t)e^{i(E+m\omega)t}\ , \qquad (2.9)$$

and by substituting Eqs. (8) and (9) to Eq. (7) and calculating it we get

$$[H + m\omega]\ \Phi_m(r_k) + \sum_i (e_i\vec{P}_i\vec{A}_i/2iCM_i)[\Phi_{m+1}(r_k) - \Phi_{m-1}(r_k)] = E_A\ \Phi_m(r_k) \qquad (2.10)$$

This schrödinger equation can be solved by expanding the wave function in term of some function, but the dimension of the matrix to be solved becomes very large for the large A value of a strong laser intensity.

To estimate the laser field strength required to enhance of $dt\mu$ molecular formation, we study the case when the laser field is linearly polarized in the direction of the hydrogen molecular axis. The study considers the case when the laser field affects only the electronic motion of the molecule; the study neglects the effect of nuclear motion resulting from perturbation of the electronic potential caused by the change in electron motion.

When the vector potential is in the direction of the diatomic nuclear axis, the electronic wave function can be expanded as the following polynomial functions:[17]

$$\Phi_m(r_{e1},r_{e2}) = \sum_{l,n,j,k,p} C_{mlnjkp}[l,n,j,k,p] \qquad (2.11)$$

where [l,n,j,k,p] stands for

$$(1/2\pi)\ \exp[-\delta(\lambda_1 + \lambda_2)](\lambda_1^l\lambda_2^n\mu_1^j\mu_2^k\rho^p + \lambda_1^n\lambda_2^l\mu_1^k\mu_2^j\rho^p) \qquad (2.12)$$

λ, μ are elliptic coordinate of electron and ρ is the interelectronic distance dividied by the internuclear distances R of (d-d) or ($dt\mu$-d).

Due to the limitation of the machine memories of CRAY computer, the following ten terms are taken as the Hilbert space expansion, [l,n,j,k,p] = [0,0,0,0,0], [0,0,0,2,0], [0,0,1,1,0] [1,0,0,0,0] [1,0,2,0,0,] [0,0,1,0,0], [1,0,1,0,0][0,1,1,0,0] [1,1,0,0,0] and [2,1,0,0,0].

Figure 1 shows the rate of resonance $dt\mu$ formation as function of laser intensity. Curve 1 is a one photon (m = 1) exchange that balances the energy conservation for the formation process. The frequency ω corresponds to the energy difference of $[(dt\mu)$-d-2e) and D_2 molecules ground state. The curves 2, 3, and 4 are the cases of two, three and four photons, respectively, which satisfy the energy conservation condition. In the low intensity region, the formation rates using the m photons increase as the m power of the intensity (I^m). As is the case for regular photon excitation and de-excitation, the cross-section of one photon process is independent of the photon intensity. The transition probability is proportional to:

$$|\langle \Psi_f | \vec{A}\vec{\nabla} | \Psi_i \rangle|^2, \tag{2.13}$$

where Ψ_i and Ψ_f are, respectively, the wave function of the initial and the final state so that the reaction rate increases A^2 which is proportional to laser intensity I. In the case of m photons, transition probability in the perturbation calculation is expressed as:

$$\propto |\langle \Psi_f | \vec{A}\vec{\nabla} | \Psi_k \rangle \langle \Psi_k | \vec{A}\vec{\nabla} | \Psi_{k'} \rangle \ . \ . \ . \ \langle \Psi_{k^{m-1}} \quad | \vec{A}\vec{\nabla} | \Psi_k \rangle|^2 \tag{2.14}$$

so that the reaction rates become proportional to $A^{2m} \propto I^m$.

Due to the limitation of the matrix dimension which can be handled by the computer in a reasonable time and of the computational method, only the case of up to ±5 photons exchange is calculated here. When the laser intensity increases above that shown in Figure 1, more than 5 photons are excited and de-excited, and the accuracy of the calculation deteriorates.

In the one photon exchange, the high reaction rate requires higher laser intensity than an exchange of multiphotons, but for low rate reaction rate, the required intensity is lower than the multiphoton case. Therefore, the laser intensity and frequency that enhances the formation rate depends on the specified formation rate. In the case of $dt\mu$ formation rate, in which the largest formation is preferable, the enhancement using many multiphoton with low frequency laser is more efficient than the laser with high frequency; however, as shown in the figure, there seems to be a threshold laser intensity with the low frequency laser below which there is no enhancement of the rate of formation. Although we studied cases of multiphoton exchange up to m = ±5, the figure suggests that the intensity required from a laser with low frequency to achieve a $dt\mu$ formation rate of $\lambda_{dt\mu} = 10^{10}$ sec^{-1} is about 10^8 watt/cm^2. This value is 1/30 times smaller than that given in our previous study.[11]

When the laser irradiates the D-T mixture, not only will the rate of $dt\mu$ formation be enhanced or decreased, but other side processes such as muon transfer, the excitation of the molecule's rotational motion associated with the muon catalyzed fusion, will be changed. These reaction rates depend on the laser frequency and the intensity in the same way as they do for dtμ formation. Figure 1 suggests that the competing phenomena for muon catalyzed fusion might be controlled by the proper choice of the laser's characteristics.

In this paper, we have treated the laser field as a classical field, so that the laser is completely coherent, and has no noise or other statistical nature. The formulation obtained for the single frequency mode laser can be easily extended to the case of P multi-frequency mode. The statistical nature of the laser can be treated as the ensemble average of the above transition probability.

2. dtμ Molecular Formation in the Clustered Ion

When the tμ atom collides with the clustered deuteron ion, the additional deuterons ions of the clustered ion play the role of third body as well as the deuteron participating for dtμ formation. The dtμ molecular formation rate is strongly affected by this spectator-participator deuteron nuclei.

Figure 2 shows the configurations of some clustered ions. In this paper the dtμ formation rate in simplest clustered ion, D_3^+ [14, 15] ions, is calculated. This ion has been extensively studied and the calculated vibrational modes show the reasonable agreement with the experimental data.

The dtμ molecular formation rate in clustered D_n^+ ions can be expressed by adding the (n-1) deuteron and (n-2) electrons to the formula for the D_2 molecule.

The interaction term can be expressed by the interaction energy between dtμ dipole and the other charged particle. In this study, the electronic contribution is neglected and only the interactions with deuteron nuclei are taken into account.

The wave functions of the d_n^+ ion and the $[(dt\mu)_8 d_{n-1}]^+$ ion are calculated by using the electronic potential[18] obtained by the Hartree Fock method.

The kinetic energy part of the Hamiltonian can be expressed with the normal coordinate S_i, t_i and r, which are defined in (3.1), but the parts associated with the t_i and r are related to the translational motion and the rotational motion, they are neglected in this study.

$$\begin{bmatrix} \Delta x_1 \\ \Delta y_1 \\ \Delta x_2 \\ \Delta y_2 \\ \Delta x_3 \\ \Delta y_3 \end{bmatrix} = \frac{1}{\sqrt{3}} \begin{bmatrix} 0 & 0 & -b & a & -a & -b \\ 0 & 1 & -a & -b & -b, & a \\ 1 & 0 & 0 & -1 & 1, & 0 \\ 0 & 1 & 1 & 0 & 0 & -1 \\ 1 & 0 & b & a & -a & b \\ 0 & 1 & -a & b & b & a \end{bmatrix} \cdot \begin{bmatrix} t_1 \\ t_2 \\ r \\ S_1 \\ S_2 \\ S_3 \end{bmatrix} \qquad (3.1)$$

where $a = 1/2$, $b = \sqrt{3}$.

The interaction terms of the dtμ dipole and the other deuteron nuclei

$$-e\vec{d} \cdot [\vec{R}_{31}/R_{31}^3 + \vec{R}_{21}/R_{21}^3] \quad (3.2)$$

are also expressed by this normal coordinate, where d is the dipole moment of dtμ molecule and R_{31} and R_{21} are, respectively, the distances of deuteron 3-1, and 2-1. The wave functions of the deuteron nuclei and dtμ nuclei are expanded by the polynomial of S_i with gaussian weighting function as

$$\phi(S_1,S_2,S_3) = \sum_{\ell,m,n} a_{\ell,m,n}\phi_\ell(S_1)\phi_m(S_2)\phi_n(S_3) \quad (3.3)$$

$$\text{where } \phi_k(S_i) = S_i^k\, e^{-\alpha_i S_i^2}, \quad (k = 1,m,n \text{ for } i = 1,2,3) \quad (3.4)$$

The dtμ molecule is formed by the interaction energy of the dtμ dipole and the other deuteron nuclei, so that the two deuterons will assist in molecular formation in the case of the D_3^+ ion, instead of the one deuteron that assists in the D_2 molecule. The nuclear interaction energy is proportional to the inverse square of the equilibrium distance and for the D_3^+ ion, the equilibrium internuclear distance R_{eq} is a little larger ($\approx 1.66\ a_0$, where a_0 = electron Bohr radius) than the distance in the D_2 molecule where $R_{eq} \approx 1.4\ a_0$. Thus, the magnitude of the interaction energy is roughly $2/(1.66/1.4)^2 = 1.42$ times larger than that of the D_2 molecule, and the cross-section will be increased about $(1.42)^2 \approx 2$ times if they satisfy the energy conservation requirement.

The calculation was performed using the 15-dimensional Hilbert space. Table II shows the eigen value of the D_3^+ ion and $(Dt\mu)D_2^+$ ion and the transfer matrix element of dipole interaction. The 3rd and 4th columns of the table are the dipole transfer integral in the cases when the dipole polarization is directed as shown in Figure 4. The fifth column shows the absolute value dipole integrals averaged over the polarization angle. Due to this dipole integral value, the dtμ molecular formation rate from D_3^+ ion has a different temperature dependence from the D_2 molecule, as shown in Figure 3.

These vibrational frequency modes have strong isotopic dependence, thus the temperature dependence of dtμ formation rate can be changed by using a different isotopic composition. Also, it might be possible to produce a large increase in the formation rate, using a suitable clustered ion in the desired temperature range. Figure 5 shows the dtμ molecular formation rate for the D_3^+ ion which was calculated using the vibrational motion excitation. The formation rate is higher than for the diatomic case in the low temperature region.

The triple nuclei molecules like $Li^7{}_3$ and $Li^6{}_3$ molecule present interesting problems in physics. Recently, the Berry phase

Fig. 1. dtμ formation under Irradiation

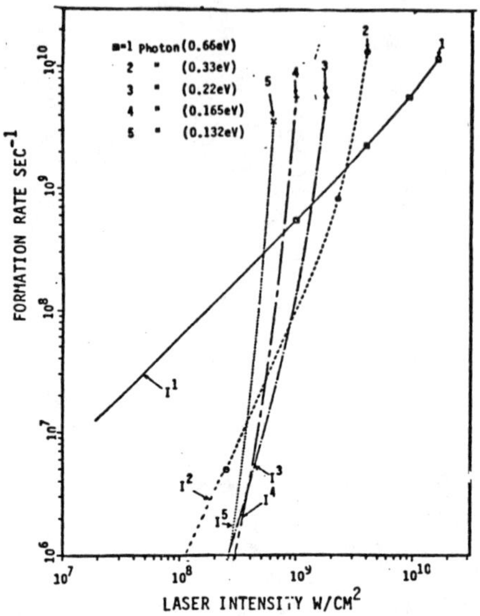

Fig. 2. Dn$^+$ clusters configuration

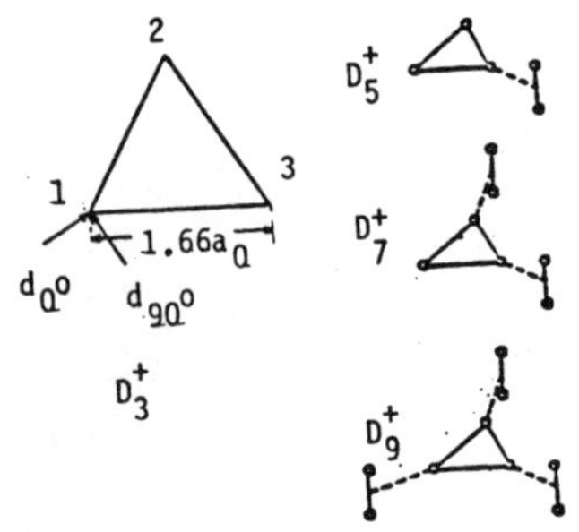

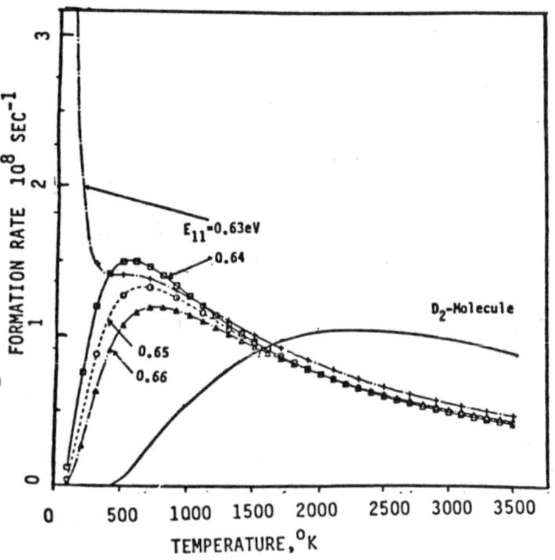

Fig. 3. dtμ molecular formation from D_3^+ clustered ion (E_{11} is the dtμ (J=1, V=1) energy level 0.63 ~ 0.66 eV)

Table I The energy levels of D_3^+ and $dt\mu d_2^+$ and the dipole element integral (Energy is unit of eV and D in unit of amu)

n	$E_{D_3^+}$	$E_{dt\mu\ D_2^+}$	$E_{dt\mu,D_2^+} - E_{D_3^+}$	$D_{0°}$	$D_{90°}$	D_{av}
1	0.3887	0.3442	-0.0445	-5.34 (-2)	-3.282 (-2)	4.430 (-2)
2	0.6251	0.5380	0.1493	-3.496 (-2)	-1.479 (-1)	1.074 (-1)
3	0.6346	0.5558	0.1671	1.872 (-1)	-1.129 (-1)	1.545 (-1)
4	0.7160	0.6321	0.2434	-1.979 (-1)	-1.566 (-1)	1.784 (-1)
5	0.8731	0.7304	0.3417	-4.49 (-3)	-1.462 (-2)	1.075 (-2)
6	0.8750	0.7623	0.3736	-1.742 (-2)	-1.129 (-2)	1.468 (-2)
7	0.8892	0.7817	0.3930	-1.518 (-2)	-9.707 (-3)	1.274 (-2)
8	0.9652	0.8573	0.4686	4.339 (-3)	1.570 (-2)	1.151 (-2)
9	0.9941	0.8827	0.4940	5.086 (-3)	-1.033 (-2)	8.1425 (-3)
10	1.0329	0.9647	0.5760	1.446 (-2)	1.173 (-2)	1.316 (-2)
11	1.0769	0.9771	0.5884	-2.483 (-3)	2.631 (-2)	2.558 (-3)
12	1.1538	1.0235	0.6348	-9.093 (-3)	-7.29 (-3)	8.241 (-3)
13	1.2710	1.0951	0.7064	-1.275 (-2)	+6.479 (-3)	1.011 (-2)
14	1.2840	1.1462	0.7575	1.704 (-2)	2.418 (-3)	1.217 (-2)
15	1.3812	1.2476	0.8589	8.449 (-3)	2.683 (-3)	6.270 (-3)

factor[19] was discussed in many quantum system. The molecular Aharonov-Bohm effect,[20] which is one of the Berry phase factors, produces a different eigen frequency from the one which does not take into account the Berry phase factor. It would be interesting to study whether or not the Berry phase factor has an effect on the dtμ formation rate from the triple nuclei ion of D_3^+.

4. The Effect of Protonium Additive on the Muon Catalyzed Fusion Cycle

4.1 Population of dμ Ground State

As discussed by Jändel et al.,[21] the energy difference between excited dμ atom and tμ atom is only $48.17/n^2$eV (where n is the principal quantum number of the excited state), and the binding energy of hydrogen molecule is about 4.6eV. Therefore, the muon transfer process from dμ to tμ is suppressed because the excess energy of this transition is not sufficient to cause dissociation of the molecule, and this results in the bottleneck of the muon catalyzed fusion cycle.

The energy difference for the muon transfer from pμ to tμ, or dμ, is much larger than the one between dμ and tμ. Therefore, the transfer is not suppressed below n = 6 for tμ, and n = 5 for dμ. If the muon transfer rate from pμ to tμ is much greater than that to dμ, then the most of the muons captured by protonium are transferred to tμ. To evaluate this, let us consider the ratio of these transfer rates. Roughly speaking, the transfer rate is expressed as the square of the transition matrix element times, the phase space, which is available for this transfer. The available phase space is approximately 3/2 power of the energy difference of the above excited level minus the molecule's dissociation energy. If we assumed that the isotropic effect of the transition matrix element is very small, and the molecular dissociation energy is not taken into account, then the transfer rates of pμ → tμ and pμ → dμ are, respectively, 7.4 times and 4.6 times of the transfer dμ → tμ. Therefore, the ratio of muon transfer rate pμ → tμ to pμ → dμ is 1.61 and only 62% of the muon capture by proton is transferred to the triton, even when there are large amounts of protonium are in the target mixture.

When we take into account the molecular dissociation energy, this ratio becomes large at high n states and a small percentage of muons captured by proton are transfered to the deutron.

Table II shows the ratio of the muon transfer of pμ → tμ and pμ → dμ for n = 2,3,4 and 5. For the n = 5 state, the ratio becomes 6.77, and at n = 6, only transfer to triton occurs. Therefore, if the population of the n = 5 and 6 of pμ is high, then the transfer to tμ is much greater than the one to dμ, and thus can avoid the bottle-neck problem. But the Auger transition from n = 5 to n = 4 is 50ϕ 10^{-11} sec^{-1} and it is much larger than the transfer rate pμ → tμ (17.ϕC 10^{11} sec^{-1}) (Fig. 4); thus the population of pμ n = 5 cannot be so large, and the effect of the addition of protonium on the fusion cycle is not very effective.

4.2 Resonance Muon Transfer

It must be pointed out that there is the possibility of resonance enhancement of muon transfer when $d\mu$ collides with T_2 or TD molecules. It was argued that the muon transfer from deuteron to triton in the collision of the highly excited state $d\mu$ to the molecules will be suppressed, due to insufficient energy to dissociate the molecule.[21] However, even then, the excess energy is enough to excite the vibrational and rotational mode of the molecule.

Analogous to the $dt\mu$ resonance formation,[22,4,5,6] the resonance muon transfer from $d\mu$ to $t\mu$ in the collision with tritium molecule can be expressed by the formula of:

$$\lambda_{d\mu \to t\mu} \propto \left| \langle \psi^{+}_{f,n}(t\mu)\, \psi^{+}_{f,\nu,k}(T,D) \right| H_{int} \left| \psi_{i,n}(d\mu)\,\psi_{i,o}(T_2) \rangle \right|^2$$

$$\delta\left[E_{d\mu,n} + E_{T_2,0} - \left(E_{t\mu,n} + E_{DT,\nu,k}\right)\right] \qquad (4.2.1)$$

In Eq (1), $\psi_{i,n}(d\mu)$ and $\psi_{i,o}(T_2)$ are the wave functions of the initial $d\mu$ atom in n^{th} excited state and the T_2 molecular state in the ground state: $\psi_{f,n}(t\mu)$ and $\psi_{f\,\nu,K}$ (T,D) are wave functions of final $t\mu$ n^{th} excited state and the DT molecule in the vibrational ν and rotational K state. The H_{int} is the interaction energy between the two states.

4.3 $dt\mu$ Molecular Formation by the D-H and D-T Molecules

When the proton is added to the DT mixture in the form of a D-H molecule, $dt\mu$ molecular formation rate shows very different characteristics than when the proton is added as a H_2 molecule.

D-H and [($dt\mu$)-p-2e] molecules have the smaller reduced nuclear masses than D-D and [($dt\mu$)-d-2e] molecules, so that the energy difference of excited [($dt\mu$)-p-2e] is wider than that of the [($dt\mu$)-d-2e] molecule. Table II shows the energy level differences between the [($dt\mu$)-p-2e] vibrational excited states and (D-H) ground state and those between the [($dt\mu$)-d-2e] and D_2, which are calculated using the Morse potential approximation for the hydrogen electronic internuclear potential. The formula of energy level E_ν for the νth excited states in the Morse potential can be expressed in:

$$E_\nu = \frac{\hbar}{a_o}\left(\frac{2D}{\mu}\right)^{1/2}\left(\nu + \frac{1}{2}\right) - \frac{1}{a_o^2\, 2\mu}\left(\nu + \frac{1}{2}\right)^2 - D\,, \qquad (4.3.1)$$

where D is the depth of the Morse potential, μ is the reduced mass of participating nuclei including the attached muon and a_o is the Bohr radius. In Figure 5, the resonance production rate of mesic molecular $dt\mu$ from D-H molecule, calculated by using the simple Vinitsky's formula of Eq. (4), which neglects the rotational motion

n	$\frac{p\mu \to t\mu}{p\mu \to d\mu}$	Energy Difference (Unit of eV) $p\mu \to t\mu$	$p\mu \to d\mu$	$d\mu \to t\mu$
2	1.70	45.8	33.5	12.0
3	1.89	20.3	14.9	5.35
4	2.43	11.4	8.38	3.01
5	6.78	7.32	5.36	1.93
6	∞	5.08	3.72	1.33

Table II Energy differences of [(dtμ)-p-2e] [dtμ)-d-2e], from the ground state of vibrational excitation [(dtμ)-t-2e] molecules D-H, D-D and D-T

Table III Ratio of Muon Transfer Difference of pμ→tμ, pμ→tμ to pμ→dμ and Energy pμ→dμ and dμ→tμ.

ν	[(dtμ)-p-2e] - (D-H)	[(dtμ)-d-2e] - (D-D)	[(dtμ)-d-2e] - (D-T) Vinitsky's	[(dtμ)-t-2e] - (D-T)
0	-0.0249	-0.0317	-0.026	-0.0355
1	0.302	0.282	0.286	0.239
2	0.772	0.585	0.588	0.506
3	1.143	0.878	0.88	0.765

Fig. 4. Muon transfer between protonium, deuterium and tritium, and Auger transition of n = 5→4 and 4→3. φ is density factor to liquid hydrogen density, X is the suppression factor discussed in reference 21.

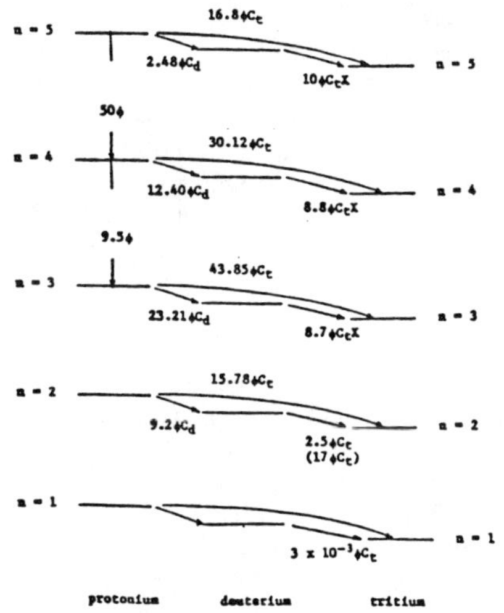

Fig. 5. dtμ molecular formation rate from D-H, D_2 and D-T molecules
(Energy level dtμ (v=1, j=1))

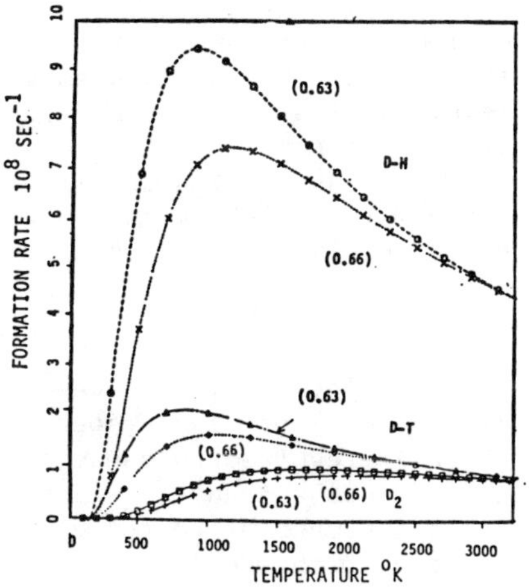

of the molecule and the finite distance of the muon from the dt nuclei, are shown as the function of the target temperature:

Because of the smaller reduced nuclear mass of the [(dt μ)-p-2e] molecule than that of [(dt μ)-d-2e] molecule, the energy level interval of the former is wider than the latter, and when the dt μ (J=1, v=1) state energy level is 0.63 ~ 0.66 eV, the energy conservation condition is satisfied by the excitation of the $\nu = 2$ level instead of the $\nu = 3$ level, as in the case of the D-H molecule. The transfer matrix element from ground state to excited ν state decreases as the value of ν increases. The value of transfer matrix for $\nu = 2$ state is about 4 times of that for $\nu = 3$ state.

Leon[6] showed that the rotational motion excitation plays a very important role for dt μ formation when the vibrational $\nu = 2$ levels are just below the threshold energy. Due to the excitation of rotational motion, the matrix element from $\nu = 2$ can contribute, and this gives a large formation rate. This contribution should be studied, and a precise determination made of the vibrational and rotational excitation energy and the dipole interaction matrix element, including the electron screening effect.

In the case of D-T molecule,[23-24] its small nuclear reduced masses compared with the D_2 molecule produces a smaller excitation level interval, as discussed by Leon.[6] The energy differences of excited vibrational level of (dt μ-t-2e) molecule from the ground level of D-T molecule also are shown in Table III. The energy level of $\nu = 2$ is too low compared with the binding energy E_{00} of dt μ (J = 1, v = 1) state, and so the contribution due to rotational motion excitation becomes small. Thus only the $\nu = 3$ state contributes to the dt μ formation. Compared to the D_2 case, the $\nu = 3$ level is closer to the dt μ shallow binding energy E_{11}, which means that dt μ formation has its maximum at lower temperature. Figure 5 shows the dt μ molecular formation rate as a function of the temperature without taking into account the changes in rotational and other levels due to fine structure splitting, etc.[9,25,26,27]

CONCLUSION

The interaction between laser and electron, which was neglected in the last paper, is strong and the laser intensity which is required to enhance the formation of dt μ can be reduced by taking into account this interaction. It is estimated from the limited number of photon exchange that by using multi photon exchange with the low frequency laser, the dt μ formation rate of $10^9\ \text{sec}^{-1}$ can be achieved with an intensity of 10^8 watt/cm^2. To verify this by numerical calculation requires solving the matrix equation with the very large number of dimensions.

The study suggests that the multiphoton mechanism of laser irradiation might be used for controlling the fusion cycle.[28-29] The effect of laser irradiation in changing the motion of the nuclei due to the internuclear electron potential is neglected here; a positive or a negative contribution depends upon the phase difference between nuclear and electronic motion. This fact should be studied in a future paper.

The study of dtμ molecular formation using the clustered ion shows that the temperature dependence of dtμ formation is quite different from that of the diatomic molecule, due to the many modes which exist in the clustered ion compared with the diatomic molecule, also there is the possibility of coherent enhancement.

The effects of addition of protonium to the deuterium tritium mixture target in the dtμ fusion cycle are studied in the context of a possible reduction in the population of dμ ground state and a change in the formation rate of dtμ by the D-H molecule from the one by the D_2 molecule. The effect of the protonium additive in reducing the dμ ground state population is not very marked, because of the limited muon transfer rate ratio of pμ → tμ to pμ → dμ; only a small percentage of reduction can be achieved even when there are large quantities of protonium present in the target. Relating to the bottle-neck problem of high population of the dμ ground state, the muon can be more effectively transfered as resonance process in the same way as resonance dtμ molecular formation in the molecular state target. This rate is very large especially in the energy range where the suppression of muon transfer is due to insufficient energy to cause the molecule to dissociate. Thus, the bottle-neck problem relating to muon suppression of transfer of dμ to tμ might not exist. By adding protonium to the DT target in the form of the D-H molecule, the dtμ molecular formation rate will be enhanced in the high temperature mixture above 1000°C.

In a similar way the characteristics of dtμ formation rate in the D-T molecule also change.[23-24] Simple calculation using Vinitsky's formula indicates that this also enhance the formation. But the formation rate is very sensitive to the data used in the calculation, and a more detailed calculation is required before a final conclusion can be drawn.

ACKNOWLEDGMENT

The author would like to express his gratitude to M. Leon and E. Vesman for their valuable discussions and to M. Jändel, B. Müller, J. Rafelski, and S. Jones for sending their papers to me prior to its publication. This research is supported by the U.S. Department of Energy, the Division of Advanced Energy Project.

REFERENCES

1. V. Bystrilsky et al., JETP 53, 877 (1981) and Phys. Letters 94B, 475 (1980).
2. S. Jones et al., Phys. Rev. Letters 56, 588 (1986).
3. W. Breunlich et al., Phys. Rev. Letters 53, 1137 (1984).
4. Vinitsky et al., Zh. Eksp. Teor. Fiz. 74, 849 (1978) [Sov. Phys. JETP 47, 444 (1978)].
5. S. S. Gerstein and L. I. Ponomarev, Phys. Lett. 72B, 80 (1977).

6. M. Leon, Phys. Rev. Lett. 52, 605 (1984).
7. J. Cohen and R. Martime, Phys. Rev. Lett. 53 738 (1984).
8. L. I. Menshikov and L. I. Ponomarev, Phys. Lett. 167B, 141 (1986).
9. Yu. V. Petrov, Muon Catalyzed Fusion 1, 219 (1987).
10. M. Leon, Muon Catalyzed Fusion 1, 163 (1987).
11. H. Takahashi, Muon Catalyzed Fusion 2.
12. M. T. Bower, L. Bass, and K. R. Jennings, J. Chem. Phys. 81, 280 (1984).
13. M. Jändel et al. CERN report.
14. G. D. Carney, Molecular Phys. 39, 923 (1980).
15. G. D. Carney and R. N. Porter, J. Chem. Phys. 65, 3547 (1976).
16. L.I. Menshikov and L.I. Ponomarev. Pisma Zh. Eksp. Teor. Fiz. 39, 542 (1984) (JETP Letter 39, 663 (1984).
17. H. M. James and A. S. Coolidge, J. Chem. Phys. 1, 825 (1933).
18. R. N. Porter, R. M. Stevens, and M. Karplus, J. Chem. Phys. 49, 5163 (1968).
19. M. V. Berry, Proc. Roy. Soc., London, A392,45 (1984).
20. C. A. Mead, J. Chem. Phys. 49, 23 (1980).
21. M. Jändel, S.E. Jones, B. Müller and J. Rafelski. CERN-Preprint CERN-TH-4856/87.
22. E.A. Vesman, Pisma. Zh. Eksp. Teor. Fiz. 5, 113 (1967) (JETP Letter 5, 113 (1976).
23. D.V. Balin et al. Muon Catalyzed Fusion 1, 127 (1987).
24. W.H. Breunlich et al. Muon Catalyzed Fusion 1, 29 (1987).
25. A.D. Gocheva, V.V. Gusev, V.S. Melezhik et al. Phys. Letter 153B, 349 (1985).
26. A.M. Frolov and V.D. Efros. J. Phys. B18, 265 (1985).
27. L.N. Bogdanova, V.E. Markusin, V.S. Melezhik and L.I. Ponomarev. Zh. Eksp. Teor. Fiz. 83, 1615 (1982).
28. K. Nagamine, Research Proposal VT-MSL (1985).
29. V. Smilga and V. V. Filchenkov, Zh Eskp Feor. Fiz 85 124 (1983).

CONSIDERATIONS FOR μCF EXPERIMENTS WITH METASTABLY SPIN-POLARIZED D AND T

A. Honig, N. Alexander and S. Yucel
Physics Department, Syracuse University, Syracuse, NY 13244

ABSTRACT

Highly spin-polarized D and T can be expected to modify muon catalyzed fusion (μCF) processes in important ways. Under appropriate conditions, the polarized reservoir induces repolarization of the muon during each fusion cycle, resulting in repeated selective hyperfine state occupation. The dependence of mesomolecular capture rates, cascading rates and fusion rates on spin (hyperfine state) can lead to different branchings in the evolutionary kinetics and possibly to modified fusion outcomes. In order to carry out most experiments on spin-dependence or to utilize polarized fuels, it is necessary for the polarized material to be removed from its production environment and retain its high polarization for a time period matched to the duration of the experiment. We describe here production of removable and transportable metastable highly spin-polarized D in HD and o-D_2, with polarization lifetime in the solid and liquid phase sufficient for application to μCF experiments. For spin-polarized DT, HT, and ortho-T_2, the polarization procedure, size of sample and duration of polarization are all less favorable than for D, and they have not as yet been developed to the extent of HD and D_2. Nevertheless, several usage modes for DT, HT + HD, and D_2 + ortho-T_2 appear feasible for μCF, including both external usage variants and in-situ experiments with the sample remaining within the dilution refrigerator. The latter relaxes the metastability requirement on the polarized tritons, but restricts the sample triton content to less than 1 Ci because of heat load from the triton beta decay. The proposed investigations exploit a truly new parameter in the μCF process, and in view of the current advanced state of development of metastably highly spin-polarized D, the time for these studies is opportune.

INTRODUCTION

New discoveries and steady advances during the past decade have brought μCF to the point where it is no longer fantasy to contemplate net power production based upon it[1]. There are still gaps in understanding, and some rates for the many processes which can occur are not firmly known, either because experiments have not directly accessed them or because of inconsistency among different measurements. There is additionally the fundamental problem of the effective sticking probability of the muon to the helium nucleus resulting from fusion. For DT fusion, this sticking imposes an upper limit on the number of fusions which can be catalyzed per muon, at present approximately 200, a number which must improve by at least a factor of 5 for hope of practical direct power production. In order to improve the understanding of μCF, investigations have been conducted in which basic experimental conditions such as temperature, pressure, density, phase, composition of hydrogen isotopes and geometry have been varied to their limits, leading to separate determination of the rates for many of the competing kinetic processes and to an improved value of the number of catalyses per incident muon.

An additional experimental parameter, high spin-polarization of the D and/or T reactants in μCF, is the subject of this report. Interest in spin dependence is not new, as is evident from the numerous reports on the role of hyperfine states in the complex kinetics[2-4]. However, scant attention has been given to theoretical analysis in the presence of a reservoir of highly polarized reactants. This is forgivable since even though considerable experience with highly spin-polarized materials exists, highly spin-polarized pure "hydrogens" have not been available; furthermore, the difficulties in separating polarized material from its complex production environment for use in a reactor or even test set-up have appeared as very formidable. Our principal purpose here is to call attention to some general reasons why spin-polarized fuels may be expected to be important for improving understanding and performance of μCF systems, and to show that highly polarized D, and soon T, removable from their production apparatus and interfacable with both small scale experimental tests and in principle with large μCF reactors, are now very much in the realm of the possible. Hopefully, this will stimulate theoretical interest to explore the benefits expected from investigations with polarized reactants.

We first inquire whether it is likely that μCF processes will be strongly influenced by spin-polarized reactants, and secondly whether spin-polarized fuels can be produced which can interface with a μCF experiment or reactor. With regard to the first question, spin dependence has already been addressed in connection with the role of hyperfine states[2-4] in the kinetics of initial mu-atom capture, cascading to the ground state of muonic atoms, spin evolution of the ground state of the muonic atoms, similar processes for the mesomolecule, and finally fusion. In fact, there is an extensive literature[5] on the problem of depolarization of an incoming polarized negative muon, partly through atomic cascade processes and partly through hyperfine interactions with the reservoir of unpolarized nuclei. This latter feature can be turned around to suggest that a reservoir of polarized nuclei can polarize the muon, and keep it polarized throughout its lifetime. Muon repolarization experiments have in fact been carried out on heavy muonic atoms with polarized nuclei[6]. In the μCF case under consideration here, it must be determined whether polarization or depolarization processes prevail in a given situation[7]. For excess D over T, depolarization should take place primarily during the cascade down to the 1s ground state of the Dμ atom, after which partial repolarization should occur through Tμ formation by isotopic exchange. If the lifetime of the resultant Tμ atom hyperfine state exceeds the TμD mesomolecular formation time, we can anticipate the possibility of initial capture states and subsequent mesomolecular cascading routes differing from those with unpolarized μ. There is fairly substantial evidence that the initial resonance capture into the mesomolecule favors a particular hyperfine component. Thus, control over the spins could increase that capture rate, which could be very useful if the effective sticking probability problem is independently solved, or decrease the capture rate, thereby possibly accentuating an alternate fusion channel more favorable for the sticking probability. For example, Kulsrud[8] has pointed out that diversion of fusion from the principal (J=0,v=1) fusable state to the (J=1,v=0) fusable state would lower the sticking probability because of smaller overlap of the helium and muon wavefunctions after the fusion. Without considering the polarization of the μ, Kulsrud noted[8] that antiparallel polarizations of the D and T in TμD could inhibit fusion from the principal (J=0,v=1) fusable state, since the fusion proceeds primarily through the spin 3/2 state, thereby possibly enhancing fusion from the more favorable (from the standpoint of sticking probability) (J=1,v=0) state. Although the relevant rates are not all experimentally known and the repolarization of the μ will affect the effective D and T polarization through the hyperfine coupling, it is nevertheless an interesting example which illustrates how a critical element of μCF may be strongly affected by spin. We have referred mostly to DT fusion, but the experimentally simpler case of fusion with polarized D_2 can also provide important information on the mesomolecular states and the DD nuclear fusion reaction.

The second question, namely the feasibility of employing polarized fuel, is our next consideration. Each type of fusion mode involves very different interfacing requirements. In some respects, μCF, with the special constraint of short muon lifetime together with the appreciable muon range in the hydrogens, presents more difficulties than either inertial or magnetic confinement fusion. Moreover, the interfacing problems for carrying out tests to improve understanding of the fundamental processes are quite different from those incurred by projecting operation of a reactor employing polarized fuels. We will be in a better position to consider these problems after we discuss the polarization methods, the special attributes of the polarized systems and the current status of our polarized hydrogens effort, and defer this second question until later.

POLARIZED D

Achievement of high equilibrium nuclear polarization requires strong magnetic fields, low temperatures and presence of sufficient interaction between the nuclear spins and the thermal bath, generally taken as the lattice vibration modes, to enable the spin system to approach equilibrium in a reasonable time. Fig. 1 is a plot of the Brillouin function which gives the equilibrium polarizations of the proton, deuteron and triton in heteronuclear molecules as a function of B/T, where B is the external magnetic field in Tesla and T is the temperature in mK. Large bore 17 Tesla magnets for dilution refrigerators are now commercially available.

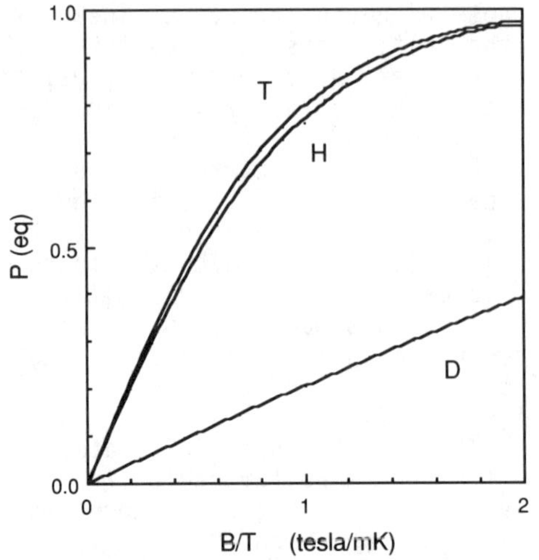

Fig.1. Equilibrium polarization of H, D and T in herteronuclear molecules.

Dilution refrigerators can typically cool to a few mK, but the heat removal capacity decreases rapidly with decreasing temperature, as does the thermal contact of the sample with the refrigerator. Since our samples all must have some rotationally excited J = 1 species present (e.g. ortho-H_2 or para-D_2) to assure contact of the nuclear spins with the thermal bath[9], there is always some internal heat generated from the (slow) ortho-para conversion of those J = 1 molecules. Taken together with the poor thermal conductivity and Kapitza resistance at the low temperatures which makes temperature equilibration between the sample and the refrigerator more difficult, one arrives at a low temperature limit of about 10 mK for these hydrogens. We have recently obtained a temperature of 15 mK with a pure o-D_2 sample[10], the lowest temperature thus far obtained for solid hydrogens, and with effort it probably can be brought down near 10 mK for samples of exceptionally high purity. Thus, a useful reference point for maximum equilibrium polarization of solid hydrogens is near B/T = 1.5, which may be stretchable to B/T = 2 in the near future. It is apparent that the polarizations at B/T = 1.5 of 91% and 93% for H and T, respectively, are very favorable, but the 30% D polari-

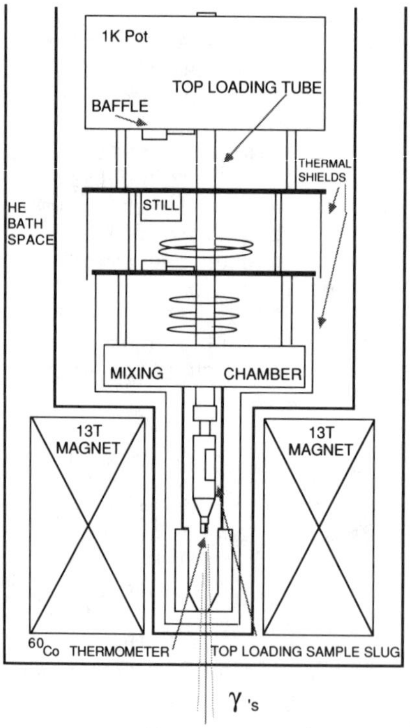

Fig. 2. Schematic of top-loading dilution refrigerator.

zation is inadequate. In earlier work, however, we adapted[11] dynamic polarization methods, originally used to enhance nuclear polarization by coupling nuclear magnetic moments to much larger electronic moments, to enhance D polarization through its intermolecular dipolar coupling to protons. The dynamic polarization is achieved through radio-frequency driving of partially forbidden transitions, which does not encumber the apparatus, and thus one can obtain useful D polarizations comparable to those of H or T. The apparatus for achieving these results is shown schematically in Fig. 2. The 77K thermal shield can and outer vacuum can are not shown. Also omitted are the sample cell and NMR coils. The feature we wish to emphasize is the top-loading tube, which is provided with cold shields and circulating liquid helium and allows sample entry and retrieval at a temperature of 4K. By incorporating a fractional Tesla compact magnet, we shall see that a metastably spin-polarized sample can be removed from the dilution refrigerator while retaining its high polarization, and subsequently be stored, transported and utilized.

We have seen that <u>contact</u> of the spin system with the thermal bath and <u>isolation</u> of the spin system from the thermal bath play central roles respectively in the polarization process (at high field and very low temperature) and in the utilization process (at modest holding fields and accessible 4K temperature). Contact of the spin system with the bath is characterized by a spin-lattice relaxation time, T_1, which is the time for the polarization of a spin system to relax to 1/e of its initial value upon removal of the magnetic field, or the time for an unpolarized system to reach 63% (1 - 1/e) of its equilibrium polarization upon turning on the magnetic field. The spin-lattice relaxation[9] occurs in the case of solid hydrogens through the J = 1 molecules, whose molecular rotation couples both to the lattice and to the nuclei. The relaxation rate depends strongly on concentration of J = 1 molecules, magnetic field and temperature. We choose HD as our prototype polarizable "hydrogen" solid, since the discussion with it can be transferred directly to other heteronuclear hydrogens such as HT and DT (except for the internal heating problems generated by the triton beta decay) and it plays an essential role in the polarization of ortho-D_2 and in some methods of polarization[12] of HT, DT and ortho-T_2. The HD, being a heteronuclear molecule, does not have symmetry restriction on its nuclear spin and rotational wavefunctions, and at temperatures below 4K, is virtually entirely in the zero rotational state J = 0. Thus, pure HD should have no relaxation to the lattice at liquid helium temperatures, and in fact, in the purest samples we have prepared, proton relaxation times, T_1^H, of the order of a day, and deuteron relaxation times, T_1^D, of several days, have been measured. Reduction of these long relaxation times is achieved by adding J = 1 impurities, such as ortho-H_2 or para-D_2. These convert to the J = 0 state very slowly, with time constants[13] in solid HD at 4K of about 6.4 days and 18 days, respectively. The nuclei on the J = 1 molecules are the ones which relax to the lattice, and since they precess at the same rate as their corresponding nuclei on the HD molecules, they share their

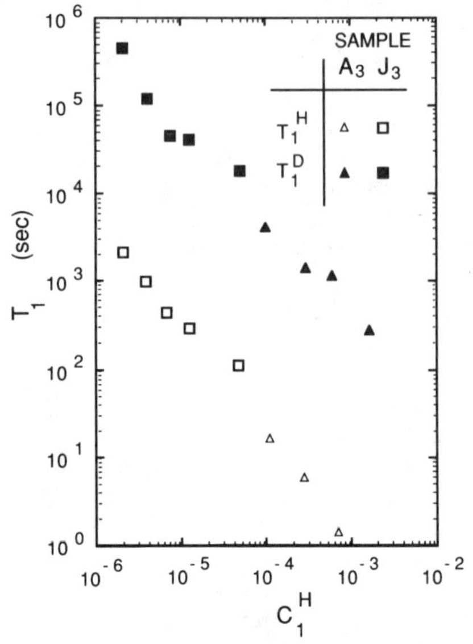

Fig. 3. Proton and deuteron spin-lattice relaxation time vs. $o\text{-}H_2$ concentration, in HD. T = 4.2K, f = 6.3 MHz. $p\text{-}D_2$ concentration equals $6X10^{-4}$ and $2X10^{-4}$ for A_3 and J_3, respectively.

magnetization growth or decay with the otherwise isolated nuclei of HD through an adiabatic spin-spin interaction at a relatively rapid rate, allowing the bulk of the H and D on the HD molecules effectively to relax at a rate proportional to the ratio of the magnetic heat capacities of the J = 1 and J = 0 molecules. In Fig. 3, relaxation times of H and D, taken from data[13] of Mano and Honig, are shown as a function of $o\text{-}H_2$ concentration, c_1^H (subscript denotes J = 1 molecules). It is crucial to note that T_1^D is more than two orders of magnitude longer than T_1^H. This is partly because the D on HD relax via the residual $p\text{-}D_2$ impurities, whose relaxation rate depends upon c_1^H but whose concentration (and magnetic heat capacity) remains constant as c_1^H increases. By further reduction of the residual c_1^D, we believe this ratio can be made even larger. At 4K with $c_1^H = 3x10^{-4}$, T_1^H is a few seconds and T_1^D exceeds a half-hour. We shall see that as the temperature decreases and the field increases, both necessary for attaining high polarization, the spin-lattice relaxation times get very much longer. Thus, only by starting with a fairly short T_1^H at 4K and 0.1 Tesla can one end up with a T_1^H which is not prohibitively long at 20 mK and 13 Tesla, our polarizing temperature-field conditions. The fact that T_1^D becomes enormously long at the polarizing conditions is of no consequence since we do not depend at all upon spin-lattice relaxation to polarize the D but rather use dynamic polarization to transfer polarization from the H to the D spins. Thus, we have the situation where polarization production is driven entirely through the H spins, and polarization retention at experimentally accessible fields and temperatures resides entirely in the D system. The H polarization is quickly lost at 4K and low fields, but the highly polarized D spins, with their long T_1^D, remain polarized and can be removed and utilized.

We proceed to describe in more detail the polarization process in terms of seven distinct stages.

1. Cold Insertion

The sample is condensed in the top-loading tube and placed in contact with the refrigerator, after which the top-loading tube is

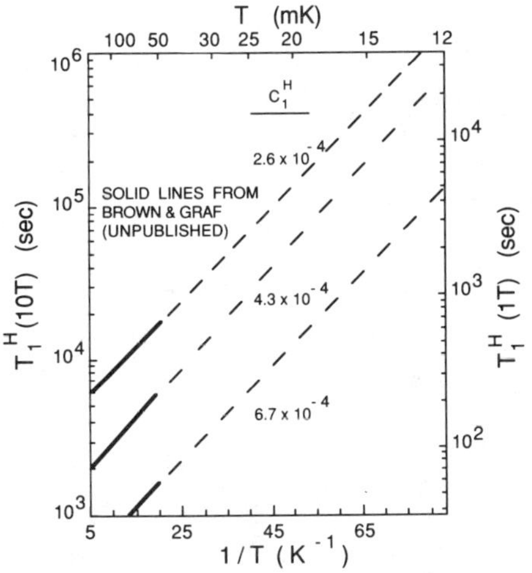

Fig. 4. Temperature dependence of proton spin-lattice relaxation time at 10 Tesla. See Ref. 14.

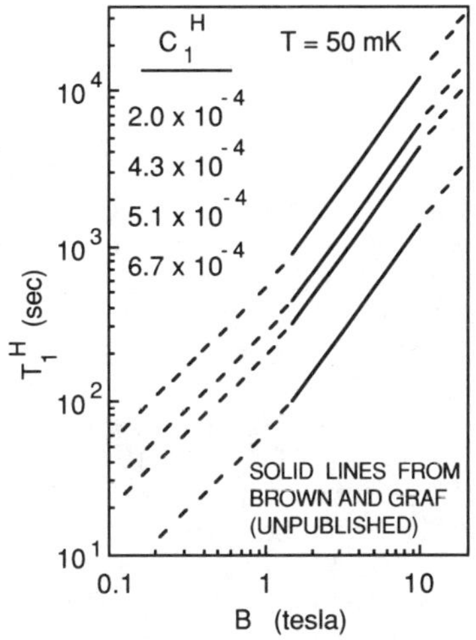

Fig. 5. Magnetic field dependence of proton spin-lattice relaxation time at 50mK. See Ref. 14.

removed via a standard left hand - right hand screw-thread system. (In fact, in most of our current investigations, cold insertion with the top-loading tube is not employed, since cold-retrieval is unnecessary and it is simpler to condense the sample through a capillary connection while the refrigerator is cooling toward 4K.)

2. Cool Down

During cool down from 1K to the mK region, T_1^H increases. In Fig. 4, an approximate representation of the dependence of T_1^H on temperature is indicated for two different magnetic field values. It should be noted that the relaxation time dependences on T and B are themselves dependent on c_1^H and c_1^D, which are often not exactly known either because of imperfectly known initial conditions or back-conversion processes. The solid lines approximately represent data from the work of Brown and Graf[14] on doubly distilled HD samples similar to our own. A few relaxation measurements we have made in the extrapolation region down to 20 mK support the validity of the exponential temperature dependence extrapolation.

3. Polarization of H

After the desired low temperature is reached, usually within 4 to 6 hours, the magnetic field is raised. From Fig. 4 (left side and right side ordinate scales) it is seen that between 1 and 10 T, T_1^H increases by a factor of about 25, and that for a sample of $c_1^H = 4 \times 10^{-4}$, the T_1^H at 20 mK and 10 Tesla is about 12 hours. This is a satisfactory choice of T_1^H for polarizing. About 70% of equilibrium polarization can be attained overnight at low temperature while the superconducting magnet is in persistent current mode. Use of larger values of c_1^H would shorten this 12 hour polarizing period but could result in several unfavorable consequences: insufficiently long T_1^D at 4K, excessive conversion heat which would make cooling to the 15 to 20 mK region more difficult, and too short a T_1^H in the low field region associated with the next two stages of our polarization process. A time much longer than 12 hours is also inadvisable, because during the course of all these operations, c_1^H is exponentially decreasing with a 6.4 day (temperature independent) time constant due to ortho-para conversion. Since the dependence of T_1^H on c_1^H is stronger than linear in this region, a given polarization goal can become elusive if too much time is spent in approaching it. An approximate representation of T_1^H vs. B at a temperature of 50 mK is shown in Fig. 5. The solid lines are based on the relation

$$T_1^H = T_1^H(1.5T)\, B^{1.4},$$

which corresponds approximately to measurements of Brown and Graf[14], and the dashed line extrapolations in the low field region (0.1 - 1 Tesla) are based on measurements of ours in the c_1^H region of 2 - 4 X 10^{-4} in which the field dependence is approximately $B^{1.0}$. We have also checked a few points at fields up to 12 Tesla which generally are in accord with the T_1^H dependence depicted in Fig. 5. Since in the high field region the power of B on which T_1^H depends exceeds unity, there are protocols for obtaining most rapid polarization growth which are superior to simply waiting at the highest field.

4. Transit to low (homogeneous) magnetic field

Preparatory to polarizing D via forbidden transitions (next stage; 5.), one must go to low, homogeneous fields, where the forbidden transition probability is effective at low rf power, thereby avoiding excessive heating. Our magnet system was specially designed for high homogeneity in the 0 - 1 Tesla field region, since it was anticipated that we would generally be dealing with long relaxation times which would permit measurements to be made in the low field region using field cycling techniques. The aim at this stage is to achieve field cycling to the low field region without appreciable loss of H polarization. One contends with a rapid decrease of T_1^H with decreasing B and a maximum sweep rate for the superconducting magnet.

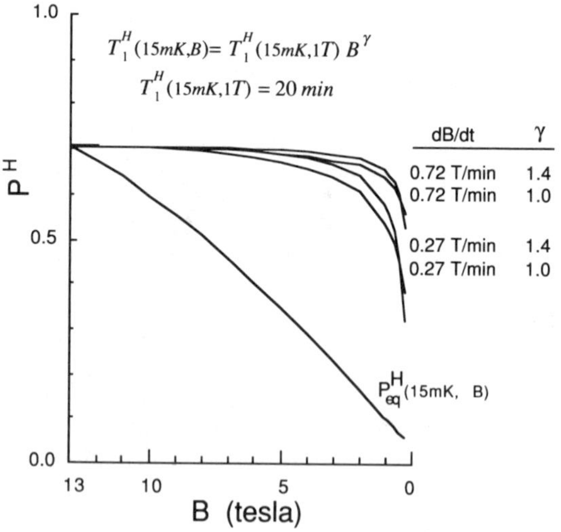

Fig. 6. Illustration of proton polarization loss from lowering magnetic field at two different sweep rates, for two different dependences of relaxation time on magnetic field.

For example, a T_1^H of 12 hours at 20 mK and 10 Tesla becomes less than 3 minutes at 0.1 Tesla, a desirable field for the next stage of the polarization process. Fig. 6 illustrates the loss of polarization during descent of the magnetic field at two different field sweep rates, the faster one representing the maximum rate for our magnet. (Note: magnets with cycling rates a factor of 5 faster have recently been developed). There is little loss of polarization down to about 1 Tesla, but below that, the polarization rapidly deteriorates. This is further exacerbated by heating which results from rapid change of magnetic field. Various procedures are available for reducing such induced current heating.

5. Polarization of D by saturation or adiabatic rapid passage of "forbidden" transitions

This process can be understood by reference to Fig. 7. The energy states are characterized by unperturbed high field magnetic quantum numbers for proton and deuteron spins on neighboring molecules, m_H and m_D, respectively. Positive proton polarization corresponds to an excess of population in the $m_H = +1/2$ state over that in the $m_H = -1/2$ state, regardless of the values of m_D, and similarly, positive deuteron vector polarization corresponds to excess population in the lower energy m_D states. Allowed proton spin transitions, of probability W_o, are vertical ones in which $\Delta m_H = \pm 1$ and $\Delta m_D = 0$, with analogous allowed deuteron transitions in the (nearly) horizontal direction. Forbidden transitions are shown as diagonal lines in Fig. 7, in which both proton and deuteron quantum numbers change. The reason transitions take place at all is that

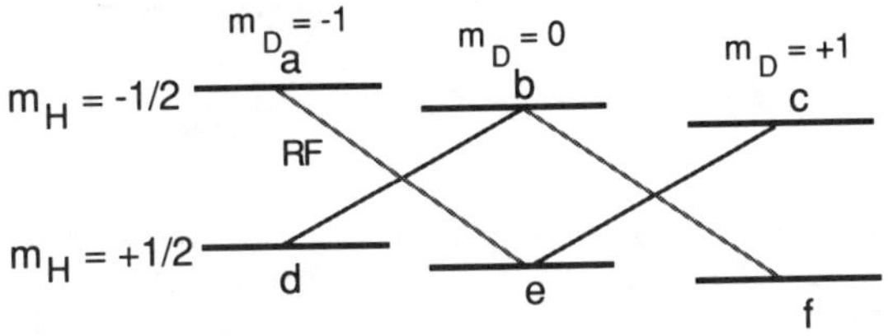

Fig. 7. Magnetic energy level scheme for HD to illustrate polarization transfer from H to D. Diagonal lines represent forbidden transitions. See text.

intermolecular dipolar interaction between the H and D admixes a small fraction, α , of neighboring m_D quantum state(s), where α is approximately $[\gamma_H \hbar / 2Br^3]$, with γ_H the proton magnetogyric factor and r the separation between H and D atoms on different molecules. This results in a forbidden transition probability, W_f, equal approximately to $\alpha^2 W_o$. At the start of this stage, we have large H polarization, P^H, and almost zero P^D, corresponding to almost the same <u>total</u> population in each of the three vertical level pairs, but large excess population in the bottom row of levels compared with the top row. It is evident that rf saturation transitions along the diagonals equalize the populations of the levels they connect; in the case of the heavy lines, this results in positive P^D, and for light lines, results in negative P^D. Adiabatic fast passage actually reverses the populations of the connected states, and results in double the polarization transfer attained with saturation. If completely efficient, which is difficult but not impossible to achieve, the resultant P^D/P^H ratio varies somewhat with the value of P^H but is always slightly higher than 2/3 of the original P^H. Two complete polarization cycles could thus bring P^D to about 90% of the P^H. For multiple polarization cycles, one requires $T_1^D > T_1^H$ at the polarizing temperature-field conditions, so that the gains in D polarization are not lost during the following cycles. In one measurement which we made at extremely long T_1, it was found that this inequality holds, but the T_1's differed only by a factor of about 2. For a specific value of forbidden transition probability, required rf power increases with B^2. Since the rf power causes deleterious heating, the advantage in effecting forbidden transitions at as low a field as possible is evident. There are fundamental limits on how low a field can be used for this process, imposed by state-mixing and resonance definition criteria, but in our case, the (considerably higher) limit is the field value where the relaxation time becomes so short that the built-up P^H collapses before there is sufficient time to transfer it to the D. That time is approximately equal to W_f^{-1}. Thus, at low field, the heating is minimized but the H polarization collapses more rapidly because of the shortened T_1^H. At higher field, the required rf power produces intolerable heating. As a result, in this early phase of our experiments, we only obtained efficient polarization transfer from H to D when T_1^H was of such duration at 0.1 Tesla that, as a consequence, it was too long at 10 Tesla to get full H polarization within a time we have thus far been willing to wait. With new superconducting rf driving coils, we expect rf heating to be reduced enough to permit effective forbidden transitions at 1 Tesla, which should allow the highest P^D to be obtained. A more complete discussion of this process can be found in Ref. 11.

6. Warm-up to 4K

The polarization of D is now "frozen" because of the very long T_1^D even at the low field. To insure that P^D undergoes minimum degradation during an extended warm-up and storage time, the field may be set at a high value which further lengthens T_1^D.

7. Cold Retrieval

This is the reverse of 1. A cold-removal top-loading tube has been designed but not as yet constructed. After removal, the sample can be inserted into a storage dewar containing a modest superconducting magnet which produces a field in the range of 3 - 8 Tesla. The temperature within the storage dewar can be reduced to 1K when necessary for longer storage times.

A polarization cycle for D should require about a day. Typical samples we use have a volume of about 0.2 cm^3, but this is principally because sample preparation is time-consuming and our present sample preparation apparatus is geared to small quantities. Samples of 10 cm^3 are perfectly feasible in our present apparatus, and there is no reason to preclude a larger refrigeration unit (at a cost of perhaps 3 times that of our present one) which could handle a liter of sample. The high ratio of T_1^H at the polarizing field to its value at the polarization-transfer low field is a problem, but raising the polarization-transfer field by use of less dissipating (superconducting) rf coils, as well as researching more favorable mixtures of hydrogens which may reduce the magnetic field dependence of T_1^H, should provide a satisfactory solution, after which we can expect deuteron polarizations with our present apparatus of 40 to 50%.

We now consider polarization of o-D_2. For fusion, polarized D with as little H as possible is often desirable, and that has motivated the efforts to be described. As is the case with HD, the rotational ground state of o-D_2 is J = 0, but unlike J=0 H_2 (p-H_2), 5/6 of the o-D_2 molecules have a nuclear magnetic moment and are polarizable. By reducing the p-D_2 (J = 1) concentration to an unprecedented low level[15], very long values of T_1^D have been achieved, thereby endowing the material with the metastability required for external use. To polarize it, however, requires protons. By admixing [0.2 HD + 3 X 10^{-4} o-H_2] mole fraction into very pure o-D_2, the sample is still 90% D. It can be polarized by the prescription given above for HD. The fact that every o-D_2 molecule does not have a nearest neighboring H is compensated by spin diffusion, but many polarizing cycles are now required because of the reduced magnetic heat capacity of the dilute HD. We have observed that the relaxation behavior of the mixture is quite different from that of HD, and it thus may be possible for shorter cycling times to be used. It is of interest that T_1^D of pure o-D_2 in the liquid state is very long[16], and it should be feasible to make the phase transition from the solid to the liquid without appreciable loss of polarization. Fig. 8 shows the deuteron relaxation time in solid and liquid o-D_2 as a function of para-D_2 concentration, c_1^D. The liquid data is from Ref. 16, and its extrapolation is shown leveled off due to a calculated collision relaxation mechanism. The points in the solid region are from Ref. 15. The additional data entry represented by the rectangular symbol in that region corresponds to our longest measured T_1^D, in excess of 8 hours. Due to back-conversion between measurements, there is some uncertainty in c_1^D for that measurement, which is represented by the span of the symbol. The interfacing problems of μCF with highly spin-polarized liquid D_2 may be the easiest to manage.

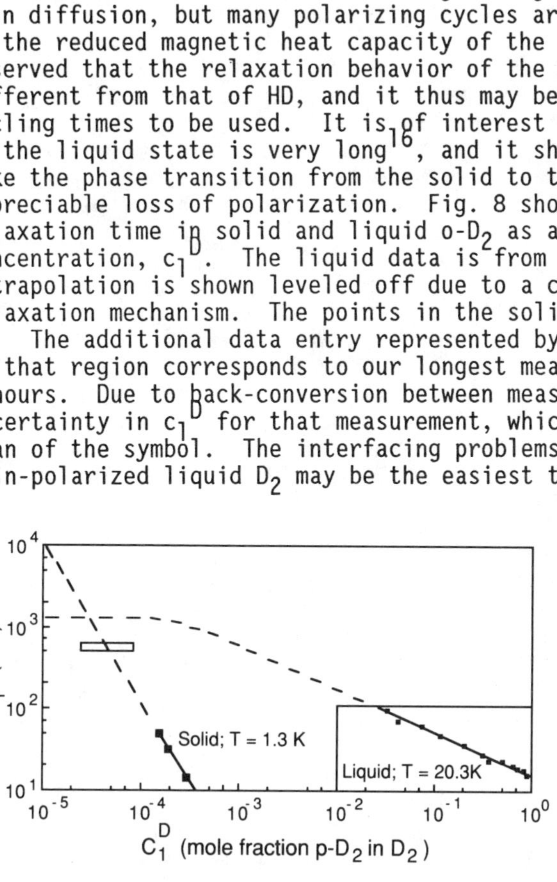

Fig. 8. D spin-lattice relaxation time vs. para-D_2 mole-fraction in solid and liquid D_2. Experimental points in liquid region are from Wang, Smith and White[16]. Solid region points are from Ref. 15 and new data.

T POLARIZATION

From the physics point of view, the fusion-eminent heteronuclear molecule DT should behave almost identically to HD, even to the point where the magnitude of the nuclear magnetic moments of the proton and triton are within a few percent of each other. From the technical aspect, they are very different due to beta decay of the triton (29 kCi/mole of T), whose 5.7 keV mean energy betas introduce a heat load of 0.98 W/mole. As a result, not much more than 3×10^{-5} moles of DT could be tolerated within a normal dilution refrigerator. Even for this 1 Ci amount, the 30 uW of generated heat would make difficult the cooling of the sample to 20 mK, where desirably high polarizations are obtainable. If this small amount of material were condensed on a spherical shell of 1 cm diameter, the sample thickness would be only about 1 μm, about 10% of the range of 5.7 keV electrons in solid hydrogens. Thus, most of the heat would be absorbed in the container walls rather than in the sample, which is some improvement. Container walls even only 1 μm thick would still absorb most of the betas, straining the cooling prospects, and while "container-less" solid hydrogens are a feasible prospect, it might be particularly difficult to produce such samples of only a few microns thickness. For dilute DT in a non-radioactive solid hydrogens matrix, more betas are absorbed in the sample itself, making the cooling problem even more worrisome. Thus, alternative methods have been sought for polarizing DT. One of these utilizes the electronic moments generated by the high energy betas in the solid to dynamically polarize the DT. This route has been discussed recently in connection with fusion experiments by Heeringa[17], and is being actively pursued by Souers[18] and co-workers at Lawrence Livermore National Laboratory. The effort to make pure DT with very low o-T_2 concentration, in order to obtain long triton relaxation times as in the HD analogous case, is complicated by the continuous generation of o-T_2 resulting from the molecular dissociation (and recombination) produced by the betas. Nevertheless, considerable progress has been made. T_1^T values up to 10 seconds have been reported[18] and some basis exists for hope of achieving 10^3 seconds at liquid helium temperatures. An alternative method for producing polarized DT or HT in the 1 - 4K temperature region has been proposed by us[12]. It utilizes epitaxial deposition of thin layers of DT or HT on blocks of metastably polarized HD or o-D_2 at temperatures in the 1 - 4K region, where the heat capacity and thermal conductivity of the polarized block are sufficient to absorb the heat from the betas without serious local heating and consequent depolarization. Polarization transfer from D to T then takes place via forbidden transitions and spin diffusion in the manner described in the section of this report on D Polarization, and is followed by ablation of the polarized layer and rapid usage, or rapid recondensation if the T_1^T permits. Analysis of the process rates and amounts of material involved leads to the conclusion that substantial throughputs of polarized DT can be achieved. A final comment is that since very pure J = 1 o-H_2 and p-D_2 (up to 99%) can be made, possibly the same or alternative techniques can be used to make 99% o-T_2. If this exhibits an ordered phase in the liquid helium temperature region, as do its isotopic counterparts, one might hope for a high polarization and a long T_1 to endure for a fair fraction of an hour despite both ortho-para conversion and dissociation due to betas. It could be polarized at the polarized HD or o-D_2 interface in the same manner as the DT or HT discussed above. The necessary D and T for fusion could then be obtained from a mixture of polarized o-D_2 and polarized pure o-T_2.

INTERFACING WITH μCF

The ingredients of polarized hydrogens have been placed on the table, and we consider ways to interface with μCF experiments. Condensed phase samples are preferable for a number of reasons. The polarized state lasts longer in the condensed phases. The shorter range of the muon in condensed samples allows a more compact geometry without need for high-pressure-resistant container walls. Lastly, continuous muon beams can be used as well as pulsed ones. Some of the problems are also easily enumerated. Even the "small" samples represent substantial amounts of polarized material. In the best cases of HD or o-D_2, polarization lasts at most a day, and generally only a few hours. Thus, many sample fillings are required to obtain sufficient data. For HT, DT, or o-T_2, whose polarization duration is not likely to exceed a half-hour in the forseeable future, even more frequent fillings are necessary. The sample size may be reduced somewhat by muon beam collimation, but that brings only limited benefits due to other problems ensuing from reduced dimensions. The principal new element with polarized condensed samples is provision for cold-entry into the fusion cell. Since this is solvable with respect to the dilution refrigerator, it should not be insurmountable with the fusion cell. Other interfacing modes include polarized frozen pellet firings into the fusion cell synchronized with a pulsed muon beam. If a high pressure polarized gas fuel is desired, a feasible means is to heat rapidly a polarized solid in the cell, either by impact of a frozen pellet on hot walls or by laser induced ablation immediately prior to the pulsed muon beam entry. The T_1's in the gas phase are only a few milliseconds, increasing with the gas pressure, but this should be adequate.

Another rather different possibility, suitable for experimental μCF tests but clearly not for reactor operation, is performing the experiments with the sample inside the dilution refrigerator. The advantage of this method is that polarization could constantly be renewed, and no change of sample is necessitated during the course of the experiment. Avoidance of top-loading procedures more than offsets the inconvenience of having the dilution refrigerator at the location of the muon beam. For polarized D, rather large samples could be accommodated. Referring to Fig. 2, we note at the bottom of the sample holder the ^{60}Co thermometer, whose increasingly anisotropic gamma emission with decreasing temperature allows absolute temperature measurements between 10 mK and about 50 mK. That the gammas are able to penetrate the three cans (2 aluminum and l stainless steel) separating the sample from the outside suggests that muons could equally gain axial access to the polarized sample. It is a simple matter to install thin wall windows of low Z material on the cans. A conically shaped split magnet can provide a reasonably unblocked acceptance angle for emitted neutrons, and electron detectors can be placed inside the cryostat. This system could be useful for polarized T as well. Even though the amount of T which can be used is small, as discussed earlier (< 1Ci), use of X'ray detection from μHe deexcitation might still be feasible and reveal how use of polarized material affects the sticking probability. Until longer T_1^T samples are developed, this mode may well be the only workable one for T. Thus, with certain constraints which have been mentioned, it appears that experimental interfacing of μCF with polarized condensed hydrogens is feasible, and new developments of uncertain outcome are not required to justify initiation of a program. There are of course still many avenues for improvement.

SUMMARY

It is likely that μCF understanding can be increased by use of polarized hydrogens. Long relaxation time polarized D_2 and HD samples can be removed from the production apparatus and externally interfaced with a μCF system, or satisfactory experiments can be done with the polarized D within the dilution refrigerator. Use of small quantities of T in DT or other T and D containing molecular mixtures is also feasible within the refrigerator, and new methods under investigation hold promise for externally usable highly polarized DT. Optimization of the polarization procedures and degree of polarization attainable will be greatly facilitated by an extended study in which the proton and deuteron spin relaxation times are mapped in the [T, B, c_1^H, c_o^H (p -H_2), c_1^D and c^{HD}] parameter space.

This work is supported by the U.S. Department of Energy under Grant # DE-FG02-84ER53179.

REFERENCES

1. For two recent reviews of the subject, see S. E. Jones, Nature 321, 27 (1986); J. Rafelski and S. E. Jones, Sci. Am. 257, 84 (1987).
2. D. D. Bakalov, S. I. Vinitskii and V. S. Melezhik, Zh. Eksp. Teor. Fiz. 79, 1629 (1980). [Engl. Tr: Sov. Phys. JETP 52, 820 (1980)].
3. L. N. Bogdanova, V. E. Markushin, V. S. Melezhik and L. I. Ponomarev, Zh. Eksp. Teor. Fiz 83, 1615 (1982). [Engl. Tr: Sov. Phys. JETP 56, 319 (1982).
4. P. Kammel et al, Phys. Lett. 112B, 319 (1982).
5. V. S. Evseev, in Muon Physics Vol. III, p. 236 - 295. [Ed: V. W. Hughes and C. S. Wu; Academic Press, N.Y. 1975].
6. R. Kadono et al, Proc. Sixth Int. Symp. Polar. Phenom. in Nucl. Phys., Osaka (1985). J. Phys. Soc. Jpn 55, Suppl.p.1038 (1985).

7. Y. Kuno, K. Nagamine and T. Yamazaki, Nucl. Phys. A475, 615 (1987).
8. R. M. Kulsrud, Nucl. Fus. 27, 1347 (1987).
9. T. Moriya and K. Motizuki, Prog. Theor. Phys. 18, 183 (1957).
10. N. Alexander, S. Yucel and A. Honig, Bull. Am. Phys. Soc. 33, 284 (1988).
11. A. Honig and H. Mano, Phys. Rev. B14, 1858 (1976).
12. A. Honig, Conf. on Intersection Between Particle and Nuclear Physics, Steamboat Springs, May 1984. AIP Conf. Proc. No. 123, 1084 (1984).
13. H. Mano, PhD Thesis, Syracuse Univ., 1978. (Unpublished). [University Microfilms, Ann Arbor, MI 48106].
14. J. A. Brown, PhD Thesis, State University of New York at Stony Brook, 1977. (Unpublished). [University Microfilms, Ann Arbor, MI 48106].
15. A. Honig, M. Lewis, Z.-Z. Yu and S. Yucel, Phys. Rev. Lett. 56, 1866 (1986).
16. R. Wang, M. Smith and D. White, J. Chem. Phys. 55, 2661 (1971).
17. W. Heeringa, Proc. Erice Workshop on Muon Catalyzed Fusion and Fusion with Polarized Nuclei, 1987. [Ed: B. Brunelli and G. G. Leotta, Plenum Press, N.Y. 1988].
18. P. C. Souers et al, Third Topical Meeting on Tritium Technology in Fission, Fusion and Isotopic Applications, Toronto (1988). [To be published in Fusion Technology].

SPIN FLIP RATES IN COLLISIONS BETWEEN MUONIC ATOMS

L. Bracci
Dip. di Fisica dell'Univ. di Pisa and I.N.F.N., Sez. di Pisa, Italy

C. Chiccoli, P. Pasini
I.N.F.N., Sez. di Bologna and C.N.A.F., Bologna, Italy

G. Fiorentini,
Dip. di Fisica dell'Univ. di Cagliari and I.N.F.N., Sez. di Pisa, Italy

V.S. Melezhik
Joint Institute for Nuclear Research, Dubna, USSR.

J. Wozniak
I.N.F.N. Pisa and Inst. of Physics and Nuclear Techniques, Cracow, Poland

ABSTRACT

We present the calculation of spin flip rates for a wide range of temperatures in the case of $(d\mu) - d$, $(p\mu) - p$ and $(t\mu) - t$ collisions. The calculations have been performed in the framework of the adiabatic representation of the three-body problem.

INTRODUCTION

We have calculated the spin flip rates of muonic atoms as a function of temperature, for $T \leq 1000^o K$ for the case of $d\mu + d$, $p\mu + p$, $t\mu + t$ collisions. To this purpose, we calculated the spin flip cross sections, up to the energy where the Maxwell distribution for the collision energy is still appreciable at these temperatures ($E \leq 1 \div 2eV$).

For the $(d\mu)$ atom the ground and excited level have $F = 1/2$ and $F = 3/2$ respectively, the threshold being $E_{th} = 0.0485eV$. For $(p\mu)$ and $(t\mu)$ atoms we have $F = 0$ and $F = 1$ for the ground and excited states, the thresholds being $E_{th}^{(p\mu)} = 0.182 \quad eV$ and $E_{th}^{(t\mu)} = 0.2373 \quad eV$ [1].

We denote by σ_{21}, λ_{21} the cross section and rate for the transitions from the excited (2) to the ground (1) state, and use σ_{12} and λ_{12} for the inverse processes.

Within our approximation the total spin S of the system of the nuclei plus the muon is conserved[2]. Consequently, for $d - d$ collisions transitions between the states of hyperfine structure can occur in both $S = 1/2$ and $S = 3/2$ channels. For $p - p$ and $t - t$ transitions are allowed only for $S = 1/2$.

The rates $\lambda^S = n_0 \sigma^S v$, where v is the relative velocity, are calculated for the density of liquid hydrogen : $n_0 = 4.25 \times 10^{22}\ atoms/cm^3$. The averaged rate (average over the energy distribution) is

$$< \lambda^S(T) >= \int n_0 \sigma^S(E) v(\frac{9E}{\pi kT})^{1/2} \frac{2}{3kT} exp(-\frac{E}{kT}) dE \qquad (1)$$

Taken into account the spin multiplicity for the total transition rate we have:

$$<\lambda_{21}>=\frac{1}{6}<\lambda_{21}^{1/2}>+\frac{1}{3}<\lambda_{21}^{3/2}> \quad ; \quad <\lambda_{12}>=\frac{1}{3}<\lambda_{12}^{1/2}>+\frac{2}{3}<\lambda_{12}^{3/2}> \quad \text{for } d-d \quad (2)$$

$$<\lambda_{21}>=\frac{1}{3}<\lambda_{21}^{1/2}> \quad ; \quad <\lambda_{12}>=<\lambda_{12}^{1/2}> \text{for } p-p \quad or \quad t-t \quad (3)$$

In addition to these rates it is interesting to have the effective rate λ_{eff}, which is connected to the depletion rate of the excited state according to

$$\frac{dn_2}{dt}=-<\lambda_{21}>n_2+<\lambda_{12}>n_1 \qquad (4)$$

where n_1 and n_2 are the populations of the ground and excited state respectively. When the populations of the ground and excited state are assumed to be statistical, we have:

$$\lambda_{eff}=<\lambda_{21}>-\frac{1}{2}<\lambda_{12}> \qquad for \quad d-d \qquad (5)$$

$$\lambda_{eff}=<\lambda_{21}>-\frac{1}{3}<\lambda_{12}> \qquad for \quad p-p \quad and \quad t-t \qquad (6)$$

RESULTS

The calculation of the cross sections was performed in the framework of the adiabatic representation of the three-body problem[3], where the complete eigenfunction is expanded over the basis of two-centre problem. We used 48 components of the discrete spectrum and 190 components of the continuous spectrum, for a range of momenta up to $k = 10$ in muonic units. Since the energies involved in the processes are small, we only considered components with $J = 0$.

For the sake of brevity we present in Figure 1 only the averaged spin flip rates as a function of temperature. An extended description of the results will appear elsewhere[4].

In all the calculations we assumed that the populations of the levels were merely statistical and omitted consideration of such processes, as back decay, which can alter this distribution. Also molecular effects and electron screening are not taken into account.

We remark that our calculations are the most extended so far performed within the framework of the adiabatic representation of the three-body problem and we could verify that our results for the spin flip cross sections are convergent when the number of equations in the considered system is increased up to the number (N=238) which we actually used.

The above remark is particularly significant for the $p-p$ case, where a marked sensitivity to the number of equations had been noticed [5,6]. For $d-d$ our results are in agreement with previous calculations [7], apart from the lower energy range $E \leq .1 \quad eV$, where the discrepancy is of the order of 10 %. For $t-t$, our results for cross sections are also in good agreement with the results of ref.8.

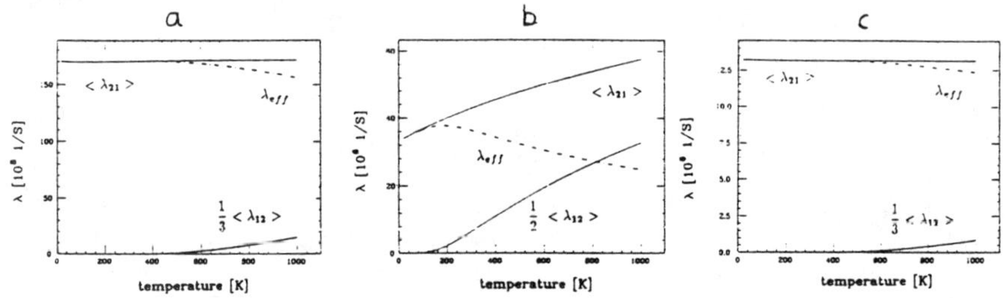

Fig. 1 The averaged spin flip rates λ_{eff}
for $p\mu - p$ (a), $d\mu - d$ (b) and $t\mu - t$ (c)

The comparison with the experimental data is fairly good for the $d-d$ case [9], whereas for $t-t$ a notable discrepancy with available data exists [10], and a further analysis is desirable. A discrepancy, however, could be a hint that the assumption of statistical population cannot be maintained [11]. As for the case of $p-p$ collisions, we are not aware of any experimental result.

ACKNOWLEDGEMENT

We gratefully acknowledge financial support from ENEA.

REFERENCES

1. M. Bubak and M.P. Faifman, JINR comm. E4-87-464, (1987).
2. S.S. Gerstein, Sov. Phys. JETP, **13**, 488 (1961).
3. L.I. Ponomarev and S.I. Vinitsky , Sov. J. Part. and Nucl., **13**, 557 (1982).
4. L. Bracci et al. (submitted to Phys. Letts.)
5. A.V. Matveenko et al., Sov. Phys. JETP, **42**, 212 (1976).
6. V.S. Melezhik et al., Sov. Phys. JETP, **58**, 254 (1983).
7. V.S. Melezhik and J. Wozniak, Phys. Lett. **116A**, 370 (1986).
8. V. S. Melezhik, J. Comp. Phys., **65**, 1 (1986).
9. P. Kammel et al., Phys. Rev. A **28**, 2611 (1983).
10. W.H. Breunlich et al., Phys. Rev. Letters **53**, 137 (1984).
11. L. I. Ponomarev, Atomkernenergie/Kerntechnik, **43**, 175 (1983).

MOLECULAR EFFECTS IN MUONIC HYDROGEN CASCADE

D. Taqqu
Paul Scherrer Institute, CH-5234 Villigen/Switzerland

ABSTRACT

Molecular effects in muonic hydrogen cascade are considered. It is shown that at liquid hydrogen densities and n≤4 the muonic atom is mostly bound to an hydrogen molecule or molecular ion. The molecular effects appreciably modify the deexcitation pathway and the μp kinetic energy.

This paper discusses the effect of molecular reactions on the atomic cascade of the muonic hydrogen isotopes. At high target densities ($\phi \approx 1$, $\phi = N/N_o$, $N_o = 2.12 \cdot 10^{23}$ mol/cm^3) the hydrogen molecules are so closely packed together that a collision process leads easily to the formation of a bound molecular state. A μp atom in an excited state interacting with a molecule via an attractive potential can excite the internal degrees of freedom of the molecule and lead to long lived intermediate state that can be stabilized in a collision with a third body. The reaction scheme is:

$$(\mu p)^* + H_2(v,r) \longrightarrow [(\mu p)^* \cdot H_2]\,(V,R) \longrightarrow (\mu p)^* + H_2(v'r') \tag{1a}$$

$$(\mu p)^* + H_2(v,r) + H_2 \longrightarrow [(\mu p)^* H_2]\,(V',R') + H_2 \tag{1b}$$

where v, v', r and r' are vibrational and rotational levels of the hydrogen molecule and V, V', R, R' are vibrational and rotational levels of the muonic atom complex. Process (1a) is a two body reaction with rate $\lambda^{(2)}(1)$ and complex lifetime $\tau(1)$ and process (1b) is the three body reaction with rate $\lambda^{(3)}(1)$ leading to a bound muonic atom complex. This complex (or the intermediate state in (1a) during its lifetime) is unstable versus the various deexcitation modes of the μp atom[1,2,3] (mostly Auger or radiative according to recent theoretical work[3]).

The Auger transition leads either to a continuum state that separates into:

$$[(\mu p)^* H_2] \rightarrow (\mu p)^{*'} + H_2^+ + e \tag{2a}$$

or to a vibrationally excited bound state:

$$[(\mu p)^* H_2] \rightarrow [(\mu p)^{*'}\, H_2^+] + e \tag{2b}$$

that can further decay via Auger effect either into[3]:

$$[(\mu p)^{*'} H_2^+] \rightarrow p + p + (\mu p)^{*'} + e \tag{3a}$$

or, if the $(\mu p)^{*'}$ is sufficiently bound and the Auger effect occurs while the protons of the H_2^+ are far from each other, into:

$$[(\mu p)^{*'} H_2^+] \rightarrow (\mu pp)^* + p + e. \tag{3b}$$

At high densities, another molecular reaction can precede Auger decay (3):

$$[(\mu p)^* H_2^+] + H_2 \rightarrow [(\mu p)^* H_3^+] + H \tag{4}$$

or follow molecular ion formation (3b):

$$(\mu pp)^* + H_2 + H_2 \rightarrow [(\mu p)^* H_3^+] + H. \tag{5}$$

These reactions supply electrons to the muonic atom complex allowing it to maintain its existence over more than two Auger transitions.

Further, if enough time is available for the bound complex to interact with the surrounding molecules, the effective binding energy of the $(\mu p)^*$ to the associated molecule or molecular ion can be appreciably increased via vibrational or rotational deexcitation:

$$[(\mu p)^* \cdot M](V, R) + H_2(v = 0, r) \rightarrow [(\mu p)^* \cdot M](V', R') + H_2(v', r'). \tag{6}$$

Greater binding means higher average kinetic energy of the protons before the Auger transition. As this kinetic energy is not changed by the Auger transition[3], process (6) can act as an accelerating mechanism.

It is clear that reaction (1) will affect the muonic atom cascade only if the reaction rate $\lambda(1)$ can compete with standard rates for external Auger effect $\lambda_e(A)$[1,3]. Processes (4),(5) or (6) also will occur only if $\lambda(4)$, $\lambda(5)$ or $\lambda(6)$ are greater than the internal Auger rates $\lambda_i(A)$ ($\lambda(3)$ or $\lambda(2)$).

An estimate of the strength of the molecular reactions (1), (4), (5) and (6) can be obtained by looking into the interactions occurring within a system quite similar to the one considered here. These are the interactions of H^+ (or H_2^+) with molecular hydrogen for which extensive experimental results are available.

Process (1) will be compared to:

$$H^+ + H_2 + H_2 \rightarrow H_3^+ + H \tag{7}$$

and $\lambda(1)$ can be estimated by noting that in both (1) and (7) only orbiting collisions can induce the reaction. For reaction (7), the orbiting cross-section $\sigma(7)$ can be computed from average polarizability α of the hydrogen molecule:

$$\sigma_o(7) \simeq \pi\rho_o^2(7) = 2\pi\sqrt{e^2\alpha/E} \tag{8}$$

where E is the collision energy.

For the μp system:

$$\sigma_o(\mu p) = \sigma(1) = \pi\rho_o^2(\mu p), \tag{9}$$

where $\rho_o(\mu p)$ will be taken in first approximation to be identical to the critical impact parameter computed by Men'shikov and Ponomarev[4] for excited μp levels with negative $\Delta = n_2 - n_1$ (n_1, n_2 are the parabolic quantum numbers).

For the two body reaction rate one has

$$\lambda^{(2)} = k^{(2)} N = \sigma_o\, v\, N. \tag{10}$$

The three body rate constant can be expressed as[5]:

$$k^{(3)} = k^{(2)} \tau\, k'$$

[n,Δ]	$\lambda^{(2)}(10^{12}s^{-1})$	$\lambda^{(3)}(10^{12}s^{-1})$	$\tau(10^{-13}s)$
[2, −1]	2.0 ϕ	0.75 ϕ^2	2.0
[3, −1]	4.7 ϕ	11 ϕ^2	4.7
[3, −2]	6.3 ϕ	25 ϕ^2	6.3
[4, −2]	7.0 ϕ	34 ϕ^2	7.0
[5, −2]	7.4 ϕ	42 ϕ^2	7.4
[10, −4]	11 ϕ	150 ϕ^2	11

Table 1: Two body reaction rate $\lambda^{(2)}$, three body reaction rate $\lambda^{(3)}$ and intermediate state lifetime for molecular reaction (1) computed according to relations (8)-(13) for various excited state [n,s] of the μp atom at thermal collision energies.

where τ is the lifetime of the non-stabilized complex and k' is the rate constant for collision with the third body. As τ is not directly measurable it is deduced from the measured values of $k^{(3)}$ the computed $k^{(2)}$ (relations (8) and (10)) and the estimated $k^{(')}$. For reaction (7), the obtained $\tau(7)$ is of the order of 10^{-11} s at thermal energies; this high lifetime is characteristic of this kind of collisions with strongly interacting collision partners. Extrapolation to reaction (1) requires an estimation of τ and k'. For the attractive potentials in $(\mu p)^*(\Delta < 0)$ levels enough vibrational and rotational levels close to the continuum are present to allow a relatively high value of τ. For k' one can make the conservative requirement that the third body interacts directly with the $(\mu p)^*$. $k'(7)$ will therefore be scaled according to the orbiting impact parameter such that:

$$k'(1) = k^{(2)}(1) = k^{(2)}(7) \cdot \left(\frac{\rho_o(1)}{\rho_o(7)}\right)^2. \tag{11}$$

For τ a similar scaling will be adopted:

$$\tau(1) = \tau(7) \cdot \left(\frac{\rho_o(1)}{\rho_o(7)}\right)^2. \tag{12}$$

This results in:

$$\lambda^{(3)}(1) = k^{(3)}(7)(N_o\phi)^2\left(\frac{\rho_o(\mu p)}{\rho_o(7)}\right)^6. \tag{13}$$

Table 1 gives $\lambda^{(2)}(1)$ and $\lambda^{(3)}(1)$ for some $(\mu p)^*$ levels computed from (8), (9) and (10), (13) using $k^{(2)}(7) = k' = 2.4 \cdot 10^{-9}$ cm^3 s^{-1} and $k^{(3)}(7) = 3 \cdot 10^{-29}$ cm^6 s^{-1} [6]. At all densities for which $\lambda^{(3)}(1) > \lambda^{(2)}(1)$ the reaction rate is $\lambda^{(2)}(1)$ and a stable complex is formed. For $\lambda^{(3)}(1) < \lambda^{(2)}(1)$ the ratio $\lambda^{(3)}(1)/\lambda^{(2)}(1)$ gives the probability that stabilization occurs during a collision. Even without stabilization levels with external Auger rates $\lambda_e(A)$ smaller than $\lambda^{(2)}(1)$ will have their Auger decay rate boosted up during the lifetime of the complex. The average internal Auger rate $\lambda_i(A)$ has to be considered ($\lambda_i(A) \gg \lambda_e(A)$) and for long enough complex lifetime ($\tau(1) > 1/\lambda_i(A)$), the effective Auger rate becomes

$$\lambda(A) = \left(\frac{1}{\lambda_i(A)} + \frac{1}{\lambda^{(2)}(1)}\right)^{-1}. \tag{14}$$

Comparison of $\lambda_e(A)$[1,3] with $\lambda^{(2)}(1)$ shows that at high n (n>10), $\lambda^{(2)}(1) \gtrsim \lambda_e(A)$ and with Auger rates as given by (14) there is an appreciable decrease in the cascade time which seems to be required in order to explain the short cascade time ($\tau_c < 80$ ns) measured at 1 torr[7].

In the region of highest Auger rates $5 \leq n < 10$, $\lambda_e(A) > \lambda^{(2)}(1)$ and little time is available for the molecular effects to influence the cascade processes.

As $\lambda_e(A)$ falls by orders of magnitudes from n=5 to n=2 we have at n=4 already $\lambda^{(2)}(1) \gg \lambda_e(A)$ for $\Delta < 0$ levels. As a consequence stable complex formation most often precedes Auger decay. Furthermore, where the n=4 level has been reached via reaction (2b) or (3b) and not via (2a) or (3a), the molecular ion formed ($[(\mu p)^* H_2^+]$ or $(\mu pp)^*$) will undergo the very fast ionic reactions (4) or (5) and turn into the molecular ion complex $[(\mu p)^* H_3^+]$ before the n=4 $\rightarrow$ 3 Auger transition takes place. This results from the fact that the rates $\lambda(4)=4.10^{13}\phi$ s^{-1} (as deduced from $k[H_2^+ + H_2 \longrightarrow H_3^+ + H] \approx 2 \cdot 10^{-9}$ cm^3 s^{-1} [8]) and $\lambda(5)$ (as deduced from $k(7)$) exceed appreciably the internal Auger rates at n=4.

After Auger decay to n=3 the μp will rapidly find itself either in the [Δ=-1] or the [Δ=-2] Stark level, bound either to H_2 or H_3^+ (or even H_5^+ [9]). In this bound state, complete thermalization is very fast but also the vibrational deexcitation process (6) will take place (especially in the molecular ion case) increasing the average internal kinetic energy of the μp atom. The rate $\lambda(6)$ can be quite high in the highest vibrational levels but, as it is sensitive to the relative vibrational spacings ΔE_v of the colliding partners, it can decrease rapidly with the increase in binding energy E_B[10]. The vibrational state at Auger decay time needs to be known in order to infer the relative population of the resulting 2p and 2s and their kinetic energy distributions. These are shown in fig. 1(a) for an equal initial [Δ=-1] and [Δ=-2] population in a $[(\mu p)^* H]$ molecule with E_B=0.5 eV. The predominant 2s population formed results from the fact that the initial [Δ=-1] and [Δ=-2] levels decay in respectively 62% and 98% of the cases [11] into the final [Δ=-1] Stark state which the adiabatic continuation of the 2s level ("$2s$") in high fields. Besides a 30% bound fraction, important μp acceleration is obtained with a non negligible high energy tail extending to many eV. The high energy component is transferred mostly into the 2p state via Stark mixing collisions [4,12,13] and at $\phi \ll 1$ decay into high energy ground state μp's . At $\phi \cong 1$, slowing down [4,13] is very effective, most μp's reach $E < 0.3$ eV and more than half of the population is found in the 2s state.

Moreover, the remaining 2p population can be efficiently transferred into the "$2s$"state via the molecular reaction

$$\mu p(\Delta = 0) \; + \; H_2(v = o) \longrightarrow [\mu p(''2s'') \cdot H_2(v = 1)]$$

whose transient formation rate has been computed[4] to exceed $3 \cdot 10^{11} \phi s^{-1}$. This transient complex can be at $\phi \approx 1$, stabilized by collision with a third body into a loosely bound $[\mu p(''2s'')\, H_2]$ vibrational level from which the return into the 2p state will be highly improbable.

The long lived 2s state itself does not require high density to undergo 3-body collisions and form a bound $[\mu p(''2s'')\, H_2]$. This occurs according to formula (13) (see Table 1) at a rate of $7.5 \cdot 10^{11}$ ϕ^2s^{-1} which exceeds the $4 \cdot 10^9$ ϕs^{-1} [4,14] collision induced decay rate of the 2s state down to ϕ=0.01. Incidentally, it is interesting to note that even at $\phi \approx 0.001$, the 2s lifetime is appreciably shortened by transient complex formation. This may explain the negative results[15,16,17] of the experiments searching for long-lived 2s.

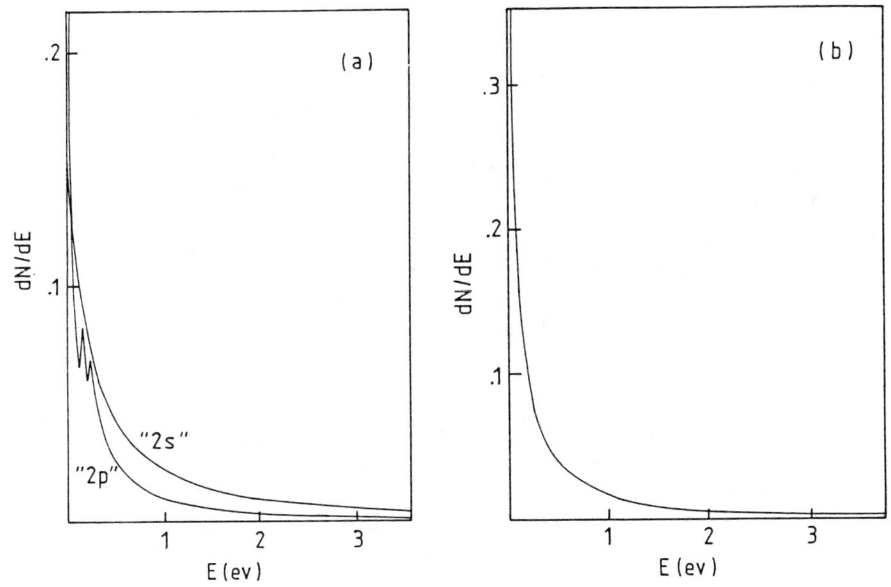

Figure 1: Kinetic energy distribution of μp atom after a bound $[(\mu p \cdot H)]$ to free transition. (a): initial states n=3, 50% Δ=-1 and 50% Δ=-2, 0.5 eV binding energy; final states: "$2s$" and "$2p$" adiabatic states; the transition is almost pure Auger. (b): initial state: "$2s$", 1 eV binding energy; final state: 1s state; the transition is about 85% radiative.

At high densities the molecular complex undergoes vibrational deexcitation and here again the question is how deeply bound it will be at decay time. On the one hand, a very high number of molecular collisions occur at $\phi \approx 1$; on the other hand, the more the μp gets bound the more its motion decouples from collisional disturbances. As a trade-off it may be assumed that deexcitation stops when vibrational energy transfer will require more than one vibrational energy quantum of the H_2 molecule. Analysis of the vibrational spectrum in the $[(\mu p) \cdot H]$ case gives a binding energy $E_B \approx 1$ eV and a resulting μp(1s) kinetic energy distribution as shown in fig. 1(b).

The appreciable 1s acceleration that results from excited bound state formation is one of the consequences of the molecular reactions that is highly relevant to μCF research. Another one, the reduction the isotopic charge exchange probability at high ϕ (see also ref. [18] and [19] where molecular effects have been discussed in this context) will be considered in more detail in a separate paper[20].

References

[1] H.A. Bethe and M. Leon, Phys. Rev. 127, 636 (1962).
[2] L. Bracci and G. Fiorentini, Nuovo Cimento 43A, 9 (1978).
[3] L.I. Men'shikov, Muon Catalyzed Fusion 2, 173 (1988).
[4] L.I. Men'shikov and L.I. Ponomarev, Z. Phys. D2, 1 (1986).
[5] P.M. Miller, J.P. Moseley, D.W. Martin and E.W. McDaniel, Phys. Rev. 173, 115 (1968).
[6] E. Graham IV, D.R. James, W.C. Keever, I.R. Gatland, D.L. Albritton and E.W. McDaniel, J. of Chem. Phys. 59, 4648 (1973).
[7] J. Böcklin, Dissertation No. 7161, ETH Zürich, 1982.
[8] L.P. Theard and W.T. Hunters Jr., J. Chem. Phys. 60, 2840 (1974).
[9] R. Johnsen, Chou-Mou Huang and M.A. Biondi, J. Chem. Phys. 65, 1539 (1976).
[10] C.R. Blakely, M.L. Vestal and J.H. Futrell, J. Chem. Phys. 66, 2393 (1977).
[11] H.A. Bethe and E.E. Salpeter, Quantum mechanics of one - and two - electron atoms, Springer-Verlag, Berlin-Göttingen-Heidelberg, 1957.
[12] G. Kodoski and M. Leon, Nuovo Cimento 1B, 41 (1971).
[13] G. Carboni an G. Fiorentini, Nuovo Cimento 39B, 281 (1977).
[14] R.O. Müller, V.W. Hughes, H. Rosenthal and C.S. Wu, Phys. Rev. A11, 1175 (1975).
[15] H. Anderhub, H. Hofer, F. Kottmann, P. LeCoultre, D. Makowiecki, O. Pitzurra, B. Sapp, P.G. Seiler, M. Wälchli, D. Taqqu, P. Truttmann, A. Zehnder and Ch. Tschalär, Phys. Lett. 71B, 443 (1977).
[16] P.O. Egan, S. Dhawan, V.W. Hughes, D.C. Lu, F.G. Mariam, P.A. Souder, J. Vetter, G. zu Putlitz, P.A. Thompson and A.B. Denison, Phys. Rev. A23, 1152 (1981).
[17] F. Kottmann, Dissertation Nr. 7179, ETH Zürich, 1982.
[18] L.I. Men'shikov and L.I. Ponomarev, JETP Letters 42, 13 (1985).
[19] M. Jändel, S.E. Jones, B. Müller and J. Rafelski, CERN-TH.4856/87
[20] D. Taqqu, to be published in Muon Catalyzed Fusion 4.

MOLECULAR EFFECTS IN NUCLEAR SCATTERING: HYPERFINE QUENCHING IN $d\mu + D_2$ COLLISIONS

O. K. Baker
Los Alamos National Laboratory, Los Alamos, N.M. 87545

ABSTRACT

A better understanding of hyperfine effects in dμ scattering from molecules of hydrogen isotopes will be helpful in untangling the intricacies of the muon catalyzed fusion process. This paper presents calculations of hyperfine quenching cross sections and rates for dμ + D_2 scattering in the absence of $[(dd\mu)d2e]$ mesomolecular formation.

The resonant formation of $[(dd\mu)d2e]$ plays an essential role in the ability of a single muon to catalyze several fusions in muon catalyzed $dd\mu$ fusion. M. Leon has shown that this resonant molecular formation is crucially affected by the hyperfine state of the $d\mu$ atom when it impinges upon a D_2 molecule[1]. The process, which he labelled resonant hyperfine quenching, in which the $d\mu$ atom undergoes a spin $F = 3/2$ to $F = 1/2$ hyperfine transition has a strong temperature dependence between 0 and 300K. The molecular formation rate is seen to be greater if the $d\mu$ atom is in the triplet state ($F = 3/2$) than if it is in the singlet state ($F = 1/2$) over this temperature range thus affecting the muon catalyzed fusion cycle. In this paper, the energy dependence of the $d\mu$ hyperfine transition cross section and the temperature ($= 3kT/2$ where T is the temperature and k is Boltzmann's constant) dependence of the hyperfine transition rate are calculated for *nonresonant* scattering, that is, in the absence of the $[(dd\mu)d2e]$ molecular formation.

The scattering of a slow $d\mu$ atom by a D_2 molecule can be described in a manner similar to that of neutron scattering from D_2 molecules[2-4] because the $d\mu$ atom is a small, tightly-bound, neutral system. The evaluation of the cross section for the nuclear hyperfine transition involves the integral for J to J' transitions for the D_2 molecule.

$$\sigma_Q = \frac{k_2}{k_1}\left(\frac{\mu_M}{\mu_A}\right)^2 \frac{\lambda_Q^2}{(2J+1)}\frac{16}{15}\left|\int d^3r \cos\left\{\frac{(\vec{k_1}-\vec{k_2})\cdot\vec{r}}{2}\right\}\phi^*_{J'm_J'}(r)\phi_{Jm_J}(r)\right|^2 \qquad (1)$$

Here J (J') is the initial (final) rotational state (with projection quantum number m_J ($m_{J'}$)) of the D_2 molecule, k_2 (k_1) is the final (initial) momentum of the $d\mu$ atom, μ_M (μ_A) is the reduced mass of the $d\mu d$ ($d\mu D_2$) system, ϕ is the D_2 wave function, and λ_Q is the scattering length. Only low energies and temperatures will be considered. Then only the lowest vibrational states will be affected and the associated indices will be suppressed. The rotational states $0, \pm 2, \pm 4, ...$ are called ortho levels while the $\pm 1, \pm 3, ...$ are called para levels. If $|J - J'| = 1, 3, ...$ then the integral in equation (1) is suppressed (≈ 0) compared with the integral which results when $|J - J'| = 0, 2, ...$ This has the following physical interpretation: If the deuteron of the $d\mu$ atom has its spin antiparallel to the spin of the deuteron of the D_2 molecule that it collides with, the quenching probability will be suppressed as compared to the case where the two deuteron spins are parallel.

An integral similar to that of equation (1) has been evaluated in reference 3 for neutron scattering from D_2 molecules. The results can be adapted to the present case if, in the quenching cross section only ortho-ortho and para-para D_2 transitions are included. The ortho-para and para-ortho transitions are "disallowed". Then the cross section for nonresonant hyperfine quenching in $d\mu$ scattering from D_2 molecules is

$$\sigma_Q = \frac{k_2}{k_1} \cdot (\frac{\mu_M}{\mu_A})^2 \cdot \lambda_Q^2 \cdot \frac{16}{15} \cdot (\frac{2J'+1}{2J+1}) \cdot 6\pi \cdot \frac{p^o}{p} \cdot (\frac{2}{\xi})^2 \cdot A_{JJ'} \qquad (2)$$

where the new symbols are as defined in reference 3. Taking the statistical average of the scattering lengths a from reference 5 with $\lambda_Q = 2\pi a$, the cross section plotted in Figure 1 results. The solid line is the calculated quenching cross section which includes only the ortho-ortho and para-para transitions, i. e. the "allowed" transitions, while the dashed line is the sum of <u>all</u> transitions including para-ortho and ortho-para or "allowed" transitions.

From this cross section the temperature dependence of the quenching rate shown in Figure 2 can be calculated:

$$\Lambda_Q = \frac{N_o \int_0^\infty \sigma_Q \cdot v \cdot e^{-\beta E} v^2 \cdot dv}{\int_0^\infty v^2 \cdot e^{-\beta E} \cdot dv} \qquad (3)$$

where $N_o = 4.25 \; x \; 10^{22} \; cm^{-3}$. In this calculation a correction (denoted by P) has been made for the temperature dependence of the rotational levels via

$$F_J = (2s+1)(2J+1)e^{-\beta \Delta E_{JJ'}} \qquad (4a)$$

$$P = \frac{F_J}{\sum F_J}. \qquad (4b)$$

ΔE is the energy difference between rotational levels and s is the $d\mu$ spin. Again the full curve includes only ortho-ortho and para-para transitions whereas the dashed curve includes the sum of "allowed" and "disallowed" transitions. The calculated quenching rate agrees with the experimental point of Kammel et. al.[6] at 30K. It is not clear why the calculation of the 3/2 to 1/2 transition for $d\mu + d$ atomic scattering should agree with the experimental value or with the present calculation, both for $d\mu + D_2$ molecular scattering. This question is still being investigated by the author.

In conclusion, the nonresonant hyperfine quenching cross section and rate have been calculated for low energies and temperatures respectively. The calculated rate is in good agreement with a measurement at 30K.

The author gratefully acknowledges helpful discussions especially with M. Leon, who suggested the calculation, and with M. Paciotti and J. Cohen.

REFERENCES

1. M. Leon, Phys. Rev. **A33**, 4434 (1986). See also ref. 6
2. J. Schwinger and E. Teller, Phys. Rev. **52**, 286 (1937)
3. M. Hamermesh and J. Schwinger, Phys. Rev. **69**, 145 (1946)
4. J.A. Young and J.U. Koppel, Phys. Rev. 135, **A603** (1964)
5. V.S. Melezhik and J. Wozniak, Phys. Lett. **B116**, 370 (1986)
6. P. Kammel et. al., Phys. Rev. **A28**, 2611 (1983)

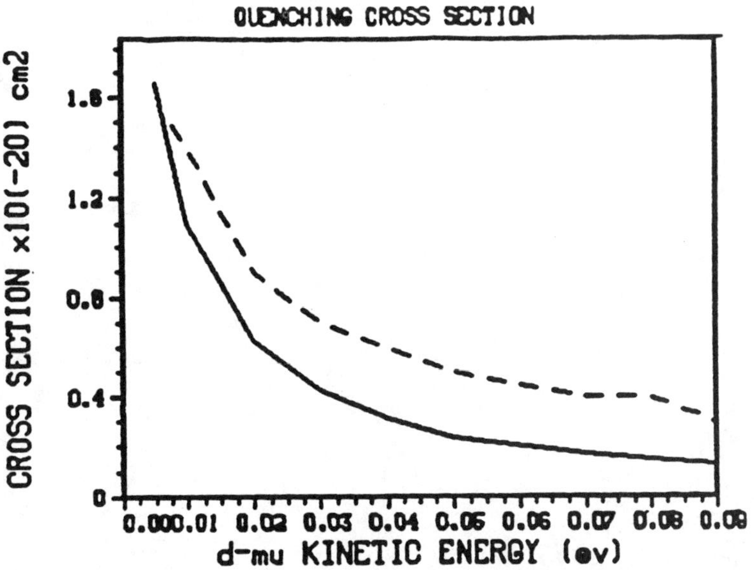

Figure 1. Nonresonant hyperfine quenching cross section as a function of incident $d\mu$ atom kinetic energy. The full curve is the cross section calculated for D_2 ortho-ortho and para-para transitions while the dashed curve includes the previously mentioned transitions plus the ortho-para and para-ortho transitions.

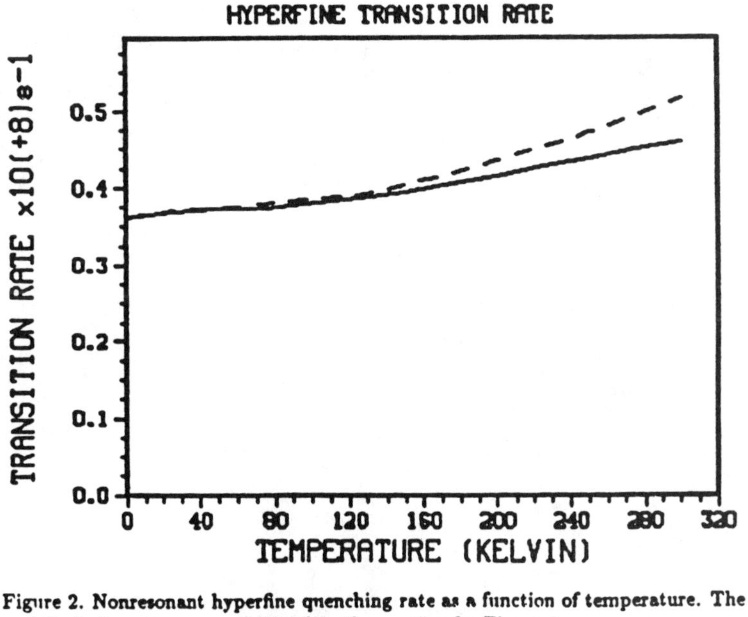

Figure 2. Nonresonant hyperfine quenching rate as a function of temperature. The full and dashed curve are as explained in the caption for Figure 1.

A Direct Process for Meso molecule Formation at High Temperature

A. M. Lane

Theoretical Physics Division, Harwell Laboratory,

UKAEA, Harwell, Oxon OX11 ORA.

There has been recent interest in the subject of meso-molecule $(dt\mu)$ formation at high temperature (leV), and a number of complex three-body mechanisms have been examined. We point out that, at leV, simple two-body dissociation $(t\mu)+D_2 \rightarrow [(dt\mu)e]+(de)$ is the dominant process with rate ≈ $6.4x10^9$ s^{-1} at liquid hydrogen density.

Discussions of muon-catalysed fusion usually focus on the low-temperature region ($T \lesssim 1000°K$) where it is believed that the dominant process for the necessary meso-molecule formation is a resonant one (Jones 1986a,b; Breunlich et al. 1987). Specifically, in the case of a muon in a chamber of D_2, T_2, DT molecules, the muon forms a $(t\mu)$ atom which then joins a deuteron in a D_2 molecule to form a resonant quasi-molecule "D_2" in which a deuteron is replaced by a $(dt\mu)$ meso-molecule:

$$(t\mu) + D_2 \rightarrow "D_2" \equiv [(dt\mu)_{11}dee] \qquad (1)$$

The $(dt\mu)$ is in the rotation-vibration state $(J\nu)=(11)$, bound by 0.6 eV, while the "D_2" is in the second vibrational state at 0.6 eV excitation energy.

The resonance process has been calculated up to 5000° (Cohen and Leon 1985), and it is found that it remains strong. Formation of "DT" becomes comparable with "D_2", and the rate of $(dt\mu)$ formation is about $4\times10^8 s^{-1}$ at liquid hydrogen density.

Recently attention has been drawn to the region of 12,000° ($kT \sim 1eV$). At such temperatures, D_2 molecules begin to dissociate. The dissociated fraction rises from 6% at 0.5 eV to 56% at 2 eV, being about 33% at 1 eV. (Here, and later, it is assumed that the density is that of liquid hydrogen, 4.25×10^{22} deuterons/cm^3.) The question has been raised (Menshikov and Ponomarev 1987) of the rate and mode of $(dt\mu)$ formation at such temperatures. Obviously the resonant process must decline with increasing dissociation. For 33% dissociation at 1 eV, the above rate becomes $2.7\times10^8 s^{-1}$ from this source. Menshikov and Ponomarev have investigated a number of non-resonant three-body processes involving D atoms from the 33% of dissociated D_2 and find a rate $10^9\ s^{-1}$. The object of this letter is to point out that a considerably higher rate arises from the 67% of surviving D_2 molecules through a simple direct process:

$$(t\mu) + D_2 \rightarrow \left[(dt\mu)_{11}e\right] + (de) \qquad (2)$$

The threshold for this is obtained from the dissociation energy 4.6 eV of D_2 to be 4 eV (for the D_2 ground state; for excited states it will be smaller). At kT = 1 eV, about 2.5% of collisions have relative energy above this value.

Now we estimate the cross-section for process (2), which can be schematised as:

$$3 + (12) \rightarrow (31)+2 \qquad (3)$$

where 3 stands for $(t\mu)$, and the deuterons are 1,2. We suppress reference to electrons, their presence being represented in $D_2 \equiv (12)$ by a screened potential between 1 and 2. We see that (3) is analogous to a stripping reaction in nuclear physics in which (12) is a deuteron, and (31) a composite of the target nucleus 3 and the stripped proton 1. In both cases, forces are short-range (non-coulomb); further the object (12) is relatively lightly bound and diffuse, while the force V_{13} that causes the transition is relatively strong and of short-range. In nuclear physics, plane-wave Born approximation gives a reasonably accurate guide (with distorted wave effects causing a reduction in cross-section by a factor of 3 or less).

Proceeding with plane-wave Born approximation, the cross-section per deuteron is:

$$d\sigma = \frac{d\Omega}{4\pi^2} \left(\frac{m_f}{\hbar^2}\right)^2 \frac{v_f}{v_i} \left|M(k_i,k_f)\right|^2 \tag{4}$$

where

$$M(k_i,k_f) \equiv \langle \phi_{13} e^{i\underline{k}_f\cdot\underline{r}_{2(13)}} |V_{13}| \Phi_{12}\, e^{ik_i\cdot\underline{r}_{3(12)}} \rangle \tag{5}$$

$\underline{r}_{i(jk)}$ is the coordinate of particle i relative to the centroid of particles j,k. m_f, v_f, k_f ($\equiv m_f v_f/\hbar$) are the final relative mass, velocity and wave-number while subscript i labels corresponding initial quantities. Introducing the Fourier transforms of the states ϕ_{13}, Φ_{12} of the composites (13), (12):

$$\phi_{13}(\underline{k}) = (2\pi)^{-3/2} \int \phi_{13}(\underline{r}_{13}) e^{i\underline{k}\cdot\underline{r}_{13}}\, d\underline{r}_{13} \tag{6}$$

$$\Psi_{12}(\underline{K}) = (2\pi)^{-3/2} \int \Phi_{12}(\underline{r}_{12}) e^{i\underline{K}\cdot\underline{r}_{12}}\, d\underline{r}_{12} \tag{7}$$

we find:

$$M = (2\pi)^3 \Psi_{12}(\underline{K}) \phi_{13}(\underline{k}) \left(|E_{13}| + \frac{\hbar^2 k^2}{2m_{13}}\right) \tag{8}$$

with

$$\underline{K} \equiv \underline{k}_f + \underline{k}_i \frac{m_2}{m_1+m_2} \tag{9a}$$

$$\underline{k} \equiv \underline{k}_i + \frac{m_3}{m_1+m_3} \underline{k}_f \tag{9b}$$

$$m_{13} \equiv m_1 m_3 (m_1+m_3)^{-1} \tag{10}$$

Turning to the evaluation of ϕ_{13}, we note that, for the lightly-bound "11" state, most of the norm of ϕ_{13} is in the region of asymptotic tail, $\phi_{13}(\underline{r}_{13}) = u(r_{13})Y_1(\Omega_{13})$ with (Menshikov 1985):

$$u(r_{13}) = 0.57\ K^{3/2}\ ((\kappa r_{13})^{-1} + (\kappa r_{13})^{-2})\exp(-\kappa r_{13}) \tag{11}$$

where κ is the binding wave-number $\approx 1.86\text{x}10^9\ \text{cm}^{-1}$. Following nuclear physics, we assume that (6) is dominated by the tail region, so integrate with (11) over the tail only to get:

$$\phi_{13}(\underline{k})(|E_{13}| + \frac{1}{2}\frac{\hbar^2 k^2}{2m}) = \frac{\sqrt{6}}{2\pi}\frac{\hbar^2 R^2}{2m_{13}}(u\ j_1'(z)\ u'j(z))_{r_{13}=R} \tag{12}$$

$$= \frac{\sqrt{6}\hbar^2}{4\pi m_{13} R^{\frac{1}{2}}}\ 0.57\ x^{-1/2}\ e^{-x}\ [(1+x)\sin Y + x^2 j_1(Y)]$$

where $j_1(z) \equiv z^{-2}\sin z - z^{-1}\cos z$, $z \equiv kr_{13}$, $Y \equiv kR$, $x \equiv KR$ and the prime means d/dr_{13}. R is the value of r_{13} where the formula (11) begins to apply. Menshikov (1985) gives $R=7a_\mu$, implying $x=0.33$. For energy close to threshold, $k \approx k_i = 0.58\times10^{10}$ cm^{-1}, giving Y=1.04, and (12) equals $0.94\ \hbar^2/m_{13}a_o^{\frac{1}{2}}$. Near threshold, $d\sigma$ is isotropic, and (4) with (8) becomes:

$$\sigma = \left(\frac{v_f}{v_i}\right) 0.77 \times 10^{-12}\ \text{cm}^2\ \overline{\left(|\Psi_{12}(K)|a_o^{-3/2}\right)^2} \qquad (13)$$

The crucial question is the magnitude of the last factor. The bar indicates average over the Boltzmann distribution of vibration-rotation states of the target D_2. Let us first evaluate for the ground state, $(J\nu)=(00)$. Using the oscillator form for $\Psi(\underline{r}_{12}) = (4\pi)^{-\frac{1}{2}}\chi_o(r_{12})$ with:

$$\chi_o(r_{12}) = \frac{2}{\pi^{\frac{1}{4}} b^{\frac{1}{2}} r_{12}} \exp\left[-\frac{1}{2}\left(\frac{r_{12}-r_o}{b}\right)^2\right] \qquad (14)$$

and the $\ell=0$ part of $\exp[i\underline{K}.\underline{r}_{12}]$, viz $(\sin Kr_{12}/K\ r_{12})$, (7) gives:

$$\left(|\Psi_{12}(\underline{K})|a_o^{-3/2}\right)^2 = \frac{b}{\pi^{3/2}K^2a_o^3} \sin^2 Kr_o\ e^{-(Kb)^2} \qquad (15)$$

With the $\sin^2 Kr_o$ factor replaced by average value $\frac{1}{2}$, $K = \frac{1}{2}k_i = 0.29\times10^{10}$ cm^{-1}, $b = 1.14\times10^{-9}$cm, Kb = 3.3, then (15) $\approx (12500)^{-1}\exp(-10.9) \approx 1.5\times10^{-9}$. This makes (13) equal $(v_f/v_i)\ 10^{-21}$cm^2. This estimate is relevant to the situation at low temperatures where a $(t\mu)$ atom, newly converted from $(d\mu)$, has enough energy for reaction (2). It begins with about 19eV kinetic energy (Ponomarev 1983), so experiences about three scattering collisions with cross-section $5\times10^{-19}cm^2$ (Melezhik, Ponomarev and Faifman 1983) before the energy is degraded below that required for (2). The cross-section just quoted means that there is <0.5% change of $(dt\mu)$ formation occurring.

We now estimate the last factor in (13) for temperature ~ 1 eV. For collisions near threshold (9a) gives $K \approx \frac{1}{2}\underline{k}_i$, so it is convenient to evaluate (7) with direction $\underline{k}_i$ as the quantisation axis for rotational angular momenta of $\Phi_{12}^{JM}=\chi_v(r_{12})Y_{JM}(\Omega_{12})$:

$$\Psi_{12}^{JM}(\underline{K}) = a_o^{3/2}\sqrt{2J+1}\,\delta_{MO}\,I_{Jv} \quad (16)$$

where:

$$I_{Jv} \equiv (2\pi^2 a_o^3)^{-\frac{1}{2}} \int_o^\infty \chi_v(r) j_J(Kr) r^2 dr \quad (17)$$

The Boltzmann average over the rotational spectrum $E_J = 3.7\ J(J+1)$meV then gives:

$$|\Psi_{12}^{JM}(\underline{K})a_o^{-3/2}|^2 = \sum_J P_J I_{Jv}^2 / \sum_J P_J \quad (18)$$

where, for kT = 1 eV:

$$P_J \equiv (2J+1)\exp(-0.0037J(J+1)) \quad (19)$$

Evaluating (17) for v=0, using (14), we find that I_{Jo} has a maximum at J=22 of 0.28×10^{-4}, with half-maximum values at j≈17,25 and negligible values at J=14,27 and beyond. Inserting in (18) we find 0.59×10^{-5}, where the reduction by a factor 5 from the maximum value of $I^2_{J\nu}$ arises because, at kT=1eV, the exponential factor in (18) is only 0.15 for J=22. In passing, we note that if we had evaluated (18) by the fixed rotor angle method, valid for higher T such that the exponential factor is ≈ 1, then:

$$|\Psi_{12}(\underline{K})a_o^{-3/2}|^2 = (2\pi a_o)^{-3}\int d\Omega \left|\int_o^\infty \chi(r)e^{ikr\cos\theta}r^2dr\right|^2 \left(\sum_J P_J\right)^{-1}$$

$$\approx \frac{.0037}{kT}\,\frac{r_o^2}{4\pi K a_o^3} \approx \frac{2.8\times10^{-5}}{kT} \qquad (20)$$

agreeing with the value from (18) if T is such that the exponentials in the numerator are ≈ 1 for J ≤ 25.

With (18) equal to 0.59×10^{-5}, cross-section (13) becomes $(v_f/v_i)4.5\times10^{-18}\text{cm}^2$. Taking the Boltzmann average of this, the rate of $(dt\mu)$ formation is:

$$\lambda = 1.9\times10^5\text{cm}^{-1}\ \langle v_f\rangle\ F$$

$$= 6.2\times10^5\text{cm}^{-1}\ \left(\frac{kT}{m_f}\right)^{\frac{1}{2}}Fe^{-E_t/kT}$$

$$= 9.5\times10^9\ F\ s^{-1} \qquad (21)$$

for temperature 1 eV and threshold energy E_t=4 eV. F is the fraction of undissociated D_2 at 1 eV. Taking F=0.67 gives $\lambda=6.4\times10^9 s^{-1}$, which comfortably exceeds the estimate $10^9 s^{-1}$ from three-body processes.

Note several aspects of the estimate:

(i) the role of excited vibrational states $\nu > 0$ of D_2. If we are guided by the fixed rotor angle method, excited states give essentially the same cross-section formula (19) as $\nu=0$. For the case $\nu=1$, it is easy to check that the fixed rotor result is exactly (19). In evaluating $|\Psi_{12}|^2$, for higher ν, one expects increased values because of better matching between important incident orbital angular momenta (shown by $I^2_{J\nu}$) and important rotational angular momenta (shown by weights P_J). This not only makes the (large) result (19) more valid, but increases the value of (19) through smaller K.

(ii) the electron overlap effect. On the basis of the fact that the electron states in D_2 are similar to those of D atoms (nodeless 1s states), we have put the overlap equal to unity. The actual value may be smaller, say 50%.

(iii) the large value (20) is specific to the (11) state of ($dt\mu$). For lower states, bound by order 100 eV, the exponential in (12) is small.

(iv) the large value is specific to the 1s state of the ($t\mu$) atom. For an excited state, (12) is reduced by an overlap between the state and ($dt\mu$), and this is small.

(v) the estimates can be repeated for the case ($d\mu$)+D_2, where the binding of ($dd\mu$) is 1.9 eV and the threshold is 2.7 eV. We find that the rate λ of (20) is increased by a factor of 2.

In summary, we find that, at kT = 1 eV, sufficient D_2 survive dissociation that (dtμ) formation rate is dominated by (2) which has a large cross-section. The rate is considerably larger than for three-body processes (Menshikov and Ponomarev 1987). The main uncertainties in our estimates arise from plane wave Born approximation and items (i), (ii) above, but we expect accuracy to within a factor 2. The rate of (ddμ) formation is large ($2 \times 10^{10} s^{-1}$) implying that, at kT=1 eV, any (dtμ) atoms which reach the ground state are likely to form (ddμ), with the consequent one-in-seven chance of being lost by sticking to a helium reaction product.

Work described in this Letter was undertaken as part of the underlying research programme of the UKAEA.

References

Breunlich W.H. et al. 1987 Phys.Rev.Lett. 58, 329.

Cohen J.S. and M. Leon 1985, Phys.Rev.Lett. 55, 52.

Jones S.E. 1985 Nature 321, 127.

Melezhik V.S., L.I. Ponomarev and M.P. Faifman 1983 Sov.Phys. JETP 58, 254.

Menshikov L.I. and L.I. Ponomarev 1987 JETP Lett 46, 313.

Menshikov L.I. 1985 Sov.J.Nucl.Phys. 42, 918.

Ponomarev L.I. 1983 Atomkernenergie 43, 175.

CAPTURE OF A CLASSICAL MUON BY A QUANTAL HYDROGEN ATOM

N. H. Kwong and J. D. Garcia
Physics Department, University of Arizona
Tucson, AZ 85721

James S. Cohen
Theoretical Division, Los Alamos National Laboratories
Los Alamos, NM 87545

ABSTRACT

Using a self-consistent method that treats the muon classically and the electron quantum mechanically, we have calculated the capture cross sections of μ^- by atomic hydrogen at low incident energies.

INTRODUCTION

The slowing down and capture of negative muons by atomic hydrogen is a heavily studied subject[1-3] (see Ref. 2 for a more complete list of references). Besides being of some interest to muon-catalyzed fusion, the μ^--H system poses a challenging prototype, three-body Coulomb problem, involving free-to-bound transitions. Here, conventional basis-expansion methods are hard to apply and are probably not very enlightening. Taking advantage of the massiveness of the muon, we apply a scheme in which the muon and the protons are treated classically and the electron quantally. Their respective trajectories and wave functions are governed by a set of coupled, time-dependent Hamilton-Schrödinger equations, Eqs. (1), which conserve total energy, momentum, and angular momentum. This technique provides arguably the most realistic results to date for the μ^--H problem and seems to be the only viable, yet realistic, way to attack the experimentally relevant case of μ^--H_2. In this contribution, we will briefly explain the technique and compare the results with those obtained previously by other methods.[1-3]

The motions of the muon, $\vec{X}_\mu(t)$, the proton, $\vec{X}_p(t)$, and the electron, $\psi(\vec{x},t)$, obey the following coupled Hamilton-Schrödinger equations, which we call the Classical-Quantal-Coupling (CQC) equation,

$$i\hbar\, \dot{\psi}(\vec{x},t) = \left[-\frac{\hbar^2}{2m_e}\nabla_x^2 + \frac{e^2}{|\vec{x}-\vec{X}_\mu(t)|} - \frac{e^2}{|\vec{x}-\vec{X}_p(t)|} \right] \psi(\vec{x},t)$$

$$m_i\, \dot{\vec{X}}_i = \vec{P}_i \tag{1}$$

$$\dot{\vec{P}}_i = \vec{\nabla}_{X_i} \frac{e^2}{|\vec{X}_\mu - \vec{X}_p|} - \langle \psi(\vec{x},t)|\ \vec{\nabla}_{X_i} \frac{e^2}{|\vec{x}-\vec{X}_i|}\ |\psi(\vec{x},t)\rangle \quad ,$$

where $i = \mu, p$. The electron wave function evolution properly reflects the muon and proton trajectories, which in turn are determined by a Coulomb force due to the instantaneous electron charge distribution (as well as that due to each other). While intuitively rather obvious, Eq. (1) can also be derived from a time-dependent action principle with a WKB ansatz for the muon + proton part of the variational wave function.[4] On the computer, $\psi(\vec{x},t)$ is represented on a spatial grid. A typical run starts wtih the muon sufficiently far away with specified incident energy and impact parameter. Equations (1) are then integrated in time, and the capture cross section is given by

$$\sigma_{capture}(E) = \pi\, b_{crit}^2(E) \quad , \tag{2}$$

where b_{crit} is the critical impact parameter such that the orbits for all $b < b_{crit}$ are eventually trapped, while those for all $b > b_{crit}$ escape.

Figure 1 shows the excitation and ionization of the electron by the incoming muon, which is captured in this run. It is obvious that even at this low energy, the evolution is essentially nonadiabatic except at the beginning. Interesting to note is the spiralling motion of the electron density caused by the rotating dipole field of the bound μ-p system. Figure 2 compares our cross sections with those calculated by the Classical-Trajectory Monte-Carlo (CTMC) method.[2] They agree rather well, except possibly at the low-energy end where CTMC data are not yet available. Comparisons of other aspects, such as final-state distributions and stopping power, will be carried out later. Another calculation,[3] treating both the muon and the electron quantally, but the interaction in the mean-field approximation, gives $\sigma(E = 0.01 \text{ a.u.}) = 9\ \pi a_0^2$, which is lower than our present result. There are, however, some uncertainties in the extraction of cross sections in that calculation that are not yet resolved. Finally, the early work[1] based on the concept of adiabatic ionization also underestimated σ, as expected.

REFERENCES

1. A. S. Wightman, Phys. Rev. 77, 521 (1950).
2. J. S. Cohen, Phys. Rev. A27, 167 (1983).
3. J. D. Garcia, N. H. Kwong, J. S. Cohen, Phys. Rev. A35, 4068 (1987).
4. N. H. Kwong, J. Phys. B20, L647 (1987).

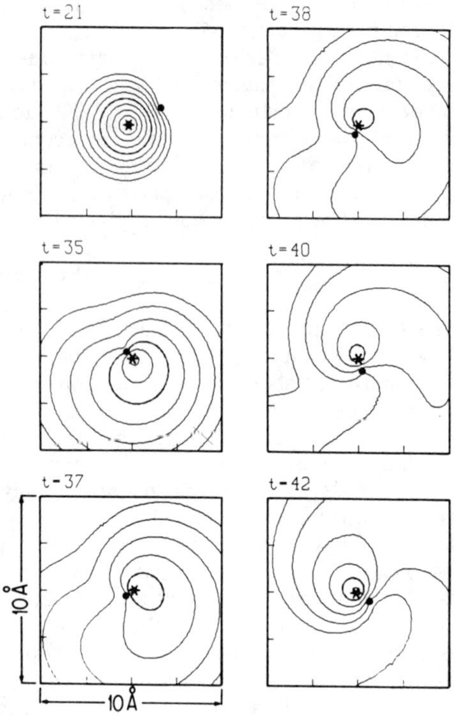

Fig. 1. Time frames showing the positions of the classical muon (·) and proton (*), and the density distribution of the quantal electron; E_{cm} = 0.165 a.u., b = 0.5 Å. At t (in a.u.) = 0, the muon was at a distance of 5 Å from the proton.

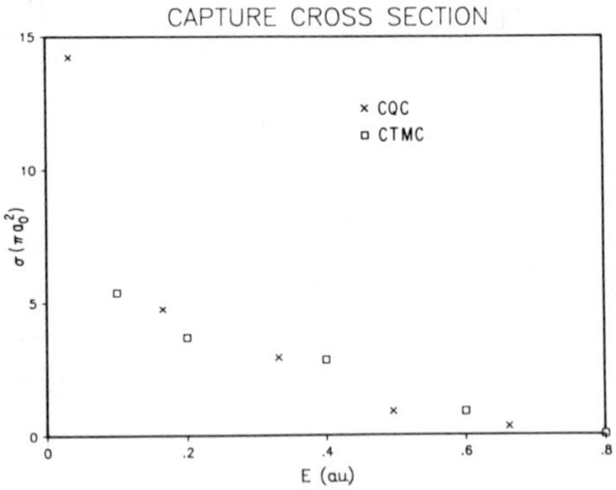

Fig. 2. Cross sections of muon capture by a hydrogen atom, as calculated by Classical-Quantal-Coupling (CQC) and Classical-Trajectory Monte-Carlo (CTMC).

MuCF WITH $Z > 1$ NUCLEI

D. Harley, B. Müller and J. Rafelski
Department of Physics, University of Arizona, Tucson, AZ 85721

ABSTRACT

We investigate the molecular processes involved in the muon catalysis of hydrogen isotope fusion with light nuclei $Z > 1$. We examine in some detail the expected fusion rates in (HZμ) molecules, and find the fusion rates to be significantly higher than previous estimates.

INTRODUCTION

MuCF reactions involving $Z > 1$ nuclei have up to now been dismissed as being unfeasible. The most obvious objection is that if the muon becomes bound to a $Z > 1$ nucleus, it is irretrievably lost from the cycle of fusion reactions. Also, as noted in the original work of Frank [1], the barrier penetration calculated in the WKB approximation results in fusion rates that are more than ten orders of magnitude smaller than those involving hydrogen isotopes only [2]. We wish to draw attention to the fact that the arguments that have been presented are not valid in the light of our recently improved insight into the dynamics of muon catalyzed fusion.

In section 1 we will first briefly consider the muomolecular requirements for a high Z nucleus + hydrogen isotope H reaction. We shall set as our target one fusion per muon, although the present uncertainty about the resonant molecular systems leaves open the potential for many more fusions. In section 2 we discuss the nuclear phenomena and Coulomb barrier penetration problem, and compute the semiclassical barrier penetration factors taking particular note of the nonadiabatic components of the molecular wavefunction. We conclude that Z+H fusion could prove to be of considerable interest, and that this process deserves further experimental and theoretical investigation [3].

1. MOLECULE FORMATION AND SCAVENGING

A major difficulty inherent in high-Z molecular processes is the realization of a high molecular formation rate, whilst avoiding the capture of the muon by Z through scavenging or transfer of the muon from H. A second problem related to the molecular processes but dependent upon the nuclear properties of the participating nuclei, is the need for a nuclear fusion rate that will dominate both the muon decay rate and the rate of molecular transfer of the muon. We shall use in the following discussion of these requirements the example Li+$d\mu$, but many points of principle apply to the other light nuclei of interest to us.

Scavenging of free muons can be reduced to manageable proportions if the relative Z abundance C_Z in the Z+H target mixture is made sufficiently small. If only one fusion per muon is desired, the abundance need be no less than $C_Z = 0.5$. Nevertheless, we wish to point out that it may be possible to support a much smaller abundance (say 1 in 2000, allowing up to 1000 free muon cycles) and still have a large molecule formation rate. This would be permitted if the Z+Hμ collision cross section was large, but also dominated by the formation of a muomolecule rather than by a direct muon transfer process. Weakly bound ^{7}Liμd molecular states [4] and near threshold resonances

[5] have already been identified. The bound states found in the improved adiabatic approximation for the ^{7}Liμd molecules are $J = 2$, $E = 2.02$ eV; $J = 1$, $E = 14.1$ eV; and $J = 0$, $E = 21$ eV, the energies of which could easily match with the energy of LiD molecular vibrations and allow molecular formation to occur. In semi-adiabatic calculations of the elastic scattering cross sections of $d\mu$ on Li by Kravtsov *et al*, it is found that cross sections of as much as 1 Å^2 are common to this system [5]. At present there is no calculation available of the resonant molecule formation rate arising from $d\mu$ collisions with LiD molecules, but we expect the molecular formation cross section to be of a similar magnitude due to the participation of Li electrons. The molecular energies stated above are much smaller than the K-electron binding energies of O(100 eV). Consequently, the cloud of the 1s electrons is effectively bound rigidly to the Li nucleus and will participate with the Li in the capture of the $d\mu$ atom. Thus although geometrically we have a relatively small Li+$d\mu$ system (of size $5a_\mu < \frac{1}{3}a_e$) inside a large electron K-shell cloud, the participation of the K electrons should assure us of a resonant μ- molecular formation cross section of atomic dimensions. Since a molecular formation rate of 10^6 s^{-1} $C_Z^{-1}\phi^{-1}$ (the approximate lifetime of the muon) corresponds to a cross section of only 10^{-6} Å$^2 C_Z^{-1}\phi^{-1}$ at $T = 500$ K, many molecule formations should be possible within the muon's lifetime. A quantitative estimate of the molecular formation rate clearly demands the inclusion of the participating electrons in calculations of Z- muomolecular resonances.

A large molecular formation cross section would be of no use if it mostly resulted in the exchange of the muon. This process is however suppressed in most high-Z systems. In the specific case of Li+$d\mu$, molecule formation is the preferred process due to the avoided crossing of the muon correlation energies in the approach of $d\mu$ to the Li$^+$-ion. The molecular orbital comprising the main component of the molecular wavefunction is the $3d\sigma$, which is the only orbital corresponding to the Li^{3+} + $(d\mu)_{1s}$ asymptotic state [6]. This orbital is always below the near degenerate (at large separations) configurations $3p\sigma$, $3s\sigma$ etc., all of which correspond to the muon being on the Li side of the molecule as the colliding nuclei separate. Transfer of the muon is therefore suppressed. Furthermore, the direct transfer of the muon to a Li $n = 3$ state is energetically forbidden, due to the participation of the electrons surrounding the Lithium atom. Should the muon transfer from the d to the Li nucleus, screening in the process one unit of nuclear charge, the two core 1s-electrons would only feel the attractive force of a quasi-helium nucleus. Hence, instead of 197 eV of binding they would have just 78 eV binding energy, a loss of 119 eV. This has to be compared with the gain of only 105 eV in the muon binding due to the change in the reduced mass of the muon. The difference, 14 eV, is greater than the any potential gain in the transfer of an electron from the (Liμ) atom to the remaining d^+ ion. Thus the configuration Li+$d\mu$ can not change to the near degenerate configuration $(\text{Li}\mu)_{3s\sigma,3p\sigma} + d$, as it would lead asymptotically to a system with energy greater than the initial energy of $(d\mu)_{1s}$ colliding with the Li electro-molecule. This is quite general for $Z \geq 3$, with the exception of the $p\mu$+Li systems, which would permit muon transfer due to an enhanced reduced mass effect. Nonetheless this transfer will be greatly suppressed due to the width and height of the dynamical barrier separating both asymptotic systems.

Once molecular formation has occurred, muons can still be transferred inside the molecular structure in competition to the nuclear fusion reaction, as the molecules of interest here will all be formed in the n = Z shell. In our example of Li+$d\mu$, the asymptotic $(d\mu)_{1s}$ state leads adiabatically into the $(\text{Li}d\mu)_{3d\sigma}$ molecular configuration. This configuration can decay either by radiative, Auger, or configuration rearrangement transitions to lower states in which the muon remains bound to the heavier Li nucleus while the molecule dissociates. The importance of these internal reactions depends on the nuclear reaction rate, which constitutes a competing reaction channel. In He it is known that radiative dissociation dominates fusion by a factor of $100 - 1000$, and some of the time constants have been measured [7], [8]. In Li the radiative transitions are much slower due to the greatly reduced transition energy and quantum selection rules. In a discussion of the Li fusion chain by Kravtsov *et al* [5], an electron Auger rate of the order of 10^6 s^{-1} is obtained. The configuration rearrangement dissociation seems to be too slow to matter [9].

In conclusion to this section, the molecular dynamics of $Z > 1$ fusion do not exclude conditions favorable for preparing Z+H for fusion. The anticipated molecular formation rate should easily be sufficient to compensate for a low abundance of the high-Z element in the reaction vessel, and both direct and molecular transfer of the muon appears to be suppressed due to the participation of the K-electrons of the high-Z element. It is however apparent that a detailed calculation of the (HZμ) molecular system including electrons and participating nuclei is needed, as the electrons and electro-molecular states are believed to play a decisive role in the capture of the Hμ atom and subsequent muo-molecule formation.

3. COULOMB BARRIER PENETRATION AND FUSION

From consideration of the muon lifetime alone, a fusion rate of at least 10^6 s^{-1} is needed in order to achieve one fusion per muon. Such a fusion rate is indeed possible if the nonadiabatic components of the muomolecular wavefunction are taken into consideration. The wave function of the bound three body system may be expressed by the adiabatic expansion

$$\psi(r,R) = \sum_n \chi_n(R)\phi_n(r;R), \tag{1}$$

where $\phi_n(r;R)$ is the two-center wave function of the muon, computed for a frozen distance R between the two nuclear centers. The coefficients χ_n(R) provide us with the amplitude of finding the muon in respective states, and are of particular interest when the two nuclei react at R$\rightarrow$ 0. The two-center wave functions become in this limit the coulombic eigenstates of the compound $Z+1$–like nucleus. In the case of a pure Coulomb problem these coefficients are determined by solving the three-body Schrödinger equation. One may not neglect the nonadiabatic coefficients, ie. those not arising from the adiabatic limit for a fixed value of $n = Z$. The pure Coulombic dtμ system is strongly nonadiabatic and the coefficients associated with muon continuum states in the range $10 - 50$ keV are not small for R$\rightarrow$ 0. For $Z > 1$ the nonadiabasy is even more pronounced since we are dealing with molecular forces arising from polarization of the muonic hydrogen atom by the Coulomb field of the Z nucleus. Specifically, the amplitudes $\chi_n(R)$ are determined by the coupled equations [10] ($c = \hbar = 1$ throughout):

$$\Big[-\frac{1}{2M_R}\Delta_R+\frac{Ze^2}{R}+\epsilon_n(R)-E\Big]\chi_n(R) = \frac{1}{M_R}\sum_k\Big[\frac{1}{2}\langle\phi_n|\Delta_R|\phi_k\rangle+\langle\phi_n|\vec{\nabla}_R|\phi_k\rangle\cdot\vec{\nabla}_R\Big]\chi_k(R) \tag{2}$$

In the first approximation, neglecting the coupling terms on the right hand side, the equation tells us how the energy of the three body system is shared between the muonic energy and relative nuclear motion. Since the total energy E of the three body system is fixed, any change in the two center binding energy $\epsilon_n(R)$ of the muon is reflected in the energy available for the relative nuclear motion; in the semiclassical approximation, we find:

$$T_{nuc}(R) = E - \epsilon_n(R) - \frac{Ze^2}{R} \tag{3}$$

In particular, when there are states ϕ_n that are more strongly bound than the adiabatic state ϕ_{n_0}, any component of the wave function found there is associated with a higher energy in the relative nuclear motion. This effect is most pronounced for the molecular ground state ie. the $1s\sigma$ state. For example, for the adiabatic state $3d\sigma$ of the Lidμ molecule, this results in a virtual energy gain of 40 keV in the relative nuclear motion, which increases the fusion rate dramatically.

The coupling terms on the r.h.s. of equation (2), which we have neglected so far, are responsible for the population of the nonadiabatic components $n \neq n_0$. As shown by Vinitskiĭ *et al* [10], the admixture of nonadiabatic components can be quite large for the loosely bound muomolecular states, contributing decisively to their binding energy. More importantly, the calculation of Bogdanova [11] has shown that for the $dt\mu$ molecule, the nonadiabatic components form a substantial part of the full wavefunction in the classically forbidden region $R \to 0$ (where $T_{nuc} < 0$), even if the wavefunction is almost totally adiabatic in the classically allowed range of R. Since the reason for this behavior is very general (that the couplings involve terms of the form $\vec{\nabla}_R \chi_k(R)$, which become large when the wavefunction is exponentially damped penetrating the Coulombic barrier), appreciable nonadiabatic components must be expected in any muomolecule in the region $R \to 0$.

We now quantitatively consider the probability of barrier penetration for the ground state $n = 1$ component. In the semi-classical nonadiabatic approximation, the tunnelling probability is given by:

$$D = 2\pi\eta e^{-2\pi\eta}, \quad \eta = \frac{1}{\pi}\int_{R_0}^{R_1} \sqrt{2\mu(V(R) - E^*)}dR \tag{4}$$

Here $V(R)$ is the H-Z barrier potential, and $E^* = E - \epsilon_n(R) + \epsilon_H$ is the energy of the nuclear system in the nth adiabatic state, with ϵ_H being the muon binding energy in hydrogen. μ is the reduced mass of Z-H, and R_0, R_1 form the boundary of the classically forbidden region. We approximate E^* by making it equal to the effective binding energy of the muon relative to the asymptotic Z-$(H\mu)_{1s}$ state as $R \to 0$:

$$E^* = R_\mu^\infty\left[\frac{(Z+\epsilon)^2}{n^2} - 1\right], \quad R_\mu^\infty = \frac{1}{2}\alpha^2 m_\mu = 2.8 \text{ KeV} \tag{5}$$

$Z + \epsilon$ $(0 \leq \epsilon \leq 1)$ is the effective charge seen by the muon throughout the fusion process, and m_μ is the muonic mass. We again emphasize that it is the energy in the relative motion of the nuclei in the fusing component of the wave function that drives the fusion, this energy arising from the large virtual Coulomb binding energy of the muon in the fusing component.

In the classically forbidden region, the Z-H potential barrier is a repulsive potential $V(r) = \alpha Z/R$, for which η is:

$$\eta = \frac{1}{\pi}\int_{V_0}^{E^*} \sqrt{(2\mu(V - E^*))}\,\frac{dV}{V^2} = \frac{2}{\pi}d\left(\frac{V_0}{E^*}\right)\sqrt{\frac{\mu}{m_\mu}}\sqrt{\frac{Z^2}{\frac{(Z+\epsilon)^2}{n^2} - 1}} \tag{6}$$

where

$$d(u) = -\frac{\sqrt{u-1}}{u} + tan^{-1}\sqrt{u-1}, \quad d(\infty) = \frac{\pi}{2}. \tag{7}$$

Equation (6) confirms that fusion will occur mostly from the $n = 1$ state, so we shall set $n = 1$ from here on (although the size of the component of the muon in the fusing state may favor a state of higher n, for large Z). The most pessimistic penetration probability for the particular case of $Z - p$ fusion ($\epsilon = 0$, for $Z \gg 1$) is then $D \simeq 10^{-8}$. We should therefore not at all confine our interest to the case $Z = 1$. We also observe that the reduced mass and finite size effects become increasingly significant with increasing Z. For $Z \geq 3$, the reduced mass of Z−H for different hydrogen isotopes can affect the penetration probability by between two to five orders of magnitude. The finite size of the fusing nuclei also has a significant impact on the penetration probability (figure 1).

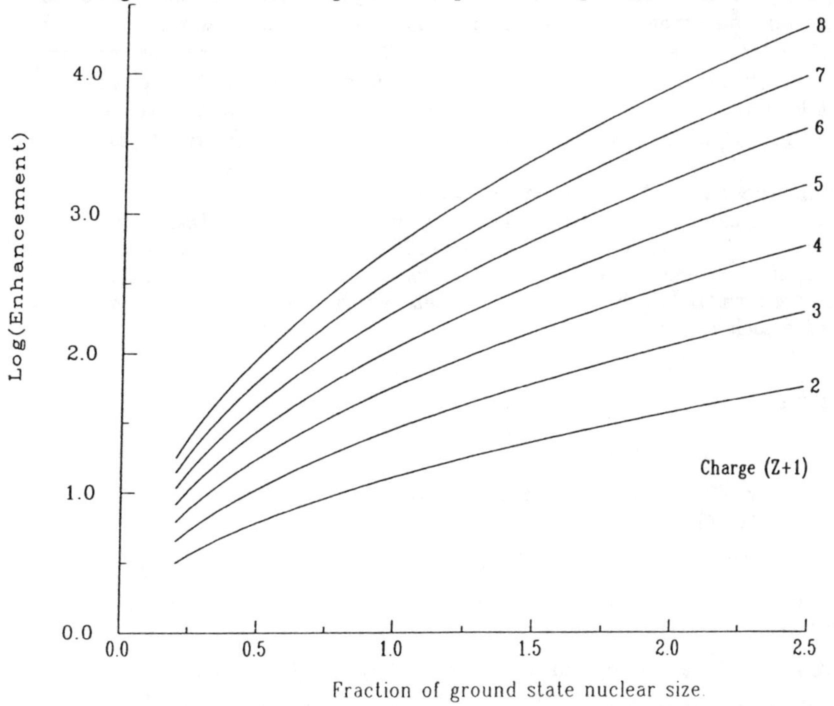

Figure 1: Fractional enhancement defined as $D(V_0)/D(V_0 = \infty)$ (orders of magnitude) in the fusion rate due to finite nuclear size for differing values of Z. The nuclear size is given as a fraction of the estimated ground state nuclear size.

The semiclassical fusion rate is obtained by multiplying the barrier penetration probability by the characteristic frequency of the nuclear oscillator. Taking a typical value $\nu \simeq 10^{18}$ s^{-1}, we can expect overall fusion rates from the fusing state of the order of 10^{10} s^{-1}. It should however be emphasized that for $Z > 1$ one can expect a substantial reduction of the amplitude of the wavefunction in the lowest molecular state $n = 1s\sigma$, which has not been accounted for in the above discussion. Nevertheless one can gain considerably from the greater nuclear size, and from the reduced mass effect for p-fusion in high-Z reactions.

4. CONCLUSION

The above qualitative discussion of high-Z systems leads us to the following conclusions:

a) The molecular resonance formation rate could conceivably be high enough to allow a significant number of fusions per muon, even for $C_Z \ll 1$.

b) The Coulomb barrier penetration is probably not as strongly suppressed as has been predicted in the adiabatic approximation. The strong virtual binding of the muon amplitude in the nonadiabatic fusing state serves to compensate the increased repulsion between the nuclei as Z is increased.

c) Nuclear finite size and isotopic reduced mass effects play a significant role in determining the fusion rate.

At present, there is still considerable uncertainty regarding the size of the nonadiabatic fusing component of the wave function in a molecular resonance, and the influence of the nuclear system on the fusion rate. The first can in principle be established by a calculation of the Coulomb three body resonance states. The nonadiabatic component leading to fusion is required to be large at $R \to 0$, where it is needed to permit tunnelling, but insignificant at $R \to \infty$, where it would give rise to molecular scavenging.

In conclusion, we believe we have shown that muon catalyzed fusion involving nuclei with $Z > 1$ cannot be totally ignored, and although extremely favorable circumstances are required to obtain fusion rates above 10^9 s^{-1} , an experimental investigation into the possibility of such fusion events is certainly justified. We are presently investigating the reactions ^{7}Li+$p\mu$, ^{7}Li+$d\mu$ and ^{10}B+$p\mu$, which we believe may be promising candidates for such studies.

References

[1] F.C. Frank, Nature 160(1947)525

[2] A.V. Kravtsov, N.P. Popov and G.E. Solyakin, JETP Lett. 40(1984)875 [Pis'ma Zh. Eksp. Teor. Fiz. 40(1984)124]

[3] D. Harley, B. Müller and J. Rafelski, in preparation.

[4] A.V. Kravtsov, N.P. Popov and G.E. Solyakin, Sov. J. Nuc. Phys. 35(1982)876 [Yad. Fiz. 35(1982)1496]

[5] A.V. Kravtsov, A.I. Mikhailov and N.P. Popov, J. Phys B 19(1986)1323

[6] K. Helfrich and H. Hartmann, Theoret. Chim. Acta 16(1970)263

[7] T. Matsuzaki, K. Ishida and K. Nagamine, Muon Catalyzed Fusion 2(1988)217

[8] A.V. Kravtsov, N.P.Popov, G.E. Solyakin, Yu.A. Aristov, M.P. Faifman and N.F. Truskova, Phys. Lett. 83A(1981)379

[9] S.S. Gerstein, Sov. Phys. JETP 16(1963)501 [J. Exptl. Theoret. Phys. (USSR) 43(1962)706

[10] S.I. Vinitskiǐ, V.S. Melezhik, L.I. Ponomarev, I.V. Puzynin, T.P. Puzynina, L.N. Somov and N.F. Truskova, Sov. Phys. JETP 52(1980)353 [Zh. Eksp. Teor. Fiz. 79(1980)698]

[11] L.N. Bogdanova, Muon Catalyzed Fusion 3(1988)359

Chapter 3
Properties of Muomolecules

A COMPARISON OF MUONIC MOLECULAR CALCULATIONS

S.A. Alexander and H.J. Monkhorst
Department of Physics, University of Florida
Gainesville, Florida 32611

K. Szalewicz
Department of Physics, University of Delaware
Newark, Delaware 19716

Abstract. For the muonic molecular ions $xy\mu$ (x,y=p,d,t) we show that the difference in the binding energies of two calculations with different sets of particle masses converges quickly to a constant. Using these constants we discuss the relative efficiency of the different basis sets which have appeared in the literature.

1. Introduction

In order to better understand how the process of muon-catalyzed fusion occurs, a large effort has been devoted to accurately computing the energy levels and associated wavefunctions of the muonic molecular ions $xy\mu$ (x,y=p,d,t) [1-26]. Although the energy of each system depends on the masses of the particles, several different estimates of the masses have been used in these calculations. In most cases these differences influence the final binding energies by less than 0.001 eV. As the accuracy of these calculations has increased, however, the proliferation of mass sets has made it difficult to directly compare the advantages and disadvantages of the different types of basis sets which have been used. In an earlier paper [25] we noticed that the difference in the binding energies of two tdμ calculations with different tritium masses converges to a constant much faster than do the binding energies themselves. This is because the Hamiltonian for one mass set can be written as a similar Hamiltonian for another mass set plus an additional term. Diagonalizing these two slightly different Hamiltonians using the same basis set and then subtracting the results is thus almost the same as evaluating a small perturbation. In this paper we compute the binding energy conversion constants for the three most popular sets of masses in use today. These particle masses are listed in Table 1. With these conversion constants we can then make an accurate comparison of the quality of several recent variational calculations.

2. Calculations

Table 2 shows that the conversion constants converge quite rapidly. We illustrate this point using a state whose binding energy converges very quickly, ttμ(00), and a state whose binding energy converges slowly, tdμ(11). For simplicity, we shall refer to each bound state as xyμ(mn) where the label m denotes the angular-momentum (0=S,1=P) and n specifies whether the state is in the ground state (n=0) or the first excited state (n=1). In these calculations we use basis sets of explicitly correlated Slater geminals. The nonlinear parameters are chosen using the random tempering method described in Ref. 25. The computed conversion constants for all the muonic systems are given in Table 3. These values show that even a relatively small change in the masses can influence the binding energies by as much as 0.0005 eV.

In Table 4 we summarize a number of muonic molecular calculations. This list includes the binding energy obtained by each study, the type of wavefunction, the number of basis functions and the set of particle masses used. In a few cases the particle masses differ slightly from those found in Table 1. These exceptions are clearly marked. Although Table 4 is arranged in chronological order, it should be noted that some of the earlier values are clearly more accurate than some of the later ones (compare, for example, the tdμ(00) calculations of Ref. 18 and 22). All of the entries in Table

4 are taken directly from their proper reference; no attempt has been made to compute the binding energy in a consistent manner. Since the binding energy depends on both the Rydberg constant and the reduced mass, the precision with which this calculation is computed can effect the final value. In some studies, e.g. Refs. 18,25 and 26, the muonic energy is truncated to the number of significant digits. In other studies, e.g. Ref. 10, machine precision is used. In most cases, however, the difference is only about 1 μeV.

In many ways Table 4 is similar to the one assembled by Puzynin and Vinitsky [27]. Our list includes several more recent calculations but is limited to published variational studies. This criteria unfortunately excludes a large number of results which appear only in various internal reports [28] as well as many important adiabatic and perturbation calculations which have helped to renew theoretical and experimental interest in muon-catalyzed fusion [3,4]. Since the particle masses are very different from those in use today, we have also omitted the early variational studies of these ions [1,2]. Obviously our goal is not to assemble a complete list of binding energy calculations. Instead, we wish to use this list to examine the quality of different wavefunctions. We feel this objective can be most easily accomplished using this particular set of calculations.

In our study of the various wavefunctions we shall look at both the convergence of the binding energy and at the number of basis functions needed to obtain a particular value. One way to measure convergence is by increasing the basis set in some systematic manner. Because we have restricted our study to variational calculations, the energy must always decrease. By comparing the size of this decrease with the size of the increase in the number of basis functions, we can get a rough idea of the convergence. Another way to estimate convergence is by performing an extrapolation. This technique, however, does not appear to be very reliable. Often the converged results which had been predicted by extrapolation were found to be quite different from those found directly. Using the 'direct' method of estimating convergence, all of the muonic ions in Table 4 appear to have converged to about 1 μeV with at least one type of wavefunction. Because differences in the particle masses can change the binding energies by much more than this amount, a knowledge of the proper conversion constants is essential for any sort of accurate comparison.

As a point of reference, we add the conversion constants to the binding energies in Ref. 25 and 26. These calculations use a basis set of Slater geminals. When these results are compared with calculations using a generalized Hylleraas basis [8,12,18,20], we see, as expected, very little difference in the converged binding energies. For some of these states the convergence of the Slater geminal basis is slower, tdμ(00) and tdμ(10), for some states it is slightly faster, tdμ(01), and for others much faster, tdμ(11). For other properties such as the sticking fraction, however, the generalized Hylleraas basis seems to converge much more rapidly [26].

When these two methods are compared with a Gaussian basis [23,24], many more differences are apparent. Because he uses Jacobi coordinates, Kamimura's calculations are much less linearly dependent than either Slater geminal or generalized Hylleraas calculations. Because of this, the Gaussian calculations could be done with 64 bit precision whereas the other calculations had to be done in 128 bit precision (double precision on the ETA-10 and CYBER 205 and extended precision on the IBM 3090). This difference significantly decreases the amount of CPU time used for such calculations.

In comparison to these other methods, the molecular type wavefunctions used in Refs. 10,14,17 and 22 seem slightly less accurate. One reason may be that the largest of these calculations used the method of "dimensional regularization" to handle the problem of linear dependence in the diagonalization. This procedure shifts the diagonal of the overlap matrix by some small constant value. In both Refs. 14 and 17 the accuracy of this method is estimated to be about 0.0001 eV.

The values in Table 4 contain a few anomalously low results. These entries are lower than the presumably converged binding energies in Ref. 25. The first is the ppμ(00) calculation of Frolov [11]. This calculation, however, used a muon mass of 206.76864 m_e which may account for the discrepancy. The second and third anomalies are the dpμ(10) and tpμ(10) calculations of Frolov [16]. Here the masses are exactly equal to those in mass set #3 so this cannot be the problem. Although Frolov's results were the first to achieve accurate estimates of the binding energies of muonic ions, in Ref. 25 we showed that several discrepancies exist between his published results and his description of their calculation. The cause of this disagreement could not be determined and is probably also responsible for the discrepancy here. The final anomaly is the ttμ(10) calculation of Vinitsky et al [14]. This disagreement could be due to their use of dimensional regularization. Since only the values of the overlap matrix are changed, the results need no longer be variational. The size of the difference between Vinitsky's result and the value in Ref. 25, about 0.0001 eV, is consistent with this view.

Although many of the entries in Table 4 use a basis of Slater geminals, the nonlinear parameters in these wavefunctions have been chosen by a number of different tempering schemes. Each scheme is a method of producing a large number of nonlinear parameters from a small set of optimized 'tempering' parameters. For a particular system, however, one method of tempering may produce converged results faster than another. In Refs. 25 and 26 we used a tempering scheme similar to one examined by Poshusta [29]. This method produced very accurate binding energies but some states required a large number of basis functions. When compared to the tempering schemes used by Frolov and by Petelenz and Smith, our method seems to converge slightly slower for most states. The reason for these differences lies in the distribution of the exponents. These can be said to fall in the range [0,Y,X] where X and Y are values determined by the tempering formulas. In Refs. 25 and 26 our method heavily samples the interval [0,Y]. In most of his calculations Frolov uses a tempering scheme similar to the one developed by Thakkar and Smith [30]. This method places most of the exponents in the interval [Y,X]. In contrast, Petelenz and Smith used an exponential distribution which sampled the region [0,X] more or less evenly. In most muonic systems very small exponents do not appear to contribute substantially to the energy. Any method which heavily samples this region must thus converge slowly. For the very weakly bound dtμ(11) and ddμ(11) states, however, small exponents seem to be essential. For this reason the binding energies in Ref. 25 and 26 converged faster than any other published study.

3. Conclusions

We have computed the differences in the binding energies for all the muonic molecules xyμ where x,y = p,d,t. The accuracy of these conversion constants is about 1 μeV. Since these constants do not require large calculations, we suggest that any future calculations which use different masses include conversion constants so a comparison can be easily made with the literature.

Acknowledgements

We would like to thank Dr. V.H. Smith Jr. and Dr. L.I. Ponomarev for bringing several binding energy references to our attention. In addition, we gratefully acknowledge a grant of computer time from Florida State University on their CYBER 205. We also wish to thank the staff of the Northeast Regional Data Center for their generous support and help in running our program on the University of Florida IBM 3090. This work has been supported by Control Data with a PACER fellowship and by the Division of Advanced Energy Projects at the Department of Energy

References

1. B.P Carter, Phys. Rev. A 141,863 (1966) and references therein.
2. L.M. Delves and T. Kalotas, Aust. J. Phys. 21,1 (1968) and references therein.
3. S.I. Vinitsky, L.I. Ponomarev, I.V. Puzynin, T.P. Puzynina, L.N. Somov and M.P. Faifman, Zh. Eksp. Teor. Fiz. 74,849 (1978) [Sov. Phys. JETP 47, 444 (1979)].
4. A.D. Gocheva, V.V. Gusev, V.S. Melezhik, L.I. Ponomarev, I.V. Puzynin, T.P. Puzynina, L.N. Somov and S.I. Vinitsky, Phys. Lett. B 153,349 (1985).
5. A.M. Frolov and V.D. Efros, Zh. Eksp. Teor. Fiz. 39,449 (1984) [Sov. Phys. JETP Lett. 39,544 (1984)].
6. A.K. Bhatia and R.J. Drachman, Phys. Rev. A 30,2138 (1984).
7. A.M. Frolov and V.D. Efros, J. Phys. B 18,L265 (1985).

8. C.Y. Hu, Phys. Rev. A 32,1245 (1985)
9. A.M. Frolov and V.D. Efros, Yad. Fiz. 41,828 (1985) [Sov. J. Nuc. Phys. 41,528 (1985)].
10. S. Hara, T. Ishihara and N. Toshima, J. Phys. Soc. Japan 55,3293 (1986).
11. A.M. Frolov, Zeit. fur Phys. D 2,61 (1986).
12. C.Y. Hu, Phys. Rev. A 34,2536 (1986).
13. V.D. Efros, Zh. Eksp. Teor. Fiz. 90,10 (1986) [Sov. Phys. JETP 63,5 (1986)].
14. S.I. Vinitsky, V.I. Korobov and I.V. Puzynin, Zh. Eksp. Teor. Fiz. 91,705 (1986) [Sov. Phys. JETP 64,417 (1986)].
15. A.M. Frolov, Yad. Fiz. 44,589 (1986) [Sov. J. Nuc. Phys. 44,380 (1986)].
16. A.M. Frolov, Zh. Eksp. Teor. Fiz. 92,1959 (1987) [Sov. Phys. JETP 65,1100 (1987)].
17. V.I. Korobov, I.V. Puzynin and S.I. Vinitsky, Phys. Lett. B 196,272 (1987).
18. K. Szalewicz, W. Kolos, H.J. Monkhorst and A. Scrinzi, Phys. Rev. A 35,965 (1987).
19. P. Petelenz and V.H. Smith Jr., Phys. Rev. A 36,4078 (1987).
20. C.Y. Hu, Phys. Rev. A 36,4135 (1987)
21. C.Y. Hu, Phys. Rev. A 36,5420 (1987).
22. S. Hara, T. Ishihara and N. Toshima, Muon Catalyzed Fusion 1,277 (1987).
23. M. Kamimura, Muon Catalyzed Fusion 3,335 (1988).
24. K. Kamimura, "Non-Adiabatic Coupled-Rearrangement-Channels Approach to Muonic Molecules" accepted Phys. Rev. A.
25. S.A. Alexander and H.J. Monkhorst, "High Accuracy Calculation of Muonic Molecules Using Random Tempered Basis Sets" accepted Phys. Rev. A.
26. S.E. Haywood, H.J. Monkhorst and S.A. Alexander, "Sticking Fraction Calculations of tdμ and ddμ Using Random Tempered Basis Sets" submitted Phys. Rev. A.
27. I.V. Puzynin and S.I. Vinitsky, Muon Catalyzed Fusion 3,307 (1988).
28. Many of these are cited in Ref. 27.
29. R.D. Poshusta, Int. J. Quantum Chem. 13,59 (1979).

30. A.J. Thakkar and V.H. Smith Jr., Phys. Rev. A 15,1 (1977); 1,16 (1977); 1,2143 (1977).

Table 1. The three most commonly used sets of constants in muonic binding energy calculations. The particle masses are in m_e and the Rydberg constant is in eV.

mass set	#1	#2	#3
m_t =	5496.918	5496.899	5496.918
m_d =	3670.481	3670.481	3670.481
m_p =	1836.1515	1836.1515	1836.1515
m_μ =	206.7686	206.7686	206.769
Ry =	13.6058041	13.6058041	13.6058041

Table 2. Comparison of the convergence of the difference in the binding energies for the ttμ(00) and tdμ(11) states. K is the number of basis functions used. BE(x) is the binding energy computed with mass set #x. Δ12 = BE(1)-BE(2) and Δ13 = BE(1)-BE(3). Values are in eV.

ttμ(00) state.

K	BE(1)	BE(2)	BE(3)	Δ12	Δ13
400	362.909773	362.909466	362.910304	0.000307	-0.000531
300	362.909773	362.909465	362.910303	0.000308	-0.000530
200	362.909762	362.909455	362.910292	0.000307	-0.000530
100	362.907302	362.906994	362.907832	0.000308	-0.000530
50	362.851059	362.850752	362.851590	0.000306	-0.000531

tdμ(11) state.

K	BE(1)	BE(2)	BE(3)	Δ12	Δ13
600	0.655734	0.655709	0.655589	0.000025	0.000145
500	0.647577	0.647551	0.647431	0.000026	0.000146
400	0.635979	0.635954	0.635834	0.000025	0.000145
300	0.562898	0.562873	0.562753	0.000025	0.000145
200	-0.003916	-0.003937	-0.004043	0.000021	0.000127

Table 3. Converged binding energies differences for all of the muonic ions xyμ. Δ12 = BE(1) - BE(2) and Δ13 = BE(1) - BE(3) where BE(x) is the binding energy computed with mass set #x. Values are in eV.

System	Δ12	Δ13	System	Δ12	Δ13
ppμ(00)	0.000000	-0.000289	ppμ(10)	0.000000	0.000124
dpμ(00)	0.000000	-0.000134	dpμ(10)	0.000000	0.000227
tpμ(00)	-0.000048	-0.000095	tpμ(10)	0.000020	0.000241
ddμ(00)	0.000000	-0.000439	ddμ(10)	0.000000	-0.000121
ddμ(01)	0.000000	0.000129	ddμ(11)	0.000000	0.000116
tdμ(00)	-0.000031	-0.000395	tdμ(10)	0.000056	-0.000109
tdμ(01)	0.000026	0.000183	tdμ(11)	0.000025	0.000145
ttμ(00)	0.000307	-0.000531	ttμ(10)	0.000295	-0.000076
ttμ(01)	0.000452	0.000091	ttμ(11)	0.000470	0.000175

Table 4. Comparison of the binding energies (in eV) of several variational calculations on the muonic ions $xy\mu$ (x,y=p,d,t). The letter under the reference column indicates the type of calculation performed, the number indicates the set of masses used and the value in parentheses is the year of publication.

ppu(00)	ppu(10)	dpu(00)	dpu(10)	tpu(00)	tpu(10)	Reference
253.15240 K=140	107.26563 K=250	221.54923 K=250	97.42990 K=250	213.83975 K=250	99.10104 K=250	A3 (1984) Frolov [5]
253.152 K=125	107.266 K=286	221.54 K=440	97.493 K=440	213.829 K=440	99.119 K=440	B2 (1984) Bhatia [6]
253.146 K=300	107.203 K=300					C2 (1986) * Hara [10]
253.152360 253.152613 K=225	107.265870 107.265757 K=300					A1 (1986) ** A3 Frolov [11]
				213.83992 K=300	99.11493 K=350	A3 (1986) Efros [13]
	107.26568 K=438		97.49774 K=698		99.12608 K=698	C3 (1986) Vinitsky [14]
253.152615 K=300	107.26583 K=350					A3 (1986) Frolov [15]
		221.54593 K=350	97.49822 K=450	213.84026 K=350	99.12632 K=450	A3 (1987) Frolov [16]
253.15230 K=140		221.54797 K=140		213.83882 K=140		A2 (1987) Petelenz [19]
253.152 K=300	107.203 K=300					C2 (1987) Hara [22]
253.152332 253.152332 253.152621 K=500	107.265971 107.265971 107.265847 K=800	221.549410 221.549410 221.549544 K=800	97.498160 97.498160 97.497933 K=1400	213.840179 213.840227 213.840274 K=1000	99.126501 99.126481 99.126260 K=1400	A1 (1988) A2 A3 Alexander [25]

ddu(00)	ddu(01)	ddu(10)	ddu(11)	Reference
325.07213 K=140	35.83658 K=140	226.68157 K=140	1.78123 K=250	A3 (1984) Frolov [5]
325.070 K=203	35.815 K=161	226.662 K=286	1.862 K=364	B2 (1984) Bhatia [6]
		226.680528 K=350	1.971121 K=350	A3 (1985) Frolov [7]
		226.68053 K=350	1.96988 K=350	A3 (1985) Frolov [9]
325.073 K=300	35.843 K=300	226.670 K=300	1.955 K=380	C2 (1986) * Hara [10]
		226.68053 K=438	1.97465 K=1286	C3 (1986) Vinitsky [14]
325.07398 K=300	35.84421 K=300	226.68176 K=350		A3 (1986) Frolov [15]
			1.97475 1.97465 K=1286	C1 (1986) C3 Korobov [17]
325.07352 K=140	35.84327 K=140			A2 (1987) Petelenz [19]
		226.68175 K=728	1.97453 K=916	D3 (1987) Hu [21]
325.073 K=300	35.843 K=300	226.670 K=300	1.955 K=380	C2 (1987) * Hara [22]
325.073540 325.073540 325.073979 K=500	35.844360 35.844360 35.844231 K=500	226.681678 226.681678 226.681799 K=800	1.974817 1.974817 1.974701 K=1200	A1 (1988) A2 A3 Alexander [25]

ttu(00)	ttu(01)	ttu(10)	ttu(11)	Reference
362.90040 K=140	83.72202 K=140	289.13931 K=250	45.19359 K=250	A3 (1984) Frolov [5]
362.900 K=203	83.630 K=161	289.12 K=286	45.096 K=286	B2 (1984) Bhatia [6]
362.909 K=300	83.770 K=300	289.137 K=300	45.198 K=300	C2 (1986) * Hara [10]
		289.14195 K=438	45.20552 K=607	C3 (1986) Vinitsky [14]
362.91029 K=300	83.77109 K=300	289.1419 K=350	45.2057 K=350	A3 (1986) Frolov [15]
362.90945 K=140	83.76996 K=140			A2 (1987) Petelenz [19]
362.909 K=300	83.770 K=300	289.137 K=300	45.198 K=300	C2 (1987) * Hara [22]
362.909770 362.909463 362.910301 K=500	83.771216 83.770764 83.771125 K=500	289.141783 289.141488 289.141859 K=600	45.205856 45.205386 45.205681 K=1200	A1 (1988) A2 A3 Alexander [25]

A. Slater geminal wavefunction.
B. Hylleraas wavefunction.
C. Molecular wavefunction in spheroidal coordinates.
D. Generalized Hylleraas wavefunction.
E. Gaussian wavefunction in Jacobi coordinates.

n.a. Not available.
* The proton mass in these calculations is m_p=1815.151 m_e.
** The muon mass in these calculations is m_μ=206.76864 m_e.

tdu(00)	tdu(01)	tdu(10)	tdu(11)	Reference
318.31805 K=250	34.82381 K=250	232.42049 K=250	0.5231 K=375	A3 (1984) Frolov [5]
319.062 K=440	34.573 K=440	232.416 K=440	0.224 K=440	B2 (1985) Bhatia [6]
		232.447561 K=350	0.607189 K=400	A3 (1985) Frolov [7]
319.117 K=380	34.776 K=380	232.436 K=300	0.628 K=500	D2 (1985) Hu [8]
319.13419 K=500		232.46867 K=740		D2 (1985) Hu [12]
		232.47155 K=698	0.65889 K=1495	C3 (1986) Vinitsky [14]
319.14007 K=400	34.83404 K=400	232.47157 K=525		A1 (1987) Frolov [16]
			0.65968 K=2084	C3 (1987) Korobov [17]
319.139752 K=1158	34.834465 K=1995	232.471537 K=1072	0.66001 K=3063	D2 (1987) Szalewicz [18]
319.13747 K=140	34.75356 K=140			A2 (1987) Petelenz [19]
319.14010 K=695	34.83332 K=780		0.658025 K=1102	D3 (1987) Hu [20]
319.139 K=398	34.830 K=398			C2 (1987) * Hara [22]
319.1395 n.a.	34.8336 n.a.	232.47112 K=1696	0.65933 K=1696	E1 (1988) Kamimura [23]
319.139606 K=1442	34.834372 K=1442	232.471506 K=2662	0.660104 K=2662	E1 (1988) Kamimura [24]
319.139722 319.139753 319.140117 K=1400	34.834491 34.834465 34.834308 K=1400	232.471594 232.471538 232.471703 K=1600	0.660172 0.660147 0.660027 K=2000	A1 (1988) A2 A3 Alexander [25]
			0.660178 0.660153 0.660033 K=2600	A1 (1988) A2 A3 Haywood [26]

SOME PROPERTIES OF THREE BODY RESONANCES OF dtμ RELATED TO MUON-CATALYZED FUSION

P. Froelich
Department of Quantum Chemistry
Uppsala University
Box 518, 75120 Uppsala, Sweden

K. Szalewicz
Department of Physics
University of Delaware
Newark, Delaware 19716, U.S.A.

H.J. Monkhorst
Quantum Theory Project
University of Florida
Gainesville, Florida 32611, U.S.A.

W. Kolos and B. Jeziorski
Warsaw University
Pasteura 1, Warsaw, Poland

ABSTRACT

The recently discovered resonant states in the dtμ molecule are investigated, with emphasis on the properties related to the muon catalyzed fusion. For the lowest resonance of s-symmetry above the (dμ+t) threshold we have calculated the energy, life time, geometry, fusion rate, sticking fraction and collisional formation rate. The results are compared with the corresponding properties for the bound states. Also, estimations of muon cycling rates and the fusion yield have been obtained for the situation corresponding to resonantly enhanced fusion in flight.

I. INTRODUCTION

The exotic molecule dtμ has been known to possess only five bound states [1]. In a recent publication we have announced the existence of <u>three -body resonances</u> of dtμ above the (dμ+t) threshold, which complement the <u>bound states</u> of dtμ [2]. These resonances are of Feshbach type, i.e. they exist only due to the three-body correlational effects, and can not be observed in the Born-Oppenheimer approximation. So far we have found two such resonances: one of p and the other of s-symmetry, both above the (dμ+t) threshold. The exact energies and widths of these resonances are given in Table 1. However we can not preclude the possibility of yet lower resonances, below the (dμ+t) threshold, where only one fragmentation mode is possible. In fact, we have seen an indication of one such resonance of s-symmetry at 5.4 eV above the (tμ+d) threshold.

<u>Table 1</u>

Resonance energies and widths for tdμ

Symmetry	Resonance energy (eV) above tμ	above dμ	Width (eV)	Basis[1)]
s	54.35	6.31	0.74	(11,11,16,18)≡1158
p	54.63	6.59	2.04	(8,8,12,16)≡1204

1) The notation for the basis is (k,l,m,ω) where k,l,m are maximum powers used in the basis set [2], and ω is the maximum of $k_i + l_i + m_i$.

In the present work we will concentrate on the states of s-symmetry, because fusion in such states is much more rapid than in states of p-symmetry due to the absence of the centrifugal force, and because their parity matches the parity of the $J^{\pi}=3/2^{+}$ nuclear resonance. The (Coulombic) resonances were calculated by means of the analytical continuation of the stabilization graphs; we refere for details to ref. 2. The lower part of the discretized continuous spectrum of dtμ together with the stabilization graphs are presented on figures 1 and 2.

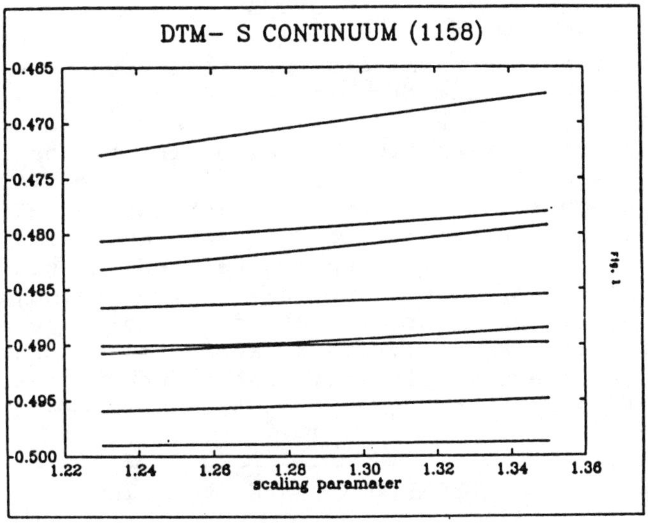

Fig. 1. Discretized continuum of dtμ as function of scaling parameter (s-symmetry, obtained with 1158 basis functions).

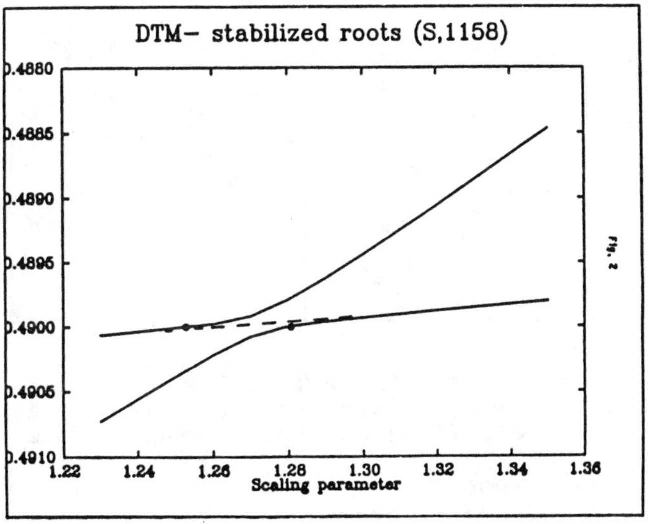

Fig. 2 Avoided crossing of stabilized roots (enlargement from fig. 1)

II. GEOMETRY

We started the analysis of the resonant states by determining their geometry. The expectation values of the distances d-t, d-μ and t-μ have been calculated with respect to the wavefunction corresponding to the resonant energy E_r, see fig 3. This calculation has confirmed the quasi-bound character of the resonant state. The meta-stable molecule turned out to have an elongated molecular shape, with the muon situated in between the nuclei. For the sake of comparison, we have also calculated the geometry of the bound states of dtμ (there are only two such s-states in dtμ). This analysis indicated that disregarding the overall size, the resonant state is not so much different from the two bound states below (see fig 3).

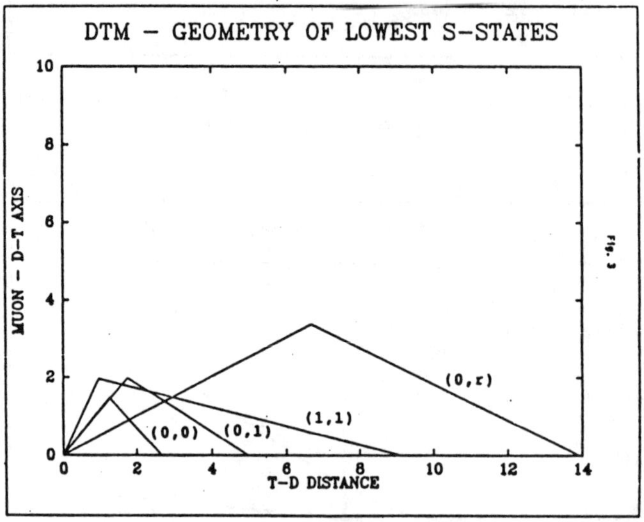

Fig. 3. Geometry of the three lowest states of s-symmetry in dtμ. From smallest to largest are: ground state (0,0), excited bound state (0,1) and the resonant state (0,r).

Thus the resonance can be regarded as a consecutive (second) excited state of the s-symmetry series. It is interesting to notice the difference between the resonant state and the (1,1) state of p-symmetry just below the threshold, with the muon situated much closer to the triton than to the deuteron. The exact d-t-μ distances are tabulated in Table 2.

Table 2

Geometry of tdμ in few lowest states

State (J,ν)	Energy	Basis	Distances[2)]		
			d-t	d-μ	tμ
(0,0)	-319.13975216	1158	2.65	2.04	1.95
(0,1)	-34.83446501	1027	4.97	3.79	2.64
(1,1)	-0.6471	1276	9.05	8.33	2.22
(0,r)	54.35	1158	13.92	7.99	7.49

2) Distances are in muonic units of length.

III. FUSION RATE

Because of the quasi-bound nature of the resonant state, it is interesting to know how fast it can fuse. In this respect we recall that the fusion rate is dictated by the efectiveness of the molecular confinement, provided in this case by the muon-induced chemical bonding. It is interesting to note that both the space and time aspects of the confinement are reflected in the resonant wave function. In other words, the resonant wave function contains the information about the life-time and the quasi-bound character of the state.

The expression for fusion rate is usually given in a factorized form [3]. The fusion rate is proportional to the probability of merging the two nuclei within the dtμ molecule, and to the probability of the nuclear reaction itself:

$$\lambda_f = T_0 |f(r_{dt})|^2_{|r_{dt}=0} \qquad (1)$$

where $f(r_{dt})$ is the pseudowavefunction describing the relative motion of the t and d nuclei (in the purely coulombic field) averaged with respect to the position of the muon:

$$f(r_{dt}) = \left[\frac{\langle \psi(r_{dt}, r_{t\mu}, r_{d\mu}) | \psi(r_{dt}, r_{t\mu}, r_{d\mu}) \rangle_{t\mu, d\mu}}{4\pi \ r_{dt}^2} \right]^{1/2} \qquad (2)$$

The bracket $\langle \ | \ \rangle$ in the above expression denotes integration of the total wave function for $dt\mu$ (obtained <u>without</u> nuclear forces) with respect to the muonic coordinates $d\mu$ and $t\mu$. The quantity $|f(0)|^2$ in equation (2) measures the probability of finding d and t at $r_{dt}=0$. It can be interpreted as the (coulombic) enhancement factor for the probability of nuclear reaction, relative to the situation when only nuclear forces are considered.

The proportionality factor T_0 is derived from the transition amplitude for the nuclear reaction itself. We adopt the same value for T_0 as used by Bogdanova <u>et al.</u>[3]

$$T_0 = 1.2 \cdot 10^{-14} \ cm^3 \ sec^{-1} \qquad (3)$$

To obtain the coalescence probability for t and d nuclei, we have first calculated the pseudowavefunction $f(r_{dt})$, describing the relative d-t motion. This function is plotted on fig. 4.

Figure 4.

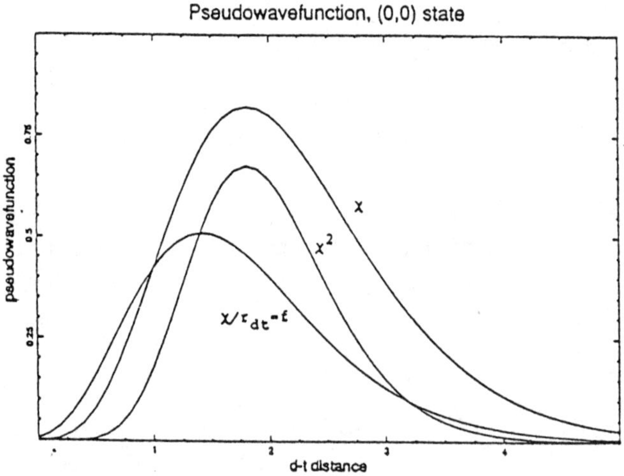

Fig. 4. Pseudowavefunction describing the relative d-t motion in $dt\mu$, averaged with respect to the position of the muon. Its value at $r_{dt} = 0$ gives a Coulombic enhancement factor for the nuclear fusion reaction.

As is easy to see, the pseudowavefunction describes the vibrational d-t motion, and is finite at the $r_{dt}=0$. Its magnitude at the coalescence point is related to the fusion rate, according to equation (1). The fusion rates for the (0,0) and (0,1) bound states are tabulated in Table 3.

To obtain the fusion rate for the resonant state, a more elaborate procedure had to be applied. We have calculated the d-t coincidence probability along the two branches of the stabilization graph, and then continued it analytically to the complex value of the dilation parameter α, corresponding to the resonant energy E_r. This procedure does not require the explicit knowledge of the true resonant wave function. The result is presented in Table 3.

Table 3.

Fusion rates for lowest states of s symmetry

State (J,ν)	Energy[3)]	Basis[4)]	f(0)	Fusion rate (sec^{-1})
(0,0)	-0.55885435	638	$0.9954 \cdot 10^{-3}$	$0.63 \cdot 10^{12}$
(0,1)	-0.50642402	638	$0.9080 \cdot 10^{-3}$	$0.53 \cdot 10^{12}$

3) 1.m.u. = 5422.5347 eV

4) with the cusp condition on r_{dt} imposed.

IV. STICKING FRACTION

With the knowledge that the resonant state can fuse with considerable speed, we have calculated the associated sticking fraction. It is defined as the probability of muon capture to any bound state of muonic ion $\alpha\mu^+$

$$w_s^0 = \sum_f |P_{fi}|^2 \tag{4}$$

where P_{fi} is the probability amplitude of muon transfer from the bound state of $dt\mu$ to the bound state of $\alpha\mu$, under a sudden perturbation. This probability amplitude is given (in sudden impulse approximation) by the overlap between the initial state of $dt\mu$ and the final state of $\alpha\mu$-n system, taken at the point of coalescence of d,t,α and n:

$$P_{fi} = \int \psi_f^*(\bar{r}_{\alpha\mu}) e^{-i\bar{p}_n \cdot \bar{r}_n - i\bar{p}_{\alpha\mu} \cdot \bar{r}_{\alpha\mu}} \psi_i(\bar{r}_{t\mu}, \bar{r}_{td}) \Big|_{\bar{r}_\alpha = \bar{r}_n = \bar{r}_t = \bar{r}_d} d\bar{r}_\mu d\bar{r}_\alpha \tag{5}$$

The application of the above formula to the case of sticking from the bound states of $dt\mu$ is described in reference [4]. Calculation of sticking from the resonant state is based on the same expression, but the procedure had to be extended in view of the fact that instead of using the resonance wave function itself as the initial state, we have rather used the scattering states from its vicinity.

Similarly to what has been done for the case of the fusion rate, the sticking fraction for the resonant state has been extracted from the values obtained for scattering states. We have calculated the sticking fractions along the two energy branches presented on fig. 2. The results of this calculation are presented on fig. 5. We notice, that sticking as a function of energy varies very rapidly in the vicinity of the resonant energy, i.e. at E_r=-0.48998 m.u. This behaviour is not unexpected and can be easily understood in view of the fact that the initial state itself varies very fast around the resonance. In fact, the initial state (being the scattering state) must change phase when its energy passes the resonant energy. This will obviously cause a sudden change of the overlap integral given by eq. (5).

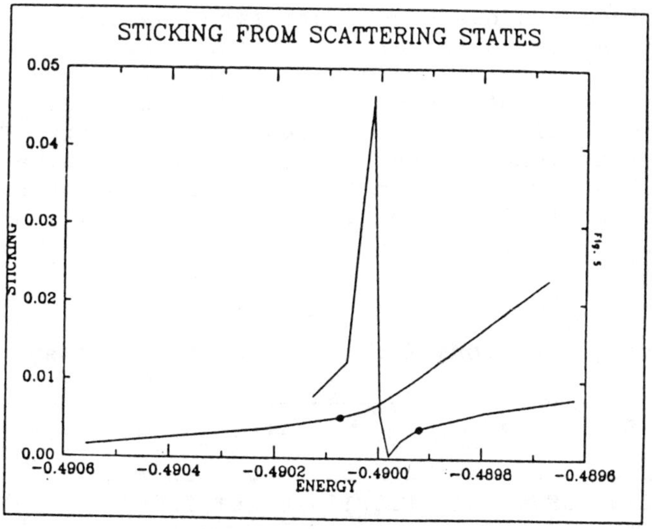

Fig. 5. Sticking fraction as function of the energy of the dtμ molecule, in the vicinity of the resonance. Note the abrupt change of the sticking fraction at the resonant energy. The two branches of sticking correspond to the two branches of the stabilization graph presented on Fig. 2. The two dots indicate points corresponding to the α_r value associated with the resonant energy E_r. Analytical continuation from the vicinity of these points gives sticking value which coincides with the intersection of the two sticking branches. This intersection occurs precisely at the resonant energy.

Incidently, such behavior of the sticking fraction convincingly confirms the very existence of the resonance, and the values obtained for its position and width. Indeed, the most rapid change of the sticking occurs precisely at the resonant energy (as obtained from analytical continuation of the stabilization graphs [2]), see fig. 5. Also, the energy interval at which the rapid change takes place is in full agreement with the value obtained for the width of the resonance (c.f. Table 1). The agreement is so good, that had we not calculated these values previously, they could be inferred now from the sticking calculation alone!

Having sticking fractions from the initial scattering states, we have obtained sticking from the resonant state by means of the analytical continuation.

This continuation has been performed as function of the scaling parameter α rather then the energy, as sticking varies less with α then it does with energy. The continuation has been done from the vicinity of the value α_r, corresponding to the resonance energy E_r. The sticking fraction obtained in this way has value 0.0065, i.e. is slightly less than the corresponding sticking for the (0,0) and (0,1) state (for comparison, all values are tabulated in Table 4).

Table 4.

Sticking fractions for lowest states of s symmetry.

State (J,ν)	Energy, eV	Basis	Sticking fraction (%)
(0,0)	-319.139752161	(11 11 16 18)≡1158	0.886
(0,1)	-34.83446501	(11 11 16 17)≡1127	0.889
(0,r)	54.578	(11 11 16 18)≡1158	0.648

We note, that sticking from the resonant state obtained in a way described above falls nearly precisely at the point of intersection of the two branches of sticking from scattering states, (see fig. 5), which in turn coincides with the energy of the resonance. At this single point, sticking from the two branches corresponds to two different α values: $\alpha_1 \approx 1.25$ for the upper branch, and $\alpha_2 \approx 1.28$ for the lower branch. As can be seen from the stabilization graph (see fig 2), these α values correspond in both cases to the same energy, which turns out to be the energy of the resonance. This provides an internal check of the consistency of the calculation. As a matter of fact, using the results of this calculation as guidance, one could obtain reasonably accurate values of sticking by calculating it with a single value of α parameter, namely with the value which on the stabilization graph corresponds to the real part of the resonant energy, regardless of the branch considered.

In a broader perspective, this conclusion throws some light on the stabilization methodology as such. It helps to find the best approximation to the resonant state, and associates it with α value corresponding to the real part of the resonant energy.

V. FORMATION RATE

Let us now consider the question about the formation rate of the resonant molecule. The resonant state can form directly during the $d\mu$-t or $t\mu$-d collisions, without intervening help from the host D_2 molecule as proposed by Vesman's formation mechanism. The cross section for the resonance formation during inelastic collision is given by[5]

$$\sigma_i(E) = \frac{\pi}{k_i^2(E)} \frac{\Gamma_i \Gamma}{(E-E_r)^2 + (\Gamma/2)^2} \tag{6}$$

where E is the total energy of the colliding system, E_r is the resonant energy, Γ is the total width of the resonance, Γ_i is the partial width for given collision channel, and k_i is the wave number corresponding to the relative kinetic energy E of the colliding fragments in this channel. In the following we will put i=d for the $t\mu$-d channel and i=t for the $d\mu$-t channel.

The resonant formation rate is given by the general expression

$$\lambda^{r,i}_{dt\mu} = \rho \, v_i \, \sigma_i \tag{7}$$

where ρ is the density of the target media (we normalize our results to liquid hydrogen density, $\rho_{LHD}=4.2\ 10^{22}\ cm^{-3}$), and v is the collision velocity. In the case under consideration, since the resonance energy is considerably high (54 eV in the d channel and 6 eV in the t channel), one needs to consider thermal averaging of the collision velocities. This means that, instead of colliding the fragments with sharply determined resonant energy, one allows for thermal spread of the collisional energy - which will obviously lower the formation rate. We assume the velocity weight function given by

$$\omega(E) = 2(E/\pi)^{1/2}\,(kT)^{-3/2}\exp(-E/kT), \quad \int \omega(E)\,dE = 1 \tag{8}$$

so that the formation rate becomes

$$\lambda_{dt\mu}^{r,i} = \int \omega(E) v_i(E) \sigma_i(E) dE \qquad (9)$$

After integration with the explicit form of the cross section given by the equation (6), the above expression reduces to

$$\lambda_{dt\mu}^{r,i} = 2\pi^2 \frac{\Gamma_i}{k_i^2(E_r)} \rho\omega(E_r) v(E_r) \qquad (10)$$

To evaluate the formation rate in each entrance channel, one should know the partial widths Γ_i. Here however, we will be satisfied with the estimation of the upper and lower limit for these formation rates, which can be derived from the knowledge of the total width. Since the resonance under consideration has only two open formation and decay channels, one has

$$\Gamma = \Gamma_t + \Gamma_d \qquad (11)$$

The upper limit for the formation rate in each channel can be easily obtained by assuming $\Gamma = \Gamma_i$; the results are presented in Table 5. Naturally, this limit is highest for the t channel, since the collision velocity in this channel is smallest. The lower limit is common for the two channels. Indeed, we can write the partial contributions to the total width as (c.f. eq. 11)

$$\Gamma = c_t \Gamma + (1 - c_t)\Gamma \qquad (12)$$

and the rates as

$$\lambda_{dt\mu}^{r,t} = \lambda_{dt\mu}^{t,max} c_t \; ; \; \lambda_{dt\mu}^{r,d} = \lambda_{dt\mu}^{d,max} (1 - c_t) \qquad (13)$$

where $\lambda_{dt\mu}^{i,max}$ are the maximal rates in each channel obtained previously. The worst case corresponds to $\lambda_{dt\mu}^{r,t} = \lambda_{dt\mu}^{r,d}$, i.e. to $c_t = \lambda_{dt\mu}^{d,max} / (\lambda_{dt\mu}^{d,max} + \lambda_{dt\mu}^{t,max})$. This lower limit for the formation rate is included in Table 5.

Table 5.

Formation rates of the resonant state

Channel	Formation rates (sec^{-1}) Upper Limit	Lower limit
t	$0.983 \cdot 10^{10}$	$0.369 \cdot 10^{9}$
d	$0.408 \cdot 10^{9}$	

One notices, that even the lower limit for the direct collisional dtμ formation is of the same order of magnitude as the one for Vesman's mechanism $(0.3-0.4 \cdot 10^{9}, ^{-1}$. Since the fusion rate is also considerable, and sticking is smaller than from the bound states, a natural question is whether the direct collisional formation mechanism of dtμ can provide a useful fusion mechanism. In this respect one has to realize that the short life time of the resonant state will have a hampering effect, because the muon will have to be cycled many times before one fusion will be obtained. On the other hand, the fact that the life time of the resonant state is shorter then the fusion time $(0.83 \cdot 10^{-12}$ sec) is not an obstacle in itself, but has to be compensated by a fast formation rate. This situation is different from the one encountered in fusion from the ground state, where the muon can "wait" for the fusion to occur in one molecule for practically indefinite time (2.2 μsec). During fusion in flight, the muon (travelling on d or t nuclei) has to form many diffrent dtμ molecules and in this way "accumulate" the time necessary for fusion. One could say that it behaves like a bee sampling honey and pollinating flowers.

VI. CYCLING RATES AND FUSION YIELD

Let us estimate the cycling rate and fusion yield for this case. Within its life time, the muon can participate in forming N dtμ molecules:

$$N = \tau_0 \cdot \lambda \tag{14}$$

where τ_0 is the muon life time and λ is formation rate. The muon spends in the $dt\mu$ molecule total time $t=\tau \cdot N$, where τ is the life time of the resonant state. Assuming fusion rate $\lambda_f=0.75 \cdot 10^{12}$ sec^{-1} (see Table 3, the exact fusion rate is under investigation), one obtaines fusion time $t_f=1.33 \cdot 10^{-12}$ sec, i.e. on average there may be $n_f=t/t_f$ fusions per muon. This number can be made more accurate by calculating the muon cycling rate (average time between fusions) including sticking probability. The cycling rate is $\lambda_c=n_f/\tau_0$, and the corresponding neutron yield is given by

$$Y = \frac{\lambda}{\lambda_0 + \lambda_c w_s^0} \tag{15}$$

where w_s^0 is the sticking from the resonant state (see Table 4), and $\lambda_0=\tau_0^{-1}$. The upper and lower bounds for neutron yield (corresponding to upper and lower bounds for formation rate from Table 5) are given in Table 6.

Table 6.

Cyclig rates and neutron yield

Formation rate (sec^{-1})	Cycling rate (sec^{-1})	Fusion yield (neutrons/muon)
upper bound $0.983 \cdot 10^{10}$	$0.66 \cdot 10^{7}$	13
Lower bound $0.369 \cdot 10^{9}$	$0.25 \cdot 10^{6}$	1

VII. DISCUSSION

The estimated neutron yield obtained for resonantly enhanced fusion in flight is smaller then the one obtained via Vesman's mechanism, at least for the resonant state which was investigated in this work, with incomplete kinetics, and without the de-excitation mechanisms taken into account. This result is mainly due to the short life time of the resonant state (there are two open decay channels), and to relatively low formation rate (caused by the high energy). The situation would improve if one could cycle the muon in a non-thermalized way. We are presently evaluating different mechanisms which can enhance the cycling rate. Would that be successful, then much better fusion yield could be obtained, particularly since the effective sticking fraction in the tepid plasma [2,7] is practically vanishing [8]. At the plasma temperature corresponding to the resonant collisional energy the stopping power of the free electrons becomes very small, which allows for almost complete reactivation.

We presume, that much higher neutron yield could be obtained for the suspected lower resonant state at 5.4 eV, mentioned earlier. This is because the required (resonant) collision energy is much smaller, and because we expect longer life time for this state which lies below the $d\mu$ + t threshold, and therefore can decay to only one channel. The investigation of this resonant state and its properties is under way.

ACKNOWLEDGMENTS

This work has been supported in parts by grants from the Swedish National Science Research Council (E-EG 4916-104), and from the Division of Advanced Study Projects of the U.S. Department Of Energy (grant DE-FG05-85ER13447).

REFERENCES

[1] See eg. L.I. Ponomarev and G. Fiorentini, Muon Catalyzed Fusion 1, 3, (1987).

[2]. P. Froelich and K. Szalewicz, Muon Catalyzed Fusion 2, ..., 1988.

[3] Bogdanova et. al., Sov. Phys. JETP 54, 442, (1982).

[4] S.E. Haywood, H.J. Monkhorst and K. Szalewicz, to appear in Phys Rev. A.

[5] A.M. Lane, preprint from Theoretical Physics Division, Oxfordshire, England.

[6] M.P. Faifman et. al., Dubna preprint E4-86-541.

[7] P. Froelich and K. Szalewicz, Physics Letters A, in press.

[8] L.I. Menshikov, preprint from the Institut of Atomic Energy IAE-4589/2, Moskva 1988.

THEORETICAL DESCRIPTION OF MUONIC MOLECULAR IONS

Krzysztof Szalewicz[a], Steven A. Alexander[b], Piotr Froelich[c], Susan E. Haywood[b], Bogumil Jeziorski[d], Wlodzimierz Kolos[d], Hendrik J. Monkhorst[b], Armin Scrinzi[a], Cliff Stodden[b], Alka Velenik[b], and Xiaolin Zhao[a]

[a]Department of Physics and Astronomy, University of Delaware, Newark, DE 19716

[b]Quantum Theory Project, Departments of Physics and Chemistry, University of Florida, Gainesville, FL 32611

[c]Department of Quantum Chemistry, Uppsala University, Box 518, 75120 Uppsala, Sweden

[d]Quantum Chemistry Laboratory, Department of Chemistry, University of Warsaw, Pasteura 1, 02093 Warsaw, Poland

ABSTRACT

An accurate theoretical description of the muonic molecular ions is critical for understanding the phenomenon of muon catalyzed fusion. We review recent calculations on the tdμ ion which have substantially improved our knowledge of the processes involving this ion.

I. INTRODUCTION

Although the muon catalyzed fusion (μCF) cycle consists of several stages, the formation and destruction of muonic molecular ions is the central event of the whole process. These positive ions contain two hydrogen isotopic nuclei and a muon. Due to the small size of this system, the nuclei fuse almost immediately once the ion is formed, without need for high temperature and pressure. The simplicity of these ions makes it possible to perform highly accurate ab initio calculations of their energy levels and properties. tdμ is the most important one of the muonic molecular ions since the most efficient fusion cycle involves this system. The latest experiments have shown that with this cycle one muon may catalyze more than 150 fusions in its lifetime[1-3]. This value is two orders of magnitude larger than for all other hydrogen isotopic ions. Therefore, in this review we will restrict ourselves to tdμ.

II. COULOMBIC ENERGY LEVELS

Because of the large mass of the muon (207 times the mass of the electron) the muonic molecular ions are intrinsically nonadiabatic systems. The first calculations for muonic molecular ions were performed by Kolos, Roothaan, and Sacks[4] in 1960. These calculations used a basis set of generalized Hylleraas functions. Several other papers appeared in the 1960s (for a more exhaustive list of references on muonic molecular ions see Ref. 5) with the most accurate results obtained by Carter[6] with Slater geminals. In 1967 Vesman[7] proposed a resonance mechanism of muonic ions formation. This mechanism, however, requires the existence of weakly bound states of these ions. Such states had not been seen in the variational calculations of that time. In the 1970s Ponomarev, Vinitsky, and coworkers developed several non-variational methods to find these states. The states were first seen[8] in 1978. The first reliable calculation was published by Vinitsky et al.[9] in 1980 using a method of expansion in adiabatic states. This calculation showed that only ddμ and tdμ have states bound by about 1 eV and that in both cases these states are the first excited levels in P symmetry (rotational quantum number J = 1). We will denote the states of the muonic molecular ions by (Jv), where v enumerates consecutive states of a given symmetry. Since 1984 a series of variational calculations confirmed the correctness of Vinitsky's et al. results. In recent papers the energy of the (11) state of tdμ has been obtained to an accuracy better than 0.1 meV.

The variational method solves the Schrodinger equation $H\Psi = E\Psi$ for a muonic molecular ion by expanding the wave function Ψ in terms of some set of basis functions

$$\Psi = \Sigma_1^N c_i \chi_i(r_1, r_2, r_3)$$

where r_k denotes the coordinates of kth particle. Energy levels and coefficients c_i are found by the diagonalization of an N by N Hamiltonian matrix. An important feature of the method is that these energy levels are upper bounds to the exact energy levels. This feature makes a comparison of different calculations straightforward.

Despite the simplicity of the muonic molecular ions, large basis sets (N of the order of 3000) are needed to achieve the required accuracy for the weakly bound states. Thus, the recent calculations for muonic ions employ much larger basis sets than present calculations for other small systems such as He, H^-, H_2, or HeT^+. The necessity of using this size expansions is mainly due to the highly diffuse and nonadiabatic character of these weakly bound states. The size of the Hamiltonian matrix which has to be diagonalized makes the calculations rather time consuming, particularly since this matrix is neither sparse nor diagonally dominant and therefore special efficient techniques for diagonalizing such matrices can not be utilized. The major problem prohibiting enlarging N is due, however, not to limitations of computer resources but rather to linear dependencies in the basis set. These linear dependencies require the use of an extended arithmetic precision (~30 decimal digits). This problem is more fully discussed elsewhere in these Proceedings[10].

Several types of basis sets have been used for calculations on muonic molecular ions. Here we discuss these basis sets for the case of S-symmetry states. For other symmetries the basis functions contain additional (angular) factors which are not relevant for the present discussion. Generalized Hylleraas functions have the form

$$\chi_i = r_{t\mu}^{k_i} r_{d\mu}^{l_i} r_{td}^{m_i} e^{-\alpha_i r_{t\mu} - \beta_i r_{d\mu} - \gamma_i r_{td}}$$

where r_{xy} denotes the distance between particles x and y. The word "generalized" indicates the presence of the factor with the exponent γ, which is absent in the traditional Hylleraas basis. In the simplest and most popular application, the powers k_i, l_i, and m_i are different for different i while the exponents α_i, β_i, γ_i are the same. In some calculations, however, a few sets of exponents were used and sometimes functions with different exponents but the same set of powers of the interparticle distances were allowed. The exponents were optimized using modest size basis sets and then the same exponents were kept for calculations with larger N. The most straightforward choice for the powers is to limit them to some chosen value and such that $k_i+l_i+m_i \leq \omega$ where ω is another chosen number. The convergence can be improved, however, if one makes a selection of these functions and includes only those which contribute to the energy at a given level[5].

Another popular choice is the basis of Slater geminals. The functions of this set are basically of the same form as above. However, rather than choosing different powers with the same set of exponents, these functions have all the powers equal zero (for S states) but all the exponents are different. One could optimize these exponents (3N for a basis of N functions) but this procedure would be too time-consuming. Instead, triples of numbers are chosen by a random number generator and only a few parameters describing the mapping to the exponent space are optimized. This method of selecting the exponents is called random tempering[10].

If one chooses to describe the system in elliptical coordinates, the following basis set can be introduced

$$\chi_i = R^{k_i} \xi^{l_i} \eta^{m_i} e^{-\alpha_i R - \beta_i \xi}$$

where $R = r_{td}$, and $\xi = (r_{t\mu} + r_{d\mu})/R$, $\eta = (r_{t\mu} - r_{d\mu})/R$ are the elliptical coordinates. In this basis set the choice of powers and exponents is similar to the Hylleraas basis.

Instead of exponents linear in the interparticle distances one may take a square dependence, which leads to a Gaussian basis set

$$\chi_i = r_{t\mu}^{k_i} r_{d\mu}^{l_i} r_{td}^{m_i} e^{-\alpha_i r_{t\mu}^2 - \beta_i r_{d\mu}^2 - \gamma_i r_{td}^2}$$

In the form given above this basis has not been used in muonic molecular calculations since it does not seem to offer any particular advantages. It becomes attractive, however, if one considers the Jacobi coordinates. In this coordinate system the distance R between two particles is one coordinate while the position vector **r** of the third particle is measured from the center of mass of these two particles. Thus, there are three such coordinate systems depending on which two particles define R, which leads to three sets of basis functions. Jacobi coordinates are very convenient for describing the dissociation of a system. In principle, they can be used with the exponential functions linear in the distances but the resulting integrals are quite complicated. In contrast, with Gaussian form the integrals are simple. Such a basis set of Gaussian geminals in Jacobi coordinates has been used by Kamimura[11,12]. Like the Slater geminal basis, this set contains a large number of exponents. The exponents were chosen, however, by even (rather than by random) tempering, i.e. the consecutive exponents are in the form ab^i where a and b are optimized. An important advantage of Kamimura's basis is that it allows to perform calculations in the regular arithmetic precision (~15 decimal digits) even with N of the order of 2000.

Table I. Binding energies of the (11) state of tdμ (in meV)

Who	Year	Method	N	Energy
Vinitsky et al.[9]	1980	non-variational	-	640
Bhatia and Drachman[13]	1984	Hylleraas	440	224
Frolov and Efros[14]	1984	Slater, random tempered	375	523.1
				(600)
Hu[15]	1985	Hylleraas	500	628
Frolov and Efros[16]	1985	Slater, random tempered	400	607.2
				(655.4)
Gotcheva et al.[17]	1985	non-variational	-	656
Vinitsky et al.[18]	1986	Elliptic	1495	658.9
				(663±2)
Hu[19]	1987	Hylleraas	1102	658.03
Korobov et al.[20]	1987	Elliptic	2084	659.68
				(660.4±0.2)
Kamimura[11]	1987	Gaussian	1696	659.33
Szalewicz et al.[5]	1987	Hylleraas, multiple exp.	3063	660.01
				(660.1±0.1)
Alexander-Monkhorst[21]	1988	Slater, random tempered	2000	660.147
Kamimura[12]	1988	Gaussian	2662	660.104
				(660.14±0.02)
Haywood et al.[22]	1988	Slater, random tempered	2600	660.153

We collected the most relevant results for the (11) state of tdμ in Table 1. [Other states of this and other molecules are given in Ref. 10]. The last four entries in the Table reached the desired accuracy of 0.1 meV which, if the relativistic and other corrections were known as accurately, would allow to theoretically predict the formation rate. When one compares different calculations to an accuracy of 0.1 meV, one must take into account possible differences in the values of the particle masses and of the Rydberg constant. It is possible to correct various results for these differences (see Ref. 10). Present uncertainties of the particle masses lead to an uncertainty of the order of several μeV in the binding energy, so that a future improvement of the energies, if needed, requires more accurate masses.

In addition to a few bound states, the spectrum of muonic molecular ions contains also several three-body resonances. These resonances, found recently by some of us[23], may play a significant role in the fusion reaction in tepid ($\sim 10^5$ K) plasmas. This subject is described in another paper of these Proceedings[24].

III. CORRECTIONS TO COULOMBIC ENERGY LEVELS

As discussed above, the non-relativistic Coulomb energy of the tdμ (11) state is presently known to accuracy of at least 0.1 meV. However, to predict the formation rate one needs to add several corrections to this energy. The largest of them are the corrections for the relativistic effects and for the electromagnetic structure of the nuclei. The latest calculation of these effects by Bakalov[25] gave the value 17.6 ± 1 meV for the (11) state of tdμ. Such accuracy could be considered almost sufficient, however, successive calculations[25,26] differ quite seriously, suggesting considerably larger uncertainty. Thus, an independent calculation of these quantities would be advisable. One of us (HJM) is pursuing work in this direction using our accurate wave functions.

Another effect which has to be taken into account is the interaction of tdμ with the host molecule. Since the ion is formed as a part of a hybrid molecule [(tdμ)dee], interactions with the other d nucleus and with the electrons cause a shift in the energies of tdμ relative to the energy of the isolated ion. Since the interaction is relatively weak, the problem can be adequately treated by perturbation theory through second order using multipole expansion for the potentials. A calculation of these corrections by Menshikov[27] provided a value of 7.7 meV and no estimate of accuracy was attempted. Recently, Bakalov[25] has published a new value of 1.2 ± 0.6 meV but the details of this calculation were not given. We are presently working on a calculation of this effect using a new variation-perturbation approach and large basis sets to describe $td\mu$. Our initial results show that the cancellation of the first-order correction with a part of the second-order one, as conjectured by Menshikov[27], holds only to an accuracy of about 3 meV. Our preliminary value of the correction is about 2 meV, in good agreement with the result given by Bakalov[25].

IV. ROVIBRONIC ENERGY LEVELS OF [($td\mu$)dee]

During the formation of $td\mu$ the binding energy of this ion and the kinetic energy of the colliding $t\mu$ are transferred into rovibrational excitations of the hydrogen-like molecule [($td\mu$)dee]. Thus, one needs to know the rovibrational energy levels of this system. The details of the temperature behavior of the resonant formation rate are extremely sensitive to the exact location of these energies. Due to the smallness of $td\mu$, the hybrid molecule resembles closely an isotopic species of the hydrogen molecule with deuteron as one of the nuclei and $td\mu$ as the other. Rovibrational energies of these isotopic species differ essentially only in the nonadiabatic correction. Faifman et al.[28] used the traditional semiempirical approach of Van Vleck to calculate these corrections. Recently, a highly accurate phenomenological isotopic scaling procedure has been proposed by Schwartz and LeRoy[29]. We have used[30] the latter approach to calculate the rovibrational spectrum for all the hybrid molecules relevant for muon catalyzed fusion achieving accuracy of about 0.05 meV.

V. STICKING FRACTIONS

In the early 1960s muon catalyzed fusion research was abandoned because it was believed that, on the average, one muon may form only a single molecule in its lifetime. Jackson[31] pointed out that even if this difficulty could be overcome, another effect, the sticking of the muon to the α particle after the nuclear reaction, restricts the number of fusions to about 100 per muon. At present, thanks to the fast resonant $td\mu$ formation rate, the muon sticking has become the main bottleneck in the process.

The probability of muon sticking to the α particle after the nuclear reaction is called the initial sticking fraction ω_s^0. The experimentally measured sticking fraction is modified by the fact that the muon can be detached from the α particle in collisions during the process of slowing down in the mixture. Thus, the total sticking fraction is given by $\omega_s = (1 - R)\omega_s^0$, where R is called the reactivation coefficient. The initial sticking fraction ω_s^0 is the sum of partial sticking fractions ω_{nl} which represent probability of sticking for the different nl channels on the $\alpha n\mu$ side, where nl denotes quantum numbers of a given energy level for the hydrogen-like $\alpha\mu$ ion. ω_{nl} includes the sum over the (2l+1)-fold degeneracy of each level. The "classical" expression for ω_s^0, which is derived assuming the sudden approximation, gives sticking probability in the form of an integral dependent on the $td\mu$ wave function taken at $r_{td} = 0$. If the $td\mu$ wave function is computed with the adiabatic approximation, ω_s^0 equals 1.172% [Refs. 31-33] for the fusion taking place from the ground state of $td\mu$. The nonadiabatic value calculated first by Ceperley and Alder[34] was 0.895%. We have recently computed[33] the initial sticking of 0.8860%. By using very accurate wave functions to represent the $td\mu$ molecule, taking into account the cusp conditions at the nuclear coalescence, performing calculations up to n = 30, and including a very small contribution from higher n by using appropriate large n asymptotic formulae we have obtained at least three-digit accuracy. A slightly less converged result (0.883%) was obtained in our group with Slater geminals[22]. Our results are compared with several other calculations in Table 2. It was really astonishing that the results reported by Kamimura at this conference and obtained using a different basis set agree with ours up to 0.0001%. This suggests that the accuracy of our calculations is probably even higher than we had estimated.

Table 2. Coulombic sticking fractions (in percent) for the (00) state of the tdμ molecular ion.

	Ceperley Alder[34]	Bogdanova et al.[35]	Hu[36]	Haywood et al.[33]	Hu[19]	Kamimura[37]
ω_s^0	0.895	0.845	0.897	0.8860	0.8583	0.8859
ω_{1s}	0.689	0.6502	0.6932	0.6826		
ω_{2s}	0.099	0.0934	0.0992	0.0979		
ω_{2p}	0.024	0.0238	0.0241	0.0238		
ω_{3s}	0.030	0.0284	0.0302	0.0297		
ω_{3p}	0.009	0.0086	0.0087	0.0086		
other	0.044	0.041	0.042	0.0434		

We have also calculated the sticking fractions for the (01) and (10) states of tdμ[38]. We attempted to calculate this quantity for the (11) state but our calculations did not converge despite the use of up to 3063 generalized Hylleraas functions. Thus, we would treat with caution the value of 0.0836% reported by Hu[19] since it was obtained with a wave function much less converged than our functions. Recently, some of us[22] repeated the calculations for the (11) state using a sequence of basis sets containing up to 2600 Slater geminals. Although the energy is even better than the Hylleraas-basis result, the sticking fraction still has not converged.

To compare our sticking fractions with the experimental ones we have to take into account the effects of reactivation. The reactivation coefficient has been calculated by several authors, most recently by Cohen[39]. If we use our ω_s^0 and Cohen's R, the total sticking fraction for a density $\phi = 1.2$ of liquid-hydrogen densities is 0.57%. This value is larger than the experimental sticking of about 0.45% obtained by the PSI collaboration[2] and much larger than the experimental result of 0.35% by Jones et al.[1]. It is possible that the effects of nuclear forces, as discussed in the next Section, may resolve this discrepancy.

VI. EFFECTS OF NUCLEAR FORCES ON THE GROUND STATE ENERGY

Although the fusion of t and d is due to nuclear forces, until recently these forces have been only marginally included in the theoretical description of the $td\mu$ ion. This is because once the nuclei are close enough, with the probability of such event determined essentially by the Coulomb interactions, fusion takes place almost immediately. In 1986 Rafelski and Muller[40] suggested that nuclear forces may significantly influence the value of the initial sticking fraction. Several other papers followed[41]. Obviously, the nuclear forces change drastically the interaction potential between the nuclei at very short distances, i.e. at the distances where the fusion takes place. The $1/r_{td}$ Coulomb repulsion, which grows to infinity when r_{td} approaches zero, is replaced at the distances smaller than a few fm by a deep potential well, which can be described by an optical potential. This approach was used in papers by Bogdanova et al.[42], Akaishi et al.[43] and Hu[44]. In this way one can calculate the shift of the energy levels due to the nuclear forces and the lifetime of a state. Unfortunately, the choice of the optical potential is to a large extent arbitrary and one would have to prove that the results are independent of this choice. An alternative is offered by the R-matrix formalism[45]. In this approach one uses experimental nuclear scattering data to construct the R-matrix which gives a linear relation between the values and derivatives of the wave function on the nuclear boundary in the td and αn channels. By assuming outgoing wave boundary conditions in the αn channel, one can compute the logarithmic derivative at the channel radius a_c. For the (00) state of $td\mu$ one gets[46]

$$L_W = a_\mu \, \partial \ln \Psi(r_{t\mu}, r_{d\mu}, r_{td}) \, / \, \partial r_{td} \simeq -63 - 5i$$

where a_μ is the muonic Bohr radius. Such boundary conditions can be imposed on our basis set. After cutting out the region of nuclear interaction, $r_{dt} < a_c$, from our matrix elements, we can compute energies of $td\mu$ with the effects of nuclear forces included. A similar calculation has been done by Struensee et al.[46] in the adiabatic approximation. In Fig. 1 we present the results of our work and compare them with literature results. The energy in Fig. 1 was plotted as a function of L_W. The curve representing our values was obtained in two different basis sets. In the first one we added the necessary flexibility by including functions with γ as large as 60. In the second approach we have added terms to the basis set with r_{td}^{-m} (which would diverge at nuclear coalescence). The two calculations are not distinguishable on the graph's scale. Let us mention that the standard basis set with γ of the order of one and without the negative powers leads to completely wrong results for large $|L_W|$.

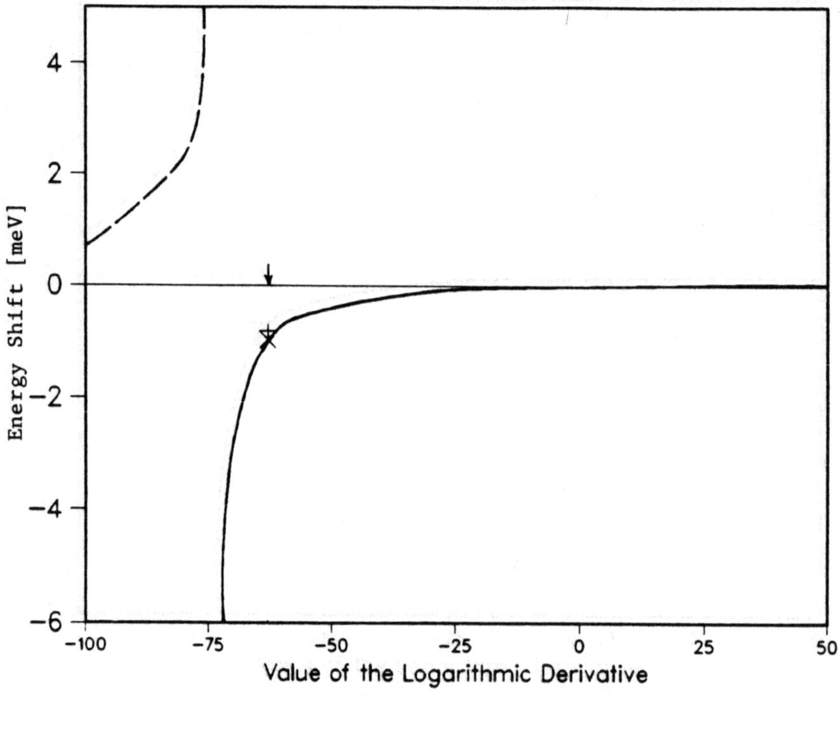

Figure 1. Energy shift for the (00) level of $td\mu$ as function of the logarithmic derivative. The values from Refs. 42 and 46 are marked by (x) and (+), respectively. (↓) shows the value of the real part of the logarithmic derivative for the outgoing wave boundary conditions.

As one may see from Fig. 1 our calculation is in excellent agreement with both the Struensee et al.[46] R-matrix calculation as well as with the optical potential work of Bogdanova et al.[42]. We wish to stress that our work represents the first high-accuracy calculation for this problem. The work of Struensee et al.[46] was done in the adiabatic approximation. In the work of Bogdanova et al.[42] the $td\mu$ wave function was represented by only a few terms in the adiabatic expansion. Our wave functions for the Coulomb value of L_W give an energy which agrees to 8 digits with the exact energy. Fig. 1 also shows that the state undergoes a drastic rearrangement at the value of L_W equal to about -72, seemingly not very far from the physical value of -63. This suggests that the process might be very sensitive to the details of the description of the nuclear forces if L_W could be changed within this range. Let us point out, however, to the talk at this conference by Ponomarev who found, using optical potentials, that the rearrangement takes place far from the physically acceptable values of the parameters for these potentials.

At this conference, Ponomarev and Kamimura reported work which estimates the effects of nuclear forces on the initial sticking. Using optical potentials and the classical expression for sticking they independently came to the same conclusion that nuclear forces increase the sticking probability by 0.03%. We are presently performing similar calculations within the R-matrix formalism.

Danos et al.[47] presented at this conference a completely novel approach to this problem which takes into account the fact that the (00) state of $td\mu$ is formed by an Auger transition from the (11) state. This approach does not assume the Siegert boundary conditions in the exit channels and abandons the classical expression for sticking. The necessity of such a treatment has been discussed at this conference. We are closely collaborating with the authors of Ref. 47 on this issue.

VII. REACTIVATION COEFFICIENT

We have mentioned above that the measured sticking fraction contains the effects of reactivation. These effects were first estimated by Jackson[31]. Initial calculations of the reactivation coefficient[32,48] gave a value of R of about 0.23 which was only very weakly dependent on the density. Recently, Cohen[39] performed a more elaborate calculation which led to R = 0.30 ± 0.05 at low density and 0.36 ± 0.05 at high density. Although the density dependence was much stronger than in the earlier work, it was still much weaker than measured by Jones et al.[1] Cohen included some new effects[39] and he is confident that all the important ones have been taken into account. In light of the accuracy being sought for the sticking fractions, Cohen's values for the reactivation coefficient have relatively large error bars due to uncertainties in the various cross sections used. There is also some evidence[49] which suggests that the scaling procedure used to obtain certain of these cross sections introduces a larger error than originally thought. It seems to us that no major change of the theory of reactivation process is possible, however, with the sticking fraction becoming the central problem of μCF, it appears worthwhile to compute the relevant scattering cross sections more accurately. We are presently working on this issue using the two-center coupled channel method of Winter[50]. This approach is able to describe ionization and charge-transfer as well as excitation processes. It provided the most accurate literature cross sections for the scattering of protons on one electron ions. More information on this work can be found in another paper in these Proceedings[49].

VII. CONCLUSIONS

We have demonstrated that several questions important for muon catalyzed fusion can be answered by performing high-accuracy quantum mechanical calculations for the muonic molecular ions. Problems concerning values of the Coulombic energies or sticking fractions have been decisively solved within the last year. Other problems such as corrections to these energy levels and sticking fractions due to the effects of nuclear forces are now being intensively examined. A careful quantum treatment of the muon reactivation is also being pursued. Solving these problems will greatly deepen our knowledge of the process and shed light on the perspectives of using muon catalyzed fusion for energy production.

ACKNOWLEDGMENTS

This work has been supported by grants from the Division of Advanced Energy Projects of the US Department of Energy. We are grateful to Dr. R. Gajewski for his constant encouragement. Partial support has also been provided by the National Science Foundation through grant CHE8505733, as well as by the Division of Sponsored Research of the University of Florida.

REFERENCES

1. S.E. Jones et al., Phys. Rev. Lett. 56, 588 (1986).
2. W.H. Breunlich et.al., ibid. 58, 329 (1987).
3. See also the results by the authors of Refs. 1 and 2 reported at this Conference and published in this volume.
4. W. Kolos, C.C.J. Roothaan, and R.A. Sack, Rev. Mod. Phys. 32, 178 (1960).
5. K. Szalewicz, H.J. Monkhorst, W. Kolos, and A. Scrinzi, Phys. Rev. A 36, 5494 (1987).
6. B.P. Carter, Phys. Rev. 141, 863 (1966); 165, 139 (1968).
7. E.A. Vesman, Sov. Phys. JETP Lett. 5, 91 (1967).
8. S.I. Vinitsky, L.I. Ponomarev, I.V. Puzynin, T.P. Puzynina, L.N. Somov, and M.P. Faifman, Sov. Phys. JETP 47, 444 (1978).
9. S.I. Vinitsky, V.S. Melezhik, L.I. Ponomarev, I.V. Puzynin, T.P. Puzynina, L.N. Somov, and N.F. Truskova, Sov. Phys. JETP 52, 353 (1980).
10. S.A. Alexander, H.J. Monkhorst, and K. Szalewicz, this volume.
11. M. Kamimura, Muon Catalyzed Fusion 3, 335 (1988).
12. M. Kamimura, Phys. Rev. A, to be published.
13. A.K. Bhatia and R.J. Drachman, Phys. Rev. A 30, 2138 (1984).
14. A.M. Frolov and V.D. Efros, JETP Lett. 39, 544 (1984).
15. C.-Y. Hu, Phys. Rev. A 32, 1245 (1985).
16. A.M. Frolov and V.D. Efros, J. Phys. B 18, L265 (1985).
17. A.D. Gotcheva et al., Phys. Lett. B 153, 349 (1985).
18. S.I. Vinitsky, V.I. Korobov, and I.V. Puzynin, Zh. Exp. Th. Fiz. 91, 705 (1986).
19. C-Y. Hu, Phys. Rev. A 36, 4143 (1987).
20. V.I. Korobov, I.V. Puzynin, and S.I. Vinitsky, Phys. Lett. B 196, 272 (1987).
21. S.A. Alexander and H.J. Monkhorst, Phys. Rev. A, in press.
22. S.E. Haywood, S.A. Alexander, and H.J. Monkhorst, Phys. Rev. A, submitted.
23. P. Froelich and K. Szalewicz, Muon Catalyzed Fusion 3, 343 (1988).
24. P. Froelich, K. Szalewicz, H.J. Monkhorst, W. Kolos, and B. Jeziorski, this volume.
25. D.D. Bakalov, Muon Catalyzed Fusion, 3, 321 (1988).
26. D.D. Bakalov, S.I. Vinitsky, and V.S. Melezhik, Sov. Phys. JETP 52, 820 (1980); D.D. Bakalov, V.S. Melezhik, L.I. Menshikov, and S.I. Vinitsky, Phys. Lett. B 161, 5 (1985).

27. L.I. Menshikov, Sov. J. Nucl. Phys. 43, 918 (1985).
28. M.P. Faifman et al., Z. Phys. D 2, 79 (1986).
29. C. Schwartz and R.J. LeRoy, J. Mol. Spectrosc. 121, 420 (1987).
30. A. Scrinzi, K. Szalewicz, and H.J. Monkhorst, Phys. Rev. 37, 2270 (1988).
31. J.D. Jackson, Phys. Rev. 106, 330 (1957).
32. S. Gerstein et al., Sov. Phys. JETP 53, 872 (1981).
33. S.E. Haywood, H.J. Monkhorst, and K. Szalewicz, Phys. Rev. A 37, 3393 (1988).
34. D. Ceperley and B.J. Alder, Phys. Rev. A 31, 1999 (1985).
35. L.N. Bogdanova et al., Nucl. Phys. A 454, 653 (1986).
36. C.-Y. Hu, Phys. Rev. A 34, 2536 (1986).
37. M. Kamimura, reported at this conference.
38. S.E. Haywood, H.J. Monkhorst, and K. Szalewicz, Phys. Rev. A, submitted.
39. J.S. Cohen, Phys. Rev. Lett. 58, 1407 (1987).
40. J. Rafelski and B. Muller, Phys. Lett. B 164, 223 (1985).
41. M. Danos, B. Muller, and J. Rafelski, Phys. Rev. A 34, 3642 (1986); H. Takahashi, J. Phys. G 12, L271 (1986); M. Danos, B. Muller, and J. Rafelski, Phys. Rev. A 35, 2741 (1987); M. Danos, B. Muller, and J. Rafelski, Muon Catalyzed Fusion 3, 443 (1988);
42. L.N. Bogdanova, V.E. Markushin, and V.S. Melezhik, Sov. Phys. JETP 54, 442 (1981); L.N. Bogdanova, V.E. Markushin, V.S. Melezhik, and L.I. Ponomarev, Sov. J. Nucl. Phys. 34, 662 (1981).
43. Y. Akaishi, M. Kamimura, and H. Narumi, Z. Phys. A 328, 115 (1987);
44. C.-Y. Hu, unpublished.
45. E.P. Wigner, Phys. Rev. 70, 15, 606 (1946); A.M. Lane and R.G. Thomas, Rev. Mod. Phys. 30, 257 (1958).
46. M.G. Struensee, G.M. Hale, R.T. Pack, and J.S. Cohen, Phys. Rev. A 37, 340 (1988).
47. M. Danos, L.C. Biedenharn, A. Stahlhofen, , this volume; M. Danos, L.C. Biedenharn, and A. Stahlhofen, NBS publication NBSIR 87-3532, National Bureau of Standards, Gaithersburg, MD (1987).
48. L. Bracci and G. Fiorentini, Nucl. Phys. A 364, 383 (1981).
49. C. Stodden, H.J. Monkhorst, and K. Szalewicz, this volume.
50. T.G. Winter, Phys. Rev. A 25, 697 (1982).

Correction to the Binding Energy of (dtμ) Meso Molecules

A. M. Lane

Theoretical Physics Division, Harwell Laboratory, UKAEA, Harwell, Oxon OX11 ORA, U.K.

The non-relativistic Coulomb binding energy of (dtμ) has been cited as 660.0 meV. Spin, finite-size and relativistic corrections, treated as perturbations, lower this by about 10%. It is pointed out that, for lightly bound systems, perturbation theory needs correction, an effect known in nuclear physics. In the present case, the change is about 0.6 meV. This is significant since energies are quoted to 0.1 meV accuracy.

In muon-catalysed fusion studies, muons introduced into a chamber containing D, T atoms or molecules quickly form $(t\mu)$ atoms, which then combine with a D_2 molecule to form a resonance complex $(D_2t\mu) \equiv "D_2"$. In this complex, the $(t\mu)$ atom is joined with a deuteron to form a $(dt\mu)$ "meso-molecule", and this replaces a deuteron in the D_2 molecule. The reaction rate for forming "D_2" resonances is one of the two crucial parameters determining the number of reactions catalysed by a muon. (The other is the "sticking probability" which is the chance that a muon is caught on an alpha-particle produced by a catalysed reaction).

The most important input parameters in the calculation of the reaction rate are the energies in the $(t\mu)+D_2$ system of the resonances $("D_2")_{K_iK}$. Here K_i,K are the rotational angular momenta of D_2, "D_2". The energy for given K_i,K is reliably calculated (in meV) as:

$$E_{K_iK} = E_{oo} + 2.4K(K+1) - 3.7\ K_i(K_i+1) \qquad (1)$$

so there is just one essential parameter, E_{oo}. The vibrational state of "D_2" is $v=2$, while that of D_2 is $v=0$, and the difference between these has been calculated as 579.4 (taking $dt\mu$ as a point; Faifman et al. 1986) or 587.1 (allowing for finite size of $dt\mu$; Menshikov 1985) so $E_{oo} = -\varepsilon_{11}+587.1$, where ε_{11} is the binding energy (in meV) of the $(dt\mu)$ system in its rotational vibrational state $(K_mv_m) = (11)$.

The essential part of ε_{11} is that arising from purely Coulombic interactions between d,t,μ, calculated non-relativistically. The most recent value is 660.0 (± 0.1) (Szalewicz et al. 1987). This value is tiny compared to the Born-Oppenheimer potential for the d-t system, typically 300 (±100) eV within a separation distance $5a_\mu$, where a_μ is the muon Bohr radius 2.56×10^{-11} cm; indeed the binding-energy of the ground state $(K_m v_m)=(00)$ is 319.1 eV. A consequence of the tiny binding energy is that a large fraction of the norm of the state is in the (p-wave) exponential tail. At this point, a subtle effect, well-known in nuclear physics, enters the picture, and this gives rise to a significant correction to a lowest-order perturbation calculation of small extra interactions. The effect comes about because the fraction of norm in the tail depends on binding-energy, so, if the binding energy is changed by a perturbation shift, the tail fraction is changed, and this, in turn, changes the size of the shift.

In the dtμ problem, there are two small shifts calculated in first-order (Bakalov et al. 1985):

(i) spin (or hyperfine) splitting. For the lowest spin states of (tμ) (S=0) and $(dt\mu)_{11}$(S=1), this gives a decrease $\delta\varepsilon_{11} = -178.0 + 142.5 = -35.5$.

(ii) relativistic and nucleon finite-size effects. These give a decrease $\delta\varepsilon_{11} = -22.0$.

Together, these amount to $\delta\varepsilon_{11} = -57.5$ and so correct the value of ε_{11} from 660.0 to 602.5, implying that $E_{oo} = -15.4$.

For any small added potential V, the normal first-order perturbation form of the energy shift is $\Delta E_1 = \langle\psi|V|\psi\rangle/\langle\psi|\psi\rangle$. We may assume that V, like the interaction between d and (tμ) is negligible for values of d-(tμ) separation, r, larger than a certain value, r=R. For r>R, ψ has the product form $\psi=\eta(\underline{r}_{t\mu})\phi(r)Y_1(\Omega)$ where $\eta(\underline{r}_{t\mu})$ is the (tμ) atom in the ground state, and φ(r) describes the d-(tμ) relative motion. For $r > 7a_\mu$, it is found (Menshikov 1985) that φ has its free-wave p-state form to within 5%. (At $r=7a_\mu$, the centrifugal potential is about 10eV, while the interaction is about 5eV):

$$\phi(\underline{r}) = \phi(R)\,\frac{(x^{-1} + x^{-2})e^{-x}}{(X^{-1} + X^{-2})e^{-X}} \tag{2}$$

where $x \equiv \kappa r$, $X \equiv \kappa R$, κ being the binding energy wave number. Normalising ψ to unity over all space means:

$$\langle\psi|\psi\rangle_R + \frac{\phi^2(R)R^3Xe^{2X}}{(1+X)^2}\int_X^\infty (x^{-1} + x^{-2})^2 e^{-2x}\, x^2 dx = 1 \tag{3}$$

where subscript R means that the r-integration is over $r \leqslant R$. Thus:

$$\langle\psi|\psi\rangle_R^{-1} = 1 + \theta^2\,\frac{(2+X)}{2(1+X)^2} \tag{4}$$

where:

$$\theta^2 \equiv \frac{\phi^2(R)R^3}{\langle\psi|\psi\rangle_R} \tag{5}$$

Since θ^2 depends only on $\psi(r)$ in the region $r<R$ of strong interaction, it may be assumed independent of binding energy, so (4) gives the dependence of $\langle\psi|\psi\rangle_R$ on binding energy. The correct energy shift due to a perturbation is:

$$\Delta E = \left(\frac{\langle\psi|V|\psi\rangle_R}{\langle\psi|\psi\rangle_R}\right) \langle\psi|\psi\rangle_R$$

$$= \left(\frac{\langle\psi|V|\psi\rangle_R}{\langle\psi|\psi\rangle_R}\right) \left(1 + \theta^2 \frac{2+X}{(1+X)^2}\right)^{-1} \tag{6}$$

where the first factor is independent of binding energy, while the last factor should be evaluated at the correct binding energy. The usual first-order form ΔE_1 corresponds to evaluating it at the unshifted energy. The correction is:

$$\frac{\Delta E - \Delta E_1}{\Delta E_1} = \frac{(3+X_o)\theta^2}{(1+X_o)\left[(1+X_o)^2 + (2+X_o)\theta^2\right]} \tag{7}$$

where δX is the change in X from its value X_o at the unshifted energy.

$$\frac{\delta X}{X_o} = -\frac{1}{2}\frac{\Delta E_1}{\varepsilon_{11}} \tag{8}$$

With unshifted energy $\varepsilon_{11} = 660.0$, and $R=79\mu$, X_o is 0.353; with $\Delta E_1 = 57.5$, δX is 0.0154 so (7) becomes $0.021\ (\theta^{-2}+1.38)^{-1}$. The value of θ^2 is obtained from Menshihov (1985), who gives (2) with:

$$\frac{\phi(R)R^{3/2}}{(X_o^{-1}+X_o^{-2})e^{-X_o}} = 0.574\ X_o^{3/2} \tag{9}$$

whence $\phi^2R^3 = 0.84$, $\langle\psi|\psi\rangle_R = 0.46$, $\theta^2 = 1.83$. The value of $\langle\psi|\psi\rangle_R$ shows that over one-half of the wave-function norm is in the tail region; this unusually large fraction gives rise to the present effect. (7) is now 0.011, so the correction $(\Delta E-\Delta E_1)$ is -0.6 meV. Thus the shift is changed from $\delta\varepsilon_{11}$ = -57.5 to -56.9, and E_{oo} is changed from -15.4 to -16.0 meV.

Finally we note another correction to the energies $E_{K_i K}$ arising from the light binding of (dtμ). This is the shift and splitting due to quadrupole coupling between (dtμ) and "D_2". The large size of (dtμ) due to its light binding gives a quadrupole moment which is large enough ($Q \sim 17\ ea_\mu^2$) to cause shifts of order 0.1 meV (after allowing for screening effects). For example, the shifts of the K=1 state coupled to (dtμ) angular momentum 1 to give total 0,1,2 are 0.11, 0.16, 0.25 meV.

Work described in this Letter was undertaken as part of the underlying research programme of the UKAEA.

References

Bakalov D.D., Melezhik V.S., Menshikov L.I. and Vinitsky S.I. 1985, Phys.Lett. 161B, 5.

Faifman, M.P., Menshikov, L.I., Ponomarev, L.I., Puzynin I.V., Puzynina T.P. and Strizh T.A., 1986, Z.Phys.D.At.Mol. and Clust. 2, 79.

Menshikov L.I. 1985 Sov.J.Nucl.Phys. 42, 918.

Szalewicz K., Monkhorst H.J., Kolos W. and Scrinzi A., 1987, Phys.Rev. A36, 5494.

INTEGRAL TRANSFORM WAVEFUNCTIONS IN THE SOLUTION OF THE QUANTUM MECHANICAL THREE BODY PROBLEM

Vedene H. Smith, Jr.
Department of Chemistry, Queen's University,
Kingston, Ontario K7L 3N6

Piotr Petelenz
Department of Theoretical Chemistry, Jagiellonian University,
ul. Karasia 3, 30-060 Cracow, Poland

ABSTRACT

The variational ansatz consisting of Slater-type geminals with random-tempered exponents was originally introduced by our group as a version of the integral transform (generator coordinate) method and applied to the solution of the eigenvalue problem for two-electron atoms. Recently, that approach has been extended to treat three-body systems where all the particles have comparable masses, e.g., in the positronium negative ion. The new version has been applied to the calculation of energies and other expectation values of the positronium and muonium ions and of muonic molecular ions. Current calculations of this group and other groups are reviewed and discussed. Special emphasis is given to deviations from the Coulombic model, in particular to the vacuum polarization correction which has recently been calculated with high accuracy for a number of muonic molecular ions.

INTRODUCTION

There has been a renewed interest in the calculation of the binding energies and other properties of three-body systems. Possibly the more intriguing of such systems to be studied recently are the muonic molecular ions such as $dd\mu^+$, $dt\mu^+$, etc., due to the involvement of some of them in catalyzing nuclear fusion[1] and the muonium[2,3] and positronium[4] negative ions due to their experimental discovery. In the case of the positronium negative ions, the essentially identical masses means that the three particles must be treated on an equal footing,[5-7] i.e. the Born-Oppenheimer approach is contraindicated.

Because of the relative simplicity of three-particle systems, one can solve the corresponding Schrödinger equation with very high accuracy. As a result, they have been the common (and logical) testing ground for a variety of aspects of the quantum mechanical description of atoms and molecules. Among the approaches that have been employed, one may cite the quasi-adiabatic approaches using spheroidal[8] and hyperspherical[9,10] coordinates, numerical Greenfunction Monte-Carlo techniques,[11,12] and variational methods wherein spheroidal trial functions[13] and those explicitly involving the interparticle coordinates. These latter trial functions have been either of Hylleraas type or of explicitly pseudo-Slater type with the exponents generated in a pseudo random manner from a set of nonlinear variational parameters (random tempering[14]).

The latter approach is essentially a version of the integral transform (generator-coordinate) method developed in this laboratory and extensively used for small atoms[15,16] and applied to a number of problems[17,18] which involve the use of Yukawa type potentials instead of Coulombic potentials. It has been recently applied with success to muonic molecule problems by several groups[19-21]. In the next section we outline this very effective method.

The standard approximation in treating muonic molecular positive ions [nucleon (1), nucleon (2), μ^- (3)] consists in the neglect of non-Coulomb contributions to the interaction potential between the particles. With this approximation, the Hamiltonian of a muonic molecule is the sum of the potential energy of the pair wise Coulomb interactions, and of the kinetic energy of the three particles involved:

$$H = \frac{-\hbar^2}{2m_1}\nabla_1^2 - \frac{\hbar^2}{2m_2}\nabla_2^2 - \frac{\hbar^2}{2m_3}\nabla_3^2 - \frac{1}{r_1} - \frac{1}{r_2} + \frac{1}{r} \qquad (1)$$

where r_1 and r_2 denote the distances between each of the two nucleons and the muon, and r is the distance between the nucleons. In the case of the muonium negative ion [e^-(1), e^-(2), μ^+(3)] and positronium negative ion [e^-(1), e^-(2), e^+(3)], the Hamiltonian is identical except for the use of the appropriate particle masses.

COMPUTATIONAL METHOD

As mentioned above, we and others have applied a modified version of the method known in atomic and molecular problems as the integral transform method[22] and in nuclear physics studies as the generator-coordinate method (GCM)[23] to the muonic molecule problem.

For the problem of finding the eigenfunctions of an **N**-particle system with Hamiltonian H, the basic idea of the method is to systematically generate trial functions by the prescription:

$$\Psi((x_1, x_2, \ldots, x_N) = \Psi(\vec{x}_N) = \int_{D_M} S(\vec{t}_M)\Phi(\vec{x}_N, \vec{t}_M)d\vec{t}_M \qquad (2)$$

where D_M is an **M**-dimensional integration domain for the parameter ($\vec{t}$ space), Φ is some known function (and can be thought of as an exact eigenfunction for some model Hamiltonian), and the weight or shape function $S(t_M)$ is to be determined. It is assumed that Ψ, Φ, and S are real.

Insertion of the ansatz (2) into the variational principle and variation with respect to $S(\vec{t}_M)$ yields a Fredholm-type integral equation which may be solved by approximate numerical integration. This produces the secular equations:

$$\sum_{i=1}^{L} W_i S(\vec{t}_i)[K(\vec{t}_i,\vec{t}_j') - EI(\vec{t}_i,\vec{t}_j')] = 0 \tag{3}$$

for $j = 1,2,\ldots,L$, where

$$I(\vec{t}_i,\vec{t}_j') = \int d\vec{x}\ [\Phi(\vec{x}_N,\vec{t}_i)\Phi(\vec{x}_N,\vec{t}_j')]\mathrm{sym} \tag{4}$$

$$K(\vec{t}_i,\vec{t}_j') = \int d\vec{x}_N[\Phi(\vec{x}_N,\vec{t}_i)\hat{H}\Phi(\vec{x}_N,\vec{t}_j')]\mathrm{sym} \tag{5}$$

$[\mathbf{f}\mathcal{O}g]\mathrm{sym} = [\mathbf{f}\mathcal{O}g + g\mathcal{O}\mathbf{f})$, and hence the Hamiltonian kernel K and the overlap kernel I are Hermitian.

The W_i and t_i are weights and abscissas, respectively, for the numerical integration. By choosing the sets $\{t_i\}$ and $\{t_j'\}$ to coincide, a convergent series of upper bounds to the true energy is obtained.

For the particular case of S states of these muonic systems we assume $\Phi(\vec{x}_N,\vec{t}_M)$ to be

$$\Phi(r_1,r_2,r,\alpha,\beta,\gamma) = (4\pi)^{-1}\exp(-\alpha r_1-\beta r_2-\gamma r_3) \tag{6}$$

This choice of Φ is equivalent to the variational treatment with a trial function of the form

$$\psi(r_1,r_2,r) = (4\pi)^{-1}\sum_{k=1}^{N} C_k(1+P_{12})\ \exp(-\alpha_k r_1-\beta\ r_2-\gamma_k r), \tag{7}$$

where P_{12} stands for the permutation operator and enters only for homonuclear muonic molecules. The C_k are linear combination coefficients, determined by diagonalizing the Hamiltonian matrix.

In effect the nonlinear parameters in the wavefunction (7) are chosen to be the lattice points of a three-dimensional quadrature formula, and the linear coefficients are found by solving the secular equation.

It is evident that we may restrict our attention to finite values of α_k, β_k, and γ_k, and so we choose the integration domain D_3 of Eq. (2) to be a paralleltope in three-space, defined by:

$$\alpha_k\varepsilon[A_1,A_2],\beta_k\varepsilon[B_1,B_2],\gamma_k\varepsilon[G_1,G_2] \tag{8}$$

for k=1,2,...,N. This parallelotope is not completely arbitrary. Constraints must be imposed because we are considering bound states and in order to ensure that all of the integrals exist. Incidentally all of the integrals required for the calculation of the energy and various other expectation values, can be done analytically.

The six parameters A_1, A_2, B_1, B_2, G_1 and G_2 which define the paralleltope D_3 are the non-linear variational parameters of the problem with the exponents α_k, β_k and γ_k generated in a pseudo-random manner from these six parameters. Several algorithms (tempering schemes) have been used for generating these exponents[15,19,20,21,24]. As an illustration of such a scheme, we list the one employed by Petelenz and Smith[20]:

$$\alpha_k = \exp[A_1 \langle \tfrac{1}{2}k(k+1)\sqrt{2}\rangle + A_2 \langle \tfrac{1}{2}k(k+1)\sqrt{7}\rangle] , \tag{9a}$$

$$\beta_k = \exp[B_1 \langle \tfrac{1}{2}k(k+1)\sqrt{3}\rangle + B_2 \langle \tfrac{1}{2}k(k+1)\sqrt{11}\rangle] , \tag{9b}$$

$$\gamma_k = \exp[G_1 \langle \tfrac{1}{2}k(k+1)\sqrt{5}\rangle + G_2 \langle \tfrac{1}{2}k(k+1)\sqrt{13}\rangle] , \tag{9C}$$

It should be noted that there exist different approaches to the optimization of the paralleltope. One may optimize it for a smaller value of N (shorter expansion length) and retain the same paralleltope for larger values of N (longer expansion lengths) or reoptimize the paralleltope for each value of N. The latter procedure will lead to more compact expansions for the same accuracy of the eigenvalue.

MUONIC MOLECULAR IONS: ENERGIES

Since Alexander et al have summarized in this volume[25] all known calculations of the energy eigenvalues of the various muonic molecular ions by this and other methods, the reader is referred to that article. In that comparison it is seen that the present method is very accurate and effective.

We turn in the next two sections to consideration of two particular expectation values for the muonic molecular ions.

MUONIC MOLECULAR GEOMETRIES

Wooley[26] has discussed the concept of the bond length in a molecule as an artifact of the Born-Oppenheimer approximation. For the muonic molecular ions where the particles have been treated on equal footing, we have determined the expectation values of various powers of the interparticle distances[27]. In Fig. 1 the expectation values of the interparticle distance are presented for the S states of the various ions. These are essentially identical with those recently calculated by Scrinzi et al[28].

As expected the excited states are larger than the corresponding ground states. Inspection of the homonuclear series indicates that both the muon-nucleon distance and the nucleon-nucleon distance increase as the nucleon mass decreases. Again this trend is to be expected - as the two nucleons become heavier the molecule becomes more compact. In the heteronuclear systems one observes that as one nucleon remains the same and the other becomes heavier, the nucleon-nucleon distance shortens as does the muon-changed nucleon distance. On the other hand, the muon-fixed nucleon distance lengthens slightly.

MUONIC MOLECULES: VACUUM POLARIZATION CORRECTION

As mentioned in the introduction, the standard treatment neglects the non-Coulomb contributions to the interaction potential such as those due to relativistic effects, finiteness of particle size, vacuum polarization, etc. The last correction, which to second order in the fine-structure constant α is the expectation value of the so-called Uehling potential,

$$V_U(r) = -2\alpha\, e^2/(3\pi r) \int_1^\infty dx(x^2-1)^{1/2}[1+1/(2x^2)]\, e^{-2\gamma xr}/x^2 \qquad (10)$$

where $\gamma = m_e c/\hbar$ and m_e is the electron mass seems to be important.

It is customary to define the vacuum polarization correction to the binding energy of the ith state of a muonic molecule as

$$\Delta E_U^i = \int\int|\Psi_i(r_1,r_2,r)|^2[V_U(r_1) + V_U(r_2) - V_U(r)]dv_1 dv_2 - \Delta E_2 \qquad (11)$$

where $\psi_i(r_1,r_2,r)$ is the wavefunction of the ith state of the muonic molecule and ΔE_2 is the vacuum polarization shift of the ground state of the (heavier) muonic atom from which the molecule originates. With this definition ΔE_U is the change of the binding energy of the molecule due to vacuum polarization.

Prior to our investigation[27,29], this correction has been evaluated[30-32] using wavefunctions from the quasi-adiabatic approach. Our initial investigation involved all the S states of these systems. We found that the earlier calculations[30] underestimated E_U by a factor of 2 to 8 depending on the state. This is probably an indication of the insufficient quality of the wavefunctions employed in that study. The present results for the states yield a correction to the binding energies which ranges from 0.25% for the ground S state of ppμ to 0.75% for the excited S state of dtμ.

MUONIUM AND POSITRONIUM NEGATIVE IONS

We mention briefly that these systems have been studied both by use of Hyleraas type wavefunctions[6,33] and by the pseudo-Slater functions with random tempering[7,19]. As in the muonic molecule problems, the latter method is very accurate and effective. Both systems are bound, the binding energy of Mu^- is very close to that of H^-, as one would expect intuitively.

We report here our calculated value from our energetically best wavefunction of the experimentally measureable two-photon annihilation rate Γ_o (2.0928 $nsec^{-1}$) which may be compared with the measured value[4] of 2.09(9) $nsec^{-1}$.

CONCLUDING REMARKS

The discretization of the integral transform (generator coordinate) wavefunction provides a very powerful variational ansatz (pseudo-Slater functions with random-temperated exponents) for the treatment of the quantum mechanical three-body problem. In this laboratory, work is in progress to evaluate the vacuum polarization and other corrections to the binding energy for all the states of the muonic molecular ions.

ACKNOWLEDGEMENTS

Support of this research by the Natural Sciences and Engineering Research Council of Canada is gratefully acknowledged.

REFERENCES

1. L. Bracci and G. Fiorentini, Phys. Reports 86, 169 (1982).
2. D.R. Harshman et al, Phys. Rev. Lett. 56, 2850 (1986).
3. Y. Kuang et al, Phys. Rev. A 35, 3172 (1987).
4. A.P. Mills, Jr., Phys. Rev. Lett. 46, 717 (1981); 50, 671 (1983).
5. Y.K. Ho, J. Phys. B 16, 1503 (1983).
6. A.K. Bhatia and R.J. Drachman, Phys. Rev. A 28, 2523 (1986).
7. P. Petelenz and V.H. Smith, Jr., Phys. Rev. A 36, 5125 (1987).
8. S.I. Vinitskii, V.I. Korobov and I.V. Puzynin, Sov. Phys. JETP 64, 417 (1987).
9. J. Botero and C.H. Greene, Phys. Rev. A 32, 1249 (1985).
10. J.G. Frey, Mol. Phys. 57, 1 (1986).
11. D. Ceperley and B.J. Alder, Phys. Rev. A 31, 1999 (1985).
12. R.G.J. Ball and B.H. Wells, J. Phys. B 21, 723 (1988).
13. S. Hara, T. Ishihara and N. Toshima, J. Phys. Soc. Japan 55, 3293 (1986).
14. Tempering is a term introduced to the quantum chemical literature by Ruedenberg to describe an algorithm for generating a set of exponents from a few parameters. [K. Ruedenberg, R.C. Raffenetti and R.D. Bardo, in Energy, Structure and Reactivity, D.W. Smith, editor (John Wiley, N.Y. 1973),p.164].

15. A.J. Thakkar and V.H. Smith, Jr., Phys. Rev. A **15**, 1 (1977); **15**, 16 (1977); **15**, 2143 (1977); J. Chem. Phys. **67**, 1191 (1977).
16. A.J. Thakkar, J. Chem. Phys. **75**, 4496 (1981).
17. V.H. Smith, Jr., and P. Petelenz, Phys. Rev. B **17**, 3253 (1978); P. Petelenz and V.H. Smith, Jr., J. Phys. C **13**, 47 (1980); **15**, 3721 (1982); Can. J. Phys. **57**, 2126 (1979); Int. J. Quantum Chem. Symp. **14**, 83 (1980); Int. J. Quantum Chem. **18**, 583 (1980); Phys. Rev. B **21**, 4884 (1980); **23**, 3066 (1981); J. Low Temp. Phys. **38**, 413 (1980).
18. D.G. Kanhere, A. Farazdel and V.H. Smith, Jr., Phys. Rev. B **35**, 3131 (1987.
19. A.M. Frolov and V.D. Efros, JETP Lett. **39**, 544 (1984); J. Phys. B **18**, L265 (1985); A.M. Frolov, Z. Phys. D **22**, 61 (1986); Sov. J. Nucl. Phys. **44**, 380 (1986); V.D. Efros, Sov. Phys. JETP **63**, 5 (1986).
20. P. Petelenz and V.H. Smith, Jr., Phys.Rev.A **36**, 4078 (1987).
21. S.A. Alexander and H.J. Monkhorst, Phys. Rev. A., accepted.
22. D.L. Hill and J.A. Wheeler, Phys. Rev. **89**, 1102 (1953); J.A. Wheeler, Nuovo Cimento Suppl. **2**, 908 (1953); J.J. Griffin and J.A. Wheeler, Phys. Rev. **108**, 311 (1957).
23. R.L. Somorjai, Chem. Phys. Lett. **2**, 399 (1968), Phys. Rev. Lett. **23**, 329 (1969), J. Math. Phys. **12**, 206 (1971).
24. S.A. Alexander, H.J. Monkhorst and K. Szalewicz, J. Chem. Phys. **85**, 5821 (1986).
25. S.A. Alexander, H.J. Monkhorst and K. Szalewicz, this volume.
26. R.G. Wooley, Adv. Phys. **25**, 27 (1976), Isr. J. Chem. **19**, 30 (1980); P. Claverie and S. Diner, Isr. J. Chem. **19**, 54 (1980).
27. P. Petelenz and V.H. Smith, Jr., Phys. Rev. A, submitted for publication.
28. A. Scrinzi, H.J. Monkhorst and S.A. Alexander, Phys. Rev. A, submitted for publication.
29. P. Petelenz and V.H. Smith, Jr., Phys. Rev. A **35**, 4055 (1987).
30. V.S. Melezhik and L.I. Ponomarev, Phys. Lett. **77B**, 169 (1982).
31. V.S. Melezhik, JETP Lett. **36**, 125 (1982).
32. D.D. Bakalov, V.S. Melezhik, L.I. Menshikov and S.T. Vinitsky, Phys. Lett. **161B**, 5 (1985).
33. A.K. Bhatia and R.J. Drachman, Phys. Rev. A **35**, 4051 (1987).
34. Y.K. Ho, J. Phys. B **16**, 1503 (1983).

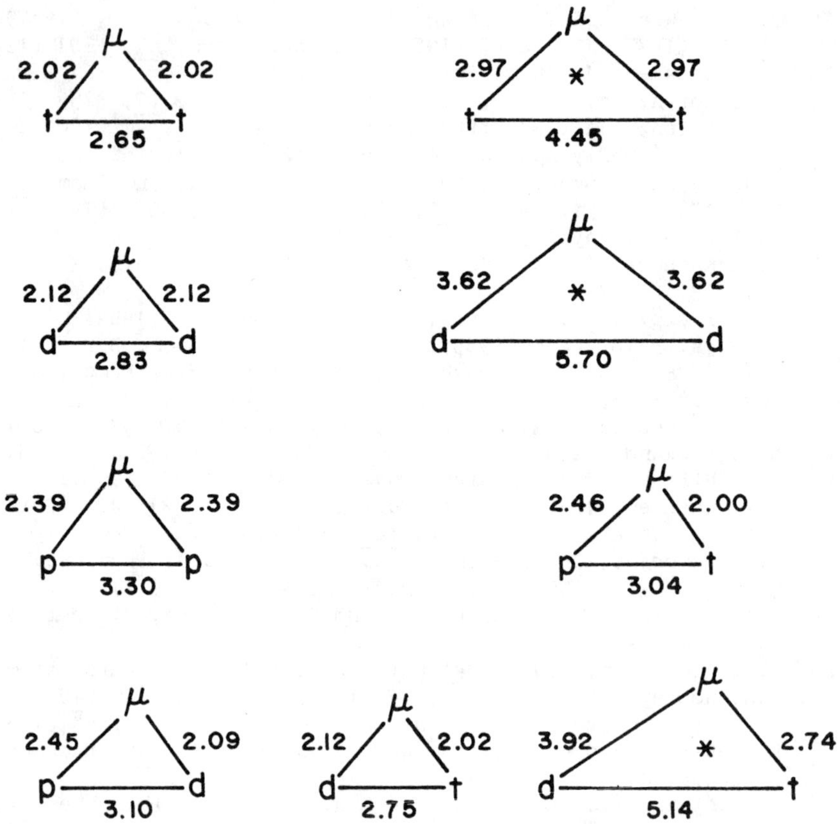

Fig. 1 - Geometries of the (J,v)=(0,0) and (0,1) states of the muonic molecular ions. The (0,1) state geometries are designated by asterisks.

DE-EXCITATION OF MUONIC MOLECULES BY INTERNAL CONVERSION

A. K. Bhatia and Richard J. Drachman
Laboratory for Astronomy and Solar Physics, Goddard Space Flight Center, Greenbelt, MD 20771

Lali Chatterjee
Physics Department, Jadavpur University, Calcutta 700032, India

The resurgence of interest in muon catalyzed fusion is largely due to the realization[1] that there exists a resonant process forming the $td\mu$ molecular ion:

$$[t\mu] + D_2 \rightarrow \{ [td\mu]^* \; d \; 2e \}^*.$$

Here the asterisks indicate that the muonic ion is in its excited J=1 state and that the deuterium molecule has been converted into a vibrationally excited molecule with one nucleus replaced by the muonic ion. The muonic ion is bound by only about 0.6 eV and it is this remarkably small binding energy that makes it possible for the vibrational excitation of the electronically bound molecule to take up the energy resonantly.

Nevertheless, it is possible for this reaction to reverse itself, returning a muonic tritium atom and a ground-state deuterium molecule.[2] If, however, the muonic molecular ion is rapidly de-excited then the system is stabilized and dt fusion will follow. The most efficient process for effecting de-excitation is internal conversion: the excitation energy is carried away by one of the molecular electrons. (We use the nuclear physics term for this process rather than the equivalent atomic terms "Auger effect" or "autoionization" and call the molecular ion the "nucleus".) Since we have previously[3] computed some fairly accurate wave functions for all the J=0 and J=1 states of the muonic molecular ions it seems appropriate to use them (suitably improved) to calculate the rate of internal conversion. We have now also carried out the first variational calculation of the J=2 wave function of the $td\mu$ ion, obtaining a binding energy of 101.59 eV. In the present calculation we have used the best available binding energies for J=0 and J=1 (Ref.4) and for J=2 (Ref.5). Instead of considering the molecular case, in which the emitted electron moves in the field of an H_2^+ ion, we have, for simplicity, assumed that the muonic ion is the nucleus of a hydrogenic atom and have calculated lifetimes and branching ratios for various transitions among the J=0, J=1, and J=2 states.

The transition matrix element has the form

$$M_{if} = \frac{e^2}{(\pi a_o^3)^{1/2}} \iiint d\vec{r}_1 d\vec{r}_2 d\vec{r} \, \frac{e^{-r/a_o} F_c^*(k,r)}{r^2} \, \psi_i(\vec{r}_1,\vec{r}_2)\psi_f^*(\vec{r}_1,\vec{r}_2) \, (\vec{d}.\hat{r}),$$

where $\vec{d}.\hat{r}/r^2 = \{(m_1-1)\vec{r}_1 + (m_2+1)\vec{r}_2\}.\hat{r}/r^2$ is the dipole term in the perturbation responsible for the internal conversion. We have used the following definitions:

$$m_1 = \frac{M_d}{M},\ m_2 = \frac{M_\mu}{M},\ \text{and } M = M_d + M_t + M_\mu.$$

ψ_i and ψ_f are the initial- and final-state wave functions and are of the form:

$$\psi_J^m = \frac{Y_J^m(\hat{r}_1)}{(2\pi)^{1/2}} f_J(r_1, r_2, r_{12}),$$

where f_J has the form $e^{-(\gamma r_1+\delta r_2)} \sum C_{abc} r_1^a\, r_2^b\, r_{12}^c$. These forms differ from those used[3] in our previous work because of the different definitions of $\vec{r}_1$ and $\vec{r}_2$; in the present case the origin is on the triton rather than on the muon. We have also restricted the form of the J>0 wave functions to reflect the fact that the rotational excitation energy of the heavy particles is much lower than the energy of the excited states of the tμ atom. $F_c(k,r)$ is the Coulomb function for the emitted electron. The transition rate can be written in the following form:

$$\lambda = \frac{2\pi}{\hbar} \frac{1}{2J_i+1} \sum_{m_i m_f} |M_{if}|^2 \rho(E)$$

$$= 5.5519 \times 10^{12} \frac{\beta F_{Coul}(k)\ I^2}{2J_i+1} \sec^{-1},$$

where β=1 for 0-1 transitions and β=2 for 1-2 transitions,

$$F_{coul}(k) = e^{-\frac{4}{k}\tan^{-1}k}\left\{\frac{2\pi}{(1+k^2)\,(1-e^{-2\pi/k})}\right\},$$

$$\text{and } I = \int d\tau \frac{f_J f_{J'}}{2r_1}\left((2m_1 + m_2 - 1)r_1^2 + (m_2+1)(r_2^2 - r_{12}^2)\right).$$

We obtain the following rates ($\sec^{-1}$) for the various transitions:

$\lambda_{11\to01} = 1.61\text{x}10^{11}$ $\lambda_{10\to00} = 8.27\text{x}10^{9}$

$\lambda_{11\to20} = 2.14\text{x}10^{10}$ $\lambda_{20\to10} = 1.28\text{x}10^{10}$

$\lambda_{01\to10} = 8.30\text{x}10^{9}$ $\lambda_{11\to00} = 3.97\text{x}10^{8}$

We find that the total decay rate of the J=1, v=1 state is $1.83\text{x}10^{11}\sec^{-1}$, which is to be compared to the rate for breakup into a tμ atom[2]. We also find that the excited J=0 state is favored by a factor of 400 compared with the ground state. This large factor can be understood by projecting out the nuclear part of the wave function and comparing the two excited states. These two states are found to be quite similar in form with a node roughly at the same point while the ground state is nodeless. This is apparently the reason why the excited J=0 state is so heavily favored, even though the energy release is so much smaller.

[1]E. A. Vesman, Pis'ma Zh. Eksp. Teor. Fiz. 5, 113 (1967) [JETP Lett. 5, 91 (1967)]; S. S. Gerstein and L. I. Ponomarev, Phys. Lett. 72B, 80 (1977).
[2]A. M. Lane, Phys. Lett. 98A, 337 (1983).
[3]A. K. Bhatia and R. J. Drachman, Phys. Rev. A 30, 2138 (1984).
[4]K. Szalewicz, H. J. Monkhorst, W. Kołos, and A. Scrinzi, Phys. Rev. A 36, 5494 (1987).
[5]S. I. Vinitskii, V. S. Melezhik, L. I. Ponomarev, I. V. Puzynin, T. P. Puzynina, L. N. Somov, and N. F. Truskova, Zh. Eksp. Teor. Fiz. 79, 698 (1980) [Sov. Phys. JETP 52, 353 (1980)].

Chapter 4
Muon Sticking
and
Muon Regeneration

COMPREHENSIVE THEORY OF NUCLEAR EFFECTS ON THE INTRINSIC STICKING PROBABILITY: I

Michael Danos*
National Bureau of Standards, Gaithersburg, MD 20899

L. C. Biedenharn**,† and Alfons Stahlhofen†
Department of Physics, University of Texas, Austin, TX 78712

ABSTRACT

A comprehensive, inherently many-body, reaction theory for an accurate calculation of the intrinsic sticking fraction, ω_s^0, for (dtμ) fusion is outlined. The non-perturbative treatment of the long-range Coulomb force and its interference with the short range nuclear force is emphasized.

1. INTRODUCTION

Since the beginning of muon-catalyzed fusion research, it has been agreed that the sticking fraction, ω_s, is one of the most important rate-limiting factors in the (dtμ) cycle.[1] An accurate calculation of the intrinsic sticking fraction, ω_s^0 — the objective of our work — has to take into account:

(i) the properties of the (dtμ) molecule
(ii) the nuclear physics involved in fusion
(iii) the subsequent fission of the intermediate states into the (αn)μ-system
(iv) the (kinematical) transformation describing correctly the final (αn)μ system.

A precision calculation of the sticking probability obviously encounters a basic difficulty:

The (dtμ) fusion involves simultaneously molecular-, atomic-, and nuclear physics in a non-perturbative way to a single overall reaction process, thus covering an energy range from (- 0.6) eV to (+ 17.6) MeV.

---

*Speaker, Part I

**Speaker, Part II

†Also Department of Physics, Duke University, Durham, NC 27706.
Work supported by the D.O.E.

Our approach is based on the conventional understanding of the μCF cycle, where the path of the muon leads from a (tμ) mu-atom (or a (dμ) mu-atom and a subsequent transfer reaction) to the molecular (1,1)- state of the (dtμ) molecule, which then decays by an electromagnetic (Auger-) transition to the fusing states $(0,0)^+$ and $(0,1)^+$.[2] The fusing states then decay via nuclear processes into a neutron and an alpha-muon system with a stuck (bound state) or unstuck (scattering state) muon. According to this framework we neglect

(a) direct fusion from all (dtμ) states with $J \neq 0$;[3]
(b) the influence of the "host-system" (i.e., the second deuteron and the electron cloud) on the static properties of the (dtμ) molecule;
(c) Coulomb excitation of the $^5He(3/2^+)$ compound system by the muon;
(d) the (Coulomb-) polarizability of the channel nuclei (i.e., d, t, ^4He).

(Note that — apart from the Auger-transition — we ignore that the (dtμ) molecule is in fact part of the muo-molecular complex ((dtμ)dee.[4]) The contributions from the assumptions (a) through (d) to the final results can be estimated by a perturbational treatment (if necessary); thus the magnitude of the uncertainties is under control.

The complete Hamiltonian for the six-body (nuclear plus muon) system can be written as

$$H = H_s + H_{int} \tag{1a}$$

where

$$H_{int} \equiv - \vec{\Omega} \cdot \langle\vec{\omega}\rangle \tag{1b}$$

is the electromagnetic interaction Hamiltonian in the dipole approximation. The Auger-transition will be treated in first order; thus the transition probability per unit time is given by the (formal) expression

$$W_f = \frac{2\pi}{\hbar} \sum_i |\langle f_i| \, \vec{\Omega} \, |\vec{d}\rangle|^2 \, \rho_i(E) \quad , \tag{2}$$

where $|f_i\rangle$, $|\vec{d}\rangle$ are the wave functions of the fusing and the molecular (1,1) ("doorway") state, computed as stationary eigenstates of the Hamiltonian H_s. (The matrix-element involving the electron of the molecule participating in the transition is written as $\langle\vec{\omega}\rangle$; $\rho_i(E)$ abbreviates the density of states at the energy E.)

It should be noted, that the total fusion rate (i.e., the sum of the "sticking" and "unsticking" fusions) is given by the Auger transition rate, eq. (2). This observation results from the fact that — due to the presence of the nuclear interactions — the fusing states ($J = 0$; $\nu = 0,1$) are non-stationary states, i.e., they decay via the nuclear reaction. (They are in fact a continuum of states, which must be treated non-perturbatively.) This observation is the starting point of our work.

We sketch in the following the computation of the fusing states $|f_i(E)\rangle$ and the asymptotic (observable) $(\alpha\mu)n$ system. This requires a detailed discussion of the incorporation of the nuclear effects, the Auger transition and the kinematics of the $(n\alpha\mu)$-channel.

2. THE $(dt\mu)$ FUSING STATES

In the center-of-mass coordinate system the configuration space is a 15-dimensional (Euclidian) space. A comprehensive treatment of the nuclear degrees of freedom is, of course, out of question since the nuclear interactions are only partly known. However, the use of Wigner's method of breaking this (high dimensional) configuration space into separate regions with effective Hamiltonians and matching conditions at the boundaries makes an accurate calculation feasible; the accuracy is limited only by uncertainties in the experimental data of the nuclear system, i.e., $^5He^*$. These data — accessible from $(n\alpha)$ and (dt) scattering experiments — are given in parametrized form by the (energy dependent) elements of the R-matrix.[5a,b]

We divide the full configuration space in three regions. In the compound nuclear region (i.e., the $^5He^*$ region, called region III) the nuclear forces are active between all nucleons. When approaching the $(dt\mu)$ region (here defined as region I), the nuclear wave function splits into the two clusters d and t, which fully separate beyond the region of the nuclear forces at the boundary (i.e., the hypersurface) $r_{dt} = a$. A similar situation occurs in region II, where the break-up leads to an alpha particle and a neutron. Here the channel boundary (i.e., the range of the nuclear forces) is defined at the "channel radius" (in Wigner's terminology) $r_{n\alpha} = b$. (Let us remark here, that the typical range of nuclear forces is of the order 3-5 fm, whereas the muonic Bohr radius is of the order 250 fm.)

It follows, that the effective Hamiltonians in regions I and II (away from the channel radii) are given by

$$H_S \to H_{dt\mu} = \sum_{i=d,t,\mu} \frac{p_i^2}{2m_i} + \frac{e^2}{r_{dt}} - \frac{e^2}{r_{d\mu}} - \frac{e^2}{r_{t\mu}} \tag{3a}$$

and

$$H_S \to H_{\alpha\mu n} = \sum_{i=\alpha,\mu,n} \frac{p_i^2}{2m_i} - \frac{2e^2}{r_{\mu\alpha}} - \frac{e^2}{r_{t\mu}} \ . \tag{3b}$$

In the neighborhood of the nuclear boundary the Hamiltonian separates to the form

$$H_s \to H_{nucl} + H_{He^*+\mu} \quad . \tag{3c}$$

Accordingly, the full wave function for the three regions can (away from the channel radii) be written in fractorized form:

$$\psi_I = \chi_d \, \chi_t \, \varphi_{dt\mu} \quad , \tag{4a}$$

$$\psi_{II} = \chi_n \, \chi_\alpha \, \varphi_{n\alpha\mu} \quad , \tag{4b}$$

$$\psi_{III} = \chi_{He^*} \, \varphi_{dt\mu} \quad , \tag{4c}$$

where χ_i ($i = d,t,n,\alpha,{}^5He$) is the internal wave function of the system i. In particular, for a given total energy E, one sees, that the wave function in the compound region III has the explicit form

$$\psi_{III}(E) = \sum_{N\ell j} \chi_{He^*}(E - \varepsilon_{N\ell j}) \, \psi^{N\ell j}_{He^*\mu}(\vec{\rho}_\mu) \quad . \tag{5}$$

Here $\varepsilon_{N\ell j}$ is the energy of the $^5He^*\mu$ hydrogenic Coulomb-system; $\vec{\rho}_\mu$ is the muon coordinate with respect ot the $^5He^*$ center-of-mass. (The corrections resulting from the difference in the $^5He^*$ center-of-mass vs. center-of-charge could be incorporated, again, perturbatively (if needed). The summation in eq. (5) abreviates a sum over bound-states (labelled by $N\ell j$) and integration over continuous-states (labelled by $n\ell j$). The factorization is (5) is exact according to the neglect of a Coulomb excitation of $^5He^*$ by the muon; cf. here assumption (c).)

The wave function in eq. (5) shows that one needs in fact the description of the nuclear compond system for all energies E^* defined by

$$E^* \equiv E_{tot} - E_{N\ell j} \quad . \tag{6}$$

This observation has some important practical consequences (see below) and is in itself essential, since the energies $\varepsilon_{N\ell j}$ are of similar magnitude as the width of the $^5He^*(3/2)^+$ resonance.

We now turn to the nuclear reaction formalism required in our treatment. The fundamental R-matrix relation is (in our context) defined by[5]

$$\begin{pmatrix} \psi_{dt}(a) \\ \psi_{n\alpha}(a) \end{pmatrix} = \begin{pmatrix} R_{11} & R_{12} \\ R_{21} & R_{22} \end{pmatrix} \begin{pmatrix} (\partial\psi_{dt}/\partial r_{dt})|_a \\ (\partial\psi_{n\alpha}/\partial r_{n\alpha})|_b \end{pmatrix}$$

$$\equiv \begin{pmatrix} R_{11} & R_{12} \\ R_{21} & R_{22} \end{pmatrix} \begin{pmatrix} \psi'_{dt}(a) \\ \psi'_{n\alpha}(a) \end{pmatrix} \quad , \tag{7a}$$

relating value and derivative of the nuclear wave functions at the channel radii. Since the problem at hand involves only two channels — (dt) and (nα) — the R-matrix relation can be re-written in propagation form

$$\begin{pmatrix} \psi'_{n\alpha}(b) \\ \psi_{n\alpha}(b) \end{pmatrix} = \begin{pmatrix} P_{11} & P_{12} \\ P_{21} & P_{22} \end{pmatrix} \begin{pmatrix} \psi'_{dt}(a) \\ \psi_{dt}(a) \end{pmatrix} \tag{7b}$$

where

$$P_{11} = -\frac{R_{11}}{R_{12}} \quad , \quad P_{12} = \frac{1}{R_{12}}$$

$$P_{21} = -\frac{R_{11} R_{12} - R_{12} R_{21}}{R_{12}} \quad , \quad P_{22} = \frac{R_{22}}{R_{12}} \tag{7c}$$

(with $R_{12} \neq 0$).

The nuclear forces obviously induce a (linear) relation between the nuclear wave function in the two channels. In other words, the intermediate (compound nucleus) region does not need to be referred to explicitly, since its effects manifest itself in relations between the logarithmic derivatives in the two channels. Hence, the boundary conditions for the (dtμ) molecule at channel radius a cannot be fixed independently from the (nα) channel: both channels must be treated in a selfconsistent manner.[6]

This fact has important consequences for the computation of the (dtμ) fusing state in region I: The R-matrix relation itself (or in its propagation form) provides a complete characterization of the nuclear reaction without requiring further "knowledge" of the compound nucleus. Furthermore, we emphasize again, that neither form of the R-matrix relation defines a logarithmic derivative of the dt channel at the matching radius; it provides only a relation between the two channels. Therefore one needs a treatment accounting simultaneously for both the (dtμ) and (nαμ) channels. It is entirely inadequate to study the (dtμ) fusing state in a stationary approximation, since this would require it to be disconnected from the (nαμ) channel. Furthermore, the coupling (connectedness) of the two channels not only accounts for the non-stationary character of the fusing states, but also determines their degeneracy and multiplicity (see below). And, most importantly, it introduces a strong energy dependence of the sticking fraction over the width of the molecular resonance.

Let us now discuss in detail the technical complications arising from this observation. In general, a system having a continuous spectrum has at each energy as many linear independent solutions as there are open channels. The (dtμ) channels are closed and the wave function vanishes accordingly for $r_{dt} \to \infty$. (This is in fact the only fixed, but still incomplete, boundary condition of our system.)[7]

The degeneracy of the system thus is determined by the degeneracy of the ($n\alpha\mu$) side, i.e., by region II. At each given energy E_{tot}, there are the following independent (and orthogonal sets of states:

(1) a denumerable infinite set of bound ($\alpha\mu$) Coulombic - and recoiling neutron - states,

(2) a continuously infinite set of ($\alpha\mu$)-scattering states and the corresponding recoiling neutron states.

Each of these channel states ((1) or (2)) determines its particular set of boundary conditions for region I. The matching conditions for the nuclear wave functions are given by the R-matrix relation (eq. (7a) or (7b)); they have to be supplemented by the matching condition for the muon wave functions, which can formally be written as:

$$\psi_I(r_{dt} = a, \vec{\rho}_\mu) = \psi_{II}(r_{n\alpha} = b, \vec{\rho}_\mu) \quad . \tag{8}$$

This condition - valid in view of our assumptions, i.e., no Coulomb excitation of the $^5He^*$ nucleus by the muon - propagates the muon wave function over region III.

The infinite number of degenerate states in the ($n\alpha\mu$) region leads obviously to a corresponding degeneracy of the ($dt\mu$) region, which follows from the Hamiltonian dynamics of the system. These states are orthogonal in the combined configuration space I, II, III. In order to proceed we require a criterion for selecting a particular linear combination from this infinitely degenerate set. This criterion is indeed provided by the physical reaction process leading to the ($dt\mu$) and subsequently ($n\alpha\mu$) state. In this process the Auger operator $\vec{\Omega}$ (c.f., eq. (2)) acts on the (1,1) state $|\vec{d}\rangle$:

$$|D\rangle = \vec{\Omega}\ |\vec{d}\rangle \quad . \tag{9}$$

The state $|D\rangle$, eq. (9), is neither an eigenstate of the Hamiltonian (1), nor is it normalized. But it is unique since according to our assumption the doorway state $|\vec{d}\rangle$, once populated, has available only the electromagnetic channels for its decay. According to assumption (a), $|\vec{d}\rangle$ exists only in region I. Thus it can be computed as a standard bound-state Coulomb three-body problem. Since the energy of the Auger electron is a continuous parameter we now understand that the states $|f_i(E)\rangle$ in eq. (2) indeed exist for all energies E, the index i reflects the degeneracies of the ($n\alpha\mu$) channel. In other words, at each energy E of the entire system the (infinite) set of fusing states $|f_i(E)\rangle$ forms a complete orthonormal set[8] where i counts all the states at that energy. We now introduce the projection operator on an energy E, given by

$$P(E) = \sum_i |f_i(E)\rangle \langle f_i(E)| \quad . \tag{10a}$$

The state $|D\rangle$, eq. (9), now can be decomposed into the components of good energy

$$|D(E)\rangle = P(E)|D\rangle$$

$$= \sum C_i(E)\ |f_i(E)\rangle \qquad (10b)$$

where

$$C_i(E) = \langle f_i(E)|D\rangle$$

$$= \langle f_i(E)|\ \vec{\Omega}\ |\vec{d}\rangle\ . \qquad (10c)$$

The immediate consequence is that any other linear combination of the states $|f_i(E)\rangle$ which is orthogonal to the dipole state $|D(E)\rangle$ has a vanishing dipole moment and cannot be reached by the Auger transition, eq. (2). Indeed, if

$$|\tilde{f}(E)\rangle = \sum B_i(E)\ |f_i(E)\rangle$$

then the orthogonality relation implies

$$\langle D(E)|\tilde{f}(E)\rangle = 0 = \sum_i C_i^*(E)\ B_i(E)\quad ;$$

the Auger matrix element is in this case

$$\langle \tilde{f}(E)|\ \vec{\Omega}\ |\vec{d}\rangle = \sum B_i^*\ \langle f_i(E)|\ \vec{\Omega}\ |\vec{d}\rangle$$

$$= \sum B_i^*(E)\ C_i(E) = 0\quad .$$

This proves the above assertion.

The first consequence is the fact that the formation of the mesomolecule selects a unique combination of components at each energy E. Therefore, this fusing state retains a memory of the molecular doorway state $|\vec{d}\rangle$ as well as the transition mechanism. These subtle inter-relations have potentially a great influence in the sticking amplitude which probes the muonic wave function as the nuclei interact.

The Auger process, like every dipole process, emphasizes the large distances. Thus the particular linear combination of components leading to a non-vanishing matrix element maximizes the muonic wave function at large distances from the nucleus, thus emphasizing muon states easily accessible to stripping reactions.

This observation now provides the criterion needed for the selection of a particular linear combination of states. A variational calculation of the ($dt\mu$) fusing state(s) at the (continuous) energy E is feasible in the following procedure:

1) choose a sufficiently large and flexible set of basic states in region I (which could include irregular functions),

2) maximize the overlap of the variational wave function $|\psi\rangle$ with $|D\rangle$ and

3) simultaneously minimize the expression $\langle\psi|\ (H-E)^2\ |\psi\rangle$ in region I (for $r_{dt} > a$.

Here $|\psi\rangle$ is an initially arbitrary linear combination over the functions of the basic set; H is the effective (dtμ) Hamiltonian given in eq. (1) and E is some chosen energy (not determined variationally). This approach approximates in an optimal way the dipole state $|D(E)\rangle$. (Recall that the actual value of E is the energy of the (1,1) doorway state less the energy of the electron emitted in the Auger transition; it therefore is indeed a continuous parameter.)

3. THE FINAL n($\alpha\mu$) - SYSTEM

The (numerically determined) wave function of the fusing state(s) provides via the nuclear, and "molecular" matching conditions (eqs. (7b), (8)) the boundary conditions for the (n$\alpha\mu$) region. Thus, one obtains at each energy E_{tot} a wave function coupling the molecular (1,1) state with the (n$\alpha\mu$) system. For the calculation we still need to consider the compound region III in order to determine the energy splitting between the nucleus and the muon; the nuclear energy (denoted as $E^*_{N\ell j}$ in eqs. (5),(6)) is required for the R- (or P-) matrix. (The energy dependence of the nuclear reaction, manifesting itself in the energy dependence of the R-matrix parameters, is the only "remnant" of the compound system entering the calculation. This dependence is no contradiction to Wigner's R-matrix theory, since the complication is caused by the energy-sharing between nuclear and muonic degrees of freedom.)

We now obtain the correct values of the boundary condition, needed for the matching procedure, by analyzing the three-body wave function ψ_I at the boundary $r_{dt} = a$ in terms of Coulombic wave functions $\psi^{N\ell j}_{He^*\mu}$, as in eq. (5):

$$\chi^{N\ell j}_{dt}(E^*,r_{dt}=a) = \langle\psi^{N\ell j}_{He^*\mu}(\vec{\rho}_\mu)|\ \psi_{dt\mu}(\vec{\rho}_\mu,r_{dt}=a,E)\rangle \tag{11a}$$

$$(\chi^{N\ell j}_{dt})'(E^*,r_{dt}=a) = \langle\psi^{N\ell j}_{He^*\mu}(\vec{\rho}_\mu)|\frac{\partial}{\partial r_{dt}}\ \psi_{dt\mu}(E,\vec{\rho}_\mu,r_{dt})\Big|_{r_{dt}=a}\rangle\ . \tag{11b}$$

The projection involves also the angular parts of $\vec{r}_{dt}$.

Since the wave functions $\psi_{He^*\mu}$ form a complete set, one achieves the matching of the muon wave function (eq. (8)) by means of eqs. (11). (Recall that the actual wave function $\psi_{dt\mu}$ at the matching radius $r_{dt} = 0$ is now a sum over products of (dt) nuclear wave functions (evaluated at the matching radius) and $^5He^*$-μ Coulomb functions; the sum is over all muon states.) By means of eq. (7b) one obtains now without further complications the complete boundary conditions for the wave function of region II at the matching radius $r_{n\alpha} = b$. In other words: The values of the wave functions χ_{dt} (and χ'_{dt}), as computed by eqs. (11), depend on the total energy as well as the muonic energy (labelled by $N\ell j$). Furthermore, the P-matrix depends on $E^* = E_{tot} - \varepsilon_{N\ell j}$.

Hence the transfer of the boundary conditions from region I ($dt\mu$) to region II ($n\alpha\mu$) results in fact in a set of boundary conditions, i.e., one condition for each Coulomb component, labeled by $N\ell j$, of region III as obtained from the unique fusing state $|D(E)\rangle$.

It follows, that the expansion of the three-body wave function into the components of region III wave functions (as formally written in eq. (5) and explicitly achieved by eqs. (11)) is needed in order to determine $E^* \equiv E^*_{N\ell j}$, required for propagating the wave function from region I to region III by means of the P-matrix.

The $(\alpha n)\mu$ wave function obtained by the matching procedure has the form

$$\psi_{IIN} = \sum_{N\ell j} \chi_{II}^{N\ell j}(r_{n\alpha} = b)\ \psi_{(n\alpha)\mu}^{N\ell j}(\vec{\rho}_\mu) \quad ; \tag{12a}$$

it is valid only in the neighborhood of the channel radius $r_{n\alpha} = b$. The general (physical) wave function in region II is determined from the effective Hamiltonian and reads (omitting the internal wave function)

$$\psi_{II} = \sum_{\{\beta\}} A_{\{\beta\}}\ \psi_n^{\ell_n k}(\vec{\rho}_n)\ \psi_{\alpha\mu}^{n\lambda\kappa}(\vec{\rho}_\mu) \quad , \tag{12b}$$

where $\{\beta\}$ is the set $\{n\lambda\kappa, \ell_n k\}$ of the muon quantum numbers (i.e., $n\lambda\kappa$; for the scattering states $n \to \eta$) and the angular momentum quantum numbers of the neutron. The wave function (12b) is written in the set of Jacobi-coordinates describing the physical system asymptotically: $\vec{r}_{\mu\alpha}$ is the muon coordinate with respect to the alpha particle and $\vec{\rho}_n$ is the neutron coordinate with respect to the $(\alpha\mu)$ center-of-mass. In order to determine this wave function (and to use the information for the $(n\alpha\mu)$-channel contained in the R-matrix relation, i.e., eq. (7.6)), the asymptotic wave function has to be matched to the interior ("nuclear") wave function (12a), which is written in terms of the "nuclear" Jacobi-coordinates $\vec{r}_{\mu\alpha}$ and $\vec{\rho}_n$ (the muon coordinate with respect to the (αn) center-of-mass). This (unpleasant) exercise is performed by re-writing the asymptotic wave function, given in the coordinates $(\vec{\rho}_n, \vec{r}_{\mu\alpha})$, in terms of the nuclear coordinates ($\vec{r}_{n\alpha},\vec{\rho}_n$) and matching value and derivative (with respect to $\vec{r}_{n\alpha}$) of both wave functions, which are now expressed in a common coordinate set, at the matching radius $r_{n\alpha} = b$.[9] This matching procedure allows the complete determination of the final wave function.

The desired amplitude of the final states (especially of the sticking states) have to be obtained from this form by a projection; the actual calculation requires the (non trivial) evaluation of various angular, and radial, overlap integrals.

The rather lengthy expressions will not be given here, but can be found elsewhere.[10]

The essential point for our discussion is that the result for the (sticking amplitudes $A_{\{\beta\}}$ corresponding to a given sticking state $\{\beta\} = \{N\ell j,\ell_n k\}$ turns out to be of the factorized form

$$A_{\{\beta\}} \sim F_{\{\beta\}}(\quad)\,(j_L(kb)\cos\delta - n_L(kb)\sin\delta)\quad, \qquad (13)$$

where L is the relative orbital angular momentum of the ($n\alpha$) system, described by a superposition of spherical Bessel and Neumann functions and a phase shift δ. (In our case L = 2 according to the angular momentum selection rules enforced by the $^5He^*(3/2)$ resonance.)

The function $F_{\{\beta\}}$ is a complicated and more or less slowly varying function of the energy; it depends on the quantum numbers of the sticking state. The phase shift δ changes from δ_0 to $\delta_0 + \pi$ as the energy sweeps over the width of the moluclar resonance(s) of the ($dt\mu$)-system, i.e., as the energy sweeps over the energies of the (0,0) and (0,1) states. This observation is easily demonstrated by considering the (molecular) R-matrix of the complete ($n\alpha\mu$) system. In the vicinity of the molecular resonance this matrix can be extremely well described by a one-level approximation, plus a (constant) backgrown term, this leads to the result given above.

As an immediate consequence of this result, the amplitude $A_{\{\beta\}}$ goes to zero within this "line width". The actual place of this zero-crossing depends on the "background" phase δ_0, which can be different for the (0,0) and (0,1) resonances and for the different sticking states.

4. THE STICKING PROBABILITY

In order to obtain the branching fraction to the sticking states we return to a more detailed discussion of the transition probability, given in eq. (2). This relation implies that, in view of assumption (a), the width of the doorway state is given by the electromagnetic transition to the fusing states, including the two-step transition to the (2,ν) states.

Accordingly, the total width of the doorway state can be written as

$$\Gamma = \sum_{\{\beta\}} \Gamma_{\{\beta\}} \quad , \tag{14}$$

where $\Gamma_{\{\beta\}}$ abbreviates the partial widths to the different orthogonal final states. Now eq. (2) can be re-written, extracting the amplitude of the final state from the matrix element:

$$\Gamma \equiv \frac{2\pi}{\hbar} \sum_{\{\beta\}} |\langle f_{\{\beta\}}(E)| \ \vec{\Omega} \ |\vec{d}\rangle|^2 \ \rho_{\{\beta\}}(E) \tag{15}$$

$$\Gamma_{\{\beta\}} = \frac{2\pi}{\hbar} |\langle f_{\{\beta\}}(E)| \ \vec{\Omega} \ |\vec{d}\rangle|^2 \ \rho_{\{\beta\}}(E)$$

$$= \frac{2\pi}{\hbar} |A_{\{\beta\}}(E)|^2 \ |\langle f(E)| \ \vec{\Omega} \ |\vec{d}\rangle|^2 \ \rho_{\{\beta\}}(E) \tag{16}$$

Since the amplitudes of the final states satisfy the obvious relation

$$\sum_{\{\beta\}} |A_{\{\beta\}}|^2 = 1 \quad , \tag{17}$$

and the sticking probability is defined as the branching ratio

$$\omega_s = \frac{\Gamma(d+t+\mu \rightarrow (\alpha\mu)+n)}{\Gamma(d+t+\mu \rightarrow (\alpha\mu)+n) + \Gamma(d+t+\mu \rightarrow \alpha+\mu+n)} \tag{18}$$

we see, that the branching ratio into a particular state can be written as

$$B_\beta = \frac{\Gamma_{\{\beta\}}}{\Gamma} = \frac{|A_{\{\beta\}}|^2 \ \rho_{\{\beta\}}(E)}{\rho(E)} \quad . \tag{19}$$

Here $\rho(E)$ abbreviates an effective density of states. The total sticking probability, obtained by summing (19) over all sticking states, requires the evaluation of the "density of states functions" ($\rho(E)$) in eq. (19). This can be done either by using the formalism of hyperspherical coordinates[11a] or the formalism of nuclear reaction theory[11b].

5. CONCLUDING REMARKS

Apart from the technical details, a complete reaction theory for the muon catalyzed (dt)-fusion has been formulated. The theory allows the correct calculation of the intrinsic sticking fraction.

REFERENCES

1. Apart from the "sticking" problems, the formation process (especially the so-called "q_{1s} problem") appears to be an important rate limiting factor; cf., for example, the contributions presented by L. I. Ponomorev (these Proceedings).

2. The bound states of the (dtμ) molecule are labeled by $(J,\nu)^{\pi}$, i.e., orbital angular momentum (J), varitional quantum number (ν), and parity (π).

3. Chiu-Yu Hu, Phys. Rev. A36, 4135 (1987) and references therein; see also the contribution by P. Froelich (these Proceedings).

4. For preliminary estimations of the resulting (small) error see the contributions by Lali Chatterjec, A. Scrinzi; these Proceedings.

5. A review of the R-marix procedure can be found in: L. C. Biedenharn et al., "Resonance Processes with Fast Neutrons," in: Fast Neutron Physics, Part II, J. B. Marion and J. L. Fowler (eds.), Interscience Publishers, New York (1963); Applications of the R-matrix procedure to μCF are: R. E. Brown et al., Phys. Rev. C35, 1999 (1987); Phys. Rev. C36, 1220 (1987).

6. The consequences of the selfconsistency condition and the interplay between a nuclear and molecular R-matrix relation (when considering the full (dtμ) and (nαμ) channels) is discussed in detail in "Comprehensive Theory of Nuclear Effects as the Intrinsic Sticking Probability: II," these Proceedings, contribution presented by L. C. Biedenharn; in the following refered to as Part II.)

7. In part two is shown, that the non-perturbative closure of the (dtμ) channel by the muon turns the non-stationary fusing states into a (plethora of) narrow molecular resonances.

8. The normalization is defined, a priori, by the connections on the S-matrix (cf. Part II).

9. The effects of this kinematical transformation on the sticking fraction are discussed in Part II, section 4.

10. M. Danos, L. C. Biedenharn, and A. Stahlhofen, "Intrinsic Sticking in DT Muon Catalyzed Fusion: Interplay of Atomic, Molecular and Nuclear Phenomena", April 1988, to be submitted for publication.

COMPREHENSIVE THEORY OF NUCLEAR EFFECTS ON THE INTRINSIC STICKING PROBABILITY: II‡

Michael Danos
National Bureau of Standards, Gaithersburg, MD 20899

L. C. Biedenharn*,** and Alfons Stahlhofen
Department of Physics, University of Texas, Austin, TX 78712

ABSTRACT

An accurate calculation of the intrinsic sticking fraction, ω_s^0, for $(dt\mu)$ fusion requires the development of a comprehensive, non-perturbative, inherently many-body, reaction theory. We identify and discuss the key problems underlying our construction of this theory.

0. INTRODUCTION

There is general agreement[1] that an accurate calculation of the intrinsic sticking fraction, ω_s^0, is one of the more important problems in muon catalyzed fusion research. An accurate calculation means, in essence, that the theoretical basis must be comprehensive, with controlled, pre-assigned error bounds. In particular, heuristic methods such as complex (optical) potentials and the like are ruled out *ab initio*[2].

To attempt such a precision calculation for μcf is a daunting task, since the problem, as we shall discuss, is one of surprising complexity, in a way that was not at all obvious to us when we began. The basic difficulty is that the muon-catalyzed fusion process involves three distinct fields of physics— molecular, atomic, and nuclear—which (despite their disparate energy scales) are intricately interwoven, ***non-perturbatively***, into a single overall reaction process. Moreover, unlike standard (Wigner, non-perturbative) reaction theory the problem is inherently a many body problem.

Let us sketch the (d, t) fusion process; we use the standard view, as developed by Ponomarev. The muon enters a deuterium-tritium mixture, is captured into a $(t\mu)$ mu-atom (or transfers to this from $(d\mu)$) and then forms a resonant excited muo-molecular complex: $((dt\mu)^*$ dee$)^*$, with $(dt\mu)^*$ denoting the excited configuration[3] with $J = 1$ and $\nu = 1$.

‡Invited paper for the Muon Catalyzed Fusion Workshop 1988, Sanibel Island, FL, 1-6 May 1988. Work supported by the D.O.E.
*Speaker
**Also Department of Physics, Duke University, Durham, NC 27706

The $(1,1)(dt\mu)^*$ system then de-excites by the Auger process[4] into the two fusing states $(J = 0, \nu = 0)$ and $(J = 0, \nu = 1)$, which decay via nuclear processes into a neutron, an alpha and a stuck (or unstuck) muon. [The $(J = 1, \nu = 1)(dt\mu)^*$ system does not have any significant direct fusion[5]].

This is, as stated, the standard scenario as originated by Vesman, Ponomarev, and others. Where is the difficulty?

The problem comes from the fact that the $(dt\mu)$ fusing system with $(J = 0; \nu = 0, 1)$ lies in the continuum. This continuum consists of a denumerably infinite part (labelled by the bound states of $(\alpha\mu)$) and a continuum infinite part (labelled by the $(\alpha + \mu)$ scattering states). This continuum of states *at each energy* must, for controlled accuracy, be treated non-perturbatively. We emphasize that these states are many-body 'molecular' states of the six-body nuclear-plus-muon system.

In our treatment, we have identified several critical issues, five in number, which are essential to an understanding of the problem. Rather than developing here in any detail our proposed comprehensive theory—a technically very complicated and accordingly rather un-illuminating exercise —we will instead focus our discussion today on these critical issues themselves, hoping to convey thereby an intuitive insight into the true nature of the problem.

1. CLOSURE OF THE $(dt\mu)$ CHANNEL: A NON-PERTURBATIVE EFFECT

In the absence of nuclear interactions, the $(dt\mu)$ molecule has five bound states, that is, states having a finite perimeter. Now let us turn on the nuclear interactions; only the two states with $J = 0$ will be significantly affected, but the molecular states will still have a finite perimeter viewed in the $(dt\mu)$ channel.

Let us focus on the motion of the d and t. Close together, but just outside the range of nuclear forces, the (dt) wave function[6] will be governed by the long range forces. The essential point, though, is that this (dt) wave function belongs to an *open nuclear channel*, when viewed locally in this way. However, the $(dt\mu)$ molecule we noted has a finite perimeter, which means that the effect of the muon is to generate reflections in the (dt) wave function so as to *close* the channel at large distances.

This closure is a non-perturbative effect (it is well known that Coulomb bound states cannot be achieved perturbatively).

What is the energy scale for significant changes in the (dt) wave function viewed near the nuclear boundary? Since the effect is of Coulombic origin, and involves reflections from distances of the order of a few muon Bohr radii, it follows that the scale is of order millielectron-volts for a change in the reflection phase of order π. This is very small compared to the ${}^5He^*, (3/2+)$ nuclear resonance width ($\sim$ 50KeV).

Thus we conclude that in the presence of nuclear interactions the non-perturbative closure of the (dt) channel by the muon generates, in the continuum, narrow ($\sim meV$) three-body $(dt\mu)$ molecular resonances. (We have constructed simple models[7] which support this conclusion.)

Most importantly we see that these resonances are inherently non-perturbative and in consequence are very far from bound states embedded in the continuum amenable to perturbative analysis. We will see later that these molecular scale resonances can generate sharp changes in the intrinsic sticking probabilities.

2. THE CRITICAL RÔLE OF THE AUGER TRANSITION

We have seen that inclusion of nuclear interactions for the $J = 0$ molecular states opens the various $(n\alpha\mu)$ channels and results in a denumerable- plus continuum- infinity of degenerate states at each energy E. This poses our second critical issue: How is one to determine the correct 'fusing state' from this infinity of possibilities?

The answer comes directly from quantum mechanics: the Auger transition itself quite literally 'prepares' the fusing state and determines the proper linear combination of degenerate energy eigenstates. The initial state for the Auger transition, the excited $(dt\mu)^*$ state in the $((dt\mu)^*\text{dee})^*$ complex, can be considered as the discrete $(J = 1, \nu = 1)$ state (being formed by three-body collisions the back-decay can be omitted, at least for an initial calculation)[8]. Moreover, the Auger transition is electromagnetic and hence can be treated perturbatively.

Calling the $J = 1, \nu = 1 (dt\mu)^*$ state the 'doorway' state, $|\psi_d>$, and denoting the dipole operator for the Auger transition as Ω_{op}, we see that the transition creates the 'dipole' state, $|\psi_{\text{dipole}}> \equiv \Omega_{\text{op}}|\psi_d>$, in a way familiar in nuclear physics. This dipole state is *not* an energy eigenstate, but may be expanded (formally), at each energy E, over the continuum of energy eigenstates:

$$P(E)|\psi_{\text{dipole}}> \quad = \quad \sum_i C_i(E)|f_i(E)>, \tag{1a}$$

where

$$P(E) \quad \equiv \quad \sum_i |f_i(E)><f_i(E)|, \tag{1b}$$

is the projection operator on energy eigenstates; the coefficients are defined by

$$\begin{aligned} C_i(E) &= <f_i(E)|\psi_{\text{dipole}}>, \\ &= <f_i(E)|\Omega_{\text{op}}|\psi_d>. \end{aligned} \tag{1c}$$

Thus at each energy E the Auger transition prepares the proper linear combination of degenerate continuum energy eigenstates.

It is important to note explicitly that these energy eigenstates are (aside from $\exp(-iEt/\hbar)$) time-independent standing waves, as required by perturbation theory for electromagnetic transitions. (We will discuss this further in Section 5, in terms of Fermi's Golden Rule.)

3. THE INCORPORATION OF NUCLEAR EFFECTS

The strong (nuclear) interactions of the nuclear components of the $(dt\mu)$ system are, of course, not fully known. This does not prevent an accurate calculation since we may use Wigner's technique for nuclear reactions[9] to parametrize the $^5He^*$ system in the vicinity of the nuclear interaction boundary (channel radii) by means of the R-matrix, which itself is known experimentally to sufficient accuracy.[10] Nuclear interactions are small for $r_{dt} \sim 5fm$ and since this distance is small compared to the muon Bohr radius ($\sim 250fm$), the muon sees the (dt) system as, in effect, $^5He^*$. (The center-of-charge does not correspond to the center-of-mass of the $^5He^*$ system, but this effect can, if necessary, be taken into account perturbatively.)

It follows that, in the neighborhood of the nuclear boundary region, the Hamiltonian separates,

$$H = H_{\text{nuclear}} + H_{\mu+^5He^*}$$

so that the $(^5He^* + \mu)$ wave function is a sum of products:

$$\psi_{\mu+^5He^*} \quad = \quad \sum_{\{\alpha\}} \psi_{\{\alpha\}}(\vec{r}_\mu)\,\varphi_{\{\alpha\}}(\vec{r}_{\text{nucl}}), \tag{2}$$

where $\{\alpha\}$ denotes the muon quantum numbers $(N, l, j, m), \psi_{\{\alpha\}}$ are Coulomb wave functions for the muon, and $\varphi_{\{\alpha\}}$ nuclear wave functions. (All other quantum numbers are suppressed.)

The nuclear wave function $\varphi_{\{\alpha\}}$ has components in the two open channels, (dt) and $(n\alpha)$, which have differing kinetic energies corresponding to the difference in thresholds for the two channels.

It is important to note that the nuclear energy for each of the states $\varphi_{\{\alpha\}}$ differs:

$$E^*_{\{\alpha\}} \quad = \quad E - E_{\{\alpha\}}(\text{muon}). \tag{3}$$

For the nuclear system, corresponding to the Hamiltonian H_{nucl}, the Wigner R-matrix approach asserts that there is a (2×2) matrix relation between the value (V) and derivative (D) of the radial wave functions of the nuclear channels at the nuclear radius. That is:

$$\mathbf{V} = \mathbf{R}\ \mathbf{D}, \tag{4}$$

where $\mathbf{V}$, $\mathbf{D}$ are column matrices and $\mathbf{R}$ is a 2×2 matrix. The existence of this linear relation is a consequence of the fact that the kinetic energy in H_{nucl} is quadratic in the momenta, which implies (at a given energy) that we have a Wronskian relation equivalent to eq. (4).

If we now consider the six-body molecular system as an entity, it is clear that a similar Wronskian relation must exist for all the (combined nuclear plus muon) molecular channels as well. That is to say, at each total energy E there exists a (large!) *molecular* R-matrix in the form:

$$\mathbf{R}_{\text{molecular}} = \text{diag}(\mathbf{R}_1, \mathbf{R}_2, \mathbf{R}_3, \ldots), \tag{5}$$

where each block, $\mathbf{R}_i$, is a 2×2 *nuclear* R-matrix, labelled by $\{\alpha\}$, and evaluated at the energy $E^*_{\{\alpha\}}$.

This explicit construction shows a number of interesting features:

(a) The molecular R-matrix (at each total energy E) is of denumerable-plus continuum- infinite dimensionality—a consequence of the large degeneracy of the energy eigenstates.

(b) For a given energy the molecular wave function has nuclear parameters that are evaluated over an effective energy spread of the order of KeV[11].

(c) Each 2×2 nuclear R-matrix does *not* relate the logarithmic derivatives of the two channels, but rather the molecular R-matrix enforces a linear relation between $\mathbf{V}$ and $\mathbf{D}$ involving all channels at once (and not separately).

(d) The basis for the degeneracy in the closed $(dt\mu)$ channel is now evident. Since the nuclear system is *open* for $r_{dt} \to 0$, we may have *irregular* as well as *regular* radial functions in each (dt) component wavefunction. This new degree of freedom exists independently for each (Coulombic) component of the total wavefunction (in the neighborhood of the nuclear boundary). These many new degrees of freedom in the closed $(dt\mu)$ channel wavefunction directly reflect the degeneracy of the fusing molecular system.

(e) Let us remark that this large degeneracy of the closed $(dt\mu)$ channel is removed by the Auger 'state-preparation' process. Since this degeneracy exists at every energy E in the continuum, it follows from quantum-mechanical first principles, that this Auger-weighted fusing state is *extremely sensitive to any external perturbation.* This sensitivity was originally suggested[12] as a possible origin for density effects on sticking as observed in some experiments. Clearly this sensitivity can be exploited further by laser coupling[13] and by magnetic fields[12,14].

(In the interest of clarity, we have neglected in this discussion to pay proper attention to various spin, orbital, ... angular momentum couplings, and the like. These details do not affect the gist of the argument given above[15].)

4. JACOBI COORDINATES AND A KINEMATIC TRANSFORMATION

We have sketched in the preceding sections how an Auger-weighted closed $(dt\mu)$ channel wavefunction can—via the molecular R-matrix—be used to determine the corresponding open $(\alpha n)\mu$ channel wavefunction. The molecular R-matrix, for each muon Coulomb state, determines V and D for (αn) in terms of (V, D) for the (dt) channel as determined by the $(dt\mu)$ Auger-weighted solution.

The problem is that this $(\alpha n)\mu$ wavefunction is valid only in the neighborhood of the nuclear boundary. The resolution is to recognize that the natural basis for the open molecular channel is determined by the Hamiltonian which splits exactly into two parts:

$$H = H_{\text{neutron}} + H_{\alpha\mu}, \tag{6}$$

with H_{neutron} being the free neutron Hamiltonian and $H_{\alpha\mu}$ being the ^4He plus muon Coulomb Hamiltonian. The corresponding $(\alpha\mu)n$ wavefunction is once again a sum of products:

$$\psi_{(\alpha\mu)n}(E) = \sum_{\{\beta\}} \psi_{\{\beta\}}(\vec{r}_{\mu\alpha})\, \chi_{\{\beta\}}(\vec{r}_{\text{neutron}}) \tag{7}$$

with $\psi_{\{\beta\}}$ denoting $^4He + \mu$ Coulomb wavefunctions and $\{\beta\}$ labelling the corresponding bound (N) plus continuum (η) Coulomb states.

The distinction between these two forms of the same wavefunction lies in the choice of two different sets of Jacobi vectors:

$(\alpha n)\mu$ *system:*

$$\begin{aligned} \vec{r}_{n\alpha} &= \vec{r}_n - \vec{r}_\alpha, \\ \vec{\rho}_\mu &= \vec{r}_\mu - (m_n + m_\alpha)^{-1}(m_n \vec{r}_n + m_\alpha \vec{r}_\alpha), \end{aligned} \tag{8a}$$

$(\alpha\mu)n$ *system:*

$$\begin{aligned} \vec{r}_{\mu\alpha} &= \vec{r}_\mu - \vec{r}_\alpha, \\ \vec{\rho}_n &= \vec{r}_n - (m_\mu + m_\alpha)^{-1}(m_\mu \vec{r}_\mu + m_\alpha \vec{r}_\alpha). \end{aligned} \tag{8b}$$

Wavefunctions expressed in these two sets of Jacobi coordinates are, in principle, related by a kinematic point transformation. Such a transformation, although very unwieldly to implement, is nevertheless quite clear in its meaning. Since the $(\alpha n)\mu$ wavefunction is valid only in the neighborhood of the nuclear boundary (unlike the $(\alpha\mu)n$ wavefunction which is valid everywhere at or outside the nuclear boundary) the desired kinematic transformation must involve matching both the value and the (radial) derivative of the various wavefunction components at the nuclear boundary. To be precise, the matching can take place only after transforming to a common set of angular and intrinsic ('alternative', see below) functions.

In effect, the kinematics of the open channels imply a *second* transformation relating (V, D) in the two sets of Jacobi coordinates, in addition to the original R-matrix relation.

This transformation is both cumbersome and complicated; it would serve little purpose to give the details here. The essential point to note (besides the fact that the transformation is well-defined and exists) is that *this kinematic transformation, because of the linear- and angular- momentum mismatch of the recoiling* $(\alpha\mu)$ *system, strongly mixes the effective molecular R- matrix from its earlier block-diagonal form.*

In fact one readily sees that even if the original R-matrix were to involve only the $N = 1$ muon state, the kinematics would strongly mix in states of large N (and η). *The smallness of* ω_s^0 *is primarily due to this kinematic mismatch.*

5. THE DENSITY OF STATES 'PROBLEM'

We have arrived at this stage in the calculation of ω_s^0: starting with the Auger-weighted closed $(dt\mu)$ channel wavefunction (at total energy E) we have obtained the corresponding open channel $(\alpha\mu)n$ wavefunction.

It is not difficult to write out explicitly the general form of such an $(\alpha\mu)n$ wavefunction, but first we must define a suitable nomenclature[16]. An *alternative,* in the language of nuclear physics, denotes a pair: a $^4He + \mu$ system in the state (Nlj) and a neutron (spin 1/2). A *channel,* denoted c, specifies the alternative (labelled Nlj), a relative $(n - (\alpha\mu))$ angular momentum $\vec{L}$ and a channel spin $\vec{s} = \vec{j} + \vec{1/2}$; that is, $c = \{Nlj, L, s\}$. A given channel state requires specifying in addition to c the magnetic quantum numbers M and σ, belonging to $\vec{L}$ and $\vec{s}$ respectively. The wave number of the relative $n - (\alpha\mu)$ radial motion depends on the total energy E and the channel c; we denote it by k_c. Finally we couple: $\vec{L} + \vec{s} = \vec{J}$. (For the unbound $(\alpha + \mu)$ states we let $N \rightarrow \eta$, a continuous parameter.)

Hence for the most general $(\alpha\mu)n$ wavefunction (at energy E, with angular momentum J and parity π) valid outside the nuclear boundary, we have the form:

$$\Psi^{J\pi M_J}_{(\alpha\mu)n}(E) = \sum\!\!\!\!\!\!\int_c \chi_c^{J\pi M_J}\left(A_c^{J\pi M_J} I_c - B_c^{J\pi M_J} O_c\right), \tag{9}$$

where $\chi_c^{J\pi M_J}$ denotes the channel wavefunctions (the $(\alpha\mu)$ Coulomb wavefunction angular momentum coupled to the spin function of the neutron) which then is coupled to the relative orbital angular momentum L; I_c, O_c are in-going (outgoing) spherical Bessel functions for the relative radial motion of the $n - (\alpha\mu)$ system at wavenumber k_c, and $A_c^{J\pi M_J}, B_c^{J\pi M_J}$ are arbitrary numerical constants. The sum (integral) is over the channel index c.

Since the wave function $\psi_{(\alpha\mu)n}$ satisfies a Hamiltonian, the constants $A_c^{J\pi M_J}$ and $B_c^{J\pi M_J}$ are not, in fact, independent but must be related by the S-matrix. This relation is:

$$B_c^{J\pi M_J} \quad = \quad \sum_{c'} S_{c,c'}^{J\pi} A_{c'}^{J\pi M_J}. \tag{10}$$

The S-matrix has the general properties that it is unitary (flux conservation) and symmetric (time-reversal symmetry). Accordingly we may write it in the *eigenphaseshift form*[9]:

$$\mathbf{S}^{J\pi} \quad = \quad (\mathbf{U}^{J\pi})^{-1} \exp{(2i\Delta^{J\pi})}\mathbf{U}^{J\pi}, \tag{11}$$

where $\mathbf{U}^{J\pi}$ is *real and orthogonal* and $\Delta^{J\pi}$ is *real and diagonal.* The dimensionality of the matrices $\mathbf{U}^{J\pi}$ and $\Delta^{J\pi}$ is $k \times k$, where k is the number of open channels.

Introducing this form into eq. (9) we have:

$$\begin{aligned}\psi^{J\pi M_J}_{(\alpha\mu)n}(E) &= \sum\!\!\!\!\!\!\int \chi_c^{J\pi M_J}\, A_{c'}^{J\pi M_J}\left(\delta_{cc'} I_c - S_{cc'}^{J\pi} O_c\right) \\ &= \sum\!\!\!\!\!\!\int \chi_c^{J\pi M_J}\, A_{c'}^{J\pi M_J} \left(\mathbf{U}^{J\pi}\right)_c^{-1}\left(I_c - e^{2i\Delta^{J\pi}} O_c\right)\mathbf{U}_{c'}^{J\pi}, \qquad (12)\end{aligned}$$

where we used:

$$\delta_{c,c'} \quad = \quad (\mathbf{U}^{-1}\mathbf{U})_{cc'}, \qquad (13)$$

and the sum (integral) in eq. (12) is over the pair of indices c, c'. Let us now write the orthogonal matrix $\mathbf{U}^{J\pi}$ in terms of row vectors, denoted by $\mathbf{U}^{J\pi k}$ with components c, that is $(\mathbf{U}^{J\pi k})_c = U_c^{J\pi k}$. We find:

$$\begin{aligned}\psi^{J\pi M_J}_{(\alpha\mu)n}(E) =& \sum\!\!\!\!\!\!\int \tilde{A}^{J\pi k, M_J}\, \chi_c^{J\pi M_J}\, e^{i\Delta^{J\pi k}} \\ & \times \left(\cos\Delta^{J\pi k} F_c(k_c r) - \sin\Delta^{J\pi k} G_c(k_c r)\right) U_c^{J\pi k}, \qquad (14)\end{aligned}$$

(the sum (integral) is over the pair of indices c and k (a diagonal eigenphase label)). We have denoted by $\tilde{A}^{J\pi k, M_J}$ the sum:

$$\tilde{A}^{J\pi k, M_J} = \sum\!\!\!\!\!\!\int U_{c'}^{J\pi k}\, A_{c'}^{J\pi M_J}, \qquad (15)$$

where the sum (integral) is over the index c'.

After these formal manipulations we are now ready to interpret our answer. The most general $(\alpha\mu)n$ wavefunction (with sharp E, J, π, M_J), is— outside the nuclear boundary—a linear combination of eigenphase solutions— labelled by $J\pi k$—each with amplitude $\tilde{A}^{J\pi k, M_J}$. A given eigenphase solution $(J\pi k)$ is a sum of channels, each channel having the (real) amplitude $U_c^{J\pi k}$. (The states describing a given channel are normalized, asymptotically, *a priori,* to unit outgoing flux by the conventions on the S-matrix.)

This interpretation of the $(\alpha\mu)n$ wavefunction is eminently reasonable, but we have yet to interpret the constant $\tilde{A}^{J\pi k, M_J}$, which determines the amplitude of the eigenphase solution labelled by $J\pi k$ (with magnetic quantum number M_J).

If we recall that the fusing state represented in the $(\alpha\mu)n$ channel by eq. (14) is defined by the Auger process, and that this process is *perturbative* since electromagnetic (and outside the S-matrix formalism for the strong interaction), then we see that the amplitude $\tilde{A}^{J\pi k, M_J}$ *is precisely the Auger matrix*

..ement of the normalized eigenphase solution $(J\pi k, M_J)$ *with the dipole state* $\Omega_{\rm op}|\psi_{\rm doorway}>$. That is:

$$\tilde{A}^{J\pi k,M_J} = <\psi^{J\pi k,M_J}|\Omega_{\rm op}|\psi_{\rm doorway}> . \tag{16}$$

With this observation the sticking calculation can be completed, noting that the perturbative (Auger) step can be treated using the Fermi Golden Rule[17]. We find that the *transition probability per unit time,* $P_c(E)$, *from the doorway* $(J = 1, \nu = 1)$ *state to the channel* c (which designates the set of sticking and non-sticking states) at energy E *is given by:*

$$P_c(E) = \frac{2\pi}{\hbar}\rho_{el} \sum_{J,\pi,k_1,k_2} (2J+1)/(2s+1) \cdot U_c^{J\pi k_1} U_c^{J\pi k_2} \cdot$$
$$\cdot <\psi^{J\pi k_1}|\Omega_{\rm op}|\psi_{\rm doorway}> \cdot <\psi^{J\pi k_2}|\Omega_{\rm op}|\psi_{\rm doorway}> \cdot$$
$$\cdot \sin\Delta^{J\pi k_1} \sin\Delta^{J\pi k_2} \cos\left(\Delta^{J\pi k_1} - \Delta^{J\pi k_2}\right), \tag{17}$$

where $\rho_{el}(E)$ is the phase space density of states for the electron multiplied by the absolute square of the electronic Auger matrix element.

We have written eq. (17) in more generality than is probably needed, in order that the coherent interference of overlapping molecular eigenphase resonances would be explicitly in evidence. It is hoped that nature will be kind enough not to yield such complications, and that the diagonal terms with one or more separated resonances will suffice.

Note that the strong interaction "density of states" is automatically taken into account by the reaction theory developed above[17]. Note, too, that the Auger overlap (matrix element with the dipole state) appears in a form with explicit (continuum) normalization conditions on the eigenphase wavefunctions, defined *ab initio* from conventions on the S-matrix.

6. CONCLUDING REMARK

In the preceding sections, 1 through 5, we have discussed the five key points underlying our development of the theory required for an accurate calculation of ω_s^0. We believe this theory provides a feasible way to carry out such a calculation, and Hendrik Monkhorst has discussed at this conference ways to implement the calculation by variational techniques.

7. ACKNOWLEDGEMENTS

We would like to acknowledge, with thanks, the many helpful discussions we have had with our colleagues Hendrik Monkhorst, Susan Haywood, Krzysztof Szalewicz, Armin Scrinzi, and Berndt Mueller; in addition we wish to thank Ryszard Gajewski for his interest and encouragement.

REFERENCES

1. L. I. Ponomarev, Muon Catalyzed Fusion **3** (1988) 629.
2. See, for example, the contribution by G. Hale, these Proceedings.
3. Detailed discussions of the formation process and resulting questions can be found in: M. Jändel et al., "Vacuum Polarization Splitting of the Hydrogenic M Shell and Muon Dynamics in Catalyzed Fusion", CERN preprint (Sept. 1987), CERN-TH 4856187; L. I Menshikov, "Mechanisms of Mesic Molecules $dt\mu$ and $dd\mu$ Formation in Gas, Kurchatov preprint (1988), IAE-4606/2.
4. Here we make the approximation that the influence of the "host- system" (the second deuteron and electron) can be neglected, or if required, treated perturbatively.
5. Chiu-Yu Hu, *Phys. Rev.* **A36** (1987) 4135 (and references therein).
6. We mean the (dt) components of the three body wave function. As discussed below the $(dt\mu)$ wave function is a sum of factors in the neighborhood of the nuclear boundary.
7. M. Danos, L. C. Biedenharn and A. Stahlhofen, "Model for the Nuclear Effects of the $(dt\mu)$ Fusing State", *NBS Publication NBSIR* 87-3532 (April 1987).
8. N. T Padial et al, *Phys. Rev.* **A37** (1988) 329 (and references therein), M. Leon, "The Collisional Broadening of Mesic Molecular Resonances", Los Alamos preprint LA-UR 86-3894 (1986).
9. L. C. Biedenharn et al., "Resonance Processes with Fast Neutrons," in: **Fast Neutron Physics**, Part II, J. B. Marion and J. L. Fowler (eds.), Interscience publishers, New York (1963).
10. R. E. Brown et al, *Phys. Rev.* **C35** (1987) 1999; *Phys. Rev.* **C36** (1987) 1220; cf. also M. C. Struensee et al, *Phys. Rev.* **A37** (1988) 340.
11. This is, in part, the origin of the detuning effect discussed previously; M. Danos, B. Müller, J. Rafelski, *Phys. Rev.* **A35** (1987) 2741.
12. L. C. Biedenharn et al., μcf Program Review Meeting, National Bureau of Standards, Gaithersburg, MD, April 29/30 (1987). (The use of magnetic fields was suggested in this report also.)
13. See, for example, the contributions by S. Eliezer, P. Kulsrud, L. I. Ponomarev and H. Takahashi these Proceedings.
14. See the contribution by J. Rafelski et al, these Proceedings.
15. The details can be found in: M. Danos, L. C. Biedenharn, and A. Stahlhofen, "Intrinsic Sticking in DT Muon Catalyzed Fusion: Interplay of Atomic, Molecular and Nuclear Phenomena", April 1988, to be submitted for publication.
16. See reference 9.
17. T. A. Griffy and L. C. Biedenharn, *Nucl. Phys.* **15** (1960) 636; G. C. Phillips, T. A. Griffy and L. C. Biedenharn, (i) *Nucl. Phys.* **21** (1960) 327; (ii) *Zeit f. Physik* **205** (1967) 420.

EFFECT OF NUCLEAR INTERACTION ON MUON STICKING TO HELIUM IN MUON CATALYZED d-t FUSION

M. KAMIMURA
Department of Physics, Kyushu University, Fukuoka 812, Japan

ABSTRACT

We briefly survey the non-adiabatic coupled-rearrangement-channels method which gives accurate energies of the $dt\mu$ molecule with a very short computational time. A complex nuclear potential between d and t is determined so as to fit the cross sections of the resonant $d+t\rightarrow\alpha+n$ reaction. The interaction is incorpolated in the three-body Hamiltonian which is diagonalized within the basis functions spanned over the three rearrangement channels with their Jacobian coordinates. Fusion rates are estimated from the imaginary part of the complex eigenenergies. The muon sticking probabilities of the J=0 states are found to be increased by 4~5 % due to the effect of the nuclear interaction. A formulation to include explicitly the coupling of the $dt\mu$ and $\alpha n\mu$ channels is presented.

I. INTRODUCTION

Accurate solution of the $dt\mu$ molecular states is one of the most important tasks in the physics related to the muon catalyzed d-t fusion.[1] Especially, the energy of the J=v=1 state, ε_{11}, is to be obtained within an accuracy of at least 1 meV compared with the total binding energy of 2.7 keV, since the resonant formation rate of the $dt\mu$ molecule[2] is so sensitive to ε_{11}. It can be stated that calculation of such an accurate eigenenergy of the non-relativistic, Coulomb three-body Hamiltonian has been accomplished in Refs.3, 4, 5 which gave ε_{11}=660 meV within 0.5-meV difference to each other.

It should then be aimed to estimate the various (relativistic etc.) corrections on ε_{11} (Ref.1 reports $\Delta\varepsilon_{11}$=23.2 meV) with the use of such an accurate wave function as well as to calculate the initial sticking probabilities, ω_S^0, of the ground and excited states. Also required is to incorpolate the nuclear interaction into the three-body calculation explicitly and evaluate the fusion reaction rates and the sticking probabilities as well as nuclear energy shifts.

I shall report recent results of my calculations along this line, showing that the method of non-adiabatic coupled rearrangement channels proposed by the present author[5] is a powerful tool to perform such investigations.

II. METHOD OF COUPLED REARRANGEMENT CHANNELS

In the coupled-rearrangement-channels method of Ref.5, the three-body wave function is expanded in terms of the basis functions spanned over the three rearrangement channels. Muon is treated on an equal footing with two nuclei since the muon mass is not much smaller than the nuclear masses, and all the three channels are described with the use of their Jacobian coordinates (Fig.1).

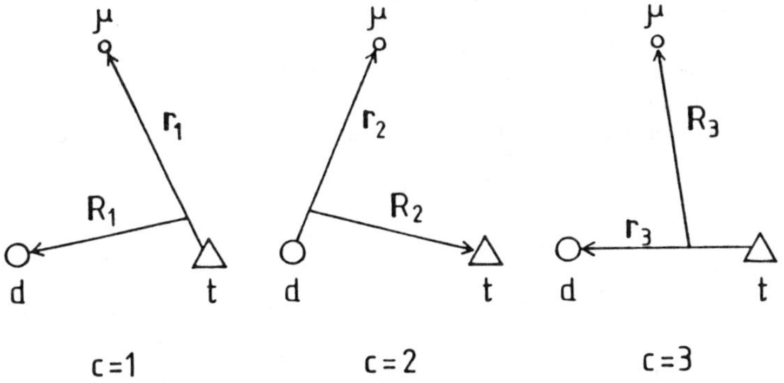

Fig. 1. Three rearrangement channels of the dtμ system and their Jacobian coordinates.

Reason for taking such an framework for the dtμ system is as follows: The fact that the state with J=v=1 lies by only 660 meV below the $(t\mu)_{1s}$-d breakup threshold suggests that the three-body wave function of the state has a large component corresponding to the $(t\mu)_{1s}$-d configuration. Similarly, the $(d\mu)_{1s}$-t configuration is likely to be important because the state is located slightly (48 eV) below the $(d\mu)_{1s}$-t threshold (cf. the $(t\mu)_{2s,2p}$-d threshold appearing 2 keV higher). The (dt)-μ component is necessary for determing the amplitude in the nuclear interaction region precisely.

Following Ref.5, we recapturate the coupled-channels method. The three channels (tμ)-d, (dμ)-t and (dt)-μ are referred to as channels c=1, 2 and 3, respectively, and their Jacobi coordinates (r_c, R_c) are defined as in Fig.1. The non-relativistic, Coulomb-interacting three-body Hamiltonian is

$$H = -\frac{\hbar^2}{2m_c}\nabla^2_{r_c} - \frac{\hbar^2}{2M_c}\nabla^2_{R_c} - \frac{e^2}{r_1} - \frac{e^2}{r_2} + \frac{e^2}{r_3}, \qquad (c=1,2 \text{ or } 3) \qquad (1)$$

where m_c and M_c are the reduced masses associated with the coordinates r_c and R_c, respectively. The total wave function with J and its z-component M, Ψ_{JM}, are expanded in terms of basis functions spanned over the three channels:

$$\Psi_{JM} = \sum_{c=1}^{3} \sum_{i_c \ell_c I_c L_c} A^{(c)}_{i_c \ell_c I_c L_c}\, \phi^{(c)}_{i_c \ell_c}(r_c)\, \chi^{(c)}_{I_c L_c}(R_c) \left[Y_{\ell_c}(\hat{r}_c) Y_{L_c}(\hat{R}_c) \right]_{JM} . \qquad (2)$$

Here, ℓ_c (L_c) stands for the angular momentum of the relative motion associated with the coordinate r_c (R_c), and the bracket $[\quad]_{JM}$ represents the vector coupling of the two spherical harmonics. In Eq.(2), ℓ_c and L_c are restricted as $0 \le \ell_c \le \ell_c^{max}$, $|J-\ell_c| \le L_c \le J+\ell_c$ and $(-)^{\ell_c+L_c} = (-)^J$. The numbers i_c and I_c specify the radial dependences of ϕ and χ, respectively (see below). For the three-body system, the expansion (2) must be the most general form of expansion with a finite number of basis. The expansion coefficients and the eigenenergies are be determined as usually by the Rayleigh-Ritz variational method.

Form of the radial functions ϕ and χ is taken as (the index c omitted)

$$\phi_{i\ell}(r) = r^{\ell} \exp\{-(r/\bar{r}_i)^2\} , \quad \bar{r}_i = \bar{r}_1 a^{(i-1)} \quad (i=1\sim n),$$
$$\chi_{IL}(R) = R^{L} \exp\{-(R/\bar{R}_I)^2\} , \quad \bar{R}_I = \bar{R}_1 A^{(I-1)} \quad (I=1\sim N), \tag{3}$$

The geometrical progressions for $\{r_i\}$ and $\{R_I\}$ are found to be useful in optimizing well the ranges with a small number of free parameters. Use of Gaussian tails, rather than exponential ones, makes the transformation between the three sets of the Jacobian coordinates very simple, and therefore closed forms of the Hamiltonian matrix elements are obtained straightforwardly.

It is to be noted that Gaussian-type basis functions have often been utilized in variational calculations of atomic and molecular problems,[8] but the present framework with the use of the Jacobian coordinate system is quite different from them.

Advantage of taking our framework with the three rearrangement channels is that the non-orthogonality among the basis functions is not exceedingly large, and therefore the eigenvalue problem hardly suffers from ill-conditioned diagonalization even for a few thausands basis functions in double-precision (14~16 decimal digits) calculations. Furthermore, owing to the explicit use of the Jacobian coordinates, the present coupled-rearrangement-channels approach was easily extended in Ref.6 to reaction problems such as $(d\mu)_{1s}+t \rightarrow (t\mu)_{1s}+d$ with the scattering boundary condition satisfied correctly.

All the numerical calculation was performed in double precision (14~16 decimal digits). Through the check of error of the whole calculation, use of quadruple precision (31~33 decimal digits) was found not to be necessary in our method for obtaining the accuracy of 10^{-6} eV in the energies ε_{Jv}. Therefore, the whole calculation can be performed on a vector processor which works in double precision, and therefore the computation time is very short. For example, when the number of the basis functions, N_b say, is 2000 for J=1, the time for calculating all the matrix elements is about 30 sec and that for the eigenvalues and eigenvectors of the lowest ten solutions is about 100 sec. (Note that most calculational methods for muonic molecules needs quadruple-precision computation on a scalar processor when $N_b \gtrsim 500$, which is due to exceedingly large non-linearity between basis functions employed.)

Such a short computation time with our method enables us to optimize the nonlinear variational parameters very carefully. For J=1, fifteen cases of N_b were chosen. For each case of N_b the optimization was made so as to obtain the best value of ε_{11}; the total number of such trial-and-error calculations for those fifteen cases of N_b was about one hundred.

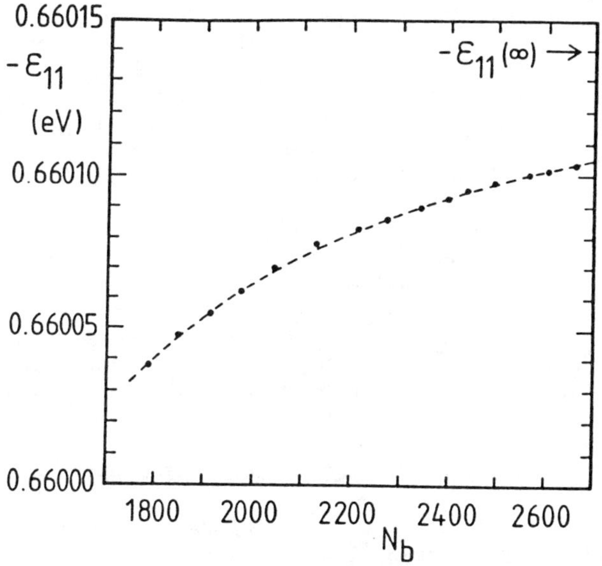

Fig. 2. Convergence of the calculated energy ε_{11} of the dtμ molecule with respect to the number of the basis functions, N_b. The dashed curve corresponds to the function $\varepsilon_{11}(N_b)=\varepsilon_{11}(\infty) + C\cdot N_b^{-\alpha}$ with $\varepsilon_{11}(\infty)=-0.66030$ eV, C=38500 eV and $\alpha=2.638$. Physical constants are taken as $m_\mu=206.7686m_e$, $m_d=3670.481m_e$, $m_t=5496.918m_e$ and $R_y=m_e e^4/2\hbar^2=13.605804$ eV, which are often used in the literature. This figure is taken from Ref. 5.

TABLE I. Corrections on ε_{11} of the dtμ molecule.

corrections		present work[a]	Ref. 1
		meV	meV
Vacuum polarization		16.7	7.3
Relatic correction	(recoil)	2.7	2.2
	(shift)	-1.8	-2.1
Finite nucleus size		10.4	14.6
Finite molecular size		1.2[b]	1.2
Nuclear interaction		$\lesssim 10^{-3}$	0.01
total		29.2	23.2

[a]Ref. 7. [b]Taken from Ref. 1.

The best values of ε_{11} versus N_b thus obtained are shown in Fig.2; for example, ε_{11}=-0.660038 eV with N_b=1789 and -0.660104 eV with N_b=2662. The dependence of ε_{11} on N_b in Fig. 2 is very smooth, which shows a high accuracy of the present calculation. The distribution of ε_{11} versus N_b is well followed by the dashed curve which is given by the function $\varepsilon_{11}(N_b)=\varepsilon_{11}(\infty) + C\cdot N_b^{-\alpha}$ with $\varepsilon_{11}(\infty)$=-0.66014 ± 0.00002 eV (C=38500 eV and α=2.638).

As for other states, we obtained ε_{00}=-319.139606 eV and ε_{01}=-34.834327 eV (N_b=1442), ε_{10}=-232.471506 eV (N_b=2662) and ε_{20}=-102.643337 eV (N_b=1566). All the ε_{Jv} are in agreement with those in Refs. 3 and 4 within 1 meV (ε_{20} is not given there).

In order to guide the experiments related to the energy of the weakly bound state with J=v=1, various kinds of corrections are to be considered.[1] With the use of the present wave function of the J=v=1 states, the correctiones were calculated[7] and are listed in Table I together with the results reported in Ref.1.

III. NUCLEAR INTERACTION AND FUSION RATE

In order to incorpolate a nuclear interaction in the three-body Hamiltonian, we first determine the interaction so that it explains the experimental data[9] of the resonant ($I=3/2^+$) d+t→α+n reaction for $E_{cm}\lesssim$150 keV. Here, we take a complex potential model for the d+t system; the absorption cross section provides the reaction cross section of the d+t→α+n reaction since there is no other open channel at the energies concerned. (A method to include the α+n+μ channel explicitely in the coupled three-body framework will be discussed in section V.)

The Schrödinger equation to describe the d+t system may be

$$[\, T_{r_3} + V(r_3) + iW(r_3) + V_{Coul}(r_3) - E_{cm} \,]\, u(r_3) = 0 \quad . \tag{4}$$

The geometry of the complex nuclear potential are assumed to be of the Woods-Saxon shape:

$$V(r)=V_0/[1+\exp\{(r-r_R)/a_R\}] \quad , \quad W(r)=W_0/[1+\exp\{(r-r_I)/a_I\}] \quad . \tag{5}$$

Although the potential should, in principle, be taken state-dependently, we simply assume here the same one not only for the $3/2^+$ (ℓ=0) resonant state but also for other non-resonant partial waves; there is little information on the latter partial waves in the energy region concerned, and their effects on the fusion rates and the muon sticking probabilities of the dtμ molecular states with J=0 are considered to be negligible small.

Since the d-t interaction is not well known from the microscopic viewpoint, we examine several different types of the potential (5) which all fit the d+t→α+n data very well down to the lowest experimental energy, E_{cm}=8 keV (Fig.2). The reaction cross section was calculated by

$$\sigma_{dt\to\alpha n}(E_{cm}) = \frac{2I+1}{(2I_d+1)(2I_t+1)}\,\frac{\hbar^2}{k^2}\,(\,1 - |S_{\ell=0}(E_{cm})|^2\,) \quad , \tag{6}$$

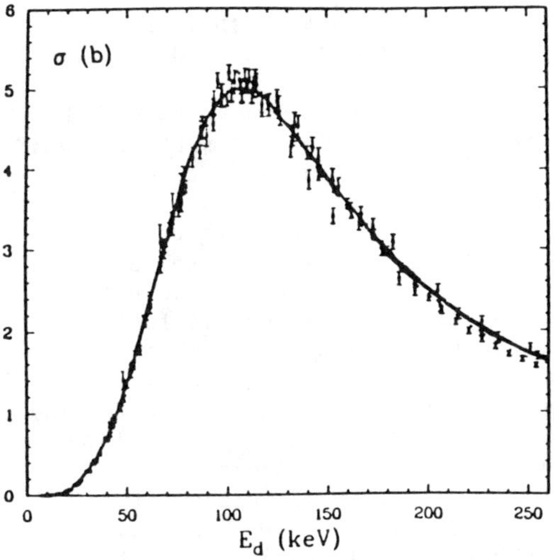

Fig. 3. Experimantal[9] and calculated cross sections of the d+t→α+n reaction. E_d=5/3·E_{cm}. The solid curve is given by the use of the complex nuclear interaction B in Table II; use of the other interactions results the same quality of fitting to the data.

TABLE II. Parameters of the nuclear potentials A~E whose geometries are defined in (6).

nuclear potentials	V_0 (MeV)	r_R (fm)	a_R (fm)	W_0 (MeV)	r_I (fm)	a_I (fm)
A	-58.52	2.5	0.3	-0.37	2.5	0.3
B	-38.01	3.0	0.5	-0.37	3.0	0.5
C	-28.27	3.0	1.0	-0.80	2.0	1.0
D	-16.04	5.0	0.3	-0.28	2.5	0.3
E	-13.19	5.0	1.0	-0.40	3.0	1.0

where I_d(=1), I_t(=1/2) and I(=3/2) are the spins of d, t and the $3/2^+$-resonant state, and $S_{\ell=0}(E_{cm})$ is the S-matrix at the c.m. energy E_{cm}(=$\hbar^2 k^2/2m_3$). Parameters of the so-determined potentials, A~E say, are listed in Table II; the Coulomb potential $V_{Coul}(r)$ between d and t is calculated with the finite sizes of the nuclei taken into account. The potential A is the deepest but narrowest among them, and E is the sharrowest but the widest; the radius of the potentials is considered to be in a reasonable range of the value.

If the calculated results of the fusion rates and the sticking probabilities are significantly dependent on the choice of the potential, we should investigate the potential much further. But, if not, I think that it will not be very important to search more realistic one extensively.

Now, the Coulomb potential e^2/r_3 in the three-body Hamiltonian (1) is replaced by $V(r_3)+iW(r_3)+V_{Coul}(r_3)$, and the Hamiltonian and norm matrices are to be diagonalized with the basis functions spanned over the three rearrangement channels.

In order to make the total wave function correlate properly with the short range nuclear interaction, we have to prepare carefully the trial basis functions in and outside the nuclear interaction region. We have precise experiences for it; for instance, as for the bound state of the three nucleons which are interacting with a realistic nucleon-nucleon interaction having shoft-core and non-central potentials, use of the present method of coupled rearrangement channels was found[10] to reproduce the binding energy given by the exact calculation with the Faddeev theory.

In the case without the nuclear interaction, we took, in Ref.5, N_b=1442 for the basis functions of J=0 states. But, since the calculation of muon sticking probability needs a very accurate three-body wave function, we carefully increased the number of the basis functions to N_b=2490 in order to describe well the nuclear interaction effect on the wave function. For J=1, a case of N_b=1789 was examined in Ref.5, but here N_b=2480 is taken (sticking calculation is not made yet).

Diagonalization of the energy- and norm-matrices are performed with the use of those basis functions for the five cases A~E of the nuclear interaction. The nuclear shift and width, $\Delta E_{Jv}-i\Gamma_{Jv}/2$, are obtained by comparing the eigenenergy with that of the calculation without the nuclear interaction. The fusion rate is estimated from the inverse life time:

$$\lambda^f_{Jv} = \Gamma_{Jv}/\hbar \quad . \tag{7}$$

The calculated ΔE_{Jv}, Γ_{Jv} and λ^f_{Jv} are listed in Table III. It is striking that the fusion rates for the J=0 states are almost the same between the five cases of the nuclear potentials A~E; namely, $\lambda^f_{00}=(1.25\pm0.03)\times10^{12}$ $\sec^{-1}$ and $\lambda^f_{01}=(1.05\pm0.03)\times10^{12}$ $\sec^{-1}$. We may then speculate that, use of any complex nuclear potential will generate very similar values as long as the potential reproduces the reaction cross section $\sigma_{dt\to\alpha n}(E_{cm})$ for $E_{cm}\lesssim150$ keV.

In the literature[11,12,13] fusion rates for the J=0 states were estimated by using the traditional method

$$\lambda^f_{J=0,v} = \lim_{v\to0} (v\sigma_0 C_0^{-2}) \times \int |\Psi_{J=0,v}(r_3,\ R_3=0)|^2\ dr_3 \tag{8}$$

with the usual notation; Ψ_{Jv_2} is the purely Coulomb wave function. Ref.11 reported $\lambda^f_{00}=1.2\times10^{12}$ $\sec^{-1}$ and $\lambda^f_{01}=1.0\times10^{12}$ $\sec^{-1}$, while Ref.13 gave $\lambda^f_{01}=0.73\times10^{12}$ $\sec^{-1}$ and $\lambda^f_{01}=0.60\times10^{12}$ $\sec^{-1}$; the result of Ref.11 agrees with ours, but note that the wave functions in Ref.13 must be considerably more accurate than those employed in Ref.11. For a consistent check of the method of (8) from the viewpoint of our approach, it is necessary to calculate Eq.(8) by

TABLE III. Enegry shift and width, $\Delta E_{Jv} - i\Gamma_{Jv}/2$, and fusion rates, $\lambda^{f}_{Jv}=\Gamma_{Jv}/\hbar$, of the dt$\mu$ molecule due to the nuclear potentials A~E. Number of the basis functions is N_b=2490 for J=0 and 2480 for J=1; in the case of no nuclear interaction, ε_{00}=-319.139702 eV, ε_{01}=-34.834452 eV, ε_{10}=-232.471470 eV and ε_{11}= -0.660070 eV.

nuclear pot.	ΔE_{Jv}	Γ_{Jv}	λ^{f}_{Jv}
J=0, v=0	(10^{-4}eV)	(10^{-4}eV)	(10^{12}sec^{-1})
A	-7.72	8.30	1.26
B	-7.01	8.42	1.28
C	-8.84	8.26	1.26
D	-8.00	8.00	1.22
E	-10.4	8.16	1.24
J=0, v=1	(10^{-4}eV)	(10^{-4}eV)	(10^{12}sec^{-1})
A	-5.87	6.98	1.06
B	-6.46	7.08	1.08
C	-7.40	6.94	1.05
D	-6.69	6.72	1.03
E	-8.69	6.86	1.04
J=1, v=0	(10^{-7}eV)	(10^{-7}eV)	(10^{8}sec^{-1})
A	-0.18	0.87	1.32
B	+0.10	1.16	1.77
C	+1.89	2.35	3.56
D	+5.48	0.95	1.45
E	+6.15	2.88	4.38
J=1, v=1	(10^{-7}eV)	(10^{-7}eV)	(10^{8}sec^{-1})
A	-0.05	0.34	0.51
B	+0.13	0.45	0.69
C	+0.75	0.92	1.40
D	+2.16	0.37	0.57
E	+2.41	1.12	1.71

using our cross section (6) at $E_{cm} \to 0$ and our wave function $\Psi_{J=0,v}(r_3, R_3=0)$ in the case of no nuclear interaction. This is, however, not performed yet.

The results for the J=1 states in Table III are seen to be rather dependent on the choice of the nuclear potentials A~E. This is because the potentials are not constructed to describe the P-wave behaviour of the d+t system; precise P-wave information is not available in the low energy region of the $d+t \to \alpha+n$ reaction. The deviation in $\lambda^f_{J=1,v}$ is due to both the S- and P-wave contributions. It may be said, however, that $\lambda^f_{10} \simeq (1\sim4)\times10^8$ sec^{-1} and $\lambda^f_{11} \simeq (0.5\sim2)\times10^8$ sec^{-1}; According to the similar method to Eq.(8), Ref.11 gave $\lambda^f_{10}=1.3\times10^8$ sec^{-1} and $\lambda^f_{11}=0.52\times10^8$ sec^{-1}, while Refs.12 and 13 reported $\lambda^f_{10}=1.95\times10^8$ sec^{-1} and $\lambda^f_{11}=1.88\times10^8$ sec^{-1}, respectively; these are consistent with ours; but, the $\sigma_{\ell=1}(E_{cm}\to0)$ employed in Refs.11-13 does not seem very reliable due to lack of the low energy data for $\ell=1$. The P-wave d-t interaction is subject to further investigations both theoretically and experimentally.

IV. STICKING PROBABILITY

According to the sudden approximation, the sticking probability for a transition to a $(\mu\,^4He)_{n\ell}$ atom is given by

$$\omega_{n\ell} = 4\pi(2\ell+1)\left| \int R_{n\ell}(r) j_\ell(qr) \Phi_{J=0,v}(r) r^2 dr \right|^2 , \qquad (9)$$

with

$$\Phi_{J=0,v}(r_3) = \Psi_{J=0,v}(r_3, R_3=0) \Big/ \left[\int |\Psi_{J=0,v}(r_3,R_3=0)|^2 dr_3\right]^{1/2} . \qquad (10)$$

$R_{n\ell}(r)$ is the radial function of the $(\mu\,^4He)_{n\ell}$ atom and $qa_\mu=0.58460$ (Ref.15), a_μ being the muon Born radius. The total (initial) sticking probability is

$$\omega^0_S = \sum_{n\ell} \omega_{n\ell} . \qquad (11)$$

The calculated ω^0_S and $\omega_{n\ell}$ for the J=v=0 state in the case of no nuclear interaction are listed in Table IV together with the literature data. It is remarkable that complete agreement is seen between the result of the present work and that of Ref. 15; their three-body wave functions must be very similar to each other.

The calculated ω^0_S and $\omega_{n\ell}$ in the case of including the nuclear interaction B is given in the last columns in Table IV. As listed in Table V for the nuclear interaction A~E, we obtain $\omega^0_S=(0.924\pm0.005)$ % for the J=v=0 state and $\omega^0_S=(0.926\pm0.005)$ % for the state with J=0,v=1. The nuclear interaction increases*) the

---

*) In this workshop L.I. Ponomarev reported that they[16] incorpolated a complex nuclear d-t potential into their method of the adiabatic representation of the three-body problem and obtained $\omega^0_S=0.93$ % for the J=v=0 state which is consistent with our result. Their rather small-sized basis functions, however, result $\varepsilon_{00}=-317.815$ eV and $\omega^0_S=0.90$ % in the case of no nuclear interaction; these are considerably worse than those obtained in the present work.

sticking probabilities of the J=0 states by 4~5 %; dependence on the choice of the potential geometry is rather small.

We investigate the contribution of the imaginary part iW of the d-t nuclear interaction the sticking probability. If we cut the imaginary part off in the three-boy Hamiltonian with the real part unchanged, the dtμ wave function of the J=v=0 state results

TABLE IV. Muon sticking probabilities ω_S^0 (%) of the J=v=0 state of the dtμ molecule.

J=v=0	no nuclear interaction				with nuclear interaction(B)
	Ref. 14[a]	Ref. 12[b]	Ref. 15[c]	present work[d]	present work[e]
ω_S^0	0.845	0.897	0.8860	0.8859	0.9261
1s	0.6502	0.6932	0.6826	0.6825	0.7141
2s	0.0934	0.0992	0.0979	0.0978	0.1021
2p	0.0238	0.0241	0.0238	0.0238	0.0248
3s	0.0284	0.0302	0.0297	0.0297	0.0310
3p	0.0086	0.0087	0.0086	0.0086	0.0089
3d	0.0003	-	0.0002	0.0002	0.0002
4s	0.0121	0.0128	0.0127	0.0127	0.0132
4p	0.0037	0.0039	0.0039	0.0039	0.0040
4d+4f	0.0003	-	0.0001	0.0001	0.0001
5s	-	0.0066	0.0065	0.0065	0.0068
all other	0.024	0.018	0.0200	0.0200	0.0208

[a] ε_{00}=-319.15 eV (by the method of adiabatic representation).
[b] ε_{00}=-319.13419 eV (N_b= 500); Ref. 13. gives ω_S^0=0.8583 (N_b=695).
[c] ε_{00}=-319.139752 eV (N_b=1995).
[d] ε_{00}=-319.139702 eV (N_b=2490). The case of N_b=2140 gives ω_S^0=0.8860.
[e] ε_{00}=-319.139702 eV + the energy shift in Table III (N_b=2490).

Table V. Effect of the nuclear interaction on the muon sticking probabilities ω_S^0 (%) of the dtμ moleclule. The case of without and with nuclear interactions A~E are shown.

	ω_S^0 (%)	
	J=0, v=0	J=0, v=1
no nuclear interaction	0.8859	0.8896
with nuclear interaction		
A	0.929	0.931
B	0.926	0.928
C	0.921	0.923
D	0.924	0.926
E	0.921	0.923

ω_S^0=(0.929±0.006) %; the role of the imaginary part is about 10~15 % of that of the whole nuclear interaction (cf. (0.929-0.924)/(0.924-0.886)=0.13). This is reasonable from the consideration that, in the d+t→α+n reaction, although the role of the imaginary part is large around the resonance, but it is small (perturbative) in the zero-energy region where the d and t interact in the dtμ molecule.

As mentioned before, the imaginary part incorpolated in the d-t interaction is a substitute of the α-n channel. we may then speculate that if we could explicitly include the αnμ channel in a coupled (dtμ+αnμ) three-body calculation, effect of the dtμ-αnμ coupling on the sticking probability would be some 10~15 % of the effect of all the nuclear interactions. Compared with the difficulty of such a coupled three-body calculation, the effect of the αnμ-dtμ coupling might be rather small (probably, within the ambiguity of the nuclear interactions employed in the calculation). Nevertherless, we intend, in the next section, to formulate the explicit dtμ-αnμ couling in the sticking calculation, since it is a fundamental and interesting problem from the few-body therォretical viewpoint.

V. COUPLING OF dtμ AND αnμ CHANNELS

Outline of our idea to include the dtμ-αnμ coupling in the sticking calculation is as follows. The Shrödinger equation, Hamiltonian and the wave function for the systems may be written as

$$(H - E)\Psi = 0 \tag{12}$$

$$H = \begin{pmatrix} H_{dt\mu} & V_{dt\mu,\alpha n\mu} \\ V_{\alpha n\mu,dt\mu} & H_{\alpha n\mu} \end{pmatrix} , \quad \Psi = \begin{pmatrix} \Psi_{dt\mu} \\ \Psi_{\alpha n\mu} \end{pmatrix} \tag{13}$$

with

$$H_{dt\mu} = T_{r_c} + T_{R_c} + V_{d\mu} + V_{t\mu} + V_{dt} , \tag{14}$$

$$H_{\alpha n\mu} = T_{r_{c'}} + T_{R_{c'}} + V_{\alpha\mu} + V_{\alpha n} . \tag{15}$$

Equation (12) is to be solved under the boundary condition of $\Psi_{dt\mu} \to 0$ in the asymptotic region and of the outgoing wave in the $\alpha n\mu$ channel:

$$\Psi_{\alpha n\mu} \to \sum\!\!\!\!\!\!\int_{\beta} T^{(\beta)} \phi^{(\beta)}_{\alpha\mu}(r_{\alpha\mu})\, u^{(+)}(K_\beta, R_{\alpha\mu,n}) \quad \text{for } R_{\alpha\mu,n} \to \infty , \tag{16}$$

where β indicates the quantum numbers of the $\alpha\mu$ atomic states including continuum states. $u^{(+)}$ stands for the outgoing wave between $(\alpha\mu)_\beta$ and n with a momentum K_β. The sticking probability is then expressed by

$$\omega^0_S = \sum_{\beta}^{\text{bound}} |T^{(\beta)}|^2 \Big/ \sum\!\!\!\!\!\!\int_{\beta} |T^{(\beta)}|^2 . \tag{17}$$

Since calculation of $T^{(\beta)}$ by (12)~(16) is of course extremely difficult, we consider an approximation bellow. As long as Ψ is the exact solution, the amplitude $T^{(\beta)}$ is equivalently written as

$$\begin{aligned} T^{(\beta)} &= \langle e^{-iK_\beta \cdot R_{\alpha\mu,n}} \phi^{(\beta)}_{\alpha\mu}(r_{\alpha\mu}) \mid \overleftarrow{H} - E \mid \Psi \rangle \\ &= \langle e^{-iK_\beta \cdot R_{\alpha\mu,n}} \phi^{(\beta)}_{\alpha\mu}(r_{\alpha\mu}) \mid V_{\alpha n\mu,dt\mu} \mid \Psi_{dt\mu} \rangle \\ &+ \langle e^{-iK_\beta \cdot R_{\alpha\mu,n}} \phi^{(\beta)}_{\alpha\mu}(r_{\alpha\mu}) \mid V_{\alpha n} \mid \Psi_{\alpha n\mu} \rangle . \end{aligned} \tag{18}$$

If we assume that $\Psi_{\alpha n\mu}=0$ and $K_\beta=K_{1s}$ and $V_{\alpha n\mu,dt\mu}$ is of zero-range type, Eq.(18) leads to the usual expression (9) for ω^0_S.

We consider a more advanced approximation in which the coupling of the $dt\mu$ and $\alpha n\mu$ channels is taken into account to a certain extent. It is to be noted that, owing to the product $V_{\alpha n}\Psi_{\alpha n\mu}$, we need, in Eq.(18), the knowledge of $\Psi_{\alpha n\mu}$ only for $r_{\alpha n} \lesssim r^0_{\alpha n}$ where $r^0_{n\alpha}$ is the range (~3 fm) of the α-n interaction. For this internal region of $r_{\alpha n}$, $\Psi_{\alpha n\mu}$ may be expanded in terms of the same type of the basis functions in Eqs.(2) and (3) but for the (αn)-μ channel with the specific angular momentum $\ell_c=2$ (spin 3/2) to describe the

$3/2^+$ resonance of the α-n channel; the other angular momentum may be negligible. The Gaussian basis functions of Eq.(3) with $\{r_1 \simeq 0.5$ fm, $r_n \simeq 10$ fm, $n \simeq 20\}$ will be sufficient to simulate the scattering α-n wave function at least for $r_{\alpha n} \lesssim 5$ fm (though they vanish asymptotically).

The coupling interaction $V_{dt\mu,\alpha n\mu}$ may be replaced by $V_{dt,\alpha n}$, since muon does not seem to affect the nuclear rearrangement process significantly. Some sets have been found[17] of the interactions $\{V_{dt}, V_{\alpha n}, V_{dt,\alpha n}\}$ which reproduce the experimental data[9] of the resonant d+t→α+n reaction concerned. The diagonalization of the coupled three-body Hamiltonian (13) is then practically possible within our present technique. Replacing the exact wave function in Eq.(18) by the approximate one thus obtained, we can calculate $T^{(\beta)}$ and ω_s^0 with the $dt\mu$-$\alpha n\mu$ coupling taken into account to a certain (rather sufficient) extent; but some improvement of the prescription to perform the integration in Eq.(18) will be required.

VI. SUMMARY

We briefly reviewed the method of coupled rearrangement channes which gives accurate energies and wave functions of the $dt\mu$ molecular states with a very short computational time. We then calculated important corrections on ε_{11} with the use of the wave function of the J=v=1 state; the correction amounts to 29.2 meV which is to be compared with the previously obtained[1] value of 23.2 meV.

We then incorpolated a complex nuclear potential between d and t into the three-body Hamiltonian; the potential was determined so as to fit the observed cross section[9] of the resonant d+t→α+n reaction. By diagonalyzing the Hamiltonian with a sufficient number of the basis functions, we calculated the nuclear energy shifts and widths of the J=0 and J=1 states as well as the fusion rates which are $\lambda^f_{00}=(1.25\pm0.03)\times10^{12}$ sec^{-1}, $\lambda^f_{01}=(1.05\pm0.03)\times10^{12}$ sec^{-1}, $\lambda^f_{10}=(1\sim4)\times10^{8}$ sec^{-1} and $\lambda^f_{11}=(0.5\sim2)\times10^{8}$ sec^{-1} ; the range of the values are due to dependence onthe choice of the nuclear potentials.

The probabilities of muon sticking to helium in the reaction $(dt\mu)\rightarrow(\alpha\mu)_{n\ell}$ +n were calculated in the cases with and without the nuclear interaction: the interaction increases ω_s^0 of the J=0 states by 4~5 %; namely $\omega_s^0=(0.924\pm0.005)$ % for J=v=0 and (0.926±0.005) % for J=0, v=1 in the presence of the nuclear interaction, while $\omega^0=0.8859$ % for J=v=0 and 0.8896 % for J=0, v=1 with the purely Coulomb wave functions.

Contribution of the imaginary part of the nuclear d-t potential to the sticking probability is 10~15 % of the whole effect of the nuclear potential. Since the imaginary part is a substitute of the dt-αn coupling, this small contribution seems to suggest that the effect of the $dt\mu$-$\alpha n\mu$ coupling on the sticking probability might be rather smaller than the effect of the nuclear d-t interaction on it.

Nevertherless, we presented an approximate method to incorpolate explicitely the coupling of the $dt\mu$ and $\alpha n\mu$ channels in the sticking calculation in a practically calculable manner. This study is in progress.

The author would like to thank Prof. M. Kawai, Prof. R.C. Johnson and Dr. M. Yahiro for stimulating discussions on the formulation of the sticking calculation with the dtμ-αnμ coupling. He is also thankful to Prof. H. Narumi, Prof. Y. Akaishi and Dr. Khin Swe Myint for the collaboration in the calculation of the corrections on ε_{11}. This work was supported by the Institute of Plasma Physics, Nagoya University and by the Research Center for Nuclear Physics, Osaka University. The British Council is also acknowledged for the financial support on the author's visit in May, 1988 to the University of Surrey and the Rutherford Appleton Laboratory to make collaborations on the muon catalyzed fusion.

REFERENCES

1. L. I. Ponomarev, Atomic Physics (Proceedings of the International Conference on Atomic Physics) 10, 197 (1987).
2. S.S. Gerstein and L.I. Ponomarev, Phys. Lett. 72B, 80 (1977), M. Leon, Phys. Rev. Lett. 52, 605 (1984); 52, 1655 (1984).
3. V.I. Korobov, I.V. Pusynin and S.I. Vinitsky, Phys. Lett. 196B, 272 (1987).
4. K. Szalewicz, H.J. Monkhorst, W. Kolos and A. Scrinzi, Phys. Rev. A36, 5494 (1987).
5 M. Kamimura, Phys. Rev. A38, (1988), in press.
6. M. Kamimura, Muon Catalyzed Fusion 3, 335 (1988).
7. Khin Swe Myint, Y. Akaishi, M. Kamimura and H. Narumi, preprint.
8. For example, S.A. Alexander, H.J. Monkhorst and K. Szalewicz, J. Chem. Phys. 85, 5821 (1986); references therein.
9. N. Jarmie, R.E. R.A. Hardekopf, Phys. Rev. C29, 2031 (1984).
10. H. Kameyama, Y. Fukushima and M. Kamimura, Contribution to the Fifth International Conference on the Nuclear and Subnuclear Systems, Kyoto, July, 1988.
11. L.N. Bogdanova, V.E. Markushin, and V.S. Melezhik, Zh. Eskp. Teor. Fiz. 81, 829 (1981) [Sov. Phys. JETP 54, 442 (1981)], L.N. Bogdanova, V.E. Markushin, V.S. Melezhik, and L.I. Ponomarev, Zh. Eskp. Teor. Fiz. 83, 1615 (1981) [Sov. Phys. JETP 56, 931 (1982)].
12. C.Y. Hu, Phys. Rev. A34, 2536 (1986).
13. C.Y. Hu, Phys. Rev. A36, 4135 (1987).
14. L. N. Bogdanova et al., Nucl. Phys. A454, 653 (1986), S.I. Vinitsky et al., Zh. Eksp. Teor. Fiz. 79, 698 (1980) [Sov. Phys. JETP 52, 353 (1980)].
15. S.E. Haywood, H.J. Monkhorst and K. Szalewicz, Phys. Rev. A37, 3393 (1988).
16. L.N. Bogdanova, V.E. Markushin, V.S. Melezhik and L.I. Ponomarev, preprint.
17. M. Teshigawara, Y. Akaishi and H. Tanaka, private communications.

BOUNDARY-VALUE APPROACH TO NUCLEAR EFFECTS IN MUON-CATALYZED D-T FUSION

G. M. Hale, M. C. Struensee, R. T Pack, and J. S. Cohen
Theoretical Division, Los Alamos National Laboratory
Los Alamos, New Mexico 87545

ABSTRACT

The use of boundary-matching techniques, as contained in R-matrix theory, to describe multichannel nuclear reactions is discussed. After giving a brief summary of the application of such techniques to the nuclear reactions in the ^{5}He system, we discuss a simple extension of the description to include muons, which was used to calculate nuclear effects on the L=0 eigenvalues of the dtμ molucule in various adiabatic approximations. Next, the form of the outgoing n$\alpha\mu$ wavefunction is discussed, resulting in a new formulation of the amplitudes used to calculate the $\alpha-\mu$ sticking fraction. Possible methods of solving for these amplitudes using the boundary-value approach are suggested, and some deficiencies of the "standard" expression for the sticking amplitudes are pointed out.

INTRODUCTION

It is, perhaps, ironic that the fundamental process underlying μ–catalyzed d-t fusion, the nuclear reaction T(d,n)^{4}He, has been one of the last elements added to the description of fusion out of the dtμ molecule. In a sense, this is justified by the fact that, although nuclear effects strongly perturb the molecular wavefunction at small d-t separations, the healing distance is short enough that their effects on all the observables of the system considered up until now have been small. In this paper, we will describe how nuclear effects were imposed on the nuclear eigenvalues of the dtμ molecule, and how they could be imposed on a quantity of vital interest in the muon-catalyzed fusion cycle, the $\alpha-\mu$ sticking fraction, using an R-matrix description of the nuclear reactions in the ^{5}He system.

We begin with a brief resume of the R-matrix relations and definitions used to characterize the nuclear reactions in the ^{5}He system, followed by a summary of results from an extensive analysis of the experimental data. Next, it is shown how to generalize the R-matrix approach to include the muon, in the approximation that the nuclear and muonic degrees of freedom are separable. We then recapitulate the results of applying the theory, in the reduced R-matrix form, to study nuclear effects on the eigenvalues of the S-wave molecular states, the wavefunctions of which were calculated with a series of increasingly more accurate adiabatic approximations. Finally, we give a new prescription for defining sticking amplitudes, and indicate how they could be obtained numerically from our formalism. Some problems with the "standard" expressions used for the sticking amplitudes are also pointed out.

R-MATRIX DESCRIPTION OF THE NUCLEAR ^{5}HE SYSTEM

R-matrix theory[1] is based on the idea that a many-body nuclear system displays simpler "channel" degrees of freedom whenever the radial separation r_c of two subgroups of the system is increased beyond the channel radius, a_c. In this "channel" region, the subgroups are assumed to be bound in their ground states, with at most (point) Coulomb forces acting between them. The channel radii $\{a_c\}$ define a channel surface which encloses the "nuclear" region. On this surface are defined the channel surface states of total angular momentum (J) and parity,

$$(r_c|c) = \left(\frac{1}{2M_c a_c}\right)^{1/2} \frac{\delta(r_c-a_c)}{r_c}\left[\left(\phi_1\otimes\phi_2\right)_{sv}\otimes i^{\ell}\, Y_{\ell m}(\hat{r}_c)\right]_{JM} \quad , \tag{1}$$

in which ϕ_1 and ϕ_2 are the spin-dependent bound-state wavefunctions of the channel, M_c is the channel reduced mass, and $Y_{\ell m}$ is the spherical harmonic describing the orbital motion. The R matrix is then defined as the channel-surface projections of a resolvent (Green's function) operator,

$$R^N_{c'c} = (c'|(H_N + \mathcal{L}_B - E_N)^{-1}|c) = \sum_\lambda \frac{\gamma_{c'\lambda}\gamma_{c\lambda}}{E_\lambda - E_N} \quad , \tag{2}$$

in which the boundary operator

$$\mathcal{L}_B = \sum_c |c)(c|(\frac{\partial}{\partial r_c} r_c - B_c) \tag{3}$$

has been added to the nuclear hamiltonian H_N for total energy E_N in order to make it hermitian in the nuclear region. This allows the spectral expansion made in the last step of Eq. (2) to be defined in terms of the solutions of

$$(H_N + \mathcal{L}_B - E_\lambda)|\lambda) = 0 \quad , \tag{4}$$

with $\gamma_{c\lambda} = (c|\lambda)$. We use the reduced-width amplitudes $\gamma_{c\lambda}$ and eigenenergies E_λ for boundary conditions B_c at channel radii a_c as a parametric expansion of the R matrix, and determine their values by fitting experimental data.

The formal solution of Schrödinger's equation inside the nuclear region,

$$\psi_N = (H_N + \mathcal{L}_B - E_N)^{-1}\mathcal{L}_B\psi_N \quad , \tag{5}$$

implies the fundamental R-matrix relation,

$$(c'|\psi_N) = \sum_c R^N_{c'c}(c|(\frac{\partial}{\partial r_c} r_c - B_c)|\psi_N) \quad , \tag{6}$$

which is a matching condition that can be used to determine quantities of interest in any scattering or bound-state problem. In the case of n coupled channels, the fundamental R-matrix relation holds for each of the n linearly independent solutions ψ^j_N so that in terms of the matrices

$$U_{ij} = (c_i|\psi^j_N) \quad \text{and} \quad U'_{ij} = (c_i|\frac{\partial}{\partial r_i} r_i|\psi^j_N) \quad , \tag{7}$$

it becomes the matrix relation

$$U = R^N(U' - BU) \quad \Rightarrow \quad R^N = (U'U^{-1} - B)^{-1} \quad . \tag{8}$$

Thus, the R matrix is essentially the reciprocal of a logarithmic derivative matrix, $U'U^{-1}$, for the coupled radial solutions. Any particular solution of the coupled problem can be constructed from a linear combination of the columns of the matrix U.

If one has a multichannel problem in which only a subset of the channels is of explicit interest, the partitioned matrix technique of the reduced R matrix[2] is often useful. This method consists of writing the fundamental R-matrix relation, Eqs. (6) or (8), in partitioned matrix (P,Q, where P+Q=1) form, and assuming that in partition Q,

$$U'_Q = L_Q U_Q \qquad (L_Q = \mathcal{O}'_Q \mathcal{O}_Q^{-1}) \tag{9}$$

the logarithmic derivatives L_Q are those for purely outgoing-wave solutions $\mathcal{O}_Q$. Then the R-matrix relation for the radial solutions in the P partition can be written as

$$U_P = \{R_{PP} + R_{PQ}(L_Q - B_Q)[1 - R_{QQ}(L_Q - B_Q)]^{-1} R_{QP}\}(U'_P - B_P U_P), \tag{10}$$

which means that the motion in channel group P is described by the <u>reduced</u> R matrix

$$\tilde{R}_{PP} = R_{PP} + R_{PQ}(L_Q - B_Q)[1 - R_{QQ}(L_Q - B_Q)]^{-1} R_{QP}\,. \tag{11}$$

With these preliminary ideas established, we can proceed to a brief description of the application to reactions in the ^{5}He system.[3] A summary of the channel configuration, data included, and χ^2 values for each of the reactions is given in Table I. The important points to note are that: (1) the energy range of the analysis is rather broad, extending to energies well above and below the d-t threshold; (2) many different types of observables (cross sections, polarizations, etc.) are included in the analysis for each reaction, allowing the R-matrix parameters to be determined accurately and unambiguously; (3) the overall χ^2 per degree of freedom for the fit (1.55), while not ideal, indicates in our experience quite a good representation of the measurements for a multichannel system.

Table I Summary of the ^{5}He system R-matrix analysis

Channel	l_{max}	a_c (fm)
d-t	3	5.1
n-^{4}He	4	3.0
n-^{4}He*	1	5.0

Reaction	Energy Range	# Observable Types	# Data Points	χ^2
T(d,d)T	E_d=0-8 MeV	6	695	1147
T(d,n)^{4}He	E_d=0-8 MeV	13	1020	1423
T(d,n)^{4}He*	E_d=4.8-8 MeV	1	10	17
^{4}He(n,n)^{4}He	E_n=0-28 MeV	2	795	1160
	Totals:	22	2520	3767

parameters = 97 ⇒ χ^2 per degree of freedom = 1.55

Examples of the quality of the fits to the integrated cross sections are given in Fig. 1. The top part of the figure shows the fit to measurements of the integrated reaction [T(d,n)^{4}He] cross section, with a comparison at the right side, expressed as a ratio, to the precise new data of Jarmie et al.[4] over the region of the low-energy resonance. The lower part of the figure shows similar comparisons with the measurements of the n-α total cross section. Again, the right side shows a detail of the fit over the resonance, in this case, containing the recent total cross-section data of Haesner et al.[5] One sees that the R-matrix representation of these data is quite good.

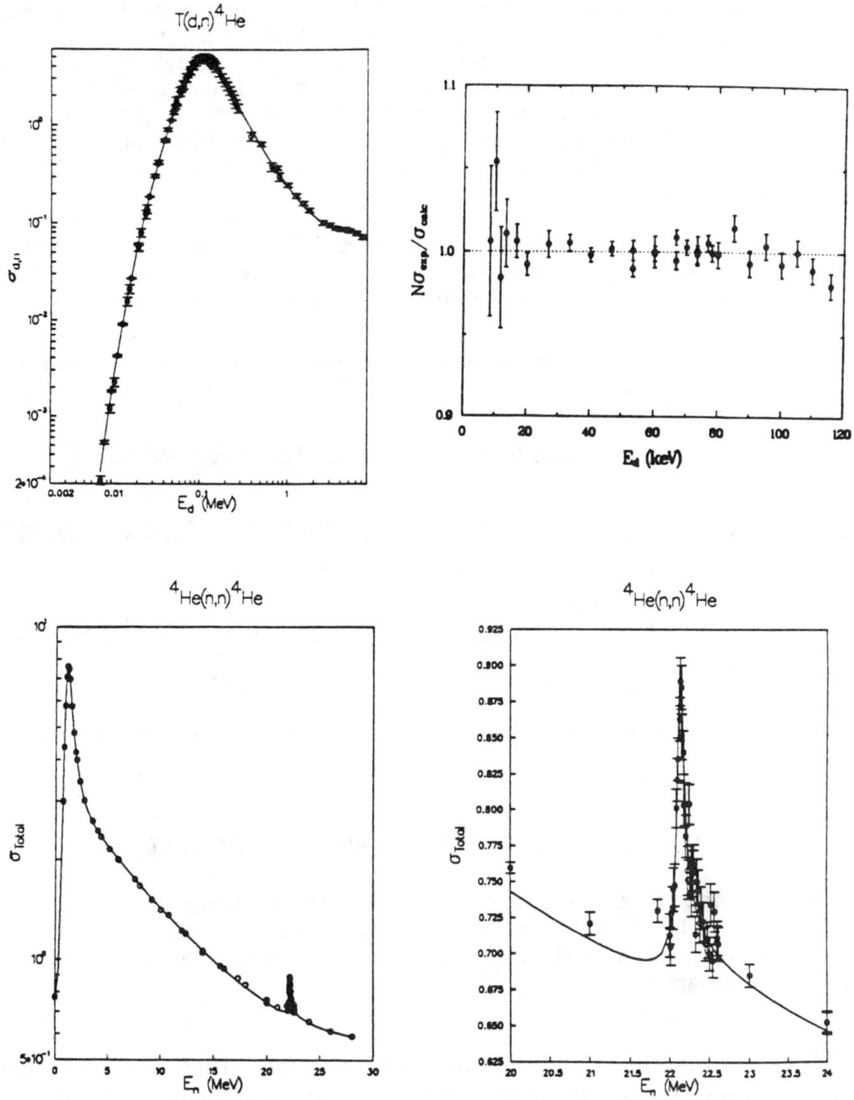

Fig. 1. Fits to measurements of integrated cross sections for the ^{5}He reactions.

At the low relative d-t energies that are of interest in cold fusion, the $T(d,n)^4He$ reaction is almost completely dominated by the resonant $J^\pi=3/2^+$ transition. The R-matrix parameters for this matrix are given in Table II. The first level is mainly associated with the low-energy resonance, the next two with higher-energy ^2D and ^4D resonances in the dt channel, and the fourth serves primarily as a background term, especially in the nα channel. Although the R matrix is clearly the relevant quantity in the muon-catalyzed fusion problem, it is of interest to mention the S-matrix pole structure[6] that results from matching to outgoing waves in the nuclear ^5He system. In addition to a "conventional" pole on the unphysical sheet (Sheet III) closest to the physical sheet of the two-channel Riemann energy surface above the d-t threshold, there is a "shadow" pole on Sheet II (the sheet on which the d-t momentum has a positive imaginary part and the n-α momentum has a negative imaginary part). The effects of the two poles are separately visible in the reaction and total cross sections, as is shown in Ref. 6. The presence of the shadow pole on Sheet II implies that the $J^\pi=3/2^+$ resonance in ^5He would occur only in the nα channel in the absence of coupling between the dt and nα channels, contradicting the usual picture that it is essentially a dt resonance.

Table II R-matrix parameters for the $J^\pi = 3/2^+$ states of ^5He. Channel labels (c) are in spectroscopic notation. Eigenenergies E_λ are center-of-mass values in MeV relative to the d-t threshold; reduced-width amplitudes $\gamma_{c\lambda}$ are also center-of-mass in units $MeV^{1/2}$.

c(J=3/2)	a_c(fm)	B_c		λ=1	λ=2	λ=3	λ=4
			E_λ	0.0837559	6.4713043	13.7357067	47.475246
^4S(dt)	5.1	-0.37	$\gamma_{c\lambda}$	1.1760678	0.0693397	-0.4955438	1.1052421
^4D(dt)	5.1	-2.00		0.1688724	-0.2729805	1.9910681	1.9847048
^2D(dt)	5.1	-2.00		-0.0484797	0.8862475	0.0958513	0.2422464
^2D(nα)	3.0	-0.59		0.3768218	-0.1562737	0.9994494	-3.8556539

R-MATRIX DESCRIPTION OF THE ^{5}HE+μ SYSTEM

The advantage of the R-matrix approach is that the addition of the muon to the system can be treated differently in the nuclear and channel regions. In the nuclear region, where the nucleons are in relatively close proximity, the total hamiltonian H is approximately separable,

$$H \approx H_N + H_\mu \quad , \tag{12}$$

with H_N the nuclear hamiltonian as before, and H_μ the hamiltonian for a muon moving about a ^5He core. The wavefunction in the nuclear region is therefore, in general, a sum of separable terms corresponding to the separation in H. However, we will in the subsequent discussion keep only the first term of this sum,

$$\Psi \approx \psi_N \phi_\mu^0 \quad , \tag{13}$$

in which ϕ_μ^0 is the ground-state μ-^{5}He wavefunction corresponding to the ground-state energy E_μ^0. This approximation joins smoothly with the adiabatic approximations to be made in the channel region, as is depicted schematically in Fig. 2. The R matrix (including the muon) is in this approximation

$$R_{c'c} = (c'|(H_N + \mathcal{L}_B + H_\mu - E)^{-1}|c) = |\phi_\mu^0\rangle R^N_{c'c}(E_N)\langle\phi_\mu^0| \quad , \tag{14}$$

with R^N given by Eq. (2) in terms of the parameters of Table II, and $E_N = E - E_\mu^0$.

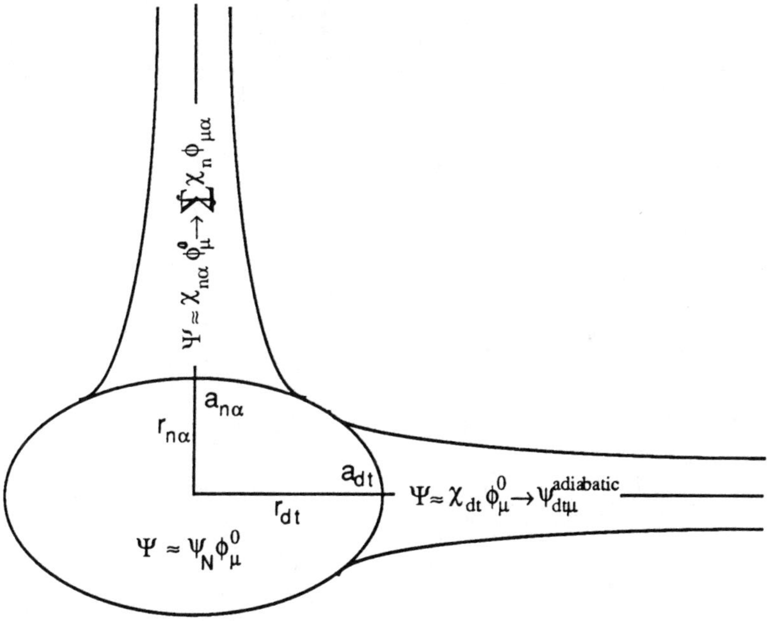

Fig. 2. Schematic of the single muon state (adiabatic) R-matrix approximation

In the channel region, the Coulomb attraction of the muon to the separated nuclear ions has a significant effect, especially in the dt channel, where it causes bound molecular $dt\mu$ states to occur. Coulomb binding of muons to alpha particles in the $n\alpha$ channel is also quite important, since this creates the "stuck" muons that are lost from the catalysis cycle. Thus, as the radial distance between the ions grows larger than the channel radius, the separable approximation made in the nuclear region breaks down, and one is faced with the difficult task (at least for $dt\mu$) of solving a three-body Coulomb problem. Exact (non-adiabatic) calculations of bound-state $dt\mu$ wavefunctions have been the subject of extensive research over the past decade.[7] However, a complication introduced by the nuclear effects is that, since the channel region excludes the origin, the wavefunctions in this region contain irregular functions in addition to the regular functions present in the bound states. In order to avoid having to deal with this kind of complexity in a fully non-adiabatic calculation, we decided to test the R-matrix approach using a sequence of increasingly more accurate adiabatic calculations for the dtμ wavefunction in the channel region.

The natural starting point was to consider nuclear effects on the eigenvalues (bound-state energies) E_b of the dtμ molecule in this approximation,[8] which allows the relative d-t wavefunction to be obtained by solving a second-order differential equation containing the muon attraction in an effective d-t potential. The solution of this equation that matches at the channel surface to values derived from the nuclear R matrix and tends asymptotically to a purely outgoing wave is a generalization of the bound-state eigenfunction sometimes called the Siegert state, which occurs for a complex eigenenergy $E_0 = E_r - \frac{1}{2}i\Gamma$. Thus, the shift $\Delta E = E_r - E_b$ and the width Γ (which is proportional to the fusion rate λ_f) are direct measures of the nuclear effect on the bound-state eigenvalues.

The matching at the channel surface is most conveniently accomplished by using the reduced R matrix described in the paragraph containing Eqs. (9)-(11). In this case, we put the ^4S(dt) channel (d) of Table II in partition P and the three remaining channels (q=1,2,3) in partition Q, in order to define the single reduced R-matrix element

$$\tilde{R}_{dt\mu} = |\phi_\mu^0\rangle \, \tilde{R}_{dt}^N \, \langle\phi_\mu^0| \, , \tag{15}$$

in which

$$\tilde{R}_{dt}^N = R_{dd}^N + R_{dq}^N(L_q - B_q)[1 - R_{qq}^N(L_q - B_q)]^{-1} R_{qd}^N \, . \tag{16}$$

Since the adiabatic solutions in the dtμ channel region near $r_{dt} = a_{dt}$ have the form

$$\Psi_{dt\mu}^{\text{adiabatic}} = \chi_{dt}(r_{dt})\phi_\mu(r_\mu) \, ,$$

in which $\phi_\mu \sim \phi_\mu^0$, we can project the fundamental R-matrix relation, Eq. (6), on $\langle\phi_\mu^0|$ to obtain

$$(dt|\chi_{dt}\rangle = \tilde{R}_{dt}^N \, (dt|(\frac{\partial}{\partial r_{dt}} r_{dt} - B_{dt})|\chi_{dt}\rangle \tag{17}$$

as a matching condition in the dt channel at $r_{dt} = a_{dt} = 5.1$ fm.

The purely outgoing-wave solutions for $r_{dt} > a_{dt}$ were calculated using three successively more realistic adiabatic approximations: the Born-Oppenheimer (BO), standard adiabatic (SA), and improved adiabatic (IA) approximations. The improvements on the familiar BO approximation involved adding diagonal pieces of the hamiltonian to the definition of the adiabatic potentials and using angular-dependent muonic wavefunctions, the details of which are contained in Ref. 8. In each case, the complex energy E_0 at which the matching equation (17) is satisfied was determined by iteration. Then, by comparing with the bound-state eigenvalues obtained by matching to regular boundary conditions at the origin, the shifts and widths arising from matching to nuclear boundary conditions at the channel radius were found.

The results for the lowest vibrational-rotational (v,L) states are given in Table III for L=0 and v=0,1. The numbers suggest that the SA and IA approximations are converging to something like the correct results, as was also indicated by a comparison of the adiabatic bound-state energies to the non-adiabatic value, and of the approximate μ-^{5}He energies to E_μ^0. The agreement with the earlier calculations of Bogdanova et al.[9] is quite satisfactory, considering that these calculations were done using an optical-potential representation of the nuclear interactions, and different approximations were made in solving for the dtμ wavefunction and the complex eigenvalue E_0. Also shown are the more recent results of Karnakov and Mur,[10] who used the single-level R-matrix representation of the reaction cross section given by Jarmie et al.[4]

TABLE III Energy shifts and resonance widths for the (v=0, L=0) and (v=1, L=0) vibrational-rotational states due to nuclear boundary conditions. Energies are in units of 10^{-4} eV.

	$\Delta E^{(v=0,L=0)}$	$\Gamma^{(v=0,L=0)}$	$\Delta E^{(v=1,L=0)}$	$\Gamma^{(v=1,L=0)}$
BO	-9.87	10.35	-7.63	8.03
SA	-8.15	8.53	-7.08	7.44
IA	-8.12	8.50	-6.75	7.10
Ref. 9	-9.6	8.2	-8.0	6.8
Ref. 10	-12.5	9.0	-10.4	7.4

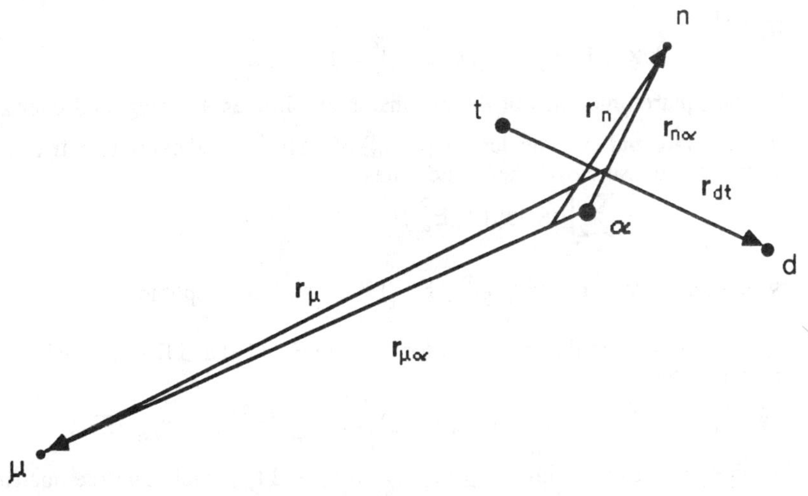

Fig. 3. Coordinates for describing the channel regions of $^5He+\mu$.

WAVEFUNCTION IN THE $n\alpha\mu$ CHANNEL REGION AND μ-α STICKING

The coordinate labeling used to describe the channel regions of the system is given in Fig. 3. At the $n\alpha$ channel surface, $r_{n\alpha}=a_{n\alpha}=3$ fm, the outgoing wave can be expressed in the separable approximation as

$$\psi_{n\alpha\mu}(\mathbf{r}_{n\alpha},\mathbf{r}_{\mu}) = \beta h_2^+(k_{n\alpha}r_{n\alpha})Y_{20}(\hat{\mathbf{r}}_{n\alpha})\,\phi_{\mu}^{\ell}(\mathbf{r}_{\mu}) \quad , \qquad (18)$$

in which β is a constant and h_2^+ is the outgoing Hankel function for ℓ=2. For $r_{n\alpha}>a_{n\alpha}$, however, since the hamiltonian is exactly separable in the variables ($\mathbf{r}_n$, $\mathbf{r}_{\mu\alpha}$), the wavefunction can be written as

$$\Psi_{n\alpha\mu}(E) = \int dE_n \int d\hat{\mathbf{k}}_n \hat{\chi}_n(\mathbf{k}_n,\mathbf{r}_n) \int d\rho(E_{\mu\alpha})\phi_{\mu\alpha}(E_{\mu\alpha}\mathbf{r}_{\mu\alpha})a_E(\hat{\mathbf{k}}_n,E_{\mu\alpha}) \times \delta(E_n+E_{\mu\alpha}-E)$$

$$= \int d\hat{\mathbf{k}}_n \int d\rho(E_{\mu\alpha})a_E(\hat{\mathbf{k}}_n,E_{\mu\alpha})\hat{\chi}_n(\hat{\mathbf{k}}_n,\mathbf{r}_n)\phi_{\mu\alpha}(E_{\mu\alpha},\mathbf{r}_{\mu\alpha})|_{E_n=E-E_{\mu\alpha}} \qquad (19)$$

in terms of the energy and momentum variables conjugate to $\mathbf{r}_n$ and $\mathbf{r}_{\mu\alpha}$. In Eq. (19), the plane-wave states $\hat{\chi}_n$ and μ-α states $\phi_{\mu\alpha}$ are taken to be energy-normalized, as is $\Psi_{n\alpha\mu}$, so that the energy-conserving delta function in the first line results naturally from the normalization conventions. The energy normalization of $\Psi_{n\alpha\mu}$ also implies the relation

$$\int d\hat{\mathbf{k}}_n \int d\rho(E_{\mu\alpha})|a_E(\hat{\mathbf{k}}_n,E_{\mu\alpha})|^2 = 1 \qquad (20)$$

for the squared modulus of the expansion amplitudes a_E integrated over all neutron directions $\hat{\mathbf{k}}_n$ and over the density $\rho(E_{\mu\alpha})$ of all bound and continuum μ-α states. This means that the sum over the bound states,

$$\omega_s^0 = \sum_b \int d\hat{\mathbf{k}}_n |a_E(\hat{\mathbf{k}}_n,E^b_{\mu\alpha})|^2 \qquad (21)$$

is the sticking fraction, and $a_E(\hat{\mathbf{k}}_n, E^b_{\mu\alpha})$ is the sticking amplitude for direction $\hat{\mathbf{k}}_n$.

One could invert the expansion for $\Psi_{n\alpha\mu}$ in Eq. (19) and find the sticking amplitude from the integral

$$a_E(\hat{\mathbf{k}}_n,E^b_{\mu\alpha}) = \int d\mathbf{r}_n \hat{\chi}_n^*(\hat{\mathbf{k}}_n,\mathbf{r}_n) \int d\mathbf{r}_{\mu\alpha}\phi^b_{\mu\alpha}(E^b_{\mu\alpha},\mathbf{r}_{\mu\alpha})\Psi_{n\alpha\mu}(E) \quad , \qquad (22)$$

but this requires knowing $\Psi_{n\alpha\mu}(E)$ for all $\mathbf{r}_n$ and $\mathbf{r}_{\mu\alpha}$ (including the nuclear region). Alternatively, one could project the expansion for $\Psi_{n\alpha\mu}$ onto a complete set of channel-surface functions

$$(\mathbf{r}_{n\alpha},\mathbf{r}_{\mu}|n\alpha\mu) = \frac{\delta(r_{n\alpha}-a_{n\alpha})}{r_{n\alpha}} Y_{\ell 0}(\hat{\mathbf{r}}_{n\alpha})\phi^j_{\mu}(\mathbf{r}_{\mu}) \quad ,$$

in which the index j ranges over all bound and continuum μ-^{5}He states, and obtain an infinite set of coupled equations for the sticking amplitudes (after a partial-wave expansion in $\hat{\mathbf{k}}_n$). If the coupling terms $(n\alpha\mu|\chi_n\,\phi_{\mu\alpha}\rangle$ on the right side of the equation vanish for high-lying states ϕ^j_μ and $\phi_{\mu\alpha}$, then the truncated system of equations can be solved by matrix inversion, knowing the left-hand side matrix elements,

$$(n\alpha\mu|\Psi_{n\alpha\mu}\rangle = \beta h^+_\ell(k_{n\alpha}a_{n\alpha})\delta_{\ell 2}\delta_{j0} \quad ,$$

vanish if the n-α angular momentum differs from $\ell = 2$ or if the μ-^{5}He wavefunction is not that for the ground state.

The sticking amplitudes and sticking fraction defined in Eqs. (19) and (21) are, in principle, quite different from the usual expressions used[11], which are based on the impulse-approximation transition matrix element

$$a_{n\ell} = \langle\phi^{n\ell}_{\mu\alpha}(\mathbf{r}_\mu)e^{i\mathbf{q}\cdot\mathbf{r}_\mu}|\Psi_{dt\mu}(\mathbf{r}_\mu,\mathbf{r}_{dt}=\mathbf{0})\rangle \quad , \qquad (23)$$

for the (n,ℓ) μ-α bound state and muon momentum $\mathbf{q}$. However, one can obtain from our framework expressions similar to those usually used by making the plane-wave approximation for $\Psi_{n\alpha\mu}$,

$$\Psi_{n\alpha\mu} \propto e^{i\mathbf{k}_{n\alpha}\cdot\mathbf{r}_{n\alpha}}\phi^0_\mu(\mathbf{r}_{\mu\alpha}) \ .$$

In this case,

$$\langle\hat{\chi}_n(\mathbf{k}_n)\phi^b_{\mu\alpha}|\Psi_{n\alpha\mu}\rangle = \delta(\mathbf{k}_{n\alpha}-\mathbf{k}_n)\langle\phi^b_{\mu\alpha}|e^{i\gamma_\mu\mathbf{k}_{n\alpha}\cdot\mathbf{r}_{\mu\alpha}}|\phi^0_\mu\rangle \quad ,$$

and the leading delta function is a poor approximation to the energy-conserving $\delta(E_n+E_{\mu\alpha}-E)$ expected, with $E = E_{n\alpha}+E^0_\mu$. The sticking amplitude is then

$$\begin{aligned} a_E(\hat{\mathbf{k}}_n,E^b_{\mu\alpha}) &= \int dE_n\delta(\mathbf{k}_{n\alpha}-\mathbf{k}_n)\langle\phi^b_{\mu\alpha}|e^{i\gamma_\mu\mathbf{k}_{n\alpha}\cdot\mathbf{r}_{\mu\alpha}}|\phi^0_\mu\rangle \\ &= \delta(\hat{\mathbf{k}}_{n\alpha}-\hat{\mathbf{k}}_n)\langle\phi^b_{\mu\alpha}|e^{i\gamma_\mu\mathbf{k}_{n\alpha}\cdot\mathbf{r}_{\mu\alpha}}|\phi^0_\mu\rangle\,\Big|_{E_n=(1-\alpha)E_{n\alpha}} \ . \qquad (24)\end{aligned}$$

In these expressions, the mass ratios $\gamma_i=\dfrac{m_i}{m_i+m_\alpha}$ and $\alpha=\gamma_\mu\gamma_n=0.0055$ are used. The matrix element in Eq. (24) is energy-conserving only for $E_{\mu\alpha}=\alpha E_{n\alpha}+E^0_\mu=86$ keV, far above the bound-state energies at which it is evaluated. However, if the delta functions are ignored, the remaining matrix element,

$$a_E(E^b_{\mu\alpha}) = \langle\phi^b_{\mu\alpha}|e^{i\gamma_\mu\mathbf{k}_{n\alpha}\cdot\mathbf{r}_{\mu\alpha}}|\phi^0_\mu\rangle \quad ,$$

is similar to the standard expression for the sticking amplitude.

SUMMARY AND CONCLUSIONS

We have discussed the R-matrix parameterization of the two-body reactions of a multichannel system, and have shown that such a description gives a detailed and accurate representation of the experimental measurements for the ^{5}He system. The "shadow" pole associated with the $J^\pi=3/2^+$ resonance in the nuclear ^{5}He system is an interesting phenomenon that could have consequences for muon-catalyzed fusion, since it occurs on the same sheet as the dtμ molecular ground state.

The R-matrix framework is well-suited also for describing the ^{5}He+μ system, since the division of coordinate space into the nuclear and channel regions allows a relatively simple separable approximation to be made in the nuclear region that is not valid in the channel regions. On the other hand, the three-body wavefunctions in the channel regions can be treated exactly, albeit with difficulty in the case of dtμ. Our calculation to test this procedure by investigating nuclear effects on the bound states of the dtμ molecule used the reduced R-matrix formalism to facilitate the matching at the channel surface and adiabatic approximations for the dtμ wavefunctions. As expected, the improved adiabatic approximation seemed to give the best results, which agreed well with calculations that had been done earlier in the Soviet Union using a generalized optical-model approach to incorporate the nuclear effects. Although in this case the use of the R matrix is less model-dependent than the use of a potential, the similarity of the nuclear shifts and widths obtained from the two approaches indicates that they result primarily from nuclear surface effects.

We have presented a more rigorous prescription for calculating sticking amplitudes and the μ-α sticking fraction than is presently in use. The solution of the equations for the sticking amplitudes from this new prescription is clearly a difficult numerical task, for which no "best" approach is yet apparent. However, it is probably worth the effort to do this more realistic calculation, in view of the serious questions that can be raised concerning the energy-conserving properties of the "sudden"-approximation matrix element normally used to calculate the sticking fraction.

ACKNOWLEDGMENTS

D. Dodder, K. Witte, R. Brown, and N. Jarmie have made valuable contributions to the work on the ^{5}He nuclear system. This effort has been sponsored by the Division of Advanced Energy Projects and the Division of Nuclear Physics of the U.S. Department of Energy.

REFERENCES

1. E. P. Wigner and L. Eisenbud, Phys. Rev. **72**, 29 (1947); A. M. Lane and R. G. Thomas, Rev. Mod. Phys. **30**, 257 (1958).
2. T. Teichmann and E. P. Wigner, Phys. Rev. **87**, 123 (1952). See also Lane and Thomas, Ref. 1.
3. G. M. Hale and D. C. Dodder, to be published. Preliminary results are described in *Nuclear Cross Sections for Technology*, edited by J. L. Fowler, C. H. Johnson, and C. D. Bowman (National Bureau of Standards, Washington D. C., 1980), NBS Special Publication **594**, p.650.
4. N. Jarmie, R. E. Brown, and R. A. Hardekopf, Phys Rev. C **29**, 2031 (1984); R. E. Brown, N. Jarmie, and G. M. Hale, Phys. Rev. C **35**, 1999 (1987).
5. B. Haesner, W. Heeringa, H. O. Klages, H. Dobiasch, G. Schmalz, P. Schwarz, J.Wilczynski, B. Zeitnitz, and F. Käppler, Phys. Rev. C **28**, 995 (1983).
6. G. M. Hale, R. E. Brown, and N. Jarmie, Phys. Rev. Lett. **59**, 763 (1987).
7. We list here only the recent, most accurate (variational) calculations: C.-Y. Hu, Phys. Rev. A **32**, 1245 (1985); **36**, 4135 (1987); A. M. Frolov and V. D. Efros, J. Phys. B **18**, L265 (1985); S. I. Vinitsky, L. I. Korobov, and V. I. Puzynin, Eksp. Teor. Fiz. **91**, 705 (1986) [Sov. Phys.--JETP **64**, 417 (1986)]; K. Szalewicz et al., Phys. Rev. A **36**, 5494 (1987); S. A. Alexander and H. J. Monkhorst, Phys. Rev. A **37** (to be published, 1988); M. Kamimura, Phys. Rev. A **38** (to be published, 1988).
8. M. C. Struensee, G. M. Hale, R. T Pack, and J. S. Cohen, Phys. Rev. A **37**, 340 (1988).
9. L. N. Bogdanova, V. E. Markushin, and V. S. Melezhik, Zh. Eksp. Teor. Fiz. **81**, 829 (1981) [Sov. Phys. JETP **54**, 442 (1981)].
10. B. M. Karnakov and V. D. Mur, Yad. Fiz. **44**, 1409 (1986) [Sov. J. Nucl. Phys. **44**, 916 (1986)].
11. L. Bracci and G. Fiorentini, Nucl. Phys. **A364**, 383 (1981).

DENSITY DEPENDENT STOPPING POWER AND MUON STICKING IN MUON CATALYZED D-T FUSION

H. E. Rafelski
Department of Physics and Astrophysics,
University of Cape Town, Rondebosch 7700, Cape Town, RSA

B. Müller
Inst. f. Theor. Physik, Universität, D-6000 Frankfurt, West Germany
Dept. of Physics, University of Arizona, Tucson, AZ 85721

ABSTRACT

The origin of the experimentally observed [1] density dependence of the muon alpha sticking fraction ω_s in muon catalyzed deuterium- tritium fusion is investigated. We show that the reactivation probability depends sensitively on the target stopping power at low ion velocities. The density dependence of the stopping power for a singly charged projectile in liquid heavy hydrogen is parametrized to simulate possible screening effects and a density dependent effective ionization potential. We find that, in principle, a description of the measured density dependence is possible, but the required parameters appear too large. Also, the discrepancy with observed (Heμ) X-ray data widens.

INTRODUCTION

In the muon catalyzed d-t fusion cycle the loss of the muon due to capture by the alpha particle formed in the nuclear reaction is an important process, since it sets an upper limit to the number of fusions induced during the muon lifetime. Several processes can reactivate the muon during the slow-down of the $(\alpha\mu)^+$ ion in the target and thus decrease the effective muon-alpha sticking fraction, allowing the muon to reenter the fusion cycle. The processes leading to muon-reactivation have been studied by several groups [2-8]. Although the theoretical reactivation rates differ amongst each other, all theoretical predictions show considerable discrepancies to experiments [1,9] at higher density (about liquid hydrogen density) and none of the theoretical calculations can find the strong density dependence of the muon-alpha sticking fraction as measured at LAMPF [1].

The reactivation probability of the muon is determined by a rather delicate balance between collisional excitation and ionization processes on one hand and the combination of electromagnetic de-excitation and slowing down of the $(\alpha\mu)$ particle on the other hand. In principle, the observed decrease of ω_s^{eff} with growing target density can have its origin in any of these contributions. The situation is complicated by the circumstance that the mechanism for a *weak* density dependence with the observed trend is well known. Its origin is revealed by the observation that the excitation rates λ_{ik}^{exc}, the stripping rates λ_i^{str} and the $(\alpha\mu)$ energy loss rate dE/dt all grow linearly with density ρ, whereas the radiative de-excitation rates λ_{ik}^{rad} do not. The identical density dependence of $\lambda_{ik}^{exc} = v\sigma_{k\rightarrow i}\rho$, $\lambda_i^{str} = v\sigma_i^{str}\rho$, and $dE/dt = -vS\rho$ effectively cancels from the rate equations when these are rewritten in terms of the kinetic energy E of the $(\alpha\mu)$, but an effective density variability is introduced in this way into the radiative rates. To wit, if P_i denote the probability that the muon is occupying the i-th level of the moving helium atom, the rate equations read:

$$\frac{dP_i}{dt} \sim \sum_k (\lambda_{ik}^{exc} + \lambda_{ik}^{rad}) P_k - \sum_k (\lambda_{ki}^{exc} + \lambda_{ki}^{rad}) P_i - \lambda_i^{str} P_i. \tag{1}$$

After the transformation to the energy as independent variable the hidden density dependence becomes explicit:

$$\frac{dP_i}{dE} \sim \frac{1}{S} \sum_k (\sigma_{k\to i}^{exc} + \frac{1}{v\rho}\lambda_{ik}^{rad}) P_k - \frac{1}{S} \sum_k (\sigma_{i\to k}^{exc} + \frac{1}{v\rho}\lambda_{ki}^{rad}) P_i - \frac{1}{S}\sigma_i^{str} P_i. \tag{2}$$

Therefore, the radiative transitions become less efficient at high density, and a larger fraction of muons can be ionized from higher states in the helium atom. This mechanism has been taken into account in previous calculations where it was found to produce a weak decrease of ω_s^{eff} with density [5,7,8].

In order to explain the *strong* density dependence reported in ref.[1] one must invoke a different mechanism. Here we make the plausible assumption that the primary sticking fraction ω_s^0 established immediately after the nuclear fusion does not depend on density, nor does the distribution of captured muons over the state of the helium atom. Any observed density dependence must then originate from the competition of density effects on the reactivation cascade. An additional, i.e. nonlinear, density dependence of the rate of energy loss of the $(\alpha\mu)$ is the most promising mechanism for an explanation of the observed phenomena: $S(\rho)$. Such non-linearities are known to exist in condensed matter targets at ion velocities below and around the equivalent K-shell velocity. As this is the region in which the stripping cross sections peak, it is not inconceivable that a strong density dependence could result from this effect.

A nonlinear growth of the stopping power with target density can have two clearly distinct origins. Firstly, the electronic structure of the target usually changes with density when the interatomic spacings are comparable to the atomic radii, and even discontinuously so when a phase transition occurs, e.g. from gas to liquid or solid. Therefore, stopping powers based on the values measured in, or calculated for, the gaseous phase [10] need not apply to liquid or solid targets. Secondly, the response of the electronic structure of the medium, of which the energy loss of the penetrating ion is one particular manifestation, may not be a linear function of the target density. Such effects arise, e.g., due to partial screening of the projectile charge by the target electrons, and are usually taken into account through a velocity dependent effective projectile charge. For singly charged ions in solid targets this effect is mostly assumed to be absent [11], but here liquid hydrogen might play a special role since it has a relatively low electron density.

BRIEF THEORY OF STOPPING POWER OF IONS

We now briefly review the stopping power of ions in matter (for more details we refer to the reviews of Fano [12] and Ziegler et al. [10,11]). The stopping power of muonic helium ions ($\alpha\mu$) in hydrogen can be derived from the proton stopping power since the stopping power of ions depends only on the charge Z_P and velocity v but not on the mass of the projectile as expressed in Bethe's stopping power formula

$$-\frac{dE}{dx} = \frac{4\pi Z_P^2 e^4}{mv^2} N Z_T \left(\ln \frac{2mv^2}{I} + \ln \frac{1}{1-\beta^2} - \beta^2 - C \right), \tag{3}$$

where m is the electron mass, NZ_T is the average electron density, and I is the mean ionization potential of the target electrons, and C describes shell corrections arising from the local inhomogeneity of the target electron distribution. This formula predicts the energy loss accurately at energies larger than 1 MeV per atomic mass unit (amu). There two physical parameters, the mean ionization potential I and the shell corrections C, successfully express the influence of the electronic structure of the target material on the rate of energy loss. At very low projectile velocities non-electronic energy-loss mechanisms become important. In the intermediate energy region (between 30 keV/amu and 1 MeV/amu) no simple theoretical description exists, as we will discuss below. However, precisely this energy range is of great importance for the muonic reactivation process, because the excitation cross sections peak around $v = 2$ (in atomic units), corresponding to an energy of 100 keV/amu.

In the framework of linear response theory, the rate of energy loss of a charged particle penetrating through matter can be expressed in terms of the dynamical structure factor [13]

$$S(q,\omega) = \sum_f \frac{1}{Z_T} |\langle f|\rho(q)|i\rangle|^2 \delta(\omega + E_i - E_f), \tag{4}$$

where i and f denote the initial and final state of the target, $\rho(q)$ is the Fourier transform of the electron density operator, and Z_T is the number of electrons per target atom. (For thermally excited targets an ensemble average over initial states must be included in the definition of $S(q,\omega)$.) The rate of energy loss is then:

$$-\frac{dE}{dt} = \int d^3p' \int \omega d\omega \int d^3q (\frac{2Z_P e^2}{q^2})^2 S(q,\omega) \delta(E' + \omega - E) \delta(p' + q - p), \tag{5}$$

where E, p and E', p' are the initial and final energy and momentum of the projectile, respectively. For projectiles much heavier than an electron the delta functions yield a connection between energy and momentum transfer:

$$\omega = vq, \tag{6}$$

where v is the projectile velocity. On the other hand, the dynamical structure factor is related to the imaginary part of the longitudinal dielectric constant:

$$S(q,\omega) = \frac{q^2}{4\pi^2 e^2} (1 + n(\omega)) \Im \left(\frac{-1}{\epsilon_L(q,\omega)} \right), \tag{7}$$

where $n(\omega)$ is the occupation probability of excited electronic states. Assuming that the target is initially in its ground state, we therefore have the relation

$$-\frac{dE}{dt} = (\frac{Z_P e}{\pi})^2 \int_0^\infty \omega d\omega \int \frac{d^3q}{q^2} \delta(\omega - qv) \Im \left(\frac{-1}{\epsilon_L(q,\omega)} \right) \tag{8}$$

which forms the basis of the Lindhard-Winther theory of electronic stopping power [14]. It can also be derived within the framework of the classical theory of electromagnetism in continuous media [15].

The Bethe formula is obtained in the limit of vanishing q-dependence of the dielectric constant. The momentum integration can be carried out, if an upper cut-off $2mv$ is introduced, yielding the stopping power formula

$$S = -\frac{1}{v\rho}\frac{dE}{dt} = \frac{2}{\pi\rho}(\frac{Z_P e^2}{v})^2 \int_0^\infty \omega d\omega \Im \left(\frac{-1}{\epsilon_L(\omega)} \right) \ln \left(\frac{2mv^2}{\omega} \right), \tag{9}$$

where ρ is the target density. Without the logarithmic term the frequency integral is related to the plasma frequency ω_p:

$$\int_0^\infty \omega d\omega \Im \left(\frac{-1}{\epsilon_L(\omega)} \right) = \frac{\pi}{2}\omega_p^2 = \frac{2\pi e^2}{m} n, \tag{10}$$

where n is the electron density in the target. Defining the average ionization potential I through

$$\int_0^\infty \omega d\omega \Im \left(\frac{-1}{\epsilon_L(\omega)} \right) \ln(\omega) = \frac{\pi}{2}\omega_p^2 \ln I, \tag{11}$$

the first term in the Bethe formula is derived. The remaining terms are relativistic corrections and shell corrections which result from the inhomogeneity of the electron density in the target [12].

For velocities in the range $(I/2m)^{1/2}$ these approximations no longer hold, and the stopping power must be calculated using a realistic description of $\epsilon_L(q,\omega)$. For gases, atomic or molecular exact dipole strength calculations are available for that purpose. Such calculations yield, e.g. $I = 15$ eV for atomic hydrogen and $I = 19$ eV for molecular hydrogen [12], showing a definite dependence of the stopping power on the chemical composition of the target. For solid targets the random phase approximation (RPA) to the homogeneous electron gas has found widespread use [16].

NONLINEAR DENSITY DEPENDENCE OF STOPPING POWER

Differences in the stopping power S between gaseous and condensed matter target of the same chemical composition have been noted by many experimenters (see e.g. the review by Thwaites [17]). Systematic investigations of this effect were carried out by Ziegler et al. [18] using alpha particles, who found S not only to decrease below theoretical expectations for small velocities, but also a marked difference in the size of this effect between gases and solids. The difference was especially large for targets with low atomic number, rising to 40% at 600 keV alpha particle energy. Very similar behavior was found by Palmer et al. for liquid targets used in biological radiation research and even for water [19]. At an alpha particle energy of 500 keV the stopping power in the gaseous phase was 39% higher than in the liquid phase.

That an explicit density dependence of S can exist even inside the liquid phase was recently shown by Both et al. [20] in an experiment making use of the Doppler-shift attenuation method. From the measurement of γ-rays emitted by moving ^{7}Li nuclei in liquid ethane (C_2H_6) they found that the stopping power decreases slightly with increasing density of the liquid. This decrease could be explained by the change in the electron density distribution inside a molecule due to proximity effects from neighboring molecules. We conclude that there is strong experimental evidence for a density dependent reduction of stopping power in the liquid phase at low ion velocities, at least for materials which exhibit a tendency to form hydrogen bonds leading to the possibility of molecular clustering in the liquid.

At this level of the formalism, nonlinear density effects are contained in the nonlinear dependence of the dielectric function ϵ_L upon density. In principle, i.e. if ϵ_L would be taken to be the true dielectric function of the target, all density dependence of this kind would be exactly described, even the discontinuous changes across a phase boundary. Unfortunately, realistic *ab initio* calculations of $\epsilon_L(q,\omega)$ do not exist for most materials, in particular, none is known to us for liquid hydrogen. Sufficient experimental information about $\epsilon_L(q,\omega)$ is also usually not available, because the polarizability can be readily measured only for values of q and ω that correspond to the photon dispersion relation and, moreover, because only the transverse and not the longitudinal component is obtained in this way. Therefore the stopping power of materials that do not either classify as dilute gases or exhibit a well-defined conduction band cannot be predicted accurately.

As eq.(9) indicates the relevant expansion parameter is $(Z_P e^2/v)$ and not $(Z_P e^2)$. At low velocities, therefore, the validity of linear response theory becomes questionable and must be replaced by a formulation that allows for the treatment of higher-order effects. The most important of these is the ability of the projectile to capture and bind electrons from the target, effectively reducing the charge seen by the target electrons. A formulation that allows for the calculation of such an effective ion charge Z_P^{eff} is furnished by the nonlinear theory of transport phenomena. In the context of which the stopping power is expressed as [21]

$$S = v_F n \sigma_{tr} = v_F n \int d\sigma (1 - \cos\theta), \tag{12}$$

where v_F is the Fermi velocity, n the target electron density, and σ_{tr} the transport cross section for momentum degradation. Here θ is the scattering angle of a target electron in the projectile rest frame, and only scattering of electrons near the Fermi surface is taken into account. The transport cross section σ_{tr} can be expressed in terms of the scattering phase shift of electrons at the Fermi surface from the projectile potential:

$$\sigma_{tr} = \frac{4\pi}{k_F^2} \sum_{l=0}^{\infty} (l+1) \sin^2(\delta_{l+1} - \delta_l). \tag{13}$$

Here it is possible to account for screening of the projectile by bound electrons by calculating the phase shifts for a self- consistent, screened projectile ion potential inside the target medium.

An analysis along these lines was performed by Echenique et al. [22], who applied the density- functional formalism to calculate the effective Coulomb potential of a slow ion travelling in a homogeneous electron gas. These authors found that the stopping power decreases below the value obtained in linear response theory when the target electron density falls below a critical value. For singly charged ions the critical density corresponds to an equivalent electron radius $r_s = (3n/4\pi)^{1/3}$ between 3 and 4. This has to be compared with an average value $\langle r_s \rangle = 3.35$ in liquid hydrogen, but it must be noted that the electron density between molecules falls to much smaller values, corresponding to $r_s \approx 9$. This indicates that the effective charge of a $(\alpha\mu)$ particle in liquid hydrogen may be less than one at low velocities.

Note that this conjecture is not based on direct experimental evidence, so far. However, ample evidence for the decrease of the effective charge exists for alpha particles, where Z_{eff} drops from two to about one in an energy region around and below 500 keV [11]. For singly charged ions the situation is still controversial. Nevertheless, the stopping power as given in Ziegler's TRIM-85 program [11] for solid hydrogen targets differs considerably from that for hydrogen gas [10], which was used in previous calculations [2,5]. This is shown in Fig. 1 depicting the stopping power function $S(v)$ as function of ion velocity v (in atomic units). At $v = 2$ the reduction agrees rather well with the empirical findings of Palmer et al. [19], indicating that it is likely to be valid also for liquid hydrogen targets.

Lacking a more reliable description of the density dependence of the stopping power function $S(v;\epsilon,\delta)$ applicable to a $(\alpha\mu)^+$ ion in hydrogen, we have carried out model calculations with a simple parametrized form of $S(v)$. The two parameters of our ansatz, δ and ϵ, describe the effects of a density dependence of the mean ionization potential I and of the effective charge Z_{eff}, respectively. If ϕ is the target density in units of liquid hydrogen density ($4.25 \times 10^{22}\text{cm}^{-3}$) the explicit forms chosen here are:

$$I_{eff} = I(1 + \phi\delta), \tag{14}$$

$$Z_{eff} = Z(1 + \epsilon\phi e^{-v^2})^{-1}, \tag{15}$$

where v is in atomic units. Note that both parameters coincide with their unmodified values at very low density. For $\epsilon = \delta = 0$ we take the function $S(v) = (Z/v)^2 S_0(v^2)$ for hydrogen gas from Ziegler [10], modifying it for other values according to the Bethe formula, eq. (3). The velocity dependence of our parametrized function

$$S(v;\epsilon,\delta) = (Z_{eff}/v^2) S_0(v^2 I/I_{eff}) \tag{16}$$

is shown in Fig. 1 for two non zero values of the parameters. The smaller set of values, $\epsilon = 0.7$ and $\delta = 0.5$ (denoted by (a) from here on), appears to mark the borderline of acceptable modification of the gas stopping power. The second, larger set, $\epsilon = \delta = 1$ (denoted by (b)), was chosen to explore the range of effects beyond these limits on a purely speculative basis, without any claim that it may be realized under experimental conditions.

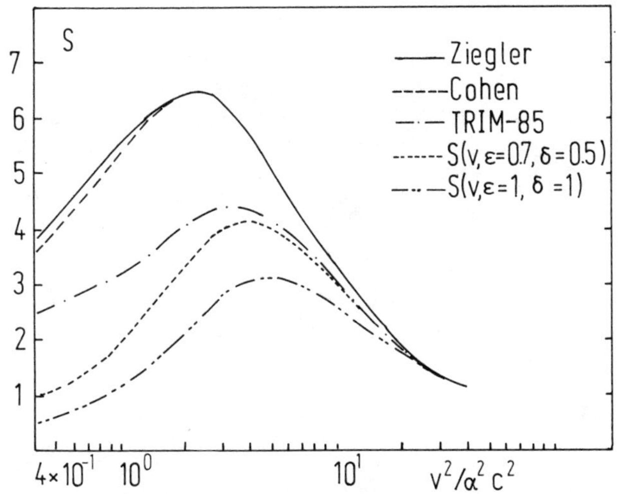

Fig. 1: Various choices for the stopping power function $S(v)$. The curves denoted 'Ziegler' [10] and 'Cohen' [5] are for hydrogen gas, the curve 'TRIM-85' [11] is for solid hydrogen. The curves labelled by values of ϵ and δ are obtained from our parametrization.

Due to the fact that the various stopping power functions deviate from one another only for small ion velocities they result only in small differences in the range of the $(\alpha\mu)$ ion emitted originally with $v = 6$. As shown in Fig. 2, the difference is less than 40% at $\phi = 1$ even for the larger parameter set (b). Precise experimental determination of the range of $(\alpha\mu)$ ions in high density targets would be useful in establishing more rigorous limits for ϵ and δ.

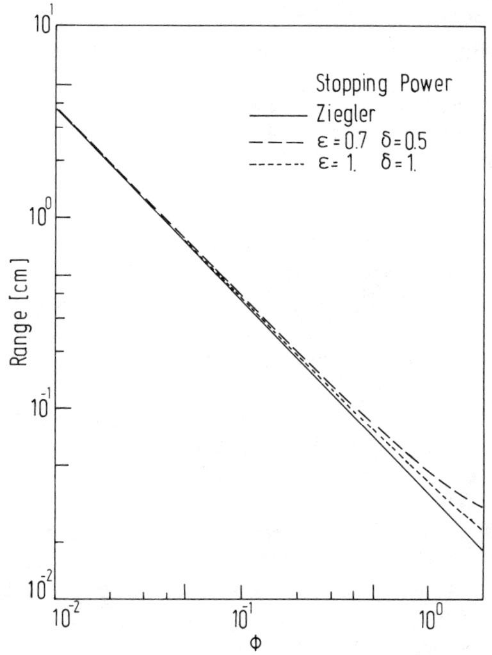

Fig. 2: Range of $(\alpha\mu)^+$ ions in liquid hydrogen at density $\phi = 1.2$. For the notation see Fig. 1.

INFLUENCE ON MUON REACTIVATION

In order to investigate the effect of a nonlinear density dependence of the target stopping power on the muon reactivation coefficient R, we have solved the rate equations, eqs. (2), which govern the reactivation cascade, with a computer code that contains several new aspects related to the excitation and ionization cross sections for the muon [8]. These are:

1. The 2s and 2p states were treated separately throughout the calculation.
2. Transitions from the 1s, 2s, and 2p states into the continuum, as well as transitions among all states up to the N-shell (n=4) were calculated in Born approximation *with* recoil and Coulomb distortion [23].
3. Transfer cross sections out of states up to the N-shell were calculated in the framework of the eikonal approximation [24].

All other rates were taken from the work of Cohen [5,25]. For the 2s-2p Stark mixing cross section we have introduced a factor Q, where $Q = 1$ corresponds to the choice of Cohen and $Q = 0.43$ to that of Markushin [7].

The effects mentioned in the second item above, i.e. recoil and Coulomb distortion, are not taken into account in the usual procedure of using scaled electronic cross section, since they depend on the nuclear masses of target and projectile as well. They are potentially important corrections, because the ratio of muon mass to nuclear mass is not small. Intuitively, one would suspect that their influence becomes more prominent at lower $(\alpha\mu)$ energies, as closer collisions are required for a muonic excitation. This expectation is borne out by our numerical results which show that the regeneration coefficient increases by 24% due to the action of both corrections in the ddμ reaction, but only by about 1% in the more energetic dtμ reaction.

The reactivation coefficient R has been calculated with the various stopping power formulas discussed in the previous section and shown in Fig. 1. The resulting muon-alpha sticking fraction

$$\omega_s^{eff} = (1 - R)\omega_s^0 \tag{17}$$

is shown in Fig. 3 for an initial sticking coefficient $\omega_s^0 = 0.88$, as favored by the most recent three-body calculations [26]. The parameter values $\epsilon = \delta = 0$ correspond to the stopping power in hydrogen gas also used by Cohen. As Fig. 3 shows, our result in this case shows the same weak density dependence as that of Cohen.

A much stronger density dependence is found for our two non trivial choices of the stopping power parameters. This is no surprise, because an explicit density dependence is introduced by them in the rate equations (2). It is also obvious, however, that the more realistic smaller parameter set (a: $\epsilon = 0.7, \delta = 0.5$) does not fully account for the density dependence reported by the LAMPF collaboration [1]. The larger parameter set (b: $\epsilon = \delta = 1$) does yield the desired steep density dependence, but we remind the reader that this choice was not considered to be very realistic. Nonetheless, if we would take the agreement seriously, it would allow us to extrapolate to an effective sticking probability of less than 2×10^{-3} at $\phi = 2.5$, reachable with the next generation of target vessels [27].

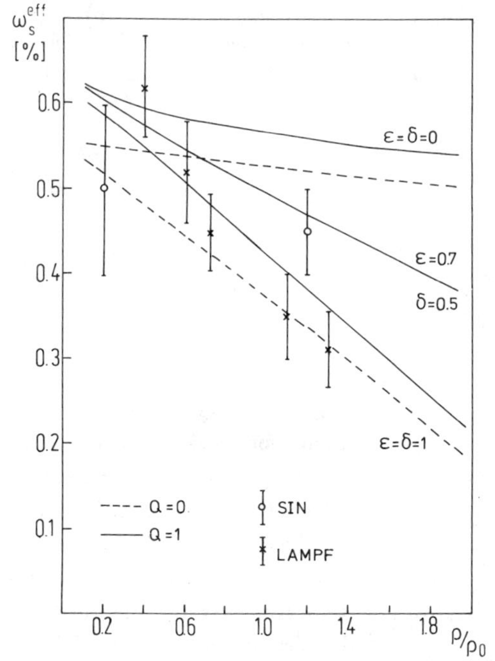

Fig. 3: Effective sticking fraction ω_s^{eff} as function of density in units of liquid hydrogen density. Our results are shown for three sets of values for the parameters ϵ and δ describing the density dependence of the stopping power. Q denotes the strength of Stark mixing between the 2s and 2p states. The experimental data points are taken from refs. [1] and [9].

Given a particular set of values for the parameters ϵ and δ our calculations predict a similar density dependence of the effective sticking fraction for the ddμ reaction. In this case the reduction compared to the standard value of ω_s^{eff} is 15% for the parameter set (a) and 30% for the set (b) at $\phi = 1.2$. Unfortunately, no direct measurement of this quantity exists for high density targets, and ω_s^{eff} cannot be deduced from the total number of fusion events in the ddμ system.

INFLUENCE ON MUONIC X-RAYS AFTER FUSION

Any modification of the stopping power also has an influence on the yield of muonic X-rays emitted after fusion. These X-rays originate from muons that are either initially captured into excited states or excited during the slow-down of the $(\alpha\mu)$ in the target. Once the muon enters into an excited state, e.g. the L-shell, the radiative de-excitation competes with all other processes leading to depopulation of the state. A simple consideration on the basis of eq.(2) shows that the K X-ray yield due to excitations into the L-shell is proportional to

$$Y(K_\alpha) \sim \frac{\sigma_{1\to2}}{S}\left(1+\frac{\lambda_2^{exc}}{\lambda_{2p}^{rad}}\right)^{-1}, \tag{18}$$

where the ratio of the K-L excitation cross section $\sigma_{1\to2}$ and the stopping power S enters. The velocity dependence of this ratio is shown in Fig. 4 for our three standard parameter sets for ϵ and δ at $\phi = 1.2$. Obviously, a reduced stopping power will result in an increase of the K_α X-ray yield. This increase will be more pronounced for the ddμ reaction, since in this case the (μ^3He) starts with a lower velocity ($v \approx 2$), whereas the ($\mu\alpha$) particle created in the dtμ reaction spends most of its time in the energy range $v > 2$ where no change occurs.

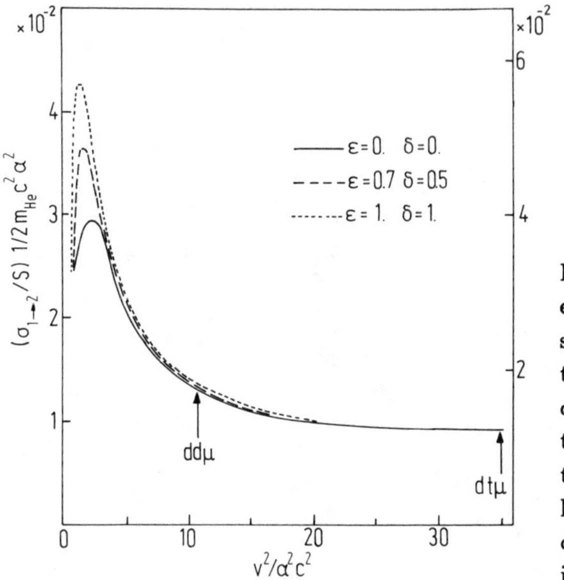

Fig. 4: Ratio of the K- to L-shell excitation cross section $\sigma_{1\to2}$ and stopping cross section S as function of the ion velocity. The scale on the left and right margin pertains to the ddμ and dtμ system, respectively. The initial velocities of the helium nuclei produced in ddμ and dtμ fusion are indicated by arrows.

The results of our calculations for the total muonic helium K_α X-ray yield emitted after ddμ and dtμ reactions are represented in Fig. 5 by crosses. The values obtained for the standard hydrogen stopping power ($\epsilon = \delta = 0$) can be compared with those of other authors, denoted by open circles. The dots labelled by 'C', 'T', and 'M' refer to references [5-7], respectively. As can be seen, our values for full Stark mixing ($Q = 1$) lie higher than those of refs. [5] and [7], but lower than those of ref. [6]. This is probably caused by the differences in the matrix elements for excitation used in the various calculations, and may be taken as an indication of the theoretical uncertainties.

It is worth noting, however, that all theoretical results are higher than the experimental values of the SIN collaboration [28] for both systems. Our results obtained for the other two parameter sets (a,b) show that this systematical discrepancy is not likely to be caused by stopping power effects: A density dependent reduction in the stopping power tends to increase the difference between theory and experiment. (Rather, the experimental data would seem to indicate that the stopping power is larger at high density, which runs counter to all reasoning.) A larger value for the Stark mixing cross section, e.g. $Q = 2$ as advocated in ref. [29], would even further increase the discrepancy. We note, finally, that the ratios of K_β and K_γ to K_α X-rays are only slightly affected by the modifications in the stopping power function.

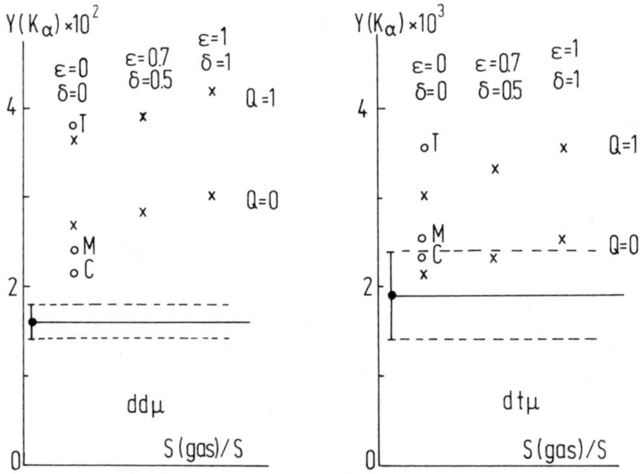

Fig. 5: Yield of muonic helium K_α radiation emitted after fusion. The crosses denote our results for various choices of the stopping power parameters ϵ and δ. The open circles denote results obtained by other authors [5-7], and the experimental data are those of Bossy et al. [28]. The figure shows that a density dependent reduction in stopping power tends to make the discrepancy between theory and experiment larger.

CONCLUSIONS

The experimental and theoretical situation yields considerable evidence that the stopping power of liquid hydrogen may be density dependent, and lower than previously expected, at ion velocities in the range $v < 2$. We have investigated the possible effects of a density dependent stopping power on muon reactivation and muonic X-ray yield in the framework of a simple, heuristic parametrization. Our reactivation calculations, which are based on a partially new set of excitation and transfer cross sections, give muon-alpha sticking fractions considerably larger than experimental results, if the standard stopping power formula is used. A strongly density dependent stopping power brings the theoretical values closer to the experimental data points, but unreasonably large values of the parameters must be introduced to explain the full density dependence observed at LAMPF [1]. Furthermore, the modified stopping power function $S(v,\rho)$ would lead to a considerable increase in the difference between experiment and theory for muonic X-rays emitted after fusion. On the basis of this evidence we conclude that an explanation of the observed density dependence of ω_s^{eff} in terms of a modified, density dependent stopping power in heavy isotope liquid hydrogen does not appear likely.

Acknowledgement: We would like to thank M. Danos, J. Rafelski, D. Trautmann, and R. D. Viollier for stimulating discussions. This work was supported in part by the U.S. Department of Energy.

REFERENCES

1. S.E.Jones *et al.*, Phys. Rev. Lett. **56**, 588 (1986).
2. L.Bracci and G.Fiorentini, Nucl. Phys. **A364**, 383 (1981).
3. S.S.Gershtein *et al.*, ZhETF **80**, 1690 (1981) [JETP **53**, 872 (1981)].
4. L.I.Menshikov and L.I.Ponomarev, Pis'ma ZhETF **41**, 511 (1985) [JETP Lett. **41**, 623 (1985)].
5. J.S.Cohen, Phys. Rev. Lett. **58**, 1407 (1986).
6. H.Takahashi, Phys. Lett. **174B**, 133 (1986).
7. V.E.Markushin, Proc. μCF-87, Gatchina, May 1987.
8. H.E.Rafelski, PhD thesis, University of Cape Town (1988).
9. W.H.Breunlich *et al.* Phys. Rev. Lett. **53**, 1137 (1984); Muon-Catalyzed Fusion **1**, 67 (1987) [Proc. μCF-86].
10. H.H.Anderson and J.F.Ziegler, *Hydrogen Stopping Power and Ranges in All Elements*, Pergamon Press (1977) [see also references therein].
11. J.F.Ziegler, J.P.Biersack, and U.Littmark, *The Stopping and Ranges of Ions in Solids*, Pergamon Press (1985) [see also references therein].
12. U.Fano, Ann. Rev. Nucl. Science **13**, 1 (1963).
13. N.R.Arista and W.Brandt, Phys. Rev. **A23**, 1898 (1981).
14. J.Lindhard and A.Winther, K. Dan. Vidensk. Selsk. Mat. Fys. Medd. **34**, 4 (1964).
15. J.Hubbard, Proc. Roy. Soc. **A68**, 976 (1955).
16. J.Lindhard, K. Dan. Vidensk. Selsk. Mat. Fys. Medd. **28**, 8 (1954).
17. D.I.Thwaites, Rad. Res. **95**, 495 (1983).
18. J.F.Ziegler, W.K.Chu, and J.S.Y.Feng, Ap. Phys. Lett. **27**, 387 (1975).
19. R.B.J.Palmer and A.Akhavan-Rezayat, J. Phys. **D11**, 605 (1978).
20. G.Both, R.Krotz, K.Lohmer, and W.Neuwirth, Phys. Rev. **A28**, 3212 (1983).
21. T.L.Ferrell and R.H.Ritchie, Phys. Rev. **B16**, 115 (1977).
22. P.M.Echenique, R.M.Nieminen, J.C.Ashley, and R.H.Ritchie, Phys. Rev. **A33**, 897 (1986).
23. D.Trautmann, F.Rösel, and G.Baur, Nucl. Inst. Meth. *214*, 21 (1983). D.Trautmann, Basel, private communication.
24. J.K.M.Eichler, Phys. Rev. **A23**, 498 (1981).
25. J.S.Cohen, preprint LA-UR-87-2225, Los Alamos (1987).
26. K.Szalewicz, Report at this conference.
27. A.J.Caffrey, Report at this conference.
28. H.Bossy *et al.*, Phys. Rev. Lett. **59**, 2864 (1987). H.Bossy, Report at this conference.
29. M.C.Struensee and J.S.Cohen, preprint LA-UR-87-4049, Los Alamos (1988).

A PROPOSED METHOD FOR REDUCING THE STICKING CONSTANT IN M.C.F.

Russell M. Kulsrud
Princeton University, Plasma Physics Laboratory
P.O. Box 451
Princeton, New Jersey 08543

ABSTRACT

We present a method of reducing the effective sticking coefficient involving reacceleration of the Heμ ion by ion cyclotron resonance. It is necessary to work with the target D-T in solid or liquid form, and the target, has to be divided into many thin rods or pellets of order 100 microns in radius. The idea is to impose a magnetic field and a rotating electric field. A balance is achieved between drag in the rods or pellets and acceleration in the region in between, so that the Heμ ion is held at a constant velocity until stripping occurs. Although stripping is greatly reduced the idea is not practical, as it stands, because of the large alpha particle heat produced in the target.

There is consequently a finite probability of either ionization by collision with target molecules or of charge transfer, thus freeing the muon for more fusions. The probability of this happening by the time that the Heμ ion slows down to zero is between 30 and 40 per cent depending on density.[1] By an inspection of Ref. 1, one finds that

$$\frac{dR}{d\varepsilon} \approx 0.1/\mathrm{MeV} \tag{3}$$

where the derivative is with respect to the drag energy lost in slowing down.

The idea described in this paper is to compensate the drag by an accelerating electric field so that the Heμ is made to move at 3.5 MeV until stripping occurs. The stopping power in a DT target at liquid hydrogen densities (LHD) is

$$\left(\frac{d\varepsilon}{dx}\right)_{drag} \approx 45\ \mathrm{MeV/cm}\ \frac{\varepsilon_o}{\varepsilon} \ . \tag{4}$$

Consequently, this method is impractical in a high density target because an electric field of the same magnitude is needed. However, if the target is filamented into many fine rods of radius a (see Fig. 1) with a small filling factor f, then it is possible to extract the Heμ ion from these target rods, accelerate it in a vacuum region in between the rods, making up for any energy lost in the rods, and to keep the Heμ ion at a constant velocity. Thus, a Heμ ion passes through many filaments so that a balance between acceleration and drag is achieved.

For this purpose there are two requirements:

First, the rods of target material cannot be gaseous DT since the very high densities needed for rapid recycling would have to be at high pressure

A liberal estimate of the minimum numbers for χ the fusions per muon necessary for pure fusion in DT muon catalyzed fusion (MCF) is obtained by taking 20 MeV/fusion, an efficiency of conversion of neutron heat to electricity of 40%, an accelerator efficiency of 80%, a beam cost of 2.5 GeV/muon and a recycling of power for production of the muon of 50%. These numbers lead to a requirement that for $\chi > 780$. It is clear that this is a very generous lower limit.

On the other hand the standard formula for χ is,

$$\chi = \frac{1}{\lambda_o/\lambda_c + \omega_S} < \frac{1}{\omega_S} \tag{1}$$

where $\lambda_o = 0.455 \times 10^{-6}s^{-1}$ is the decay rate of the muon. λ_c is the cycling rate and ω_S^{eff} is the effective sticking parameter allowing for reactivation. Experimental and theoretical values are converging on a sticking value that is, with certain exceptions greater than 0.3%. The upper limit on χ is 333, at least a factor of two below the limit indicated above.

Thus, to make MCF practical it seems necessary to either reduce the cost of the muon or to reduce the effective sticking coefficient ω_S^{eff}. A possible method to accomplish the latter involves enhancing the reactivation coefficient R where

$$\omega_S^{eff} = \omega_S^{o} \exp\text{-}R \tag{2}$$

where ω_S^o is the initial sticking probability and R is the accumulated probability for stripping, or reactivation of the muon from the $He\mu^+$, muonic ion. The $He\mu^+$ ion resulting from a sticking event, is initially moving rapidly, with an energy of $\varepsilon_o = 3.5$ MeV and velocity $v_o = 1.3 \times 10^9$cm/s.

and have to be confined by a thick wall. This wall would necessarily stop the Heμ ion before it can reach the vacuum region to be reaccelerated. This means the target DT material must be in liquid or solid form and thus at a low temperature $\lesssim$ 25°K. It must be quite thin since the average path length for a $H_e\mu$ to leave the rod starting at a random position and moving in a random direction is approximately a. Combining this with Eq. (4) we see that for a = 100 microns the mean loss is 1/2 MeV.

The second requirement is a strong magnetic field B parallel to the rod axes and of order 5-10 tesla to keep the Heμ ion in the target area. Further, the accelerating electric field E is required to rotate at the gyration frequency of the Heμ ion to keep the ion energized against the drag. If we treat the loss per passage through a single rod as small, the perpendicular motion of the Heμ ion is approximated by

$$\frac{dV}{dt} = \frac{e}{m_{He}} \underset{\sim}{E}(t) + \underset{\sim}{v} \times \underset{\sim}{\Omega} - \alpha \underset{\sim}{V} \qquad (5)$$

where α is the mean drag coefficient, $\Omega = eB/m_{He\mu}c$, and

$$\underset{\sim}{E}(t) = E_o\left[\cos(\Omega t + \phi)\ \hat{x} + \sin(\Omega t + t\phi)\ \hat{y}\right] \qquad (6)$$

where $\hat{x}$ and $\hat{y}$ are cartesian unit vectors perpendicular to the cylinder axes. Since the fractional change in energy going through a rod is small, we approximate the discontinuous drag of the discrete rod by a smoothed out average drag force.

The initial velocity in the x,y plane ranges between 0 and v_o, depending on the initial direction of motion relative to the cylinder axes. For illustration purposes we consider only the case that $v = v_o$. By axisymmetry we may assume $\underline{v} = v_o\hat{x}$ initially.

Orbits are shown in Figs. 2a - 2d for various choices of α and ϕ. If $\phi < 90°$, the acceleration of the ion by the electric field is clear. The case $\phi > 90°$ would seem to lead to deceleration, but for all except $\phi \approx 180^0$ phase stabilization rapidly occurs and v remains near v_o after several gyrations. Phase stabilization is discussed in the appendix.

The velocity dependent drag coefficient α is given by

$$\frac{d\underline{v}}{dt} = - \alpha \underline{v} = \frac{1}{m_+ v_o} \left(\frac{d\varepsilon}{dx}\right)_o \left(\frac{v_o}{v}\right)^3 \underline{v} \tag{7}$$

where m_+ is the mass of the Heμ ion, and $(d\varepsilon/dx)_o$ is the stopping power at the velocity v_o.

Because of the randomness of the phases at the time of emissions, and the distribution of initial velocities of the Heμ ion perpendicular to the rod axes the ions should be treated by a kinetic theory which also includes the distribution of R values. However, as a much simplified theory seems to represent the situation fairly well, we restrict ourselves to this simplified theory. This enables us to make estimates of the field strengths and filling factors necessary for an effective reduction of the sticking process.

We assume that ϕ, the initial phase, is zero and estimate the velocity necessary to keep the Heμ ion at v_o. We ignore any ions with velocity components along the rod. If the phase is different, a somewhat larger electric field is necessary. The size of this field relative to the zero phase field is given in Table 1. An additional complication is the velocity dependence of the drag. If the electric field is slightly too large for balance, the ion will accelerate and the drag will decrease leading to a greater imbalance and further acceleration. Similarly, for a slightly weaker field the particle will systematically decrease. Fortunately, this instability is not very fast and the ion can be kept near its initial velocity until stripping occurs. This effect is of substantial importance if one includes the variation possible in the initial perpendicular velocity. Only an accurate kinetic treatment of this problem will establish the viability of the stripping theory, but a rough treatment seems to show that this effect is not really serious. For this note we ignore these important complications.

Let us assume that we keep the Heμ ion at a constant velocity v_o. Then from Eq. (3) R will increase with time. Combining with the probability that the particle remains unstripped, exp -R, we conclude that the mean time for stripping is that time when the Heμ^+ has suffered a drag energy of 10 MeV. During the same time the ion has gained an energy 10 MeV from the electric field. Hence

$$eE_+ = \frac{10\ \text{MeV}}{(v_o t_+)} = \frac{7.7\ \text{KV/cm}}{t_+(\mu s)} \tag{8}$$

where t_+ is the mean stripping time in microseconds.

In the proposed scheme there will be several large cycles for the muon. First it will generate an average of $\chi_o = (\omega_s^o)^{-1}$ fusions before sticking. This will take a time of $1/(\omega_s^o\lambda_c)$. Next it will enter the acceleration phase and spend a time t_+ as a Heμ ion before being stripped. The large cycle will be repeated a number of times and the mean big cycling rate is thus

$$\Lambda_c = \frac{1}{t_+ + 1/\omega_s^o\lambda_c} \ .$$

There is a finite probability W that the Heμ will slow to zero or escape to the outer wall and avoid stripping during the acceleration phase. (This loss arises from particles with initial $\phi \approx 180$ and we estimate W as 10%.) Thus, the total number of fusions becomes

$$\Xi = \frac{1}{\omega_s^o}\left(\frac{}{W + \lambda_o/\Lambda_c}\right) = \frac{1}{\omega_s^o(W + \lambda_o t_+) + \lambda_o/\lambda_c} .$$

On comparison with Eq. (1) we see the effective sticking probability is reduced to

$$\Omega_s^{eff} = \omega_s^o\,(W + \lambda_o t_+) .$$

Note that the initial sticking enters so to improve on the ω_s^{eff} in ordinary targets, needing $W+\lambda_o t+ < (1-R) \approx 0.7$.

The filling factor is determined by t_+ since the Heμ ion goes a distance $v_o t_+$. The stopping power at 3.5 MeV is about 45 MeV/cm in the rods so an average thickness of 10/45 = 0.22 cm of rod material must be transversed. Thus,

$$f = \frac{\pi a^2}{D^2} = \frac{0.22}{v_o t_+} = \frac{1}{5900\ t_+(\mu s)}$$

and the rod spacing D is

$$D = 136\ \sqrt{t_+(\mu s)}\ a \approx 1.4\ \sqrt{t_+(\mu s)}\ \text{cm} ,$$

the last figure corresponding to $a \approx 0.01$ cm.

By way of illustration we assume $\omega_s^o = 0.5\%$, $\lambda_c = 500 \times 10^6$/sec, W = 0.1 and list in Table 2 the values of the effective study coefficient Ω_s^{eff}, the number of fusions Ξ, the necessary electric field E_+ the reciprocal filling factor $1/f$, and the interrod spacing for various values of t_+. The last column, for B_{++}, will be discussed presently.

Table 2

It is seen that rather large but still achievable electric fields can lead to a substantial reduction of ω_{eff}. For the cycling rate chosen in this illustration $\lambda_c = 500/\mu sec$, the number of fusions per muon can approach 1000.

The magnetic field B has not yet been determined. The electric field is independent of it provided that several gyrations of the Heμ ion occur during the stripping time. However, B does set a lower limit of the total target area. Clearly its diameter must be considerably bigger than twice the gyration radius of the Heμ ion, ρ_+ which is given by

$$2\rho_+ = \frac{52}{B_{Tesla}}$$

This in turn sets the total voltage V_+ in the rotating electric field, equal to $2\rho_+ E_+$.

There is a very important drawback to this scheme and any other scheme of power production at very low temperatures. This is presented by the He^{++} 3.5 MeV/fusion energy Q_1, delivered to the target. The minimum work per fusion w_1 to remove this energy is given by Carnot efficiency. That is

$$w_1 = \frac{T_o - T_1}{T_1} Q_1$$

where T_1 is the temperature of the target and T_o is the room temperature to which this energy must be removed. Taking $T_1 = 25^oK$ and $T_o = 280^oK$ we get

$$w_1 = 35.7 \text{ MeV/fusion}$$

clearly this exceeds the 14.6MeV fusion energy of the neutrons.

There are several schemes to overcome this difficulty. One such scheme has been suggested by E. Valeo and P. Kaw. It involves choosing the rod radii so small that the He^{++} ion can leave the rod without appreciable loss of fusion energy, and filling the interrod region with gas. If $a \approx 0.006$ cm, the mean loss is 1 MeV/He^{++}. Second, choose the magnetic field B_{++} large enough so that the He^{++} gyro-orbit is smaller than D/2 so that the He^{++} ion cannot reach the neighboring rod. As a result the residual He^{++} energy is lost to the gas and can be removed at a reasonable temperature. Unfortunately, the magnetic fields required to satisfy this condition are very large. They are given in the last column of Table 2.

In conclusion it appears technically possible to reduce the effective sticking coefficient by an arbitrarily large factor by working with a filamented solid or liquid DT target. However, associated with this method is the serious problem of the removal of the 3.5 MeV He^{++} fusion energy at very low temperatures. This problem must somehow be resolved to make the scheme viable.

I should like to acknowledge useful discussion with a large number of people who have been very helpful with criticisms and helpful suggestions. Particularly important have been those with M. Rosenbluth, H. Furth, S. Cowley, S. Jones, T. Tajira, S. Eliezer, D. Jassby and S. Cohen. Y.C. Sun has participated fully over several years in resolving the numerical situation and carrying out calculations. This work was performed under the auspices of U.S. Department of Energy Contract No. DE-AC02-76-CH03073.

Appendix - Phase Stabilization

The equation of motion Eq. (5) of the Heμ ion in the frame rotating at the gyration frequency is

$$\frac{dV}{dt} = \frac{e}{m_{He\mu}} \underline{E} - \alpha \frac{v_o^2}{v^3} \underline{v}$$

where $\underline{E}$ is a constant vector. The initial angle between $\underline{v}$ and $\underline{E}$ is ϕ. If $\phi < 90^o$ (Fig. 3a) $\underline{v}$ will be accelerated along $\underline{E}$. For $\alpha = 0$ one gets the dashed line. With drag it becomes the dotted line and converges towards the mean velocity discussed in the text. If $\phi > 90^o$ (Fig. 3b), the ion is again accelerated along $\underline{E}$. At first deceleration occurs and then reacceleration occurs with convergence towards the same mean velocity.

Because the mean balance point P_o at $v = v_o\ (\alpha\ m/eE)^{1/2}$ is an unstable point of Eq. (A-1), it is impossible to converge exactly to this point for all ϕ and fixed E. Those with ϕ near 180° approach P_o and than decrease to the origin while those ϕ near 0° pass P_o and accelerate to infinite velocity. However, if the times are longer than the normal slowing down time, nearly complete stripping is possible. These details are not included in the simple treatment in the text, which merely presents the idea and rough estimate for the mean stripping of the required electric field. A more complete kinetic treatment will be presented in another paper.

References

1. S.E. Jones, Nature 321, 127 1986.
2. J.S. Cohen, Phys. Rev. Lett 58, 1407, 1987.

Table 1

ϕ	$E/E(\phi=0)$
0°	1
30°	1.1
45°	1.2
60°	1.34
90°	2.0
120°	4.0
135°	6.8

Table 2

t_+(μs)	Ω_s^{eff}	Ξ	E_+(KV/cm)	1/f	D(cm)	B_{++}(tesla)
0.5	.16	398	15.4	2950	1.0	160
0.4	.14	432	19.2	2360	.88	178
0.3	.12	474	25.7	1770	.77	206
0.2	.10	534	38.5	1180	.63	252
0.1	.07	621	77.0	590	.44	357

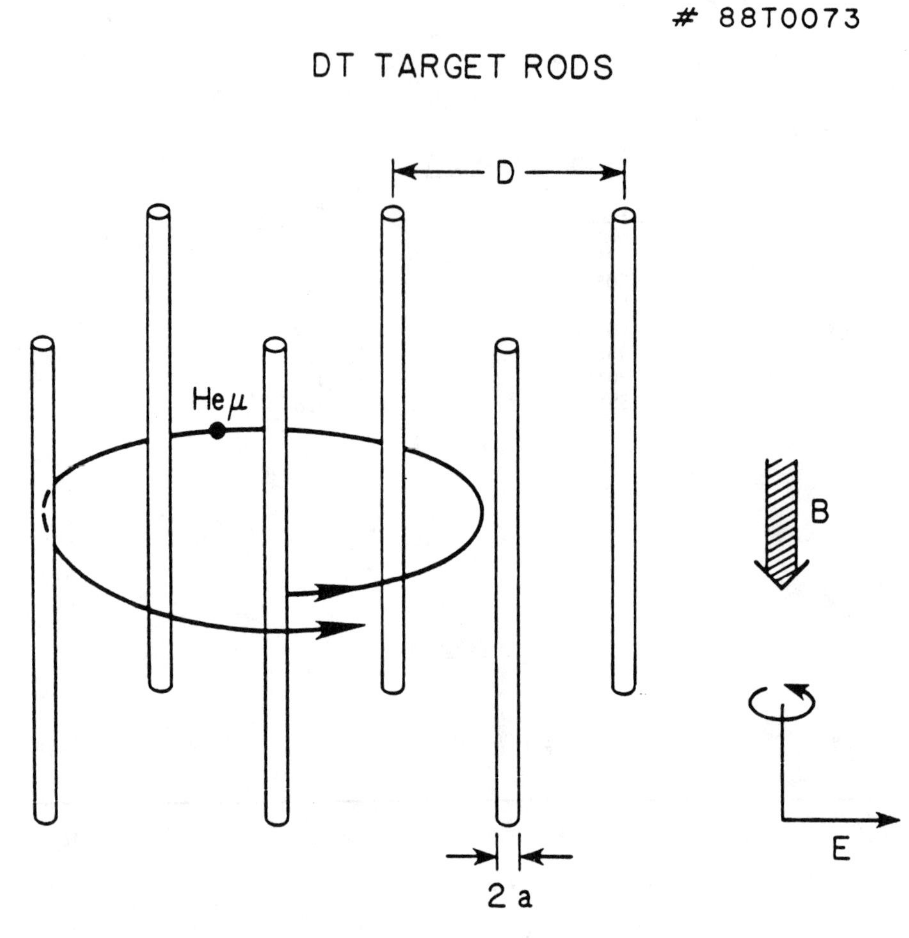

Fig. 1 The D-T target is filamented into many fine rods. A $\underline{B}$ field is imposed parallel to the rod axes. There is an electric field rotating at the gyroperiod of the Heμ ion perpendicular to $\underline{B}$. A Heμ ion is shown leaving one rod and passing through another rod.

88T1038

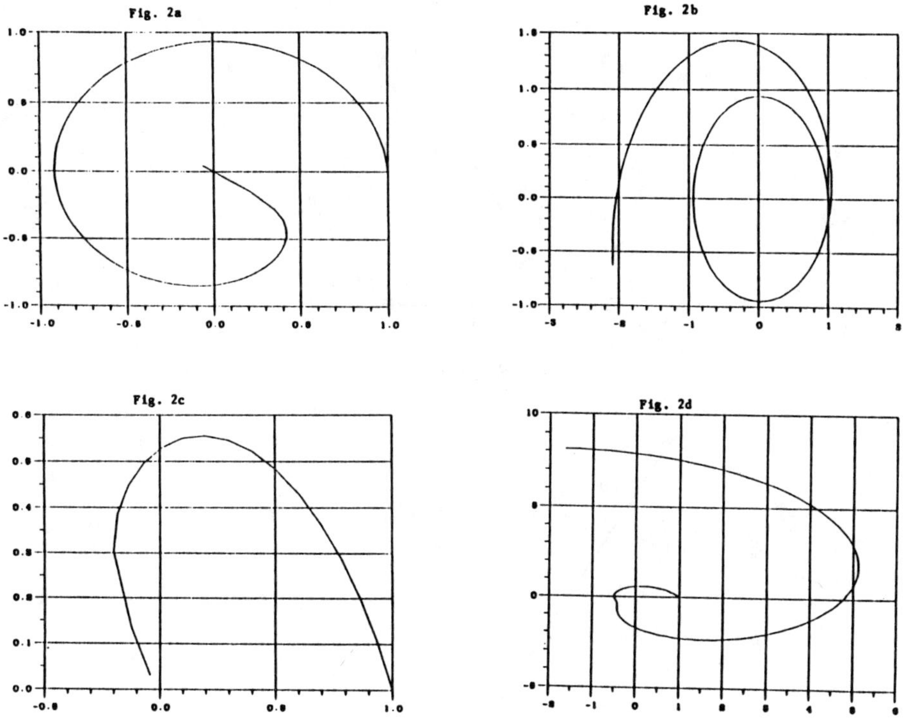

Fig. 2a-2d Orbits of the Heμ ion in velocity space. (2a) ϕ=45° and the electric field is such that deceleration occurs. (2b) ϕ=45° and acceleration occurs. (2c) ϕ=120° and deceleration occurs. (2d) ϕ=120° and acceleration occurs.

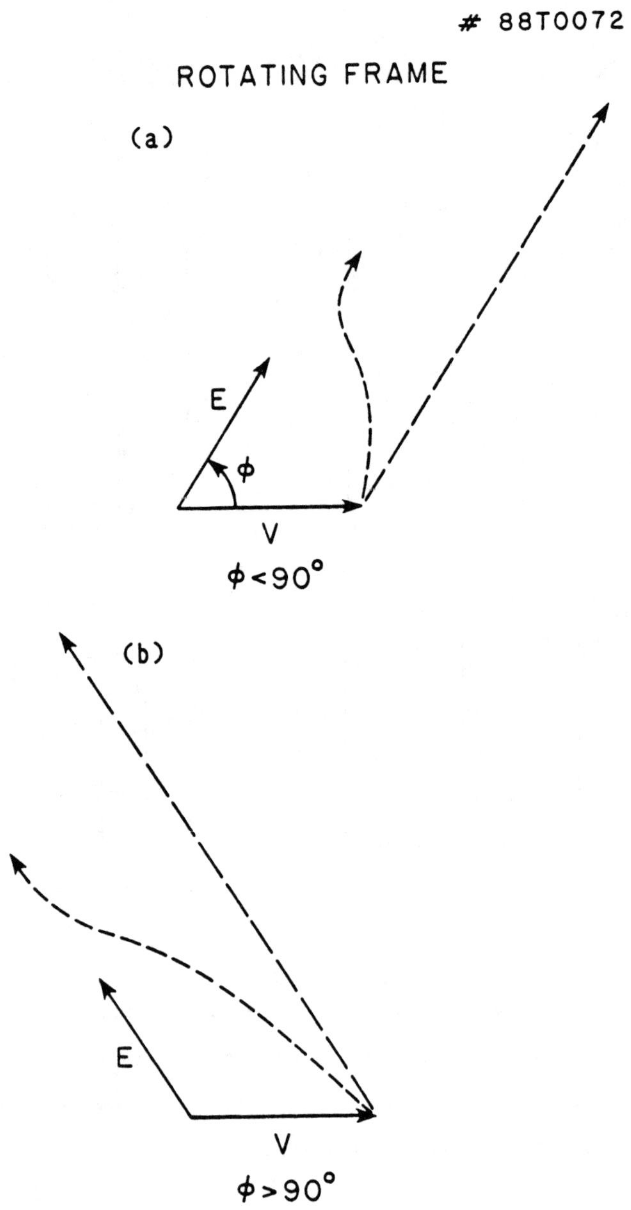

Fig. 3 Velocity orbits in the rotating frame.

A FIRST BORN APPROXIMATION CALCULATION OF THE TOTAL CROSS-SECTION FOR IONIZATION OF THE $\alpha\mu^+$ ION BY COLLISION WITH A DEUTERON

Guy C. Larson and Piotr Froelich
Department of Quantum Chemistry
Uppsala University
Uppsala, SWEDEN

The first-Born ionization cross-section for the post-fusion $\alpha\mu^+$ is recalculated "from scratch." The results are very different from those reported by other authors.

INTRODUCTION

This is a first Born calculation of the total ionization cross--section of the $\alpha\mu^+$ ion for a collision with a deuteron at "rest" and the $\alpha\mu^+$ ion moving with a speed near the post-fusion speed of 5.84 v_0 (where $v_0 \equiv \alpha c = e^2/\hbar$). The procedure is based mostly on a scaled version of H. Bethe [1].

The ionization of the alpha-mu ion via binary collisions with the deuterium nuclei is the most important mechanism in reactivation of muons stuck to the alpha particles in the chain of the muon catalyzed fusion. Ionization is responsible for some 70% of the reactivation effect [2]. The stripping rates used for reactivation calculations include both ionization and muon transfer, and the most reliable values so far are obtained by scaling of the experimental data from atomic physics. The ionization contribution to stripping is assumed to be well known (at least on the first-Born level) but we have not been able to explicitely localize the source. Instead, we have noticed that different authors use different values of the first-Born ionization cross-section. It may be useful to remove this ambiguity, and have a firm bench-mark result for the purpose of future more exact studies of reactivation.

For this reason, we have gone to the original first-Born calculations for the differential cross section (of the hydrogen atom) done by Bethe [1], and then carefully scaled them for the $\alpha\mu^+$ ion in what we believe is a very straightforward and logical way.

Note that we are using <u>muonic</u> units ($\hbar = m_\mu = e = 1$.)

PROCEDURE

The procedure can be outlined as follows.

1) Determine differential cross-section $\frac{d^2\sigma}{dQdE}(Q,E)$, where $\hbar Q$ and E are respectively the [magnitude of the] momentum transfer[1] and the energy transfer[1] to the $\alpha\mu^+$ ion.

---

[1] Note that we are using <u>relative coordinates</u>, and that these quantities are the [reference frame transformation] invariants: $\hbar Q$, the magnitude of the <u>total</u> momentum transferred to the target system [in this case the $\alpha\mu^+$ ion] (which is equivalent to the momentum transferred to the target's center of mass); and E, the energy transferred to the <u>internal</u> energy of the target system (i.e. the <u>total</u> energy transferred to the target, <u>minus</u> the kinetic energy transferred to its center of mass); in any arbitrary laboratory frame.

2) Determine (from kinematic constraints) the appropriate limits of the integral for

$$\sigma^{ion} = \iint_{EQ} d^2\sigma(Q,E) \equiv \iint_{EQ} \frac{d^2\sigma}{dQdE}(Q,E)\ dQdE \tag{1}$$

and integrate.
3) Plot the result and compare with others'.
Each of these steps will be treated in turn.
1) The differential cross-section (in first Born) for a hydrogen-like target can be written as

$$d^2\sigma(Q,E) = \frac{4\pi z^2}{(v/v_o)^2} \frac{dQ^2}{Q^4} F'\left(\frac{Q}{\alpha_*},\frac{E}{H_*}\right) \frac{dE}{H_*} \tag{2}$$

where
$v \equiv |\underset{\sim}{v}_d - \underset{\sim}{v}_{(\alpha\mu)}|$
is the relative speed of the collision,
ze is the charge on the projectile [in our case the deuteron],
H_* and α_* are scaling constants which depend on the target,
and

$$F'(q,\epsilon) \equiv \frac{\partial F}{\partial \epsilon}(q,\epsilon) \tag{3}$$

is the <u>differential inelastic-scattering form factor</u>.
To be specific, if the target is a hydrogen-like[2] system of reduced mass m_* and "nuclear" charge Ze, then

$$\alpha_* \equiv \frac{m_* Ze^2}{\hbar^2} = \frac{1}{a_*} \tag{4}$$

and

$$H_* \equiv \frac{m_* Z^2 e^4}{\hbar^2} = 2R_* \tag{5}$$

[2] i.e. in our case we mean a two-particle bound system with [what is assumed to be] only coulombic interaction, where at least one of the particles has only unitary [±e] charge.

where a_* and R_* are respectively the "Bohr radius" of the target and it's threshold energy. (For a real hydrogen atom target, $H_* = H$ is a Hartree and $a_* = a_0$ is the familiar Bohr radius.) In our case the target is an $\alpha\mu^+$ ion, so

$$m_* = m_{\mu\alpha} \equiv (m_\mu^{-1} + m_\alpha^{-1})^{-1} \tag{6}$$

and

$$Z = 2. \tag{7}$$

From Bethe [p. 32, eq. (30)][3] we derive (for a target initially in it's ground state):

$$F'(q,\epsilon) \equiv \frac{\partial F}{\partial \epsilon}(q,\epsilon)$$

$$= \frac{2^8 q^2\left(q^2 + \frac{2\epsilon}{3}\right)}{[(q+\kappa)^2 + 1]^3[(q-\kappa)^2 + 1]^3} \times \frac{e^{-\frac{2}{\kappa}\arctan\left(\frac{2\kappa}{q^2-\kappa^2+1}\right)}}{1 - e^{-\left(\frac{2\pi}{\kappa}\right)}} \tag{8}$$

with[4] $\kappa \equiv \text{sqrt}(2\epsilon - 1)$ (9)

Thus for our case we get

$$d^2\sigma(Q,E) = \frac{4\pi z^2}{(v/v_0)^2} \frac{dQ^2}{Q^4} F'(q,\epsilon) \frac{dE}{H_{\mu\alpha}} \tag{10}$$

where

$$q \equiv \frac{Q}{\alpha_{\mu\alpha}}, \tag{11}$$

---

3 Note that the unitless quantities we call q, κ, and ϵ are equivalent to Bethe's q/α, k/α, and $(1 + k^2/\alpha^2)/2$ respectively.

4 In a real hydrogen atom target, $\hbar(\kappa\alpha)$ is the magnitude of the momentum transferred to the ejected electron [in the relative coordinates of the H atom]. In a generalized hydrogen-like target, $\hbar(\kappa\alpha_*)$ is the magnitude of the momentum transferred to the ejected relative particle.

$$\epsilon \equiv \frac{E}{H_{\mu\alpha}}, \tag{12}$$

$$\alpha_{\mu\alpha} \equiv \frac{m_{\mu\alpha} Z e^2}{\hbar^2}, \tag{13}$$

$$H_{\mu\alpha} \equiv \frac{m_{\mu\alpha} Z^2 e^4}{\hbar^2}, \tag{14}$$

$$Z = 2, \tag{15}$$

$$z = 1, \tag{16}$$

and $\Gamma'(q,\epsilon)$ is given by equation (8).

2) Conservation of energy requires that

$$(\hbar K_i)^2 - (\hbar K_f)^2 = 2ME \tag{17}$$

where $\hbar\vec{K}_i$ and $\hbar\vec{K}_f$ are respectively the initial and final relative momenta (and $\hbar K_i$ and $\hbar K_f$ their respective magnitudes), and

$M \equiv$ reduced mass of projectile [deuteron]

$$\equiv [m_d{}^{-1} + (m_\alpha + m_\mu)^{-1}]^{-1} \tag{18}$$

Combining equation (17) with the definitions

$$Q^2 \equiv |\vec{K}_i - \vec{K}_f|^2 = K_i^2 + K_f^2 - 2\vec{K}_i \cdot \vec{K}_f$$

$$\equiv |\vec{K}_i - \vec{K}_f|^2 = K_i^2 + K_f^2 - 2K_i K_f \cos\theta \tag{19}$$

and

$T \equiv$ initial relative kinetic energy

$$= \frac{Mv^2}{2} = \frac{(\hbar K_i)^2}{2M} \tag{20}$$

yields

$$(\hbar Q)^2 = 4TM\left[1 - \frac{E}{2T} - \mathrm{sqrt}\left(1-\frac{E}{T}\right)\cos\theta\right] \tag{21}$$

This gives us the kinematic limits

$$(\hbar Q)^2_{\gtrless} \equiv (\hbar Q)^2_{\substack{max\\min}} = 4TM\left[1 - \frac{E}{2T} \pm \mathrm{sqrt}\left(1-\frac{E}{T}\right)\right] \tag{22}$$

which can equivalently be written as

$$(Qa_\mu)^2_{\gtrless} = 4\left(\frac{T}{H_\mu}\right)\left(\frac{M}{m_\mu}\right)\left[1 - \frac{E}{2T} \pm \mathrm{sqrt}\left(1-\frac{E}{T}\right)\right] \tag{23}$$

where

$$H_\mu \equiv \frac{m_\mu e^4}{\hbar^2} \tag{24}$$

and

$$a_\mu \equiv \frac{\hbar^2}{m_\mu e^2} \tag{25}$$

The "total [energy-dependent] integrated cross section" is

$$\frac{d\sigma}{dE}(E) = \int_{Q_<}^{Q_>} \frac{d^2\sigma}{dQdE}(Q,E)\ dQ \tag{26}$$

(Note that this integration over Q corresponds to an integration over angles.)

Since we are only concerned with the <u>ionization</u> cross-section σ^{ion}, it is clear that we are after energy transfers from the ionization threshold

$$I \equiv \frac{m_{\mu\alpha} Z^2 e^4}{2\hbar^2} = \tfrac{1}{2} H_{\mu\alpha} = \frac{m_{\mu\alpha} Z^2}{2m_\mu} H_\mu \tag{27}$$

to the maximum possible energy transfer T (the initial relative kinetic energy). The total ionization cross section is thus

$$\sigma^{ion} = \int_I^T \frac{d\sigma}{dE}(E)\, dE = \int_I^T \int_{Q_<(E)}^{Q_>(E)} \frac{d^2\sigma}{dQdE}(Q,E)\, dQdE \tag{28}$$

with $Q^2_{\gtrless}(E)$ given in equation (23). Substituting our expression for $\frac{d^2\sigma}{dQdE}(Q,E)$ [equation (10)] into this gives

$$\sigma^{ion} = \frac{4\pi z^2}{(v/v_o)^2} \int_I^T \int_{Q_<(E)}^{Q_>(E)} F'\left(\frac{Q}{\alpha_{\mu\alpha}}, \frac{E}{H_{\mu\alpha}}\right) \frac{dQ^2}{Q^4} \frac{dE}{H_{\mu\alpha}} \tag{29}$$

3) Equation (29) was evaluated (numerically) for approximately 10^2 values of v, and the results are plotted in curve 4 in Figure 1.

These results are obviously quite different from conventional results from other sources (see figure) but after carefully examining all our work, we are unable to locate any errors. We have, on the other hand, found some small errors (some of which may just be typographical) in the formulae of others' (for instance, equation (A.4) in Bracci[2] et al has a $16/m^4_{\mu\alpha}$ where it should actually have $256/4m_{\mu\alpha}$), but none sufficient to account for such a large discrepancy. We therefore believe that the discrepancy must lie somewhere in the numerical calculations (an unsuccessful effort on our part to reproduce Bracci's graph from his own equations seems to confirm this), and for that reason we have done the above integration many different times with several quite different numerical integration programs, but we still come up with the same result.

REFERENCES

1 H. Bethe, Annalen der Physik 5, 325 (1930).
2 L. Bracci and G. Fiorentini, Nuclear Physics A364, 383 (1981).
3 J. S. Cohen, Muon Catalyzed Fusion 00 (1987) MCF00060.
4 S. S. Gershtein, et al. Sov. Phys. JETP 53 (1981) 872.
5 H. Takahashi, Muon Catalyzed Fusion 1(1987)237.
6 J. S. Cohen, Proc. of International Symposium on Muon-Catalyzed Fusion, Tokyo, Japan (1986). (Preprint).

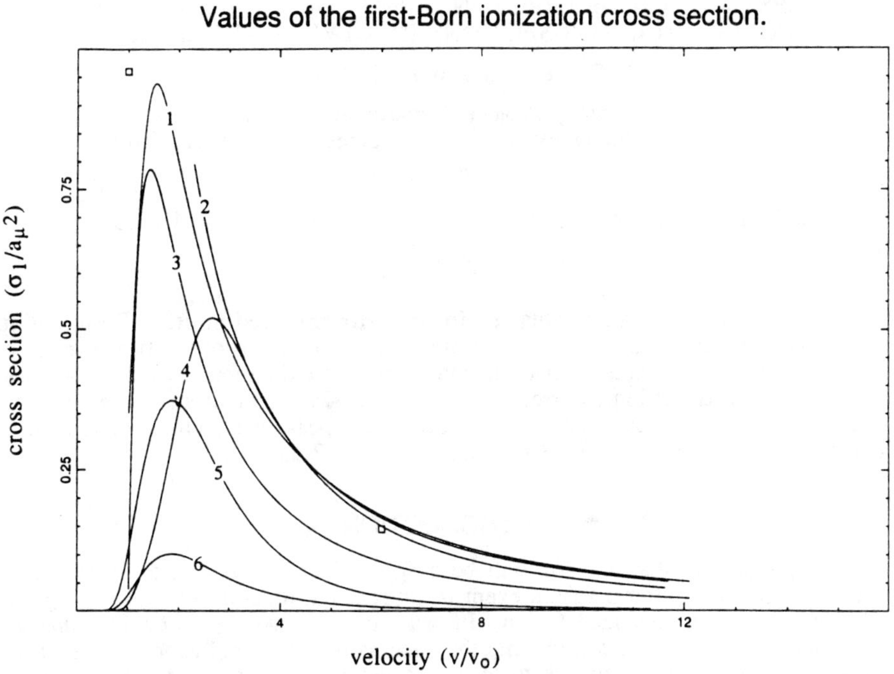

Fig. 1. Values of the first-Born ionization cross section, as cited by various authors. (All cross-sections are for collisions with $\alpha\mu^+$ in ground state. $\sigma^{tot[al]} = \sigma^{ion} + \sigma^{tr[ansfer]}$.) Curve 1 is σ^{tot}, scaled by Cohen [3] from experimental data. Curve 2 is the Born approximation for σ^{tot} used in Gershtein [4]. Curve 3 is Rutherford (Bohr model) formula for σ^{ion}, from Takahashi [5] and Bracci [2]. Curve 4 is a Classical Trajectory Monte Carlo (CTMC) calculation of σ^{ion} by Cohen [3]. Curve 5 is ours (σ^{ion}, first-Born). Curve 6 is σ^{ion} calculated using the erroneus eq (A.4) in Bracci [2]. The squares are CTMC calculations of σ^{tot} by Cohen [6].

ESTIMATION OF ERROR IN USING BORN SCALING FOR COLLISION CROSS SECTIONS INVOLVING MUONIC IONS

C. D. Stodden and H. J. Monkhorst

Quantum Theory Project, Departments of Physics and Chemistry, University of Florida, Gainesville, Florida 32611

K. Szalewicz

Department of Physics, University of Delaware, Newark, DE 19716

ABSTRACT

A quantitative estimate is obtained for the error involved in using Born scaling to calculate excitation and ionization cross sections for collisions between muonic ions. The impact parameter version of the Born Approximation is used to calculate cross sections and Coulomb corrections for the 1s→2s excitation of $\alpha\mu$ in collisions with d. An error of about 50% is found around the peak of the cross section curve. The error falls to less than 5% for velocities above 2 a.u.

INTRODUCTION

The probability that a muon will be stripped from an $\alpha\mu$ ion as it collides with other particles after a fusion event is an essential quantity in determining the efficiency of muon catalyzed fusion. In order to calculate the probability that a muon will be stripped from a muonic ion such as $\alpha\mu$ many collision cross sections are needed. It is very important to obtain accurate cross sections for all collision processes over the entire range of velocities spanned during slowdown.

In calculating the stripping from $\alpha\mu$ several authors have relied on the use of scaling collision cross sections for the p + He^{+} system in order to obtain some of the important cross sections for the d + $\alpha\mu$ system[1,2,3]. The possibility of obtaining cross sections by scaling values from other systems follows from the Born approximation expression for such cross sections[4]. It has been shown that for the collisional excitation and ionization of electrons the use of scaling gives negligible error even in regions where the Born approximation itself begins to break down. In fact in some cases the accuracy of the scaled results is actually greater than the unscaled[5]. The Born scaling technique has become a common and useful tool in obtaining cross sections for collisional excitation and ionization of an electron bound to a nucleus. It is not surprising then that this technique has been used in an attempt to obtain the most accurate cross sections for the collisional excitation and ionization of a muon bound to a nucleus. It has been hoped that the same assumptions which allow for the use of Born scaling in the case of a normal ion will be valid also for a muonic ion. The validity of these assumptions for muonic atoms has been a concern which has been raised before[6]. The purpose of the present work is to provide a quantitative estimate of the error introduced when using Born scaling to obtain collisional cross sections for muonic ions from cross sections for normal ions.

Born Scaling

The expression appropriate for scaling excitation and ionization cross sections from collisions between normal ions to collisions between a normal and a muonic ion of corresponding charges is

$$\sigma^{(\mu)}(v) = (m_e/m_\mu)^2 \, \sigma^{(e)}(v) \tag{1}$$

where m_x is the reduced mass of x in the bound system. The expression above is valid at energies greater than threshold. In effect Eq. (1) simply scales the cross section to muonic units (m_μ =1 m.u.). The assumptions implicit in this simple expression can be formulated in a semiclassical fashion in the center of mass frame as (i) the projectile follows a rectilinear trajectory during the collision and (ii) the projectile travels with a constant velocity relative to the target. The two assumptions above must be valid for those impact parameters whose trajectories contribute significantly to the relevant cross section. If assumptions (i) and (ii) are valid then the cross sections do not depend on the relative mass of the nuclei and the muonic system simply scales down according to the reduced mass of the muon. This proceedure is called Born scaling.

The most important impact parameters will be those of the same order of magnitude as the electronic (muonic) bohr radius. Because of its mass ($\approx$207m_e) the muonic Bohr orbital is about 200 times smaller than for an electron in the same system. This fact is what leads to doubts about the validity of assumptions (i) and (ii) for the muonic system.

Impact Parameter Formalism

In order to quantitatively estimate the error introduced in Born scaling we will follow the proceedure of Bates and Boyd[5]. In this treatment the nuclei are treated classically and the electron (muon) quantum mechanically. A transistion amplitude is calculated along a trajectory defined by an impact parameter and the cross section obtained by integrating over all such impact parameters as well as over the azimuthal angle. Without making any assumptions about the trajectories the transition integral along a single trajectory may be written

$$|\hbar a_{ij}(\infty)| = \left| \frac{1}{v} \int_{-\infty}^{\infty} \Lambda(R)\, U_{ji}(R)\, exp\left(\frac{-i\epsilon_{ij}\zeta}{\hbar v}\right) d\zeta \right| \tag{2}$$

where $\epsilon_{ij} = \epsilon_i - \epsilon_j$, v is the initial relative velocity and R is the internuclear distance. The expression above follows from transforming the integral from dt to dζ using the relations of standard orbit theory and the definition dζ=$v$dt . The symbols Z_t and Z_p refer to the target and projectile charges respectively and the factor Λ (defined in ref. 5) is equal to 1 for an s-s transition.

If the trajectory does not deviate too far from a straight line the matrix element can be expanded about a variable η as

$$U_{ji}(R) = U_{ji}(\eta) + \omega \left\{ 1 - \frac{\zeta}{\eta} log \left[\frac{\zeta + \eta}{\rho} \right] \right\} U'_{ji}(\eta) \tag{3}$$

where

$$\omega = Z_t Z_p e^2 / m v^2 \tag{4}$$

and η=$(\zeta^2 + \rho^2)^{1/2}$. The prime on U_{ji} indicates differentiation with respect to η and ρ is the impact parameter. The first term in Eq. (3) gives the matrix element assuming a straight line trajectory while the second approximates the correction for a Coulomb trajectory. For Eq. (3) to be valid a requirement is that $\omega/\rho \ll 1$.

The transition amplitude can now be approximated by the sum

$$a_{ij}(\infty) = a^{\circ}_{ij}(\infty) + \delta a^{\circ}_{ij}(\infty) \tag{5}$$

The cross section in the first Born approximation is then

$$Q^{\circ}_{ij} = 2\pi \int_0^{\infty} \rho \left| a^{\circ}_{ij}(\infty) \right|^2 d\rho \tag{6}$$

while the Coulomb correction is

$$\delta Q^{\circ}_{ij} = 4\pi \int_0^{\infty} \rho a^{\circ}_{ij}(\infty)\, \delta a^{\circ}_{ij}(\infty)\, d\rho \tag{7}$$

Calculations and Discussion

We will now use Eqs. (6) and (7) to estimate the error in using Born scaling to calculate cross sections for the system d + $\alpha\mu$(1s). To illustrate the error involved over a range of collision velocities we have chosen the simplest process, namely

$$d + \alpha\mu(1s) \rightarrow d + \alpha\mu(2s) \tag{8}$$

Table I. The excitation cross section and estimated error as a function of relative velocity. The estimated error is defined as $-\delta Q^{\circ}_{ij}/Q^{\circ}_{ij}$.

v(a.u.)	Q(/1.0E-23 cm**2)	Est. Error
0.75	6.83	114%
0.9	7.77	62%
1.0	7.91	44%
1.1	7.80	33%
1.2	7.53	25%
1.5	6.33	13%
2.0	4.46	6%
2.5	3.18	4%
3.0	2.35	2%
4.0	1.43	1%

From the results of Table I we can see that the Coulomb correction is substantial for velocities less than 2 a.u. Around the maximum of the cross section curve (v=1 a.u.) the error is about 50%. Based on these results we conclude that for the 1s→2s excitation of $\alpha\mu$ Born scaling is completely inadequate except for velocities above 2 a.u. According to Bates and Boyd we will obtain similar error values for the 1s→2p transitions at these velocities although the cross section curves will peak at slightly higher velocities. This is not surprising since the important values of ρ for all of these velocities are about the same as for the s-s transition (0.4 m.u).

If we compare the impact parameters and velocities with those for ionization of the muon from the 1s state by scaling the values for He^{+} + p we find that the important impact parameters are $\approx$0.7 m.u. which is only slightly larger than for the s-s transition[7]. The cross section peak occurs at about the same velocity as for the 1s$\rightarrow$2s excitation but the high energy tail is much longer. Thus we can expect that for ionization the error will again be substantial for velocities less than 2 a.u., but the overall effect will probably be smaller for ionization because of the larger area of the cross section curve at velocities greater than 2 a.u.

In conclusion we can state that these results cast doubt on the reliability of Born scaling as applied to the collisional excitation and ionization of $\alpha\mu$ at velocities less than 2a.u. A similar calculation should be performed for ionization to verify this. The effect of this type of error on the overall stripping of a muon from $\alpha\mu$ is diminished by the fact that most of the stripping takes place at velocities greater than 2 a.u. However to obtain a more accurate estimate of this effect calculations should be done above and below v=2 a. u. using Coulomb trajectories. As an alternative to Born scaling we are now in the process of calculating collisional cross sections for excitation and stripping of $\alpha\mu$ initially in the ground and first excited states using a coupled-state Sturmian approach with a very large basis. This should result in very accurate quantum mechanical cross sections[8] which may lower the error bars on the stripping probability. In light of the present results it seems advisable to use Coulomb trajectories in these calculations for those velocities less than 2 a.u. and to investigate the Coulomb effect at even greater velocities.

Bibliography

1. L. Bracci and G. Fiorentini, Nucl. Phys. A **364**, 383 (1981).
2. V. E. Markushin, Muon Cat. Fusion **3**, 395 (1988).
3. J. S. Cohen, Phys. Rev. Lett. **58**, 1407 (1987).
4. D. R. Bates and G. Griffing, Proc. Phys. Soc. A **66**, 961 (1953).
5. D. R. Bates and A. H. Boyd, Proc. Phys. Soc. A **79**, 710 (1962).
6. J. S. Cohen, Phys. Rev. A **37**, 2343 (1988).
7. T. G. Winter, Phys. Rev. A **33**, 3842 (1986).
8. T. G. Winter, Phys Rev. A **25**, 697 (1982).

Chapter 5
Muon Production

OPTIMAL ACTIVE TARGETS FOR MUON CATALYZED FUSION REACTORS

Magnus Jändel

Research Institute of Physics, S-104 05 Stockholm

Abstract

A simple but realistic design for the muon production system of a muon catalyzed fusion reactor is presented. Starting with a careful optimization of the pion production target we next consider the complete system where pion conversion losses and muon losses in the pressure vessel are taken into account. Problems and inefficiencies are identified to provide a basis for future inventions.

Introduction

Many ideas for efficient muon production such as high energy cascade production [1] [2] [3] [4], recirculating beams [5], hybrid systems [6] [7] and colliding beams [8] have recently been proposed. We shall here in some detail consider a simple but fairly realistic muon production system "the Muon Cave" (see Fig. 1). This is an extension of previous work [1] in collaboration with M. Danos and J. Rafelski. In particular previously ignored but important effects related to the construction material of the pressure vessel are considered.

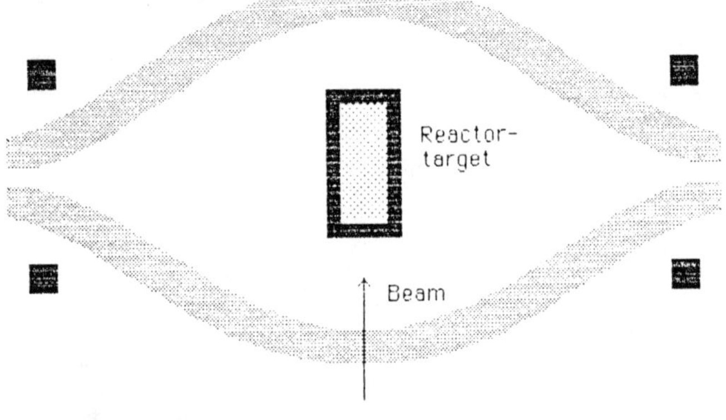

Fig. 1. The "Muon Cave" design consists of a magnetic bottle enclosing an active target. Pions emitted from the target are reflected by the magnetic fields until they decay. Muons are stopped in the target.

To calculate pion and muon yields in the "Muon Cave" configuration we have used the FLUKA87 Monte Carlo code [9] for high energy particle cascades in matter and magnetic fields. FLUKA87 in the context of muon catalyzed fusion was described at the 1987 Leningrad conference (see also A. Bertin and A. Vitales contribution to this workshop). Particle production cross sections in FLUKA87 are interpolated from data tables using the resonance production model by Ranft and coworkers [10].

In the following we first consider the geometrical optimization of high density π^- production targets. These results could be useful for exploring many different design principles. Next we analyze "the Muon Cave" design. Finally some possible ways of improving the performance of muon production systems are discussed.

Optimal Target

The optimal geometrical shape of a π^- production target is defined by the competition between pion creation and pion loss. A large target is required in order to use the full π^- production capacity of the high energy cascade while the produced pions escapes more easily in a small target. A target where the shape has been chosen to optimize the yield of emitted pions shall here be refered to as an optimal target.

To find the number of emitted π^- per beam particle for an optimal target within a reasonable amount of computer CPU time it is convenient to study the cascade in an infinite target. Using the cylindrical symmetry we then divide the target into volume bins. In each bin we count the number of: 1) created pions n_c 2) pions lost by charge exchange n_{ch} 3) pions lost by absorption n_{abs} 4) stopped pions n_s. Stopped pions are readily captured by target nuclei and are hence lost for the purpose of muon production.

The contribution from a given bin to the outgoing pion flux is $n_{em} = n_c - n_\ell$ where n_ℓ is the number of lost pions given by $n_\ell = n_{ch} + n_{abs} + n_s$. For each choice of bin structure we now obtain the optimal target shape by cutting away all bins with negative n_{em} leaving only a connected set of bins with positive n_{em}. This procedure saves computer time since the infinite target Monte Carlo simulation can be done once for a given material and beam type using an "eventtape" to register all significant events. Different choices of bin size and statistical treatment can then be applied "offline".

The energy cost per emitted π^- E_{opt} is displayed in Figs 2 - 4 for various materials. Each calculated point in the figures has a relative error of 5 - 10% and corresponds to about 5 minutes of IBM-3090 CPU time. The bin structure ranges between 8×8 and 16×16 depending on the available statistical sample.

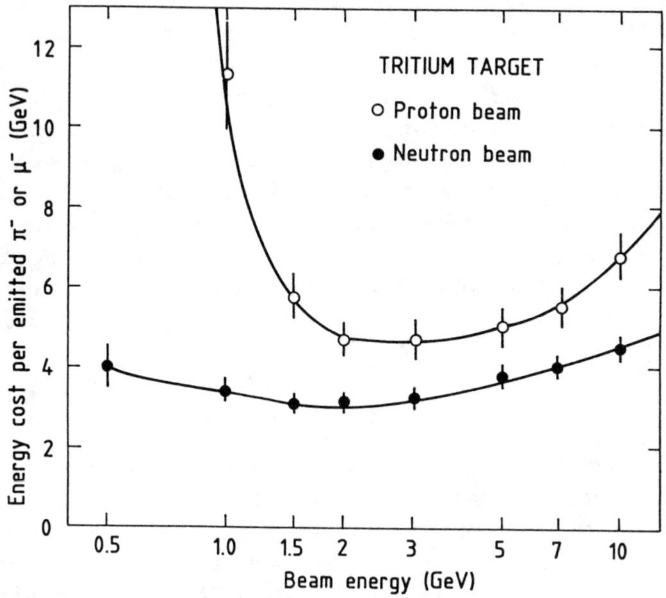

Fig.2. Energy cost per negative pion or muon emitted from an optimal tritium target as a function of beam energy.

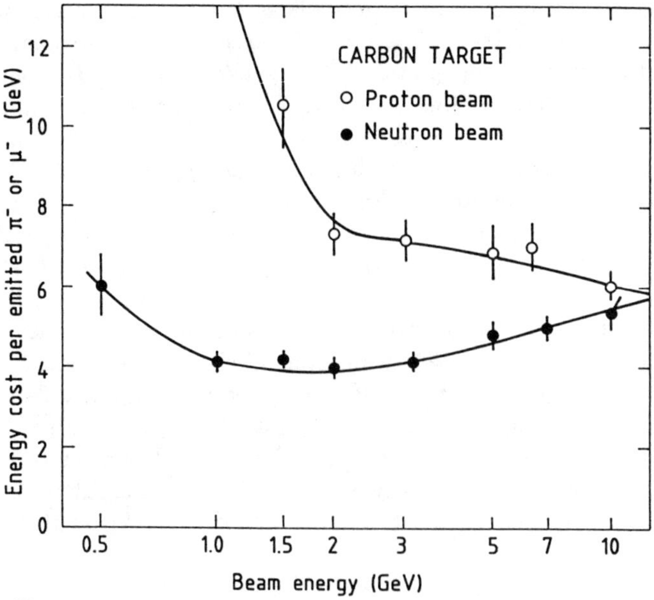

Fig. 3. Energy cost per negative pion or muon emitted from an optimal carbon target as a function of beam energy.

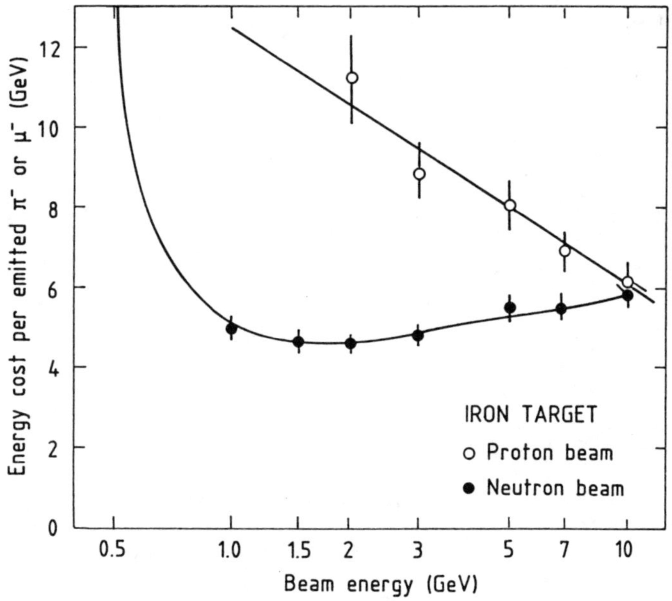

Fig. 4. Energy cost per negative pion or muon emitted from an optimal iron target as a function of beam energy.

Results for both proton and neutron beams are given in the figures. For other light-ion beams the corresponding results can be estimated by:

$$E_{opt}(A,Z) = \frac{(ZE_{opt}(1,1) + (A-Z)E_{opt}(1,0))}{A} \qquad (1).$$

The energy cost of pion production for optimal targets is usually more than a factor two larger than for infinite targets [1]. The qualitative behaviour is otherwise similar with E_{opt} increasing with increasing atomic number A of the target nuclei and a difference between proton and neutron beams that increases with increasing A and decreasing beam energy. The optimum beam energy of e.g. a deuteron beam is reached for each target material at the smallest beam energy where the difference in efficiency between a proton and a neutron beam is essentially washed out by the extensive cascade of high-energy reactions. The prefered beam energy is higher for heavier target nuclei. The lowest energy cost for a deuteron beam was 4 GeV per emitted pion in a tritium optimal target.

Pion losses in an optimal target as compared to an infinite target are displayed in Fig. 5. About 50% less primary pions are created in the optimal target. Losses due to charge exchange and absorption in flight or at rest are much smaller. Newborn pions in a heavy material are often absorbed or charge exchanged before leaving the target nucleus. A smaller fraction of the surviving pions are hence lost before leaving the optimal heavy target.

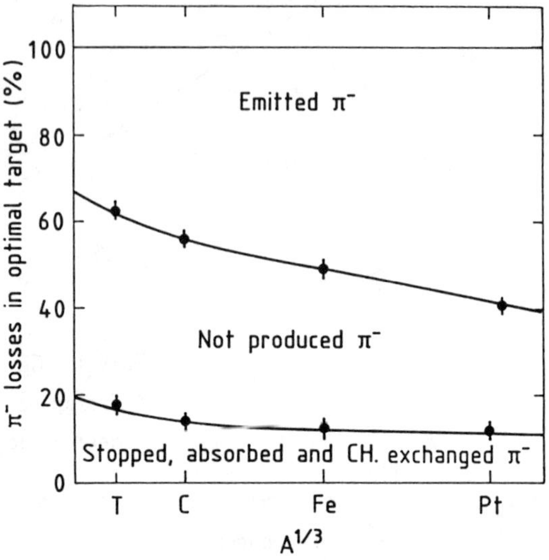

Fig. 5. Pion losses in an optimal target as a function of target material. The 100% level refers to the number of pions produced in an infinite target. The pion loss caused by not using the entire potential for pion production in a target of limited size is described as "not produced" π^-.

The volume of an optimal target depends of the density Φ (in units of liquid hydrogen density) according to

$$V_{opt}(\Phi) = V_{opt}(\Phi = 1)\Phi^{-3} \tag{2}$$

where $V_{opt}(\Phi = 1)$ is shown in Fig. 6 together with the volume of the smallest target with 10% reduced efficiency compared to the optimal case. The large errors and the substantial reduction in tritium inventory obtained with a 10% reduction in muon production clearly shows that the net contribution to the number of emitted pions is strongly peaked in a comparatively small region surrounded by a larger volume with a small but significant net production. Because of the Monte Carlo method and the finite amount of available computer time the statistical sample in the outer regions is necessarily small. The estimated total volume has therefore a much larger relative error than the pion yield from the same Monte Carlo sample. The volume of a tritium optimal target at density $\Phi = 1$ and beam energy 2 GeV is about 6.13 m^3 which means a total tritium inventory of 650 Kg at 50% tritium concentration. The small difference between the properties of deuterium and tritium in active targets is not important in this context.

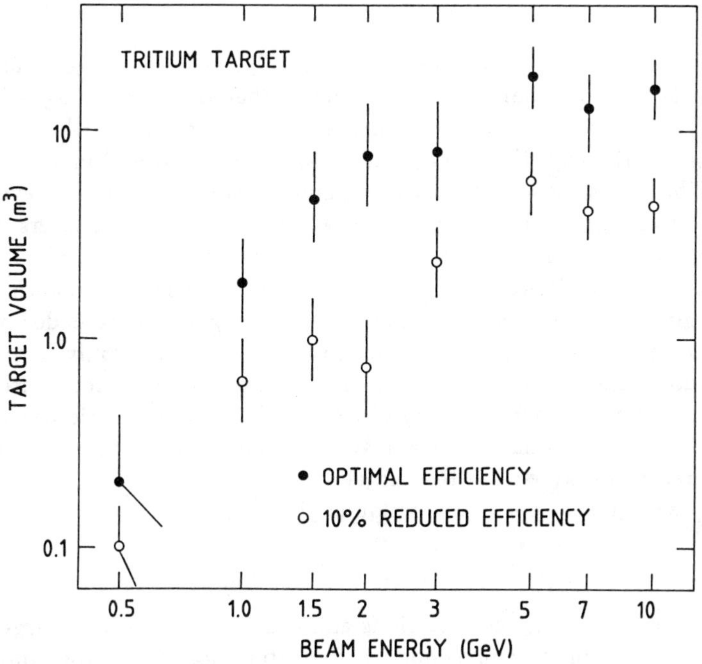

Fig. 6. Volume of an optimal $\Phi = 1$ tritium target (closed points) and of the smallest target with 10% reduced efficiency (open points) as a function of beam energy.

Pions can decay in flight within the optimal target and some of the decay muons are stopped before leaving the target. The number n_{dir} of such directly stopped muons is a interesting quantity in the design of active target reactors. The time of flight within the target is proportional to the target length which is inversely proportional to the density. At sufficiently high densities it is therefore resonable to fit the Monte Carlo results for n_{dir} to the form

$$n_{dir}(\Phi) = n_{dir}^{0}(1 - e^{-\frac{\Phi_m}{\Phi}}). \tag{3}$$

In a tritium target we get $n_{dir}^{0} = .30 \pm .04$ and $n_{dir}^{0} = .18 \pm .04$ per beam particle for a neutron and proton beam respectively. The density dependence calculated at $\Phi = 5$, 1, 0.5, 0.1 and 0.01 is in both cases well described by setting $\Phi_m = 0.012 \pm .001$ in (3). The contribution from directly stopped muons to the total yield of realistic high-density active targets is hence quite small.

Muon Cave

The "Muon Cave" design for a muon catalyzed fusion target-reactor system consists of an optimal active target in an otherwise empty magnetic bottle (see Fig. 1). Pions produced within the active target are with high probability emitted through the target surface and contained within the magnetic bottle. The pions move between the magnetic walls and decays eventually. The kinetic energy of the decay muons is lost in several encounters with the active target before the muons are stopped in the target.

In this scheme there is no need for detailed tailoring of magnetic fields in order channel the pions away from an external target and focus decay muons on the reactor vessel. Pions and muons move almost randomly in the muon cave. Muons are, however, stopped only in the reactor since there are no external targets or other auxiliary objects within the magnetic bottle. High density targets with small tritium inventory can be used since the pions are not required to decay within the target.

The energy cost per active muon is given by

$$E_{act} = E_{opt} \times \eta_{conv} \times \eta_{matr} \tag{4}$$

where E_{opt} is the energy cost of pions emitted from an optimal target, η_{conv} corrects for losses in the conversion from emitted pions to stopped muons and η_{matr} describes the loss caused by muons stopped in the construction material. Both E_{opt} and η_{matr} depends of the type and amount of construction material used in the pressure vessel. While ignoring much of the engineering details we must here consider the gross features of the reactor vessel.

The maximum internal pressure P and the tensile strength σ of a thin-walled pipe is related by

$$2RP = 2\ell\sigma \tag{5}$$

where R is the radius of the pipe and ℓ is the wall thickness. Modelling the active target as a thin-walled pipe or a set of such pipes filled with an ideal gas of pressure $P(MPa) = 0.293T(K)\Phi$. we get the relation:

$$\frac{V_{matr}}{V_{fuel}} = \frac{2\ell}{R} = \frac{0.586T(K)\Phi}{\sigma(MPa)} \tag{6}$$

between the volume V_{matr} of the construction material and the volume V_{fuel} of deuterium-tritium fuel.

The construction material degrades the performance of the active target by beeing an inferior target material for pion production and by deactivating muons. Both these effects are less severe for low-Z materials such as carbon composites than for a comparatively high-Z material like steel. The tensile strength of high-strength structural steel is about $\sigma_{steel} = 1200$ MPa.

Composites based on e.g. the Torayca T1000 carbon fibre can reach tensile strengths of at least $\sigma_{carbon} = 3850$ MPa [11]. Graphite is also a suitable material since the neutron capture cross section is small and the migration length of fusion neutrons is much larger than the 1 - 10 cm thickness of a carbon composite reactor vessel.

The energy cost for emitted pions E_{opt} is for some relevant materials given in Figs. 2-4. The realistic active target consists of both fuel and construction material and E_{opt} is hence usually 10-20% larger than for a pure DT-target. This effect has been taken into account in the Monte Carlo simulations. The detailed spatial distribution of the construction material was found to be of little importance.

The conversion of emitted pions to stopped muons is associated with both muon and pion losses. Pions are captured with the average rate λ^{π}_{cap} since they can be reflected back into the target and subsequently absorbed, charge exchanged or stopped. Muons are stopped with the average rate λ^{μ}_{cap}. This rate should be smaller than λ^{π}_{cap} since muons feel only the stopping power of the active target but not the strong interaction forces. Monte Carlo experiments show that $\lambda^{\pi}_{cap} = c\lambda^{\mu}_{cap}$ with c = 1.45 at $E_{beam} = 2$ GeV. Muons and pions decay with their natural rates $\lambda^{\pi}_{dec} = 2.6030 \times 10^{-8}s$ and $\lambda^{\mu}_{dec} = 2.19703 \times 10^{-6}s$ respectively. The competition between capture and decay gives an expression for the conversion losses:

$$1 - \eta^{-1}_{conv} \approx \frac{\lambda^{\pi}_{cap}}{(\lambda^{\pi}_{dec} + \lambda^{\pi}_{cap})} + \frac{\lambda^{\mu}_{dec}}{(\lambda^{\mu}_{dec} + \lambda^{\mu}_{cap})}. \qquad (7)$$

The capture rates are inversely proportional to the size of the "Muon Cave" and can thus be scaled to minimize the losses. The best choice is $\lambda^{\mu}_{cap} = \sqrt{\frac{\lambda^{\mu}_{dec}\lambda^{\pi}_{dec}}{c}}$. Inserting this value in (7) gives a total conversion loss of 20% in a perfectly optimized magnetic bottle and thus

$$\eta_{conv} = 1.25 \qquad (8)$$

A similar value for the conversion efficiency was obtained by Petrov and Sakhnovsky [12] in a different scheme.

We now calculate η_{matr} - the inverse fraction of muons stopped in the fuel. More than 90% of all muons have an energy larger than 50 MeV. The range of a 50 MeV muon is 11 cm in e.g. carbon and the required thickness of a single shell carbon fibre reactor vessel of tensile strength 4000 MPa and a working temperature of 500 K is only 3 cm. The average muon travels hence several times through the reactor before capture. The relevant quantity is the total average thickness of the construction material while the details of the structure are unimportant.

The fraction of stopped muons in each component would be proportional to the thickness ℓ, density ρ and stopping power S:

$$\eta_{matr} = \frac{\Phi \rho_{lhd} \ell_{fuel} S_{fuel}}{(\Phi \rho_{lhd} \ell_{fuel} S_{fuel} + \rho_{matr} \ell_{matr} S_{matr})} \tag{9}$$

From a standard stopping power formula [13] we get $\frac{S_{fuel}}{S_{carbon}} = 0.86$ and $\frac{S_{fuel}}{S_{steel}} = 1.06$ assuming a 50% deuterium-tritium mixture. A muon momentum of 150 MeV was used in the stopping power calculation. Changing the muon momentum within the intervall 150 - 800 MeV where 80% of the muons are found would lead to a maximum variation of 4% in the stopping power ratio. Inserting (6) in (9) we get

$$\eta_{matr} = (1 + a_{matr} \frac{T(K)}{\sigma_{matr}(MPa)}) \tag{10}$$

where $a_{carbon} = 4.37$ and $a_{steel} = 12.2$.

We now collect our results for E_{opt}, η_{conv} and η_{matr} to obtain the energy cost per active muon E_{act}. The result is displayed in Fig. 7. For a carbon composite reactor vessel with tensile strength 4000 MPa and a working temperature of 500 K we get $E_{act} = 7.5$ GeV. Using a steel container with tensile strength 1200 MPa we get at the same temperature $E_{act} = 15.7$ GeV.

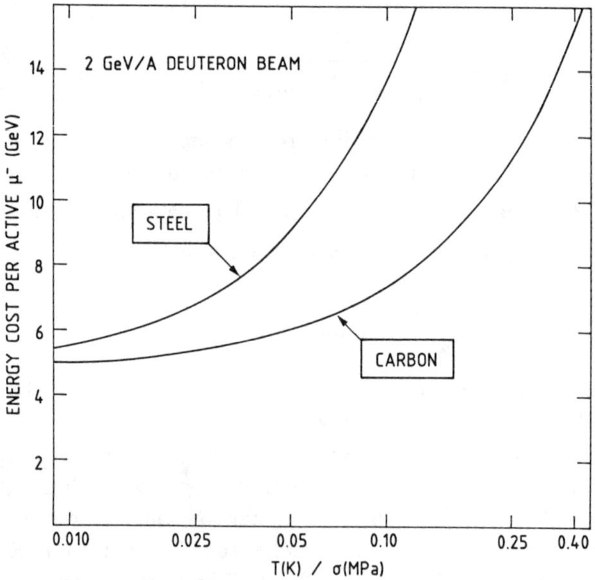

Fig. 7. Energy cost per active muon in the "Muon Cave" scheme. as a function of the fuel temperature divided by the tensile strength of the pressure vessel. Ideal gas behaviour is assumed for the fuel. The results can be used also for other equations of state as discussed in the text.

Note that E_{act} is independent of the fuel density Φ if the fuel is an ideal gas. The corrections due to the presence of the pressure vessel depend of the ratio $\frac{M_{matr}}{M_{fuel}} = \frac{V_{matr}\rho_{matr}}{V_{fuel}\Phi\rho_{lhd}}$ of the mass of the construction material to the mass of the fuel. This ratio is independent of Φ. The equation of state of hydrogen at high pressures is, however, in general not linear in Φ [14] but the performance of the Muon Cave system can always be obtained from Fig. 7 by inserting an equivalent temperature $T(K) = \frac{P(MPa)}{0.293\Phi}$.

The tritium inventory of a fusion reactor is of crucial importance both because of the limited tritium lifetime and because of safety considerations. The total volume V_{opt} of a pure deuterium-tritium optimal target of density $\Phi = 1$ is given in Fig. 6. For a beam energy of 2 GeV we get $V_{opt} = 6.1m^3$. If 10% of the efficiency is traded for reduced target size we get $V_{10\%} = 0.87m^3$. The corresponding tritium inventory for 50% deuterium-tritium mixture is $M_{opt} = 650Kg$ and $M_{10\%} = 92Kg$. The tritium inventory decreases, however, as Φ^{-2} with increasing density. Experiments are now planned at $\Phi = 2.5$. A μCF-reactor at this density would need a tritium inventory of only $M_{opt} = 104Kg$ or $M_{10\%} = 14.7Kg$. Including the effect of the construction material would further reduce the required amount of tritium.

The size of the magnetic mirror system can be estimated assuming that the pions move randomly within the confining field. The capture rate of pions is then defined only by the average density $\Phi_{av} = \frac{V_{fuel}\Phi}{V_{cave}}$ where V_{cave} is the volume between the reflecting fields and we for simplicity have assumed a pure deuterium-tritium optimal target. Conversion losses are according to (7) minimized when 12% of the emitted pions are captured before decay. Eq. (3) can now be employed to estimate the suitable average density finding $\Phi_{av} = -\frac{\Phi_m}{ln(0.12)} = 0.0057$. The volume of the "Muon Cave" is hence

$$V_{cave} = \frac{V_{fuel}\Phi}{\Phi_{av}} \tag{11}$$

At density $\Phi = 1$ and $E_{beam} = 2GeV$ we get $V_{cave} = 1075m^3$ and $V_{cave} = 153m^3$ for $V_{fuel} = V_{opt} = 6.13m^3$ and $V_{fuel} = V_{10\%} = 0.87m^3$ respectively. Chosing the 10% reduced efficiency and a high density $\Phi = 2.5$ we get a Muon Cave radius as small as 1.8 m.

Conclusions

Comparing the earlier studied dilute infinite target [1] to the present scheme (with typical parameters T = 500K and σ = 4000MPa for the carbon composite vessel) we note that the energy cost per muon is larger by a factor 4 in the more realistic scheme. We have identified the main pion and muon traps: 1) π^- not produced since the cascade partly escapes from the target. 2) conversion losses due to capture of emitted π^- and decay in flight of μ^- and 3) μ^- stopped in the construction material. In our example these losses are responsible for the increased cost of muons by factors 2, 1.25 and 1.5

respectively.

Future technological inventions might improve these numbers. Auxiliary targets protected by magnetic fields, careful tailoring of the confining fields, improved construction materials and pulsed devices where thermodynamical pressure equilibrium is not maintained during the active phase could enhance the efficiency of muon production systems.

References

[1] M. Jändel, M. Danos and J. Rafelski, Phys.Rev. C **37**, 403 (1988); M. Jändel, M. Danos and J. Rafelski, J. Muon Catalyzed Fusion, **3** 557 (1988).

[2] Yu. V. Petrov and Yu. M. Shabel'skii, Yad. Fiz. **30**, 129 (1979) (Sov. J. Nucl. Phys. **30**, 66 (1979).

[3] A. Bertin et al. Europhys. Lett. **4**, 875 (1987).

[4] L. Chatterjee, Fusion Tech. **12**, 444 (1987).

[5] S. Eliezer, T. Tajima and M.N. Rosenbluth, Institute of Fusion Studies Report IFSR 223/1986.

[6] Yu. V. Petrov, Nature **285**, 466 (1980).

[7] H. Takahashi et al. Atomkernenergie-Kerntechnik **36**, 195 (1980).

[8] G. Chapline and R. Moir, J. Fusion Energy **5**, 191 (1986).

[9] P.A. Aarnio, J. Lindgren, J. Ranft, A. Fasso and G.R. Stevenson CERN-TIS-RP/190/1987; P.A. Aarnio, A. Fasso, H.J. Moehring, J. Ranft and G.R. Stevenson, CERN-TIS-RP/168/1986.

[10] J. Ranft, Nucl. Instr. Meths. **48**, 133 (1967); K Haenssegen and J. Ranft, Nucl. Sci. Eng. **88**, 537 (1984).

[11] De Bossu et al. Proc. 8th Int. Conf. of Soc. for the Advanc. of Matr. and Proc. Eng. European Chapter May 18-21 (1987).

[12] Yu. V. Petrov and E.G. Sakhnovsky, Atomkernenergie-Kerntechnik **46**, 25 (1985).

[13] U. Fano, Ann. Rev. Nucl. Sci., **13**, 1 (1963).

[14] P.C. Souers, Hydrogen Properties for Fusion Energy, (Univ. of California Press 1986).

ON THE ENERGY COST OF PRODUCING MUONS IN A DEUTERIUM-TRITIUM TARGET FOR MUON CATALYZED FUSION

A.Bertin(*), M.Bruschi(*), M.Capponi(*), S.De Castro(*), I.Massa(***), M.Piccinini(*), M.Poli(**), N.Semprini-Cesari(*), A.Trombini(*), A.Vitale(*)and A.Zoccoli(*)

(*) Dipartimento di Fisica dell'Università di Bologna, and Istituto Nazionale di Fisica Nucleare, Sezione di Bologna, Italy

(**) Dipartimento di Energetica dell'Università di Firenze, and Istituto Nazionale di Fisica Nucleare, Sezione di Bologna, Italy

(***) Dipartimento di Fisica dell'Università di Cagliari, and Istituto Nazionale di Fisica Nucleare, Sezione di Cagliari, Italy

ABSTRACT

The costs of producing negative muons in d-t targets at different densities are determined. The advantages of a segmented target structure are discussed, alternating high-density layers to suitably dimensioned vacuum gaps.

INTRODUCTION

For a full exploitation of the possibilities of muon-catalyzed fusion (MCF)[1] in a deuterium-tritium (d-t) target,[2] based on the cyclical reaction

$$d\mu t \rightarrow {}^4He + n + \mu^- \quad \text{(N times)} \tag{1}$$

two essential requirements are to be fulfilled.

First of all, one has to maximalize the number N of fusions induced by a single muon.[3] In its turn, this demands to minimalize all muon losses due to processes which take the muon out of the fission cycle. The most important of these are due to the sticking effect[4] of the muon to the ^{4}He nucleus released in reaction (1); and to the muon transfer processes to the ^{4}He and ^{3}He nuclei present within the target.[5]

As a second point, the energy expenditure in the production of muons for the purposes of MCF has to be minimalized. So far, dedicated experimental work is missing on this problem. The first estimates of the cost C_{μ^-} of negative muons (which most effectively are obtained from decaying pions) were given by Petrov and collaborators,[6] who assigned a value of about 5 GeV to this global energy cost. Other evaluations followed,[7] yielding comparable results.

More recently,[8] we performed a calculation of C_{μ^-} by a Monte Carlo method,[9] the reliability and flexibility of which were checked while interpreting the measurements[10] of the yield Y_π of low-energy pions obtained from a 300 GeV proton beam. A further check of the method has been carried out while measuring Y_π at a 10 GeV proton beam.[11] We can then assume that the predictions obtained in this way, which are in good agreement with those provided by Jandel, Danos and Rafelski in an independent analysis,[12] are leaning on well founded experimental grounds.

In the present report, we shall discuss some extensions of the results already published,[8] examining the possibility of a particular structure for the d-t target, which in principle may help in reducing C_{μ^-} significantly.

OUTLINE OF THE METHOD

Quite generally, the energy expenditure C_{μ^-} in producing one muon through pion decay can be expressed as

$$C_{\mu^-} = \frac{C_{\pi^-}}{P_{\pi-\mu}} \tag{2}$$

where C_{π^-} is the analogous cost referring to one negative pion, and $P_{\pi^-\mu}$ is the fraction of produced negative pions which decay into muons. C_{π^-} can be defined as

$$C_{\pi^-} = \frac{TN}{N_{\pi^-}} \tag{3}$$

where T is the kinetic energy, N the number of primary beam nucleons impinging on the pion production target, and N_{π^-} the number of correspondingly produced negative pions.

The cost evaluations we wish to discuss refer to the hypothesis that negative pions are produced directly within the deuterium-tritium target, where the muons from pion decay are stopped and interact. Along this line, the results are obtained through the following steps:

i) Systematic calculations of C_{π^-} are performed, using the quoted Monte Carlo method, and varying the main parameters within the limits given in Table I.

ii)A further Monte Carlo calculation is developed, in order to determine the probability $P_{\pi^-\mu}$ of pion decay (integrated over the momentum distribution of the produced pions) within the targets considered in Table I. Starting from its initial energy, one produced pion is followed along its path inside the assumed material, until it stops or decays: the variation of its decay rate during the slowing-down process, the Coulomb multiple diffusion and straggling effects, and the nuclear interactions of the pion with a momentum-dependent cross section $\sigma_{in}(\pi^- d)$ ($\sigma_{in}(\pi^- t)$) are duly taken into account.

Table I - Main target parameters considered for the calculation of C_{π^-}.

Primary particle beam	Range of primary particle momentum p_B(GeV/c)	Element	Density range (units of ρ_o)[a]	Thickness (units of nuclear interaction length)	Diameters (cm)
Neutrons Protons	2÷8	Deuterium Tritium	0.1÷1	3	50÷700

[a] ρ_o is the atomic density of liquid hydrogen (4.2×10^{22} cm^{-3})

In the absence of specifical data on the quoted cross sections for the relevant momentum range, starting from the available results[13] on the inelastic cross-sections $\sigma_{in}(\pi^- p)$ of negative pions against protons, it is assumed that $\sigma_{in}(\pi^- d) = \alpha\ \sigma_{in}(\pi^- p)$, and $\sigma_{in}(\pi^- t) = \beta\ \sigma_{in}(\pi^- p)$, where α and β are suitable coefficients. They are obtained by considering the known ratios between the total cross sections $\sigma_{tot}(\pi^- d)$, $\sigma_{tot}(\pi^- p)$ (for the case of α); and $\sigma_{tot}(\pi^- d)$, $\sigma_{tot}(\pi^- {}^4He)$ and $\sigma_{tot}(\pi^- p)$(for the case of β). After suitable averaging and minor corrections, we obtain $\alpha = (1.5 \pm 0.5)$, and $\beta = (2.5 \pm 0.5)$.

The determination of the fraction of decaying pions ($N_{\pi^-\mu}/N_\pi$) is obtained as a function of the initial pion momentum p_{π^-}. The momentum distributions (dN_π / dp_π) of the pions are calculated separately for deuterium and tritium production targets for different primary particle beams and momentum. The integral decay probability $P_{\pi^-\mu}$ for the pions produced within the targets considered in Table I are then determined by combining the results obtained on ($N_{\pi^-\mu} / N_\pi$) and on (dN_π / dp_π). Finally, the muon costs C_{μ^-} are calculated through Equation (1). The uncertainties on the coefficients α and β induce inaccuracies of 20-25% in the obtained C_{μ^-} values.

RESULTS ON C_{π^-}

The results obtained[8] on the energy cost for producing negative pions in liquid deuterium or tritium targets can be summarized as follows:

i) C_{π^-} decreases with the target diameter D, attaining a constant value for $D \cong 3$ m.

ii)For $D \gtrsim 5$ m, the optimal momentum of the primary beams is in general $p_B \cong 3.5$ GeV/c.

iii)the minimum values obtained for C_{π^-} in these conditions are summarized in Table II.

Table II - Optimal values for C_{π^-} for 3- nuclear interaction lengths thick deuterium and tritium targets at the density of liquid hydrogen.[a]

Primary beam	p_B (GeV/c)	Target:	C_{π^-} (GeV) Deuterium	Tritium
Neutrons	3.5		1.3	1.0
Protons	3.5		1.7	1.3

[a] See reference 8. Relative inaccuracies are of the order of a few per cent.

One may comment here that the C_{π^-} values given in the Table could in principle be affected if the target density is decreased. In fact, once the average time between two subsequent inelastic interactions of the pions increases, its decay process its favoured, and the number of pion-producing nuclear interactions undergone by the pion becomes smaller. We have therefore used the Monte Carlo method to explore the variations of C_{π^-} with the target density, keeping constant the effective target thickness (3 nuclear interactions lengths) and increasing its diameter correspondingly. The results obtained are presented in Fig.1 for a typical beam-target choice, and in Table III for the different cases examined.

Table III - Variations of C_{π^-} with the target density for different (p_B = 3.5 GeV/c) beams and (3 nuclear interaction lengths thick) targets.

Density (ρ/ρ_o)	Target element:	Deuterium		Tritium	
	Beam:	Neutrons	Protons	Neutrons	Protons
1		1.3	1.7	1.0	1.3
0.5	C_{π^-}(GeV)	1.35	1.8	1.1	1.5
0.2		1.4	2.0	1.2	1.6
0.1		1.5	2.1	1.3	1.7

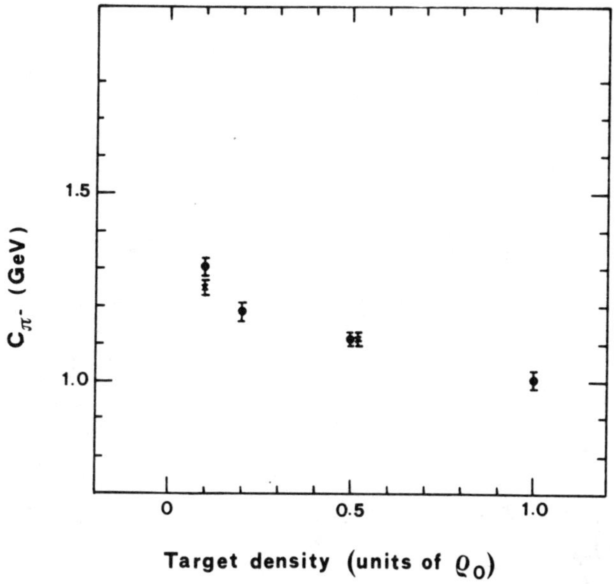

Fig.1 . Decrease of the pion cost C_{π^-} with the target density for the case of 3.5 GeV/c neutrons impinging on tritium. Full points: compact target (3 nuclear interaction lengths). Crosses: segmented target, having a total tritium thickness of 3 nuclear interaction lengths, but alternating tritium layers to suitably dimensioned vacuum gaps. (ρ_o is the density of liquid tritium).

From these findings, we can conclude that the variation of C_{π^-} with the target density is quite reduced. Indeed, because of the relativistic time dilatation, the expected effect becomes sensitive only for low-energy pions. Therefore, it has only little influence on the growth of the nuclear interaction cascade process within the production target. Furthermore, as a general trend, the C_{π^-} values turn out to be smaller for the case of neutrons impinging on tritium targets (su Table III).

RESULTS ON C_{μ^-}

Two typical momentum distributions of the negative pions produced in a tritium target are shown in Fig.2. These figures, which do not depend neither on the target diameter, nor on its density, are similar to the ones obtained[14] in the case of deuterium.

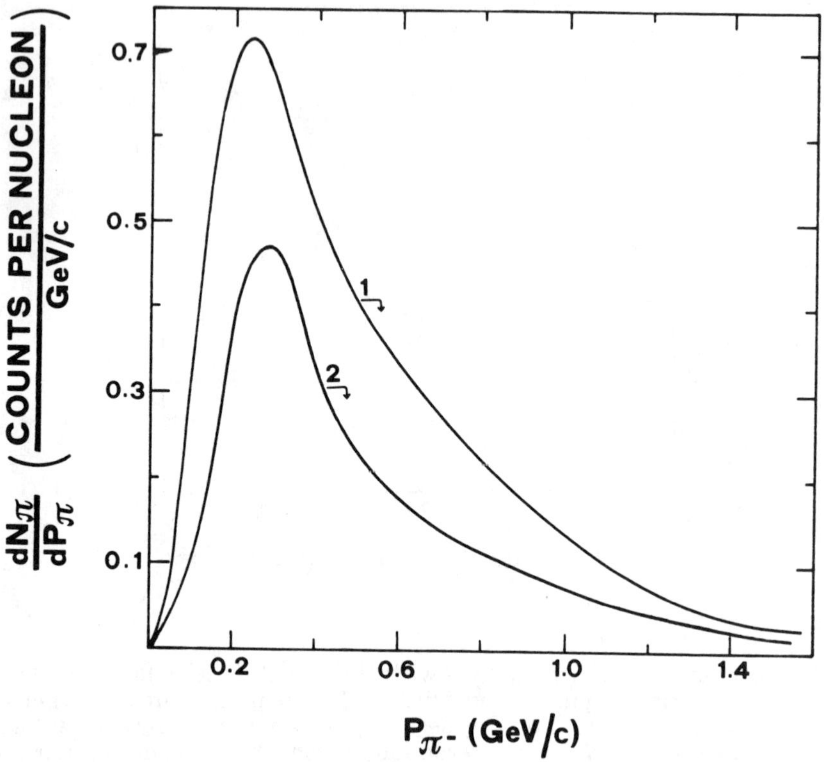

Fig.2 . Calculated momentum distributions of the negative pions produced in a liquid tritium target by 3.5 GeV/c neutrons (curve 1) and protons (curve 2).

Fig.3 shows, in its turn, the results obtained in the present calculations for the fraction of negative pions decaying (when they are produced with a given initial momentum p_{π^-}) within (3-nuclear interaction lengths thick) tritium targets having different densities. As expected, they indicate an efficiency for pion-muon conversion which is somewhat smaller than for the case of deuterium targets,[8] due to the higher nuclear interactions between pions and tritons. These results are in close agreement with those obtained by an analytical calculation we already described.[8]

By combining the results of Fig.3 with those of Fig.2, we obtained the integral probabilities $P_{\pi^-\mu}$ of pion decay within tritium targets which are presented in Fig.4. The pronounced dependence of $P_{\pi^-\mu}$ on the target density is apparent.

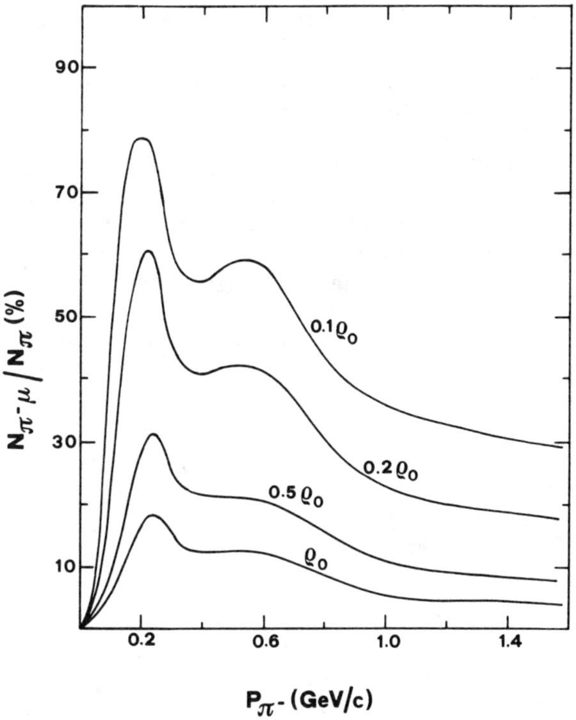

Fig.3 . Fraction of decaying negative pions as a function of their momentum, when pions are formed within (3 - nuclear interaction lengths) tritium targets having different densities (ρ_o is the density of liquid tritium).

Using then the C_{π^-} values given in Table III, and remembering Equation (2), one gets the energy costs for producing one negative muon in a tritium target which are reported in Table IV. In the Table, these values are compared to those corresponding to deuterium targets, which are quite similar to those obtained in our previous work,[8] where a constant C_{π^-} value was assumed.

DISCUSSION AND CONCLUSIONS

From the results reported in Table II, it is seen that the costs of producing muons in deuterium, tritium (and consequently in deuterium-tritium) targets turn out to be quite similar, once the primary particle beam and the target density have been chosen. The worse pion-muon conversion efficiency in tritium (with respect to deuterium) is in fact somewhat compensated by the smaller cost of producing pions (see Tables II and III).

Taken as a whole, the results of Table IV point at an effective cost C_{μ^-} of about 10 GeV for a (d-t) target at a density equal to the one of liquid hydrogen.

This cost is high for the purposes of MCF, and the naive impression one might draw from Table IV (i.e. that one might work in low-density targets to reduce such an expenditure) would in practice be erased by the opposite requirement of producing a high number N of fusions per stopping muon (which is best achieved in high-density d-t media).[1,3]

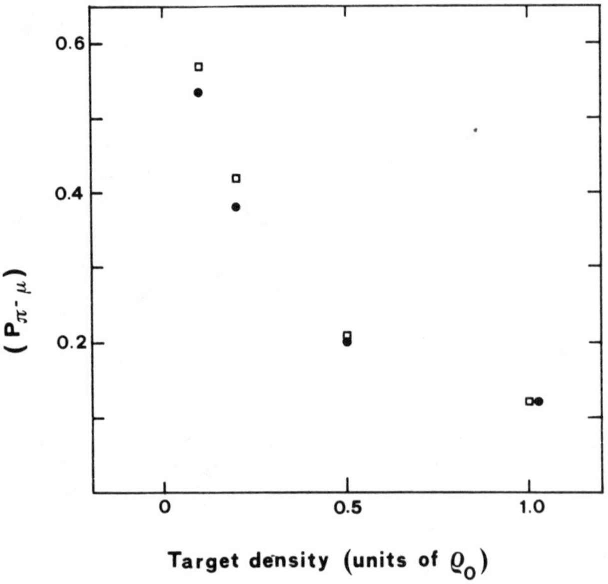

Fig.4 . Decay probability for pions produced in (3 - nuclear interaction lengths) tritium targets at different densities by 3.5 GeV/c protons (square points) and neutrons (round full points), respectively.

Keeping in mind these contradictory demands, a possibility which seems worth while being examined is the one of a segmented target, where high-density layers of deuterium-tritium mixtures are alternated to suitably dimensioned evacuated gaps. This type of structure would correspond, for a constant number of nuclear interaction lengths, to a reduced average density. It would exploit, therefore, the fact that C_{π^-} varies little with the target density (opposite to $P_{\pi^-\mu}$).

As a test of the present hypothesis, our calculations were run for two ideal segmented targets, having a total thickness of 3 nuclear interaction lengths, and effective average densities $\rho = 0.5\ \rho_o$ and $\rho = 0.1\ \rho_o$. The results obtained for C_{π^-} are presented in Fig.1. The corresponding curves for the conversion probability of pions into muons practically are identical with those presented in Fig.3 for $\rho = 0.5\ \rho_o$ and $\rho = 0.1\ \rho_o$. The advantage of the proposed target structure are then apparent.

Table IV - Calculated Cost (C_μ-) of producing muons by different 3.5 GeV/c particle beams impinging on deuterium or tritium targets.[a]

Target (3 nuclear interaction lengths thick)					
Density (units of ρ/ρ_o)	Element:	Deuterium		Tritium	
	Beam:	Neutrons	Protons	Neutrons	Protons
1		8.3	11.4	8.6	11.8
0.5		4.6	6.6	5.5	6.9
0.2	C_μ-(GeV)	2.7	4.0	3.1	3.7
0.1		2.3	3.2	2.3	2.9

[a] The inaccuracies of the presented values range between 20 and 25%.

The idea of density gradients within the target (or system of targets) which are suitable for MCF is not new: under different respects, it is present in the hybrid mesocatalytic reactor suggested by Petrov,[6] in the recirculating-beam synthesizer proposed by Eliezer et al.,[7] and in the active target arrangement conceived by Jandel, Danos and Rafelski.[12] The hypothesis considered here presents conceptual difficulties, such as the problem of the solid walls which would include the high-density d-t layers, which may well represent a significant disturbance for the stopping muons; and practical problems, such as e.g. the huge quantity of tritium contained within the considered geometries.

Nevertheless, the calculations presented indicate that, for sufficiently large target dimensions, (a) the average target density is the leading parameter on which depends the energy cost of producing muons, and (b) even a non sophisticated target geometry may allow to exploit the small variation of C_π- with density for the purpose of reducing C_μ- .

In our opinion, therefore, the hypothesis of a segmented, multilayer ((high-density) + (vacuum)) d-t target should be taken into consideration for the purpose of ameliorating the energy balance of muon-catalyzed fusion.

REFERENCES

1. See e.g. S.E.Jones, Nature 321, 127 (1986); and references therein.
2. S.S.Gershtein and L.I.Ponomarev, Phys. Lett. B72, 80 (1977).
3. See for instance: V.M.Bystritskii, V.P.Dzhelepov, Z.V.Ershova, V.G.Zinov, V.K.Kapishev, S.SH.Mukhamet-Galeeva, V.S.Nadezhdin, L.A.Rivkis, A.I.Rudenko, V.I.Satarov, N.V.Sergeeva, L.N.Somov, V.A.Stolupin and V.V.Filchenkov, Sov. Phys. JETP 53, 877 (1981); S.E.Jones, A.N.Anderson, A.J.Caffrey, J.B.Walter, K.D.Watts, J.N.Bradbury, P.A.Gram, M.Leon, H.R.Haltrud and M.A.Paciotti, Phys. Rev. Lett. 51, 1757 (1983); S.E.Jones, A.N.Anderson, A.J.Caffrey, C.Dew. Van Siclen, K.D.Watts, J.N.Bradbury, J.S.Cohen, P.A.M.Gram, M.Leon, H.R.Maltrud and M.A.Paciotti, Phys. Rev. Lett. 56, 588 (1986); W.H.Breunlich, M.Cargnelli, P.Kammel, J.Marton, N.Naegele, P.Pawlek, A.Scrinzi, J.Werner, J.Zmeskal, J.Bistirlich, K.M.Crowe, M.Justice, J.Kurck, C.Petitjean, R.H.Sherman, H.Bossy, H.Daniel, F.J.Hartmann, W.Neumann and G.Schmidt, Phys. Rev. Lett. 58, 329 (1987).
4. See for instance: L.Bracci and G.Fiorentini, Phys. Rep. 86, 169 (1982); D.Mueller and J.Rafelski, Phys. Lett. B164, 223 (1985).
5. See e.g. A.Bertin, M.Bruschi, M.Capponi, J.D.Davies, S.De Castro,I.Massa, M.Piccinini, M.Poli, N.Semprini-Cesari, A.Trombini, A.Vitale and A.Zoccoli, report presented at the present Workshop; and references therein.
6. Yu. V.Petrov and Yu. M.Shabelskii, Sov. J.Nucl.Phys. 30, 66 (1979); Yu. V.Petrov, Nature 285, 464 (1980).
7. S.Eliezer, T.Tajima and M.N. Rosenbluth, DOE-ET-53088-223 (1986).
8. A.Bertin, M.Bruschi, M.Capponi, S.De Castro, I.Massa, M.Piccinini, M.Poli, N.Semprini-Cesari, A.Vitale and A.Zoccoli, Europhys. Lett. 4, 875 (1987).
9. P.A.Aarnio, A.Fasso, M.J.Moehring, J.Ranft and G.R. Stevenson, FLUKA 86 Code: see FLUKA 86 User's Guide, CERN TIS-RP/168 provisional, 21 January 1986; Enhancements to the FLUKA 86 Program (FLUKA 87), CERN TIS-RP/190, 30 January 1987.
10. A.Bertin, M.Capponi, S.De Castro, I.Massa, M.Piccinini, M.Poli, F.Scuri, N.Semprini-Cesari, A.Vacchi, A.Vitale, E.Zavattini and A.Zoccoli; Europhys. Lett. 3, 1255 (1987).
11. A.Bertin, et al., (to be published).
12. M.Jandel, M.Danos and J.Rafelski, Phys. Rev. C37, 403 (1988).
13. Review of Particle Properties, Phys. Lett. B, 170 April 1986. See also V.Flaminio, W.G.Moorhead, D.R.O.Morrison and N.Rivoire, Compilation of Cross Sections I: π^+ and π^- Induced Reactions, CERN-HERA 83-D1, 30, August 1983.
14. A.Bertin, M.Capponi, S.De Castro, I.Massa, M.Piccinini, M.Poli, N.Semprini-Cesari, A.Vitale, E.Zavattini and A.Zoccoli, Proceedings of the Course/Workshop on Muon Catalyzed Fusion and Fusion with Polarized Nuclei, "Ettore Majorana" Centre for Scientific Culture, Erice, 3-9 April 1987, to be published.

PION (AND MUON) PRODUCTION BY ANTIPROTONS ON LIGHT AND HEAVY NUCLEI*

R.A. Lewis, E. Minor, F. Persi, G.A. Smith,
W. Toothacker and J. Whitmore
Laboratory for Elementary Particle Science
Department of Physics
Penn State University
University Park, PA 16802
(presented by G.A. Smith)

ABSTRACT

The possibility of producing pions (muons) for muon-catalyzed fusion by antiproton annihilation in nuclear targets is considered and compared with other methods involving nuclear projectiles and targets. Although the kinematical properties of pions produced by antiproton annihilation are very attractive, the energy cost per pion is not competitive with that of the other methods. This arises from the cost of producing antiprotons.

INTRODUCTION

The subject of pion (muon) production for muon-catalyzed fusion has been explored in some detail by several investigators.[1-6] Each of these studies has involved nucleon-nucleon, nucleon-nucleus or nucleus-nucleus collisions in the projectile laboratory momentum region of $\lesssim$ 5 GeV per nucleon. Because antinucleon-nucleon annihilation produces an abundance of pions (~ 4-5 at rest), it is of interest to also consider this reaction as a possible alternate source of pions (and hence muons from pion decay), and to compare it with the previously mentioned studies.

CROSS SECTIONS

Figure 1 shows proton-proton total and elastic cross sections up to 5.5 GeV/c. We observe that above ~ 2 GeV/c the difference, or inelastic cross section, is ~ 25 mb and constant. Pion production becomes abundant only after one passes through the $n\Delta$ threshold region at ~ 1.5 GeV/c. Figure 2 shows the antiproton-proton annihilation cross section up to ~ 10 GeV/c. Here we see that the annihilation cross section at the lowest momentum shown (~ 200 MeV/c) is eight times larger than the proton-proton inelastic cross section. And, at rest (zero momentum) the probability for annihilation is unity. This comparison would suggest a distinct advantage for antiproton-proton annihilation over nucleon-nucleon inelastic scattering.

*Work supported in part by the U.S. Air Force and National Science Foundation.

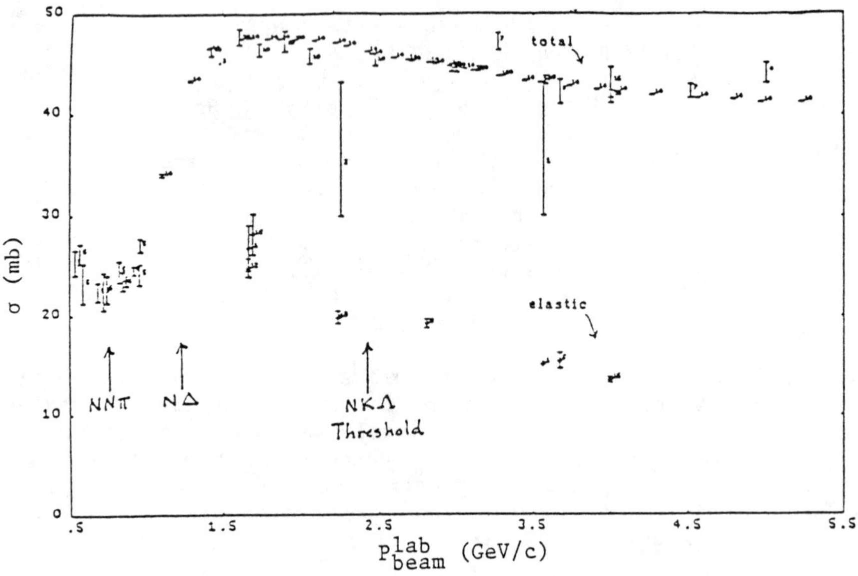

Fig. 1 - Total and elastic proton-proton cross sections up to 5.5 GeV/c momentum.

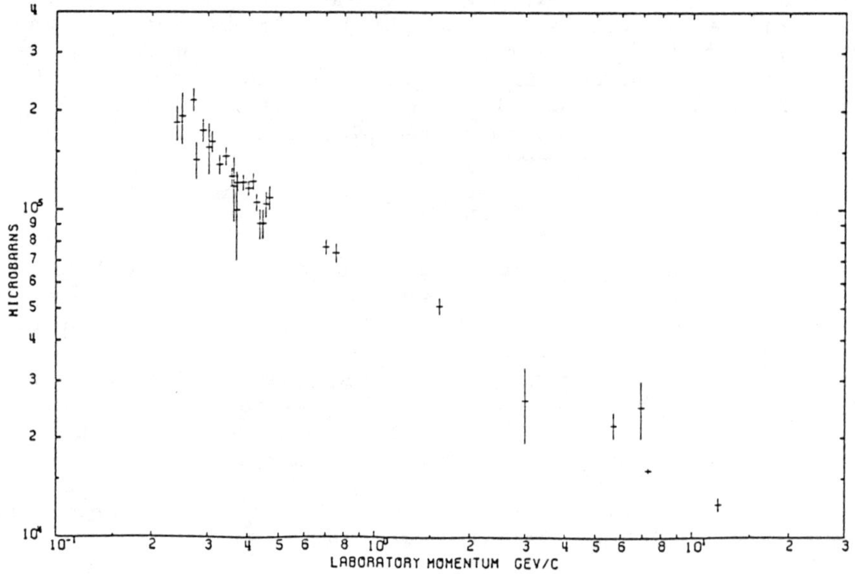

Fig. 2 - Antiproton-proton annihilation cross section up to 10 GeV/c momentum.

PION MULTIPLICITY

A quick glance at pion multiplicity data shows that nuclear projectiles and targets are preferred over nucleon projectiles and targets. This is seen in Fig. 3, where we plot the average pi-minus multiplicity per collision versus the atomic mass (A) of the target for various light ion projectiles and momenta.[7-12] We observe that for fixed projectile mass and energy the pion multiplicity increases with A, presumably due to intranuclear cascading. We also see that the multiplicity increases with projectile mass and energy at fixed A. However, the most striking feature of these data is the enormous yield of pions at 60A and 200A GeV/c. We will return to this point in a moment.

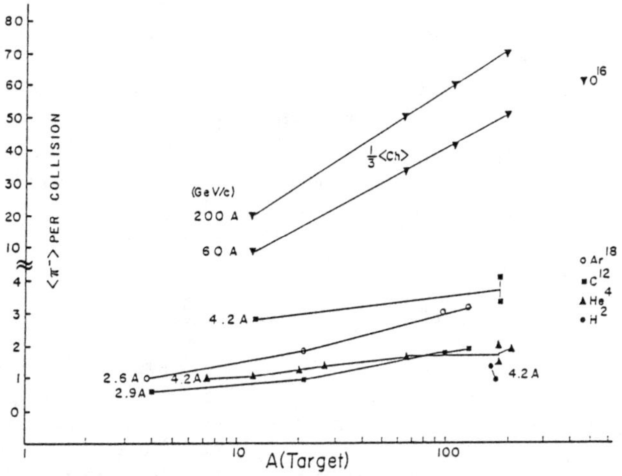

Fig. 3 - Average pi-minus multiplicity per collision versus atomic mass (A) of the target for a variety of nuclear projectiles and beam momenta (GeV/c).

A similar plot for antiproton-nucleus collisions is presented in Fig. 4.[13-17] Again, we select nuclear targets to profit from cascading effects. As in Fig. 3, the high energy data (100-200 GeV/c) give large multiplicities, but far below those for light ion-nucleus collisions in the same momentum range, and are only slightly larger (~ 1-2 units) than proton-nucleus data, which indicates that annihilation is not important at these energies.

The antiproton annihilation data at rest exhibit no significant A dependence. This may be explained if one recalls that at rest the antiproton captures in an atomic orbit, cascades down by emission of Auger electrons and X-rays, and finally annihilates on the surface of the nucleus. Consequently, many pions do not penetrate the nucleus and this, plus the lower energy of the pions (~ 200 MeV), leads to little if any extra pion production. In spite of this, however, the pion multiplicity is comparable to that from light ion-nucleus collisions at $\lesssim$ 4.2A GeV/c seen in Fig. 3.

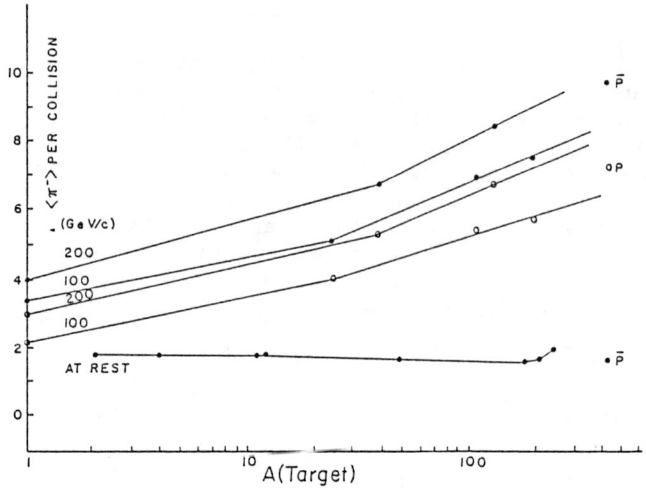

Fig. 4 - Average pi-minus multiplicity per collision versus atomic mass (A) of the target for antiprotons at a variety of beam momenta (GeV/c). Also shown are proton data at 100 and 200 GeV/c.

KINEMATIC CONSIDERATIONS

We now turn to kinematic variables to give us some guidance on the likelihood of capturing muons from pion decay in flight. In Fig. 5 we show pseudorapidity and transverse energy distributions of pions produced in light ion-nucleus collisions at 60A and 200A GeV/c momentum.[12] A straightforward calculation of the longitudinal momentum of a pion at the peak of the pseudorapidity distribution at 200A GeV/c gives ~ 6 GeV/c. Such a pion moving down a magnetic solenoid of size sufficient to trap the transverse momentum component would travel ~ 335 m in one lifetime! This is obviously too long for any practical trapping device to be used for energy production purposes.

In Fig. 6 we show the same information for antiproton-silver collisions at 100 GeV/c.[16] A similar calculation as above yields ~ 2 GeV/c longitudinal momentum, or ~ 112 m decay length. Again, this is much too large, and we therefore abandon the idea of using very high momentum light ions or antiprotons as a practical source of muons, in spite of the very large multiplicity of pions produced.

In Figs. 7 and 8 we show scatter plots of transverse momentum versus rapidity for pi-minus produced in 4.2A GeV/c carbon on carbon collisions[7,9-11] (Fig. 7) and antiproton annihilation at rest in carbon and uranium[17] (Fig. 8). Here, the decay lengths are found to be ~ 22 m and ~ 11 m respectively, which are consistent with previously discussed dimensions of magnetic traps.[1,3] In addition, an interesting feature of Fig. 7 is that one can easily translate into the center-of-mass of the carbon-carbon system by symmetrizing the density of points around zero rapidity. This requires a translation of ~ -1.25 units of rapidity, corresponding to center-of-mass momenta of 1.25A GeV/c for each carbon nucleus. With this translation it is difficult to distinguish between the carbon-carbon and antiproton-carbon or -uranium data. In other words, the kinematics of a 1.25A GeV/c carbon-carbon collider and antiproton annihilation at rest are virtually indistinguishable, and the decay length of a pion produced in either case is sufficiently short (~ 11 m) to make both candidates for practical use.

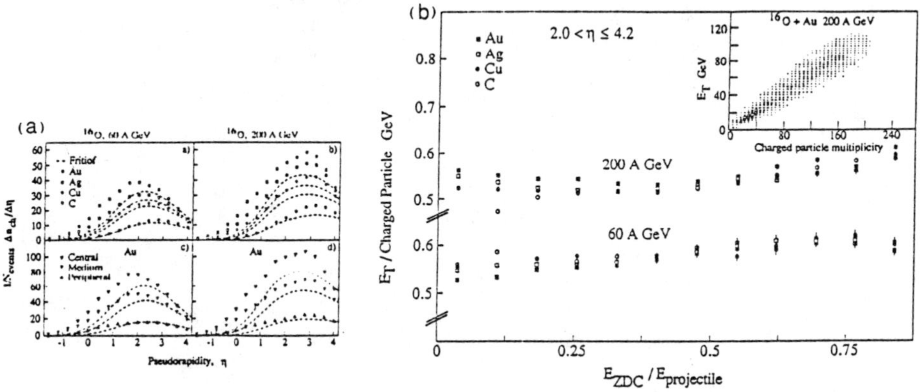

Fig. 5 - (a) Pseudorapidity and (b) transverse energy distributions of pions produced in ^{16}O collisions with various nuclear targets at 60A and 200A GeV/c. See ref. 12 for details.

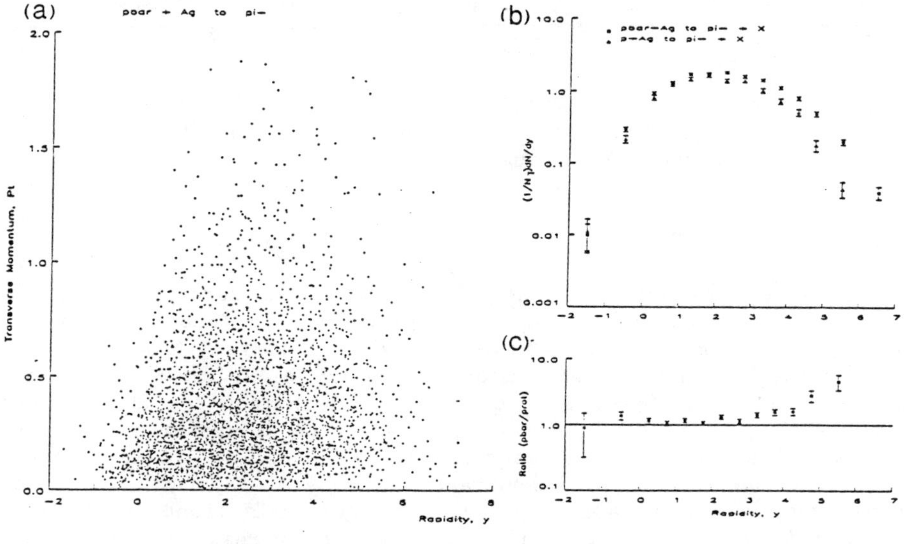

Fig. 6 - (a) Scatter plot of rapidity versus transverse momentum (b) rapidity projection and (c) ratio of antiproton to proton production versus rapidity for antiproton (proton) production of pi-minus on a Ag target at 100 GeV/c.

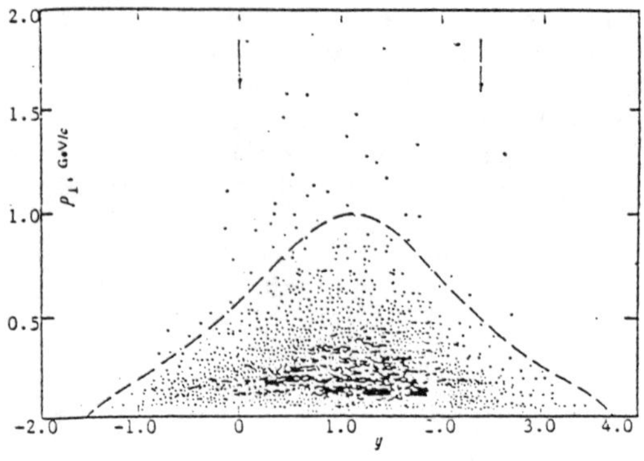

Fig. 7 - Scatter plot of rapidity versus transverse momentum of pi-minus produced in 4.2A GeV/c carbon collisions on carbon. See refs. 7,9-11 for details.

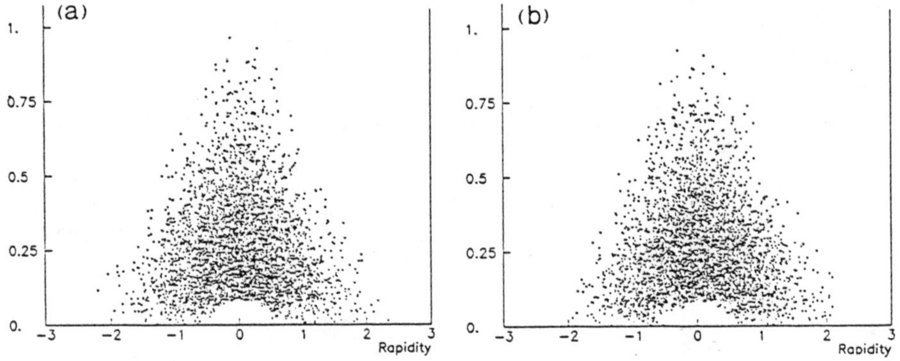

Fig. 8 - Scatter plots of rapidity versus transverse momentum of pi-minus produced in antiproton annihilation at rest in (a) carbon and (b) uranium.

COMPARISON OF THREE PRODUCTION SYSTEMS

Based on previous studies and the above considerations, we compare the basic parameters of three possible pion (muon) sources in Table 1. System A (linac beam on a very large target) has been studied thoroughly and is capable of delivering pions at 2-10 GeV per pion.[4] System B (carbon on carbon collider) requires an enormous luminosity (L) to yield pions at the same rate as system A (assuming an inclusive pi-minus cross section of 17b[7,9-11]), and is less energy efficient than system A by about a factor of 2-3. System C (antiprotons stopping in nuclear targets) is vastly inefficient compared to systems A and B, and is still so after an increase in antiproton collection efficiency by a factor of 10^2 (which at this time is due strictly to the authors' imaginations!).

Table 1

Comparison of light ion- and antiproton-induced pi-minus production schemes using nuclear targets.

A. LINAC ON LARGE TARGET

2.25A GeV/c DEUTERONS ON INFINITE DEUTERIUM-TRITIUM TARGET (BEAM DUMP)

1 mA BEAM CURRENT $\Rightarrow 0.6\times10^{16}$ d/sec * 1.5 π^-/d
$= \sim 0.9\times10^{16}\ \pi^-$/sec

$E_{IN}/\pi^- = 2\text{-}10$ GeV (DENSITY DEPENDENT)[4]

B. COLLIDER (CARBON ON CARBON)

1.25A GeV/c + 1.25A GeV/c

$L \cong 0.9\times10^{16}/17b = 0.5*10^{39}\ cm^{-2}s^{-1}$

$E_{IN}/\pi^- \simeq [14.4/3 * 1.5/3]\ E_{IN}/\pi^-$(LINAC)
$\simeq 5\text{-}24$ GeV

C. ANTIPROTONS (STOP)

E_{IN}/sec $\cong 2\times10^{12}$ p/sec $\times$ 28 GeV/p
$= 5.6\times10^{13}$ GeV/sec (CERN example)

π^-/sec $\cong 10^9\ \bar{p}$/sec * $2\pi^-/\bar{p} = 2\times10^9\ \pi^-$/sec

$E_{IN}/\pi^- = 28\times10^3$ GeV

INCREASE $\bar{p}$ ACCEPTANCE BY $\times10^2$?

($\Delta p/p$ = 100%, $\Delta\Omega$ = 1 STERADIAN)

$E_{IN}/\pi^- \Rightarrow 280$ GeV

CONCLUSIONS

Conventional high current linacs accelerating deuterons into a beam dump appear to be the most attractive scheme considered in this work, although a light ion collider is not far behind if the required luminosity can be produced. The energy cost of producing antiprotons makes them of no value in this problem.

REFERENCES

1. Yu.V. Petrov and Yu.M. Shabel'skii, Sov. J. Nucl. Phys. 30, 66 (1979).
2. H. Takahaski et al, Atomkernenergie-Kerntechnik, 36, 196 (1980).
3. Yu.V. Petrov and E.G. Sakhnovsky, Atomkernenergie-Kerntechnik, 46, 25 (1985).
4. M. Jändel et al, Phys. Rev. C27, 403 (1988) and contribution to this workshop.
5. S. Eliezer, Laser and Particle Beams, 6, 63 (1988).
6. L. Chatterjee et al, "On Heavy Ion Collisions as a Muon Source," paper submitted to this workshop.
7. A.M. Baldin, Proc. Seventh International Conference on High Energy Physics and Nuclear Structure, Zürich, Switzerland, 1977, p. 280.
8. S. Fung et al, Phys. Rev. Lett. 40, 292 (1978).
9. G.N. Agakishiev et al, Z. Phys. C12, 283 (1982).
10. G.N. Agakishiev et al, Sov. J. Nucl. Phys. 37, 559 (1983); 38, 90 (1983).
11. J. Bartke et al, Z. Phys. C29, 9 (1985).
12. R. Albrecht et al, Phys. Lett. B202, 596 (1988).
13. W. Bugg et al, Phys. Rev. Lett. 31, 475 (1973).
14. C. DeMarzo et al, Phys. Rev. D26, 109 (1982).
15. F. Balestra et al, Nucl. Phys. A474, 651 (1987).
16. J. Whitmore et al, Penn State University, preliminary results from Fermilab #597.
17. R.A. Lewis et al, Penn State University, preliminary results from CERN #PS183.

A New Concept for Muon Catalyzed Fusion Reactor

T. Tajima
Institute for Fusion Studies and Department of Physics
The University of Texas at Austin
Austin, Texas 78712

S. Eliezer
SOREQ
Yavne, 70600
Israel

and

R.M. Kulsrud
Plasma Physics Laboratory
Princeton University
Princeton, New Jersey 08544

Abstract

A new concept for a muon catalyzed pure fusion reactor is considered. To our best knowledge this constitutes a first plausible configuration to make energy gain without resorting to fissile matter breeding by fusion neutrons, although a number of crucial physical and engineering questions as well as details have yet to be resolved. A bundle of DT ice ribbons (with a filling factor f) is immersed in the magnetic field. The overall magnetic field in the mirror configuration confines pions created by the injected high energy deuterium (or tritium) beam. The DT material is long enough to be inertially confined along the axis of mirror. The muon catalyzed mesomolecule formation and nuclear fusion take place in the DT target, leaving α^{++} and occasionally $(\alpha\mu)^{+}$ (muon sticking). The stuck muons are stripped fast enough in the target, while they are accelerated by ion cyclotron resonance heating when they circulate in the vacuum (or dilute plasma). The ribbon is (eventually) surrounded and pressure-confined by this coronal plasma, whereas the corona is magnetically confined. The overall bundle of ribbons (a pellet) is inertially confined. This configuration may also be of use for stripping stuck muons via the plasma mechanism of Menshikov and Ponomarev.

I. Introduction

In order to achieve a thermodynamically meaningful energy gain by muon catalyzed DT fusion alone without resorting to breeding fissile matter by fusion neutrons, it is believed that the fusion catalysis cycling rate in a muon lifetime, X_μ, needs to substantially exceed 1000, while the so-called break even (after counting the thermodynamical efficiency of energy conversion) needs X_μ between 500 and 1000 under our current understanding of the processes of muon catalyzed fusion. Neglecting faster intermediate processes, nowadays we can identify two crucial bottlenecks for muon catalyzed fusion energy production. The onc is the $dt\mu$ mesomolecular formation process (rate $\lambda_{dt\mu}$) and the other is the muon sticking with fusion α particles (probability ω_{s0}):

$$X_\mu^{-1} = \omega_s + \lambda_{dt\mu}\tau_\mu , \tag{1}$$

where

$$\omega_s = \omega_{s0}(1 - R) , \tag{2}$$

and τ_μ is the lifetime of a muon and R is the stripping rate of muon from the $(\alpha\mu)^+$ atom.

In recent years best estimated experimental and theoretical values for ω_{s0}, R, and $\lambda_{dt\mu}$ have steadily firmed and are presently converging toward each other, as well evidenced in the presentations of the present Workshop:[1,2] $\omega_{s0} \simeq 0.008$, $R \sim 40\%$, and $\lambda_{dt\mu} \sim 10^9\,\text{sec}^{-1}$. The formation rate $\lambda_{dt\mu}$, of course, is a function of density ϕ and can be a function of temperature T as well. The exponent to ϕ may be larger than unity, although not enough is known. The exponent to T may also be positive and it may change in the regime higher than the room temperature. Again it is not for certain. In much higher temperature regime different molecular formation physics come into play and the mesomolecule can be destroyed by collisions. Some of the processes involve a plasma and three-body collisions.[3,4] These new channels of molecule-plasma reactions are potentially of immense importance, as their rates could be respectably large,[4] in spite of mesomolecule destruction. They tend to peak near temperatures of 10^1 eV. At present, however, no firm experimental investigations of these as we know of exist. Although we witness new infusion of physics in the mesomolecule formation processes and some degree of uncertainty in their rates, it is fair to say that the debate on understanding of the severest bottleneck of all, the sticking rate, has now been experimentally and theoretically settled at least to a degree accurate enough to decide the rate of energy production: $\omega_s \sim 0.0045$. This entails that even if $\lambda_{dt\mu} \to \infty$, the catalysis cycle rate X_μ is merely ~ 220. A very nice scientific progress in advancing the understanding of sticking physics. Nevertheless, a disappointing conclusion in itself.

The next question is: if the nature gives us a value of $X_\mu \sim 220$ (or lower), is it possible to enhance it by an ingenious method or two? Roughly speaking, the determination of ω_{s0} involves nuclear physics along with $dt\mu$ molecular dynamics; that of R involves atomic and molecular physics, while that of $\lambda_{dt\mu}$ includes atomic and molecular physics. Thus it seems prudent to us to try to "manipulate" or improve the processes that pertain to lower energy physics, i.e., atomic and molecular physics, as it is easier (or "cheaper") to influence lower energy physics. This is one of our guidelines. In the following we consider the enhancement of stripping by one way or another that does not involve nuclear physics.

Before we proceed to discuss our concept, we feel obligated to emphasize the famous but oft forgotten thermodynamical theorem, the second law of thermodynamics. Sometimes the muon catalyzed fusion is said to be *cold* fusion. This is appropriate only for the fundamental processes. As far as energy production is concerned, if it is too cold, no energy gain is possible to achieve (under 'normal' circumstances). Imagine that we want to let fusion happen in DT ice bars (below 25K). Per a dt fusion reaction, a neutron of 14.1 MeV and an alpha particle of 3.5 MeV are created. The neutron leaves the ice and presumably hits a surrounding blanket (or worse, a metal), while the alpha particle is likely to stay in the DT ice and heats it. If we are to insist on keeping the low temperature of the specimen, every time a fusion alpha particle heats the target, we have to cool (or extract the energy of that amount from) the material. That is, we have to remove 3.5 MeV energy per fusion. With the cooling efficiency η_c (because of the second law of thermodynamics, it is not unity and in fact a miniscule number typically of the order of) ~ 0.1, the necessary energy to keep the material at a desired cool temperature is $3.5\,\mathrm{MeV}/\eta_c$ ($\sim 35\,\mathrm{MeV}$) per fusion. On the other hand the total fusion energy production is of course only 17.6 MeV per fusion. No energy gain is possible. We shall come back to this difficulty when we discuss our concept.

II. Enhancement of Stripping

The effective sticking probability ω_s can be reduced by increasing the stripping coefficient R [see Eq. (2)]. The stripping (or reactivation) coefficient R may be written as[4,5,6]

$$R = 1 - \exp\left[-n \int_{E_i}^{E_b} \frac{\sigma_{\mathrm{st}}(E)dE}{F(E)}\right] , \tag{3}$$

where σ_{st} is the stripping cross-section, F is the stopping power of$(\alpha\mu)^+$ in matter, and E_b and E_i are the birth energy of $(\alpha\mu)^+(\sim 3.5\,\mathrm{MeV})$ and the $(\alpha\mu)^+$ ionization potential ($\sim 10\,\mathrm{keV}$). The cross-section peaks around $\sim 10\,\mathrm{keV}$ and quickly decreases below this energy. In order to make the reactivation coefficient large (i.e., close to unity), Bracci and Fiorentini[7] argued $(\alpha\mu)^+$ should be kept as long as possible in the higher velocity range where the stripping probability is higher. They suggested to apply accelerating electric fields in the matter. Unfortunately, the effective slowdown field in the solid matter is of the order of 45 MeV/cm, with which, either by dc or ac fields, breakdown of the target arises.

Kulsrud proposed[8] a clever separation of the target and the vacuum section in which acceleration of $(\alpha\mu)^+$ atoms takes place while their stripping happens in the solid (or liquid) target. The acceleration is achieved by the ion cyclotron resonance heating with the frequency $\omega = \Omega_{\alpha\mu} = eB/4m_pc$, where m_p is the proton mass. In his calculation the necessary strength of rf electric field E_{st} for stripping is

$$eE_{st} \cong 5 \times 10^4 f \text{ kV/cm} , \tag{4}$$

where f is the filling factor of the DT target. On the other hand, the breakdown field for ac field E_{bd} (which is much higher than the dc case) is approximately given[9,10] by

$$eE_{bd} \simeq m\omega c , \tag{5}$$

where m is the electron mass and ω the rf frequency. At the frequency range of $\Omega_{\alpha\mu} \sim 10^{10}$ Hz (with 1M Gauss magnetic field) the field in Eq. (4) is tolerable against breakdown. Unfortunately, his idea does not work as he put it, since the alpha heating of the target cannot be tolerated, as we discussed the cooling energy in the Introduction.

Menshikov[4] proposed an alternative idea for stripping. Instead of keeping the energy of $(\alpha\mu)^+$ high by acceleration (Refs. 7 and 8), he replaces the condensed matter by a (very high density) plasma. He points out that the plasma process significantly enhances the stripping if the temperature of the plasma is $\gtrsim 10$ eV. However, he fails to provide any plausible mechanism to confine such a high pressure plasma, except for mentioning vaguely about a shock wave.

In this paper we present a plausible confinement scheme that allows significant stripping by Kulsrud's mechanism (or perhaps Menshikov's) and simultaneously maintains favorable catalysis-reaction conditions of high densities and temperatures of the target. In principle, such a system can produce energy in a thermodynamically meaningful fashion.

III. Reactor Concept

In what follows we depict only the essence of the idea, as this paper does not have enough room to discuss all the relevant issues in detail. A detailed discussion will become available later.[11] The overall configuration is given in Fig. 1.

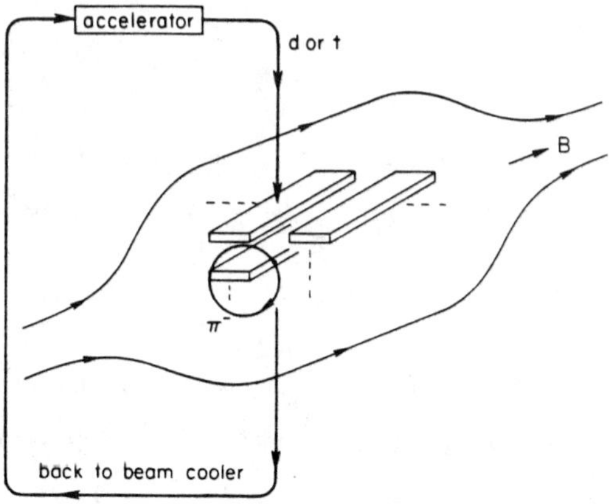

Figure 1: The overall reactor concept. The magnetic mirror axis coincides with the DT ice specimen axis. The injection direction of high energy beams of d or t is perpendicular to this. The blanket surrounds the mirror.

A simple mirror of mirror ratio R_m with magnetic field $\mathbf{B}$ generated by superconductors is surrounded by neutron blankets such as Li etc. Although the strength of the magnetic field is not uniquely fixed, we may use a typical number of 1MG. A pellet (or pellets) of DT ice is repetitively injected either mechanically or gravitationally into the mirror vessel. The pellet consists of a series of ribbons of DT ice. The minimum overall size is $\sim$(2cm)^3, which is determined by the two conditions, the inertial confinement condition of the overall pellet and the magnetic confinement condition of π^-. These DT ice ribbons fill the pellet with the filling factor f in order to accommodate sufficient acceleration time, while keeping enough matter to strip $(\alpha\mu)^+$ according to Kulsrud's mechanism of stripping. The choice of the filling factor f will be different for Menshikov's idea.[11] These DT ice specimens are penetrated by a high energy beam of d or t particles ($\sim$ 1 GeV/nucleon) injected perpendicular to the mirror axis, which coincides with the (longitudinal) axis of the specimens. Beam particles collide with the target nuclei, creating π^-'s. Pions are created *in situ* and confined by the mirror magnetic field, as in the earlier hybrid reactor concept.[12] The portion of the beam that did not suffer strong interaction will be collected, cooled, accelerated, and reused for the next pass. At this energy of the d or t beam the cooling may be appropriately provided by Budker's electron cooling method[13] or Van der Meer's stochastic cooling method.[14] The energy cost of this cooling will be studied in the future. The beam energy has to be boosted as well.

The ribbons have the following characteristics. The thickness ℓ_1 of a ribbon should be smaller than the range of $(\alpha\mu)^+$ at the 3.5 MeV energy at birth. Typically we choose 100 μm. The width ℓ_2 is typically several times ℓ_1. The length ℓ_3 is determined by the larger of the inertial confinement length and a few π^- Larmor radius:

$$\ell_3 = \max(c_s \tau_\mu \ , \ 4\rho_\pi) \ , \tag{6}$$

where c_s is the sound speed of the DT specimen and $\rho_\pi = \beta c/\Omega_\pi$. For 1 MG field and 1 eV DT specimen, these two are in the same ball park of $\sim 10^0$ cm. These ribbons are stacked up as shown in Fig. 2. The distance d_2 between two ribbons in the width direction is

$$d_2 = \min(2\rho_{\alpha\mu} \ , \ \ell_2/\sqrt{f}) \ , \tag{7}$$

and the distance d_1 between two ribbons in the thickness direction is

$$d_1 = \max\left(\ell_1/\sqrt{f} \ , \ \frac{\ell_1 \ell_2}{2f\rho_{\alpha\mu}}\right) \ , \tag{8}$$

where $\rho_{\alpha\mu}$ is the Larmor radius of $(\alpha\mu)^+$. Typical numbers for these are 0.4 cm and 0.2 cm, respectively. Radio frequency electromagnetic waves with frequency $\omega = \Omega_{\alpha\mu}$ are applied with the electric field rotating in the plane of Fig. 2.

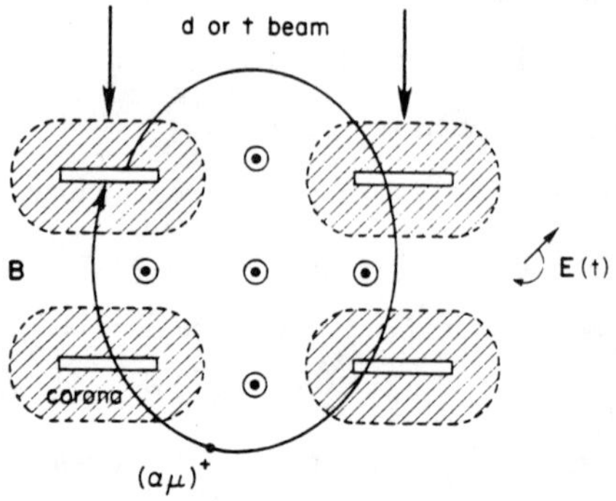

Figure 2: The array of DT ice ribbons and coronal plasma surrounded by the magnetic field.

When the beam is injected into these ribbons and fusion reactions begin, fusion alpha particles are created and all their energy is deposited in the heating of DT ice. In order to avoid the dilemma of 'cold fusion — no energy production' mentioned in the Introduction, the DT ribbons are heated up to temperature T_h which is much higher than the room temperature T_r so that we can now *extract* energy from the DT ribbon instead of investing energy to cool it. This condition is written as

$$\eta_h \Delta T_h - \Delta T_c/\eta_c > 0 \ , \tag{9}$$

where η_h and η_c are the thermodynamical efficiencies of the heat energy extraction from the hot material (T_h) into electricity and the cooling, and $\Delta T_h = T_h - T_r$ and $\Delta T_c = T_r - T_{\rm ice}$. From this T_h has to be larger than $1 - 2\,\text{eV}$ for $\eta_h \sim 0.3$ and $\eta_c \sim 0.1$. This will resolve the difficulty of Kulsrud's idea. Of course, this brings in a new difficulty of confining hot DT ribbons at nearly the solid density to cope with an enormous pressure.

In order to confine these DT ribbons, we introduce the concept of coronal confinement.

We surround each ribbon with a coat of a coronal plasma with density n_c and temperature T_c such that

$$n_h T_h = n_c T_c \ , \tag{10}$$

where n_h and T_h are the heated DT ribbon density ($\phi \sim 1$) and temperature ($\lesssim 1\,\text{eV}$). The thickness ℓ_c of the coronal plasma is

$$\ell_c \simeq \frac{1}{4} d_1 \ . \tag{11}$$

We choose the density of the corona such that the majority of $(\alpha\mu)^+$ of stripping does not take place there:

$$n_c \ll \frac{\ell_1}{\ell_c} n_h \qquad \text{and} \qquad T_c \gg \frac{\ell_c}{\ell_1} T_h \ . \tag{12}$$

Equation (12) can be written as

$$n_c \ll \frac{4\ell_1}{d_1} n_h = 4\sqrt{f}\, n_h \qquad \text{and} \qquad T_c \gg \frac{1}{4\sqrt{f}} T_h \ , \tag{13}$$

where the equality in Eq. (13) applies when the first term in the parenthesis in Eq. (8) applies. T_c typically is $10\,\text{eV}$. When the DT ribbon expands after alpha heating, the coronal pressure balances it, while the coronal pressure is now borne by the magnetic pressure outside of it. Thus, the corona is magnetically confined:

$$n_c T_c + \frac{B_c^2}{8\pi} = \frac{B^2}{8\pi} \ , \tag{14}$$

where B_c is the magnetic field penetrated into the corona and B is that of outside ($\sim$ vacuum). Details of the creation method of such a corona will be discussed elsewhere. Some possibilities include the alpha heating of the ribbon itself with or without a sponge-like surface, injection of very hot and very tenuous plasma to evaporate the surface, surface current application, laser surface ablation, etc.

The magnetic field tends to penetrate into the corona (and the ribbon) by the resistive process. The velocity of such penetration v_p is given by

$$v_p = \frac{\eta c \nabla P_c}{B^2} \sim \frac{\eta_0 c (n_h T_h)/\ell_c}{B^2 (T_c/T_0)^{3/2}} , \tag{15}$$

where η is the collisional resistivity of the corona and an explicit dependence of the coronal temperature is written in. This velocity should be sufficiently small such that

$$\ell_c / v_p > \tau_\mu . \tag{16}$$

This equation imposes a constraint on B, ℓ_c, T_h, and T_c. Besides the classical resistive penetration, there may arise anomalous penetration. A most prominent example is due to the Rayleigh-Taylor instability developing at the corona-vacuum interface, as the expanding corona decelerates due to the magnetic pressure. A simple estimate of the growth rate of the R-T instability is

$$\gamma \simeq \alpha \sqrt{g/\ell_c} \sim \alpha c_s / \ell_c , \tag{17}$$

where α is a geometrical factor ordinarily less than unity. There may be a possibility of developing an electric field and the coronal plasma penetrate through the magnetic field with $cE \times B/B^2$ drift velocity. In this case it may become necessary to short out the electric field.[15] Detailed considerations of these will be referred to later publications.[11] Finally, the collection of all these ribbons is confined inertially

$$\ell_3 / c_s \simeq a / c_s > \tau_\mu , \tag{18}$$

where a is the size of the pellet.

IV. Conclusion

We have presented a first plausible concept for fusion reactor via muon catalysis. It tries to reduce the effective sticking probability much less than 10^{-3} by the ion cyclotron resonance heating (or perhaps by the plasma stripping effects). The necessary structure of the DT fuel was elaborated. This particular structure allows the above operation, simultaneously keeping the necessary criteria for the mesomolecule formation processes. A number of parameters such as the strength of magnetic field, the filling factor etc., have yet to be optimized. It is entirely possible, however, that a comfortable (for physics) choice of parameters may lead to the requirement of a fairly large magnetic field of the order of 1 MG. Many anomalous plasma related processes may come into play. The rf power may be dissipated by extraneous resonances and other unwanted heating and thus the Q-value of rf source may not be as high as we wish. All these physical as well as technical problems can be formidable and certainly provide an immense challenge. The work is cut out for us. Nonetheless we are happy to point out that a major line of our concept has still withstood various conceivable difficulties in an essential way up till now.

One of the authors (T.T.) would like to thank Professor D. Baldwin for his interest in the project. The work was supported by the U.S. Department of Energy Contract DE-FG05-80ET-53088.

References

1. S.E. Jones, in this Proceedings.
2. C. Petitjean, in this Proceedings.
3. L.I. Menshikov and L.I. Ponomarev, JETP Lett. **46**, 312 (1987).
4. L.I. Menshikov, to be published in J. Muon Cat. Fusion.
5. S.S. Gerstein, *et al.*, Zh. Eksp. Teor. Fiz. **80**, 1690 (1981).
6. L. Bracci and G. Fiorentini, Nucl. Phys. **A364**, 383 (1981).
7. L. Bracci and G. Fiorentini, Nature **297**, 134 (1982).
8. R.M. Kulsrud, in this Proceedings.
9. R.J. Noer, Appl. Phys. **A28**, 1 (1982); J.W. Wang and G.A. Loew, IEEE Trans. Nucl. Sci. **NS-32**, 2915 (1985).
10. R.B. Palmer, in *Laser Acceleration of Particles* AIP Proc. **91**, ed. P.J. Channel (1982); C. Callan, C. Max, F. Perkins, and M. Rosenbluth, JASON preprint JSR-85-302 (1988).
11. T. Tajima, to be published.
12. S. Eliezer, T. Tajima, and M.N. Rosenbluth, Nucl. Fusion **27**, 527 (1987).
13. G.I. Budker and A. Skrinsky, Usp. Fiz. Nauk **124**, 561 (1978).
14. D. Möhl, G. Petrucci, L. Thorndahl, and S. Van der Meer, Phys. Rept. **58**, 73 (1980).
15. F.S. Felber, R.O. Hunter, N.R. Pereira, and T. Tajima, Appl. Phys. Lett. **41**, 705 (1982).

On Heavy Ion Collisions as a Muon Source

Lali Chatterjee and Dipak Ghosh

Physics Department
Jadavpur University
Calcutta - 700 032

S. D. Verma

Physics Department
Gujrat University
Ahmedabad

ABSTRACT

We explore the possibility of heavy ion collisions providing an efficient pion, and eventually muon source for research purposes and for ultimate utilisation of muon catalysed fusion.

Pion and muon beams are conventionally produced by proton impact on light nuclear targets. It has been demonstrated[1,2] that neutron rich projectiles and targets favour negative pion production so that they are suitable for π^-, μ^- beams. The usefulness of including tertiary pions by operating at slightly higher projectile energies has also been discussed.[3,4] Advantage of deuteron and triton beams and targets for efficient negative muon production has also been pointed out.

We investigate here the possibility of increasing efficiency in π^-, μ^- production by collecting pions showered out in heavy ion collisions. For a cyclotron-like machine one can accelerate heavy ions to higher total energies than lighter ones. For a machine of given radius, the gain in total energy is directly proportional to A. The question that then arises is how is our input energy related to increasing the A number to obtain a given energy E per nucleon. Even without access to a storage ring, the total energy obtained for given magnetic field B and potential diference V is proportional to q^2/A, and the energy per nucleon to q^2/A^2. one can conclude therefore that fully stripped ions with the larget (q^2/A) will be the most efficient to accelerate, as the total availabe energy for pionisation is largest.

In the ultimate analysis the cost of stripping to attain the highest q/A benefit will have to be optimised.

The number of pions produced per collision is then the relevant quantity to be compared as we increase, A number of projectile. In this context increasing A number of target will not increase costs except for individual complications of aquiring specific targets. As an example we display in Table I the mean shower multiplicities for some different A projectiles on different targets at projectile energies of 3.7 GeV/nucleon.

We must remember these are pions of both charge states and number of π^- will be half the number. However, as already discussed increasing neutron richness of projectile and target would favour π^- The possibility suggested in[5] that production of dense neutron rich matter in central collisions could produce additional pions and muons as their debris may provide an interesting advantage.

An empirical variation of $\langle n_\pi \rangle$ with A of projectile as $A^{.4}$ is observed experimentally from the data shown in Table I. QCD predicts a variation of $A^{1/3}$ in case of A-A collisions. Therefore increase of A of target along with projectile can give further enhancement of π^- yield.

It may be noted that 3.7 GeV/n Mg^{24} correspond to projectile protons of energy $\sim$ 14 GeV for which one knows the mean multiplicity to be $\sim$ 3. This also demonstrates an appreciable multiplication of the number of produced pions by using Mg^{24} instead of protons as projectiles.

One can thus conclude that significant gain in pion yield can be obtained by using a heavier projectile over conventional proton beams, colliding with heavy targets. An optimisation of the stripping cost against pion yield would determine the ideal combination. Such targets could perhaps be incorporated into the 'active' target fusion systems suggested in[4], and near thereshold pions would give slow muons directly usable for fusion.

ACKNOWLEDGEMENT

L.Chatterje wishes to thank University Grants Commission, India for financial support of the work, and the Department of Science and Technology, India for partial travel support to report the work at Muon Catalysed Fusion Workshop, Florida, USA, 1988.

TABLE - I

Interactions	A target	n_s
p-emulsion		1.63 ± 0.02
^{12}C-emulsion	A ~ 58	7.60 ± 0.13
^{24}Mg-emulsion		12.37 ± 0.22
p-emulsion		1.68 ± 0.03
^{12}C-emulsion	A ~ 14	4.21 ± 0.15
^{24}Mg-emulsion		9.20 ± 0.35
p-emulsion		1.55 ± 0.04
^{12}C-emulsion	A ~ 98	8.54 ± 0.16
^{24}Mg-emulsion		13.09 ± 0.25

REFERENCES

1. Yu. V. Petrov and Yu.M. Shabelski, Yad. Fiz. 30, 129 (1979); Sov. J. Nucl. Phys. 30, 66 (1979)
2. H. Takahashi et al, Atomkernenergie/Kerntechnik 36, 195 (1980)
3. L. Chatterjee, Fusion Technology 12, 444 (1987)
4. M. Jandel et. al. Phys. Rev. C 37, 403 (1988)
5. L. Chatterjee and J. Dey. Jadavpur University Preprint JU/NHEP/88-2

Chapter 6
Summary
and
Conclusions

FISSION: AN OBJECT LESSON FOR FUSION*

Alvin M. Weinberg

Introduction

The development of a new, and possibly hazardous, long-range energy source is beset with two political problems (as well as the many technical ones): survival and public acceptance. By survival I mean continuing support, year after year, of a very expensive enterprise whose promise always seems greater than its achievement: can this support continue long enough to allow the promised goal to be achieved. By public acceptance, I mean the reaction of the public to the new energy source, assuming that it achieves its technological goals. Both of these problems have been faced by fission power: I propose to describe the experiences of fission in confronting these issues in the hope that they might be dealt with more deftly by fusion. My account will be anecdotal and personal.

Fission and The Energy Crisis

When fission was discovered in 1938, the world hardly recognized that the energy problem would become the "moral equivalent of war." We were on the verge of World War II, and bombs, not power, were on our minds.

From Hahn and Strassman to Enrico Fermi's first reactor, hardly four years elapsed. (I still remember how skeptical I was of the whole matter until May of 1942 when the first subcritical experiments yielded a multiplication factor greater than unity.) Thus, nuclear power burst on the world in an incredibly short time.

One result of this flash of inventive genius--and luck--was to focus attention on the future of energy. In 1953, Palmer Putnam, in a study sponsored by the U.S. Atomic Energy Commission, argued that a prudent custodian of man's energy future would plan for a cumulative energy demand of 70,000 quads by 2050--(a quad being about 10^{18} joules)--several times

*Presented at Muon Catalyzed Fusion Workshop, Sanibel Island, Florida, May 2-6, 1988.

the estimated reserves of oil and gas. No wonder his report urged development of various alternatives to fossil fuel--particularly nuclear energy and solar.

Nuclear fission quickly became a panacea for the coming shortage of conventional fuel. Indeed, it was as though the discovery of an ultimate solution to the problem of energy made us aware of the problem! I can recall the extraordinary excitement I felt when, in 1943, Philip Morrison showed that the breeder reactor was really an infinite energy source. He had calculated that the energy derived from burning residual uranium in the granitic rocks comfortably exceeded the energy required to extract the uranium from the rocks. The breeder reactor was therefore a means to "A World Set Free," as H. G. Wells put it in his book of that name, in 1914. All of us in the fission community were exhilarated and euphoric: we had, at least in principle, solved mankind's energy problem for all time! What remained were a few engineering details!

Reality vs. Dreams

Alas, the world since 1943 has not developed in the simple, orderly way we had expected. Nuclear fission, let alone the breeder reactor, has not become the universal panacea we had expected it to be. What went wrong?

First was the remarkable diminution in the energy intensity of most of the world's industrial economies. Many energy analysts of the 1970s believed E/GNP was a constant of nature, and we expected large increases in energy demand on this basis. In the United States, the ratio of the yearly increase in energy to the yearly increase in GNP, $\Delta E/\Delta GNP$, fell from 1.4 between 1966 and 1970 to -0.2 from 1980 to 1985; and in 1986 we used the same amount of energy as in 1973! The figures for the rest of the world were much the same: the world war becoming more energy efficient. At the same time, the electricity to GNP ratio did seem to be constant--as GNP increased, so did electricity consumption. Since energy overall hardly increased, this implied that electricity accounted for an increasing fraction of total primary energy. In 1968, 18% of primary energy in the

U.S. was converted into electricity; today this number has increased to 36%, although the 7%/year increase in electricity fell to around 2%/year. Again, this shift to electricity seems to occur throughout the world. (All this suggests that Lenin's prediction: Communism = Socialism + Electrification is at least partly right; electrification is the wave of the future.)

Nevertheless, the euphoric projections of the 1960s for nuclear electricity expanding to about 700 GWe in the United States by 2000 never come to pass. Instead, about 120 reactors in the U.S. generate around 117 GWe; and in the entire world about 550 reactors will generate, when completed, around 400 GWe. The lower growth rate in electricity (2% instead of 7%/year) accounts for much of this lowered rate of installation of nuclear plants. But equally important has been the extraordinary public disaffection with things nuclear, especially after Three Mile Island and Chernobyl.

In retrospect, perhaps we nukes should have anticipated this turn of events. I recall Enrico Fermi, at one of the staff meetings at the wartime Metallurgical Laboratory saying that as significant as the introduction of a new source of energy--was the generation of huge amounts of radioactivity, accompanying large scale fission. He was not at all sure what the public's reaction to this totally new hazard would be. And a few years after the war, James Conant insisted that we would decide that nuclear energy "was not worth the candle" because we could find no satisfactory way of disposing of the wastes.

The public's disaffection has complicated the regulatory process--standards have continually been tightened, quality assurance has become a byword--and costs have risen exorbitantly. To be sure, much of the escalation has come from general price increases: but much of the increase is attributable to the aforementioned factors, which reflect the public's uneasiness about nuclear energy. At the moment, in the U.S., as well as other countries, there are no plans to build a new nuclear power plant.

Let me enlarge on the issue of reactor safety--and give you my personal recollections of the matter. That the 10^{10} curies of radioactivity contained in a 1000 MWe reactor posed a formidable engineering problem was recognized at the Metallurgical Laboratory from the outset. One of my first tasks at the Laboratory in 1942 was to help Edward Teller estimate the amount of C^{14} that might be released from an air-cooled graphite reactor, such as we built at Oak Ridge. But our ideas about reactor safety were primitive, as judged by the experience of today.

My main job was to help Eugene Wigner in he nuclear design of the Hanford reactors for production of plutonium. The design was dominated by the requirement that the system be chain-reacting--i.e., that the multiplication constant exceed unity. This was a rather close thing restricted as we were to using very scarce natural uranium, not entirely pure graphite, and neutron absorbing water (as coolant) and aluminum (as sheathing). Thus we chose lattice constants (i.e., ratio of graphite to uranium) which would minimize the amount of uranium needed for the reactor. That the resulting configuration was unstable against loss of coolant--i.e., that the reactor became prompt critical if all the water in the cooling tubes was voided, was never a design constraint--simply because we could do nothing about the positive void coefficient.

The DuPont Company, which actually built Hanford according to Wigner's design, provided an engineering fix for this instability. They introduced flow restricters or "gags" at the inlet of each pressure tube; the pressure drop across each gag was much higher than the pressure drop downstream in the tube itself. Were local boiling to start downstream of the gag, the overall flow through the tube would not be changed very much: the "boiling disease," as we called it, would not cause the entire tube to void, and thereby lead to melting of the fuel in that tube.

The simultaneous voiding of many channels, resulting in a prompt critical excursion, was never considered! It was not until the Hanford N reactor, which replaced the original reactors, that the nuclear design incorporated a negative, not a positive, void coefficient. But this required the use of slightly enriched uranium, something not available to us during the war! My main point is that considerations other than inherent safety dominated the early design of fission power reactors.

The pressurized water reactor, which now accounts for around 85% of all nuclear power, further illustrates the point that safety doctrine has evolved rather slowly. Eugene Wigner, even during the war, asked me to examine the nuclear properties of reactors moderated by various moderators: light water, heavy water, Be, BeO, as well as graphite. The first calculations on H_2O moderated lattice systems were made (by Bob Christy and myself) in 1942. (Wigner indeed proposed converting Pu^{239} into U^{233} in plate-type, water moderated reactors in 1943 when we first discovered that Pu^{239}, because of its spontaneous fission, could not be used in a gun-type weapon.) Thus the idea of water as a moderator was not foreign to us in Chicago; and the use of pressurized water for power was discussed in our group, especially with the late Forrest H. Murray. Our calculations on H_2O systems were first tested at Oak Ridge in 1944-45; we discovered that very little enrichment was needed to sustain a chain reaction moderated by water.

Enter Admiral, then Captain, Rickover--who spent 1946 at Oak Ridge. His mission--to build a nuclear submarine. The Oak Ridge suggestion: use pressurized H_2O as moderator and coolant (a proposal originally made to P.H. Abelson and R. Gunn of the Naval Research Laboratory)--simply because the moderation length in water is only around 6 cm, and therefore the reactor was compact--a requirement for a submarine.

Note that, as at Hanford, inherent safety was not a prime consideration--indeed, was not a consideration at all in our choice of PWR for the nuclear navy--which is surely justified for a military device where the crew would often be placed in far greater jeopardy than posed by a malfunctioning reactor.

That the LWR, originally proposed for submarines, came onto dry land, and now dominates, often seems miraculous to me. This feat is a triumph of engineering design and ingenuity. Not least is the extraordinary amount of safety that has been engineered into modern LWRs--so much so that the core melt probability of the world's existing fleet of LWRs is of the order of 10^{-4}/reactor year; and of the newer LWRs, such as Sizewell B in England, perhaps 20-fold lower than this.

These improvements in engineered safety of LWRs I regard as miraculous, particularly in view of the LWR originally being chosen not because it was safe, but because it was compact. Nevertheless, the Health and Safety Executive of the United Kingdom has recommended a limit of 100 LWRs in that country, on the ground that the overall core melt frequency (CMP) of between 10^{-3}/yr and 10^{-4}/yr is tolerable; but if the world builds, say 5000 reactors, and a priori CMP is 10^{-5}/yr, the core melt frequency becomes 5×10^{-2}/yr, at CMP=10^{-6}/yr, 5×10^{-3}/yr, which may be too high.

Are Reactors Safe Enough?

Are nuclear reactors of present type safe enough? Each of us must have his own answer to this. I believe this question can be, will be answered only by the public, not by the reactor engineering community. And, in several European countries, notably Sweden, Austria, possibly Italy, the public seems to have decided: No, reactors are not safe enough. The situation in the United States is unclear. Seabrook and Shoreham are built but not operating because of local opposition; and, especially after Chernobyl, a large minority of people object to nuclear reactors being placed near their homes.

Can nuclear energy be made acceptable by improved education of the public? I am doubtful. After all, the nuclear power community has been trying to educate the public about the virtues of fission power ever since 1945. Yet all the education has done little to persuade the public in the wake of Three Mile Island and Chernobyl. At least in the United States, I think we shall have to adopt a different strategy.

What I have in mind is the development of a new generation of fission reactors that is first of all acceptable to the articulate but skeptical elites who seem to have so much influence on the public's attitude toward nuclear energy. We must have systems that are not merely safe enough for those of us who believe in nuclear energy; we must have systems safe enough for those who are skeptical of nuclear energy.

Is this possible? Until 10 years ago I would have said no. The 10 billion curies of radioactivity and the 200 megawatts afterheat in a fission reactor made the idea of an inherently safe reactor all but impossible. The situation has changed drastically in the past few years.

Inherent safety, to greater or lesser degree, has now become a catchword within the reactor community.

Two approaches to inherent safety have emerged. The first approach is through incremental improvements in existing LWRs. This is represented by the Sizewell B reactor and by the Advanced BWR and LWR being planned for Japan. These systems incorporated many additional redundant safety systems; their designers expect the a priori core melt probability to be reduced as much as 100-fold. Most of this improvement is attributable to the addition of more active safety systems that intervene in case of malfunction; but several additional intrinsic or passive safety features are also incorporated.

The other approach is to design reactors whose safety depends not on timely actively interventions in the case of malfunction, but rather on inherent or passive characteristics of the reactor itself. In the wake of TMI-2, and particularly Chernobyl, at least 4 inherently safe reactors have been seriously proposed. Two are variants of pressurized water reactors--the Russian heat only integral boiling water reactor, and the Swedish Process Inherent Ultimately Safe Reactor (PIUS). One is a modular, metallic-fuelled liquid Metal Cooled Fast Reactor; and one is a modular High Temperature Gas Cooled Reactor.

The PIUS reactor's safety stems from its immersion in a huge concrete tank of borated water. The borated water is separated from the pure water which cools the reactor by two hydraulic locks, one at the top and the other at the bottom of the reactor. During normal operation, the hydraulic locks are kept shut by hydrostatic and hydrodynamic forces; any upset in the cooling regime--such as incipient boiling, breaks the locks allowing the borated water to flow into the reactor. The boron shuts off the chain reaction, and there is enough water in the concrete tank surrounding the reactor to remove the afterheat for a week or more.

The modular HTGR is a small helium-cooled, graphite moderated reactor. Its inherent safety depends on its uranium fuel being encapsulated in tiny spherical grains (about 500 microns diameter) covered by a pressure-tight coating of SiC and pyrolitic graphite. Thus each uranium grain is encased in its own pressure vessel; and tests demonstrate that these tiny pressure vessels retain fission products even at temperatures up to 1600°C. To keep the temperature below 1600°C. under all circumstances, the reactor is shaped like a long annular cylinder, and the thermal power is only ~300 megawatts. The combination of low power and high surface area allows the reactor to dissipate its after heat by passive natural convection without exceeding the 1600°C. temperature of the fuel even if all active cooling fails.

Of these inherently safe reactors, the Russian boiler is actually being built; and serious work is going on in the U.S., in the Soviet Union and in Germany on the mod-HTGR. The Swedish PIUS is being examined quite carefully by the Italian Ansaldo Company in collaboration with the Swedish-Swiss Company--Brown-Boveri-ASEA.

Lessons for Fusion

No one really knows whether the future of fission power will in fact require the development of inherently safe reactors. Should the world decide that greenhouse gases pose a more serious threat than the possibility of a malfunctioning reactor, fission in its present embodiment may once again be accepted by skeptical elites, and ultimately by the public. And of course, in some countries, notably France and Japan, not to speak of many Eastern European countries, fission seems to have survived TMI-2 and Chernobyl.

Nevertheless, as I look back on 50 years of fission, I am persuaded that we might have avoided the current malaise had we done two things differently: first, established inherent safety as a design criterion for power reactors at the very start; and second, sequestered our radioactive wastes permanently thirty years ago rather than leaving them in temporary storage. We did neither of these things fundamentally because we were in a hurry. From the discovery of fission to Fermi's pile, 4 years elapsed; from Fermi to Hanford, 2 years; from Hanford to Nautilus, 7 years; from Nautilus to Shippingport (60 megawatts), another 3 years; and from Shippingport to the 1000 megawatt monsters of today, about 10 years. Thus one of the things we have learned is that 30 years, which seems like a very long time, is not so long in the development of a new energy source: that our practice of scaling by factors of 10 and more before we have fully learned all the lessons on the smaller scale has caused us, especially in the United States, serious grief.

But in a way fission had no choice but to move very fast. Until the space program appeared, fission was the world's most expensive R&D enterprise. When several billions are spent each year on fission, there is enormous pressure to exhibit milestones that justify continued support of the enterprise. The fission enterprise was caught in its own euphoria of success after success--until TMI-2, Chernobyl, and the public's disaffection caught up with us.

No one can accuse fusion--whether magnetic, inertial, or muonic--of going too fast. On the contrary, I suspect that most of the fusion community must worry whether the rate of progress is sufficient to warrant the continued outlay of roughly 1-1/2 billion dollars worldwide each year.

That the energy crisis seems to have eased for the time being relieves the pressure for quick results on those seeking new energy sources. If the world continues to improve its energy efficiency, the time for replacing dwindling fossil fuels recedes. But this too adds to fusion's dilemma: if the time for replacing fossil fuel is more distant, how can one justify a massive effort today to develop alternatives for the next century?

This is the dilemma that besets the fission breeder (which after all is the direct competitor for fission as well as fusion). Our euphoric vision of a world set free by inexhaustible, autarkic fission breeders is hardly able to command the resources it did 20 years ago, when all of us expected energy to become scarce much more quickly than it has. Had the breeder turned out to be much cheaper than alternatives, economics alone would have guaranteed its support; but the breeder in present embodiment is not cheap, and economics alone cannot justify it.

For this problem of survival--how to justify vast sums for support of applied energy research that pays off far in the future, and with devices that at the moment appear to be expensive, fission has little to teach fusion--first, because fission power in its LWR version was developed so extraordinarily fast that in its first 30 years the milestones kept coming, as did the lavish support; and second, because the breeder finds itself in much the same boat as fusion. Though the breeder is inexhaustible energy, it is not cheap energy; the breeder suffers therefore from the same lack of urgency as does fusion itself.

My only advice, to both the breeder and the fusion community, is to somehow restore the sense of urgency in the quest for inexhaustible energy. We must, both of our communities, insist that 50 years passes quickly when a basic energy source is being developed; that the present seeming surfeit of energy simply cannot last, that the world continues to electrify, that long term energy sources are going to be needed; and that the greenhouse, and acid rain, can't wait. In short, that just as the exploration of space is now an accepted part of the world's long term research agenda, so the quest for non-fossil energy must also remain a part of man's agenda.

But in the matter of public acceptance, fission has much to teach fusion. Though fusion is much cleaner than fission, it is not devoid of radioactivity--100 million cuties of T^3 in a magnetically confined reactor, I don't know how much in a muon catalyzed reactor--and billions of curies in a fission-fusion hybrid. I am therefore much in favor of the consideration now going on toward developing inherently safe fusion reactors--reactors which because they are small and use materials that are only slightly activated by neutrons, are immune from melt-downs. I should think that a little thought also ought to be given to the control of muon catalyzed fusion reactors in view of the positive temperature coefficients of the muon multiplication--at least in some versions of these systems.

We fission people were extraordinarily lucky--our basic physics was almost trivially simple, and our technology went at a fantastic pace--so that by 1990 we will be providing almost 10% of the world's primary energy. But because we went so fast, we ignored Fermi's warning: we assumed that because we nukes accepted fission power, the public would accept it too. We are trying to fix things with Inherently Safe Reactors, and belated waste disposal--but whether we succeed in gaining back the public's confidence is surely a close thing. I hope this object lession from fission enables our fusion colleagues to avoid these mistakes in policy. It would be a tragedy indeed if fusion, so much more difficult technically than fission, eventually foundered not because it was technically impossible but because the public became disaffected with it, and governments stopped supporting its development.

THE CHALLENGES OF MUON CATALYZED FUSION

Johann Rafelski

Department of Physics, University of Arizona, TUCSON AZ 85721

ABSTRACT

This is a survey of new and significant developments which promise to shape the future of muon catalyzed fusion, based on presentations made at the 1988 MuCF workshop. The new ideas about muon production, muon sticking, regeneration and the muomolecular reactions represent steady progress towards understanding of the basic physics of MuCF and reaching towards 1,000 fusions per muon. I will attempt to elucidate the progress made in the theory of MuCF and to illustrate the practical importance of some of the issues raised. In many instances I will follow my personal viewpoint, justifying as much as possible the reasons behind my judgement.

THE FUNDAMENTAL CHALLENGE

The subject of Muon Catalyzed Fusion has been abandoned and resurrected already once: following the exciting discovery nearly 30 years ago many active particle physicists, including the current directors of the main particle physics Laboratories and Noble prize winners, Leon Lederman (USA-Fermilab) [1] and Carlo Rubbia (Europe-CERN) [2] have worked on MuCF and made important contributions. The interest soon subsided, as experiments found only a few fusions per muon. The renewed interest in this field is due to the path setting series of experiments carried out initially at the Los Alamos National Laboratory by S.E. Jones [3] and his close collaborators A.N. Anderson, A.J. Caffrey and M. Paciotti. Today we have 150 fusions per muon and are gathering here to find avenues leading perhaps to still many more fusions. Thank you Steve, Al, Gus and Mike for this accomplishment which has led us to gather here and have a most remarkable scientific meeting.

We have been listening in rapid sequence to presentations of results based on units GeV and meV. These very different energy units, spanning 12 orders of magnitude, involve practically all fields of modern physics which are rarely seen together. The unifying element bridging the gap between soft chemistry and hard particle processes in muon catalyzed fusion is an elementary particle akin to the electron but 207 times heavier, the muon, μ. We know little about the muon: like the electron it is electrically charged and hence subject to the electro-weak interactions. Once its mass is measured, its other properties follow, by assuming that it is otherwise identical to the electron. Its mean lifetime is 2.2 μs. The interaction which causes the muon to decay is also felt by the electron, but the latter does not decay as there is no charged particle lighter than the electron. The masses of the leptons, viz. electrons, muons, and tau particles are thought today to originate in the world spanning 'Higgs' field, whose vacuum expectation value is non-vanishing. This hypothesis has led to the resolution of the puzzle of the Weak Interactions and the correct prediction of the masses of the elementary particles (W and Z bosons) responsible for this interaction. The next challenge of fundamental particle research is probably the question of how and why the different leptons and their masses are generated. Obviously, any progress in this matter would have great impact on MuCF as well.

In principle, the mass of the muon is relatively small, 105.652 MeV, and five or six nuclear fusions would suffice to generate the energy equivalent needed to produce a muon. Each d-t fusion reaction releases 17.6 MeV energy, 80% being carried by the fusion neutron. The energy yield is further enhanced by the secondary reactions required to breed the needed tritium and it can be assumed that the effective thermal energy yield per fusion is $\sim$ 22 MeV, just 1/5 of the muon mass. However, unless we made progress in the understanding of the 'mass' we cannot hope to create muons so efficiently.

THE MUON PRODUCTION CHALLENGE

For the present, we shall only consider off-the-shelf technologies when considering the energy cost of muon production. Muons need to be formed in large quantities since even if each muon were to catalyze 1000 fusions, the thermal power produced would be 1.5 10^{-3}W per muon during its lifetime. Said differently, in order to generate the continuous thermal power of say 1.5 MW, we need a steady presence of 10^9 muons, that is a flux of $10^9/2.2 \times 10^{-6}$/s=70nA muons. Such muon flux intensities can probably be only reached in nuclear beam collisions with fixed targets. We cannot hope to make muons for less than about 1.5 GeV in beam power per muon, since in all practical approaches we will be making nearly equal number of the three pions: π^+, π^o, π^- and only the last decays into the negatively charged muons we need. Working at near the delta resonance energy of the nucleon, i.e. 300 MeV above the nucleon rest mass in its frame of reference, we maximize the ratio of inelastic to elastic cross section, such that losses to elastic 'nuclear' collisions are small, perhaps 10% of the energy on average. Similarly, choosing a target of relatively moderate Z, we assure that the stopping power losses are of the same magnitude. Clearly, we need to create a system with essentially negligible other losses, in order to be close to this very optimistic assessment of muon cost. There are two clear kinds of losses: firstly, the projectile may be lost before making the desired interaction in the target, and secondly, the produced pion may undergo an undesirable interaction in the target, before it decays into a muon.

Even with this, as we will see, very optimistic muon cost assumption of 1.5 GeV and the equally very optimistic, if not unreachable, muon yield assumption of 1000 fusions, our system would effectively multiply the beam power by only a factor 15 with the lower 'grade' thermal energy being generated. Most importantly, these figures suggest that we need a highly efficient means of converting thermal energy into the beam power, if MuCF were to become a viable energy option. In all these issues we clearly have to stretch our knowledge of accelerator technology to the limits.

Presently, the best studied systems of muon production are based on the 'active target' principle in which the fusion vessel serves at the same time as the target for a deuteron beam [4], [5]. This design suffers from the large absorption losses of stopped pions. In order to circumvent this problem, one would need to reduce the active target

D-T density to values too low to facilitate many catalyzed fusions during the muon lifetime. Hence in my view we must now turn our attention to more sophisticated systems comprising a separate target area with carefully selected combination of the projectile and target material. Still, the calculations of Magnus Jändel [4] establish upper limit of 6 GeV as a "realistic" cost in energy for production of usable muons in active targets, although the optimum system with separate target and fusion areas may be significantly more efficient, as implied above.

Considering this discussion of muon production our objective today should be the attainment of at least 1000 fusions per muon, if applications of muon catalyzed fusion as sole source of energy production is to be seriously contemplated. It is worthwhile to keep in mind that in a hybrid reactor, in which the fusion neutrons are driving a fission mantle leading to an additional energy yield of 200 MeV per neutron, thus effectively increasing the energy yield by a factor 10, the currently achieved fusion yield of more than 100 fusions per muons suffices, as was first observed by Yuri Petrov [6].

The muon production challenge can be summarized as follows: ***how does one produce μA fluxes of muons at an energy cost which is not greater than few GeV beam power per muon.***

Before proceeding with further discussion, it is interesting to pose some 'off the record' questions. How would a small MuCF pilot system with a flux of 10^{13} muons per second influence our understanding of the muon itself? Current experiments investigating muons can observe only a sample of about 10^{13} muons during one year of operation. Could they be then repeated in about one second? Should MuCF experiments have the highest priority in order to be able to move to such intense muon beams sooner?

THE NUCLEAR CHALLENGE

Nuclear physics presents itself with an equally overwhelming challenge related to the proper understanding of the fate of the muon during the nuclear fusion reaction. In particular the problem of sticking of the muon, that is of muon capture by the fusion product, the α particle, turns out to be on the forefront of nuclear reaction theory. In order to emphasize the importance of sticking, note that the fusion yield Y per muon, i.e. the number of fusions (Y) each muon can participate in, is limited by the ratio of the average cycle rate λ_c of the muon and the effective muon disappearance rate λ_μ^{eff} :

$$Y = \lambda_c/\lambda_\mu^{eff}, \tag{1}$$

where:

$$\lambda_\mu^{eff} = \lambda_\mu + \mathrm{w}_s\lambda_c. \tag{2}$$

Here $\lambda_\mu = \tau_\mu^{-1}$ is the rate of muon decay in the catalytic cycle and w_s is the mean probability per muon cycle to immobilize the muon whatever the cause. This implies that Y satisfies

$$Y < (\mathrm{w}_s)^{-1} < (\omega_s^{eff})^{-1}. \tag{3}$$

The first inequality becomes equality when $\lambda_c \gg \lambda_\mu$, which is true in favorable circumstances. The last inequality is a consequence of the fact that a priori, the quantum mechanical muon loss probability ω_s^{eff} is a lower limit on w_s. Here

$$\omega_s^{eff} = (1 - R)\omega_s. \tag{4}$$

R is the regeneration probability, which we will discuss in the next section, while ω_s is the intrinsic sticking probability. ω_s is a branching ratio of the two-body decay of $dt\mu$ to all possible decays:

$$\omega_s = \frac{dt\mu \to \alpha\mu + n}{dt\mu \to (\alpha + n + \mu), (\alpha\mu + n)}. \tag{5}$$

We begin our discussion of the sticking ω_s by recalling that the muomolecules are quite stable in the sense that the natural oscillation frequency of the muomolecule is 100 000 times faster than the (perturbative) fusion rate which is usually quoted around one meV [7]. This implies that in a *perturbative* treatment of the influence of strong interactions on the muomolecular structure only negligible influence can be found on muon sticking, as we have seen in the lecture by Gerry Hale [8]. Of course, this result is implicitly inherent to the perturbative approach. If, however, we do not presuppose the validity of the perturbative treatment of the nuclear effects a different picture emerges, as presented in the lectures by Mike Danos [9] and Larry Biedenharn [10].

Why need we proceed beyond the perturbative description? We know that muon sticking arises from a small fraction of the total amplitude; when computing the sticking amplitude, we pose the conditional question of what the muon is doing while the nuclei are close together. It does not matter at all that this condition is very rarely satisfied, indeed perhaps only once in every 100,000 molecular "oscillations". But once this condition is fulfilled, viz. the nuclei are close together, it is clear that what the muon is then doing is subject to the detailed behavior of the nuclei. The logic of this argument is such that it can be turned around: it is very fortunate that the strength of the nuclear interaction is washed out in the muomolecule by 100,000 molecular oscillations, assuring that we have a molecular structure despite the strength of the nuclear interaction. But this says next to nothing about the response of molecular components to the nuclear interaction when the nuclei in the molecule coalesce.

Thus we learn that it is entirely improper to use classical physics arguments in order to discount the influence of the nuclear interaction on the behavior of the muon. Classically, the muon can be in only one place and hence on these grounds there is only rarely any chance, perhaps 1 in 100,000, of influencing the system. This argument applies only to the semi-classical components of the molecular amplitude, but not to the typically small, truly quantum components, *i.e.* those responsible for muon sticking. One has here to realize that sticking occurs, as experiments show, in about 0.5% of all reactions: thus even the total elimination of the sticking channel from the nuclear reaction would reduce the total nuclear reaction rate by only this "perturbative" amount, and nobody will argue that such minute change in the total reaction rate due to interference of Coulomb and strong interaction effects is impossible.

The proper treatment of this problem requires a nonperturbative consideration of the d+t nuclear interaction [9]. First recall that the fusion reaction proceeds via a well-defined intermediate nuclear state with well defined quantum numbers, the strong ^{5}He$(3/2)^+$ resonance, just 50 keV above the energy threshold. This resonance filters the quantum numbers of the reaction, as it couples only to select parity and angular momentum states of the entrance and exit channel. Because of its small width of 70 keV it is also an energy filter. This is one of the important differences with respect to the *dd* fusion system, the other being the presence of a strong dipole in the *dt* case, as the center of mass and center of charge differ significantly here. Finally, the velocity of the ^{3}Heμ in dd fusion is substantially less, the reaction thus probes more classical components of the three body $dd\mu$ wave function.

THE R-MATRIX CHALLENGE

Returning now to the discussion of the $dt\mu$ system we note the remarkable coincidence that the nuclear resonance parameters are of the same order of magnitude as the binding energy of the muon to the compound ^{5}He system, 10 keV, while the kinetic energy of the muon, should it be attached to the fusion product, the α particle, is 86 keV, just the width of the intermediate nuclear state. This suggests that the energy dependence of the nuclear interaction will interfere with the muonic system, requiring an unambiguous treatment using Wigner's R matrix approach, generalized to involve the third 'Coulombic' particle with its long range interaction, the muon.

We now try to assess the likelihood that the nonperturbative treatment will lead to a revision of the sticking fraction. When considering this question it is helpful to realize that the muomolecular state $dt\mu$, which decays in the fusion reaction either into a two- or three-body final state, is actually itself a resonance, with a meV width, as determined by perturbative calculations of the rate of fusion within the s-states in the molecule. Thus considering the inverse process of neutron scattering with the $(\alpha\mu)^+$ ion, we see an elastic resonance of such a width at the appropriate energy. This consideration shows that the coupling of the muomolecular $dt\mu$ wave function component to its continuation in the $n\alpha\mu$ part of the space must involve, as function of the eigenstate energy, a rapidly changing phase. Thus in order to obtain the proper eigenstate of the system reaching into both configuration space domains one has to consider how the state has been prepared.

Because of the temperature sensitivity of the muomolecular formation rate it is believed that the fusing state is formed by the Auger transition from the $J=1$ doorway muo-electro resonance. This near threshold muonic state is subject to a greatly diminished nuclear interaction, which can be disregarded in this discussion. The Auger process thus prepares a particular distribution of the fusing state within its energy width of about 1 meV. In principle, this energy is observable by recording the energy of the Auger electron. In order to compute the distribution of the Auger transitions in energy, we must consider what the fusing state looks like for large r_μ, R_{dt}; only this part of the configuration space is relevant for the Auger transition from the loosely bound doorway resonance. As in this region the Coulomb $1/r, 1/R$ interaction is negligible in the sense that the form of the wave function is totally described once the energy is known, we have:

$$\psi^E \to N \int e^{-P_r r} e^{-P_R R} f^E(P_r, P_R) dP_r dP_R \tag{6}$$

where the weight f is constrained by the Schrödinger energy condition:

$$\frac{P_r^2}{2\mu} + \frac{P_R^2}{2M_{dt}} = -E\ . \tag{7}$$

The remaining ambiguity is associated with one continuous parameter, say φ, and it is convenient to write:

$$P_r = \sqrt{2|E|\mu}\cos\varphi$$

$$P_R = \sqrt{2|E|M_{dt}} \sin\varphi \ . \tag{8}$$

Thus, the asymptotic form of the wave function is simply

$$\psi^E \underset{\substack{r\to\infty \\ R\to\infty}}{\longrightarrow} N \int_0^{\pi/2} e^{-\sqrt{2|E|\mu} r \cos\varphi} e^{-\sqrt{2|E|M_{dt}} R \sin\varphi} g^E(\varphi) d\varphi \ . \tag{9}$$

The weight function g is controlled in part by the interaction (near) the origin, and in part by the (Auger) process which prepares the state. The index E indicates that we are dealing with a continuum wave function and hence the weight g changes rapidly within the width of 1 meV, as already argued. (There would be only one particular g if we were considering the Coulombic system, as the wave function of a given discrete state with a well behaved potential is unique, independent of the way the state is formed.) The Auger transition amplitude leading to the formation of the fusing state is governed by

$$\overline{\Omega} \equiv \langle f|\Omega|i\rangle = \int g^E(\varphi)\chi^E(\varphi)d\varphi \tag{10}$$

where χ^E, the Auger matrix element of the (1,1) muon with the exponential shown in eq. (9) is in principle a known function. The electron part of the Auger matrix element is not dependent on the energy E within the interval of meV. Hence the states of different E (within meV out of 34 eV) are populated according to the energy dependence given by the square of eq. (10), which contains the undetermined function $g^E(\varphi)$. To determine this function very complex calculations are needed, as has been elucidated in this meeting [9],[10].

This effort can now be easily motivated, noting that within the narrow range of energies we are interested in, g is a peaked function in comparison to a nearly flat χ. Approximating g by a Lorentz form and χ by a linear function in the vicinity of the right energy, we can easily investigate how far the energy structure in the Auger transition shifts the peak of the distribution away from the peak of the quasi-stationary $dt\mu$ pole: the quantity to consider, also on dimensional grounds, turns out to be the width parameter of the peaked distribution g. Thus the result is that the Auger transition prepares the fusing state some place within the perturbative width of the 'stationary' state. But within this width, as we have argued before, sticking varies considerably. Of course the distribution of the Auger amplitude in energy will average in the end over the energies smoothing out the largest effects. Even so, we can expect that sticking will be changed by a significant amount.

At this point it is necessary to recall that the study of nuclear effects on sticking has been a consequence of the observation that the muon sticking is about 35% less than the best "conventional" wisdom Coulombic value, as can be seen comparing the experimental results [3] with the calculations of Kamimura [7]. The current 'conventional wisdom' best theoretical value of the initial sticking coefficient ω_s as given in [7] is 0.93%. This value will be further reduced by regeneration. At highest available densities, ϕ =1.2 the value of sticking is 0.65%, while the average of all experiments is around 0.42%, with possible further reduction due to the improved dd corrections (see Muomolecular Challenge section) to about 0.39%. Thus at high density there seems to be a very substantial disagreement between theory and experiment on sticking, while at low densities it is difficult to make a similar assessment due to much greater error bars. As we can now better appreciate, proper understanding of such a small and seemingly insignificant, though most important effect requires a very substantial theoretical and computational effort. The nuclear R-Matrix challenge thus can be summarized as follows: ***compute non-perturbatively the behavior of a three body system undergoing nuclear reactions and involving the long range Coulomb interaction.***

CAN ONE INFLUENCE NUCLEAR BRANCHING RATIOS?

But we are not chasing here the last 10% effects: if the mechanisms behind the unexpected reduction of sticking are understood, there is potentially the possibility to conceive of ways to take advantage of the phenomena at play here. Normally, one would argue that it is next to impossible to influence the branching ratio of a nuclear reaction, which the sticking coefficient indeed is. But we just saw the sensitivity of this branching ratio to the minute, meV structure of the resonant state. This opens up the possibility, first emphasized by Larry Biedenharn [11], that we could conceivably alter the nuclear branching rates by macroscopic means, such as by varying the atomic density or by imposing an external magnetic field. The objective is to alter the density of final states involving in particular the fusion accompanying Auger electrons. So far we have not investigated the electron part of the Auger rate, assuming that the matrix element would be quasi constant over the meV energy range of interest here. Let us now see if it is possible to change, on the scale of meV, the density of states of the outgoing electrons by application of magnetic field of several Tesla. Indeed, the unit of magnetic interaction energy is:

$$\hbar\omega_c = 2\mu_e\hbar B = 0.116 \times B[\mathrm{T}]\ \mathrm{meV}\ . \tag{11}$$

Thus even rather 'conventional' 35 kGauss (3.5 T) magnetic fields have an interaction energy of 0.42 meV with the electrons. Does this imply that the Auger electrons have a structured final density of states and hence with preference populate certain energies, with the fine structure of this effect being, just as required, a good fraction of meV? To find an answer I must consider the ratio of the density of states with magnetic field, to that without the field. Considering the motion of electrons in a magnetic field chosen for simplicity to point along the z–axis we first recall the (non–relativistic) Landau orbital eigenenergies:

$$E(k_z, B_z) = \frac{\hbar^2 k_z^2}{2m_e} + (l + \frac{1}{2})\hbar\omega_c. \tag{12}$$

The total number of states up to a given energy E is obtained by integrating over all allowed k and summing over all permitted values of l up to maximum value L such that the quantity

$$\bar{k}_z = \sqrt{\frac{2m}{\hbar}}\sqrt{E - (L + \frac{1}{2})\hbar\omega_c} \tag{13}$$

remains real:

$$N(E) = 2G\sum_{l=0}^{L}\int_0^{\bar{k}_z} \frac{\hbar}{\sqrt{2m}} dk_z, \tag{14}$$

where G counts the number of phase space bins in the volume occupied by the magnetic field:

$$G = \frac{eBV}{(2\pi\hbar)^3}. \tag{15}$$

The differential of $N(E)$ gives us the desired density of states as function of E. Dividing out the usual phase space density I obtain:

$$\frac{(dN/dE)}{(dN_c/dE)} = \frac{\hbar\mu_e B}{E}\sum_{l=0}^{L}(1-(2l+1)\mu_e\hbar B/E)^{-\frac{1}{2}}, \tag{16}$$

where L is the maximum value which leaves the root real. Clearly, exactly when

$$E' = (2L+1)\mu_e\hbar B = (L+\frac{1}{2})\ \ 0.116\ \ B[\mathrm{T}]\ \ \mathrm{meV}, \tag{17}$$

we encounter a root singularity in the normalized density of states. For L not too large, the width of this singularity is similar to the magnetic energy and this new structure comprises a good fraction of the total reaction strength. But in our case, with the energy E' being of the order of 34 eV the value of L for $B = 3$ T is of the order 100,000. This implies that the peak in the distribution is very sharp, with a width 1/100,000 of the magnetic energy.

Thus the initial answer is yes, we find the required structure in the electron density of states in that a series of sharp peaks separated by O(meV) has been identified in the density of states. However, the integrated strength within the peak is greatly reduced by the fact that we are dealing with "100,000's" reoccurrence. Thus I do not expect a more refined calculation to reveal a great influence on sticking, unless perhaps the fusion were to involve a hyper-fine transition $(O(100meV))$ or proceed directly from the (1,1) resonant state, a highly unlikely possibility for the $dt\mu$ case given the present state of the understanding of MuCF. However, in the $dd\mu$ case we indeed may have direct fusions from the (1,1) resonant state and the magnetic effects could play a greater role here.

THE REGENERATION CHALLENGE

Once the muon sticks to the α–particle, it is not entirely lost from the cycle of reactions: at the initial velocity of about 6 a.u. it carries about 90 keV kinetic energy which is significantly greater than the energy needed, 11 keV, to strip it from the α–particle. Even more importantly, it takes many atomic collisions before the $\alpha\mu^+$–ion looses its 3.5 MeV kinetic energy. It is easy to see that muon stripping processes will occur in nearly 30% of all sticking reactions, because of the competition between the energy loss and stripping processes. Both these processes are governed by similar atomic cross sections, the main difference being that energy loss (stopping power) involves the atomic scale associated with the electron (a_e), while the stripping of the muon is governed by the muonic atomic scale (a_μ) which is 200 times smaller. Because the energy loss in atomic collisions is $O(10eV)$, which happens just to be one part in 200^2 of the energy carried by the $\alpha\mu^+$–ion, there is indeed a competition between these processes.

The time it takes for the $\alpha\mu^+$ occasionally created in $dt\mu$ fusion to be slowed down from its initial velocity $v_{(\alpha\mu)} = 5.82$ to $v_{(\alpha\mu)} = 0$, depends on the stopping power $S(v)$ of the target. This quantity is defined by the rate of energy loss:

$$\frac{dE}{dt} = -\rho v S(v). \tag{18}$$

Here v is the velocity of the $\alpha\mu^+$–ion. Integrating we find the slowing down time t_{stop}:

$$t_{stop} = \int_0^{E_{in}} \frac{dE}{\rho v S(v)}. \tag{19}$$

E_{in} is the initial energy and we integrate to near $E = 0$ where no stripping can occur anymore. Stopping time is inversely proportional to the density ρ, usually expressed in units of the liquid hydrogen density ρ_0:

$$\rho_0 = 4.25 \times 10^{22}\text{cm}^{-3},$$

$$\rho = \phi\rho_0. \tag{20}$$

This slowing down time is $t_{stop} \sim 4 \times 10^{-11}$s, if the stopping power of the $\alpha\mu^+$ is obtained from the scaled proton stopping power. Thus even for $\phi = 10^{-2}$ the time 4 ns needed for the muon stripping during the slowing down of the $\alpha\mu^+$–ion is of no particular significance in MuCF cycle dynamics, as it will occur in fewer than one in two hundred reactions.

In order to understand the general expressions describing the muon regeneration, it is helpful to consider first a simplified problem, assuming that the muon can only be stripped from the $\alpha\mu^+$–ion by a one step excitation processes. The cross section for this will be denoted $\sigma_{str}(E)$ and comprises both ionization and transfer components. The rate for the stripping process to occur is:

$$\lambda_{str}(E) = \sigma_{str}\rho v. \tag{21}$$

The 1s-state population P_0 of the $\alpha\mu^+$–ion thus satisfies the equation

$$\frac{dP_0}{dt} = -\lambda_{str}(E(t))P_0 \tag{22}$$

with the solution

$$P_0(t = t_{stop}) = \exp(-\int_{t=0}^{t_{stop}} \lambda_{str} dt),$$

$$P_0(t = t_{stop}) = \exp(-\int_{E_{in}}^{0} \lambda_{str}\frac{dt}{dE}dE). \tag{23}$$

Recalling eq.(19) the final population is

$$P_0(t_{stop}) = \exp(-\int_0^{E_{in}} \frac{\sigma_{str}}{S}dE), \tag{24}$$

or simply

$$P_0 = 1 - R_0 = \exp(-I),$$

where

$$I = \int_0^{E_{in}} \frac{\sigma_{str}}{S} dE. \tag{25}$$

The regeneration exponent I invokes the ratio of the ***muon*** stripping cross section σ_{str} to the ***electron*** ionization cross section weighted with the mean energy loss $\delta E \sim O(20eV)$. The simplest estimate of the exponent thus is

$$I \sim (m_e/m_\mu)^2 \times E_{in}/\delta E \sim 0.4.$$

Hence a regeneration at the level of $R_0 \sim 0.3$ can be expected.

The actual stripping processes are Coulomb ionization on target nuclei, d or t, either in a single step or by multi–step processes as internal Coulomb excitations; e.g. $1s \rightarrow 2p$ may be followed by either transfer to a hydrogenic bound state of the target (d or t) or Coulomb ionization. Note further that once the muon is in an excited state, be it due to Coulomb excitation or due to initial sticking, the variouse de-excitation processes (radiative, Auger and Coulomb induced de-excitation) compete with Coulomb stripping. This then makes the regeneration probability R density-dependent. To treat the full problem, the population equation has to be generalized to incorporate the individual population of all $\alpha\mu^+$ bound states. Such calculations generally can not explain the experimentally observed density dependence [3]. However, in the calculation of the stripping probability the essential input is the detailed stopping power [12] of the medium acting on the $\alpha\mu^+$–ion. One finds that a density-dependent reduction of the stopping power could explain the density dependence of the effective sticking without conflicts with other experimental data. However, then the X-ray transitions in the $(He\mu)^+$–ions following fusion in $dd\mu$ and $dt\mu$ become even more different from the current theoretical expectations. Thus the conclusion of the regeneration studies carried out by many groups is that it is impossible to quantitatively describe the density dependence of sticking observed so far. The results of Berndt Müller and Helga Rafelski [12] further show that it is impossible to explain the experimental results even when the stopping power is made strongly density dependent. That is, while the comparison with the $dt\mu$ sticking data improves, there is a substantial discrepancy in the X-ray yields observed in the $dd\mu$ and $dt\mu$ fusion.

An interesting aspect of the regeneration phenomenon is the possibility of exploiting it to prolong the active lifetime of the muon. One could conceive of a mechanism to replenish the energy of the $\alpha\mu^+$–ion during the time the muon is attached to the α particle [13], [14]. We need to supply to the $\alpha\mu^+$–ion energy at a rate which is nearly equal or faster than the energy loss due to the stopping power. In effect, even at $\phi = 0.01$ the rate of energy supply must be 5 MeV in 4 ns, i.e one eV in 10^{-15}s, a surely difficult if not impossible proposition.

Instead of supplying the energy to the $\alpha\mu^+$–ion we could equally change the environment in which muon catalyzed fusion takes place in order to take advantage of a possibly reduced stopping power. In his report of the work done recently by L.I. Menshikov [15], Leonid Ponomarev suggested that in a low temperature plasma of about 15 eV we could expect that a sufficient reduction of the stopping power would result leading to practically total muon stripping following on dtμ fusion. Even though there are many problems with the plasma environment for MuCF in general, as many of the chemical resonant processes are lost and new mechanisms will therefore have to be (hopefully) identified, the main problem with this beautiful idea is that more careful theoretical analysis of the initial calculations [16] reveals that any reduction of the stopping power should not be expected below $T = 300$ eV. Even at this energy, when essentially all electrons are ionized, we should not consider it a sure bet that the stopping power will be reduced, as the Coulomb coupling of electron and ion densities can lead to major collective dissipation phenomena. This is not to mention that at these temperatures, and at the needed densities of about $10^{22}/\text{cm}^3$, we are indeed considering rather extreme plasma environments, which are probably more conducive to thermonuclear fusion without the need for the catalysis!

The challenge of muon regeneration is to find a path to take advantage of the high kinetic energy available to strip the muon from the $\alpha\mu^+$*−ion.*

THE MUOMOLECULAR CHALLENGE

The need to catalyze 1000 fusions during the short life span of the muon creates the need for resonant reactions. The practicability of MuCF thus depends on the resonance condition between the weakly bound $dt\mu$ muomolecule and the particular rotational-vibrational excitation in the host hydrogenic (electro) molecule. This makes a thorough understanding of soft chemical processes in dense hydrogen necessary, along with understanding of the few-body system at the level of 1 part in 2.8 keV/0.28 meV = 10 000 000, a really challenging problem of quantum chemistry. In several lectures [17], [18], [7] the highly sophisticated variational quantum chemistry calculations carried out in US and Japan were presented. The precision reached permits us for the first time to consider *abinitio* calculation of the chemical reaction rates governing the muon cycle in the dense mixture of hydrogen isotopes. In this way bottlenecks in the muomolecular reaction chain can be identified. The best Coulomb value of the (1,1) resonant energy was given as −660.26 meV. However, this value does not include the different corrections, such as relativistic effects, QED effects (Vacuum Polarization), and the effects of the host electro-molecule.

Steve Jones [3] summarized in his lecture the surprising results found at Los Alamos about the fusion yield Y. At the present time there is no indication that there is an approach to a possible saturation of the fusion yield as function of the hydrogen density. This goes together with the beautiful nonlinear effects found by Steve [3] in the rate of muomolecule formation, which were discussed from differing perspectives by Jim Cohen [19] and Yuri Petrov [20], as presented by Leonid Ponomarev. What is very surprising is that there seem to be no adverse surprises concerning the losses of the muon, such as due to inevitable side cycles of fusion reactions, scavenging by helium etc. As visible in the fusion yield, high density favors fusions, and even seems to help reduce sticking as already discussed. At the present time it is premature to argue what the origin of these effects is, as long as we do not even understand properly the reasons behind the anomalously high fusion yields observed: clearly all loss mechanisms should reduce, not enhance the observed fusion yield, perhaps this is what is happening at low densities and we are only catching up at high densities with the true muon catalysis rates and losses, which are much better than we originally have expected.

Most notable at the meeting has been the recognition of the importance of the understanding of the dynamics governing the $1s$ population of the μd atom, commonly called Q_{1s}. Not only is it the worst bottleneck in the muon cycle reaction chain as we understand it, but it also profoundly influences the global data analysis [21]. The value of Q_{1s} was predicted by Ponomarev [22] to be rather small, as transfer processes to tritium originating in excited μd states should dominate the d–muonic cascade. But the presence of this bottleneck, viz. large values of Q_{1s}, makes a reevaluation of these muon transfer calculations necessary. Indeed, it has been pointed out by Müller [23] that one can easily reach agreement with experiment, if there is inhibition of the calculated transfer rates between muoatomic M–shells of D and T. All this implies that the theoretically expected rate of muon transfer from deuterium to tritium is reduced, in particular at high densities, thus making the corrections due to occasional $dd\mu$ formation and subsequent fusion reactions greater than theoretically expected. In the Los Alamos data analysis, this correction is made on the basis of experimental data, while the SIN/PSI analysis relied previously on the unproven theory [22]. In consequence, the SIN/PSI results on sticking at high density seem to require a further correction downward, reducing the discrepancy between experiments and increasing the difference with the conventional theory described in Steve's lecture [3].

The meeting brought to us the first results on a significant new research direction, aiming to substitute the need for high densities in muomolecular formation by powerful lasers; several effects are being identified [24], [25], [26] with the common aim of creating the resonance condition essentially exactly, the laser beam of photons serving either as a 'third' energy balancing body or as means of altering the density of states for the processes. It is too early to draw final conclusions about these efforts, but their success would alter our approaches to possible practical exploitation of MuCF, pointing perhaps towards comparatively low density (fraction of LHD, 4.25×10^{22} atoms/cc).

Another most interesting and new development concerns the study of the three body $dt\mu$ scattering resonances near to the thresholds $d+t\mu$ and $t+d\mu$, by the Florida-Sweden-Poland collaboration [27]. The results are highly encouraging, and indicate the possible existence of quasi bound molecular states near the $d+t\mu$ threshold. Presence of such resonances could facilitate MuCF processes at high temperatures, which could arise locally in the fusion target. One should note at this point that such "runaway" effects all terminate within $2.2\mu s$ with the muon decay.

This quantum chemistry challenge of muon catalyzed fusion can thus easily be put into very few words: *how to squeeze 1000 fusions into the short lifetime of the muon at possibly low density.*

CONCLUDING REMARKS

A particularly attractive feature of MuCF is the possibility of achieving a functional system in an STP (standard room, temperature and pressure) environment. The plasma approach to MuCF, as we have discovered from this workshop, is a possibility which differs little from technologically challenging thermonuclear plasma fusion projects. In the more conventional line of approach to MuCF we have seen a number of projects and developments ripening in the past year, and our general understanding of the fundamental MuCF processes has been steadily increasing, being far from complete at the present time.

Even though we have come a long way since the first MuCF workshop Steve Jones organized in Jackson Hole in 1984, with subsequent meetings at Tokyo 1986 and Leningrad 1987, we can not as yet specify the key bottlenecks in the MuCF cycle limiting the number of fusion reactions, nor have we resolved the difference in the experimental results on sticking which began emerging about four years ago. While in the past 4 years the calculations we dreamed of in Jackson Hole were performed, new and more complex needs have arisen, leaving us with even more questions open than before. We must yet understand the nuclear influence on sticking, the highly nonlinear nature of the muomolecular formation along with the questions pertinent to the Q_{1s} population of the deuteron groundstate, the 'density dependent' muon sticking, the regeneration and stopping power, and last if not indeed first, ways of efficiently making the large number of needed muons.

Acknowledgement: Throughout my involvement with MuCF as well as in deep discussions of the problem of muon sticking I have greatly benefited from numerous stimulating conversations with M. Danos. The frequent discussion of experimental results with Steve Jones have been instrumental in my understanding of the many different MuCF processes. I thank Helga Rafelski for her continuous interest and collaboration.

I would like to particularly acknowledge continuous interest and support of MuCF within the US Department of Energy, which has been the cause of the rapid recent progress.

References

[1] E.J. Bleser, E.W. Anderson, L.M. Lederman, S.L. Meyer, J.L. Rosen, J.E. Rothberg and I.T. Wang; Phys. Rev. 132, 2679 (1963).

[2] G. Conforto, C. Rubbia, E. Zavattini and S. Focardi; Il Nuovo Cimento 33, 1001 (1964).

[3] S.E. Jones; *Survey of Experimental Results in Muon-Catalyzed Fusion*; this volume.

[4] M. Jändel; *Optimal Active Targets For Muon Catalyzed Fusion Reactors*; this volume.

[5] A. Bertin, M. Bruschi, M. Capponi, S. De Castro I. Massa, M.Piccinini, M.Poli, N. Semprini-Cesari, A. Trombini A. Vitale and A. Zoccoli; *On the Energy Cost of Producing Muons in a Deuterium-Tritium Target for Muon Catalyzed Fusion*; this volume.

[6] Yu.V. Petrov; Nature 285, 466 (1980).

[7] M. Kamimura; *Effects of Nuclear Interaction on Muon Sticking to Helium in Muon Catalyzed dt-Fusion*; this volume.

[8] G.M. Hale, M.C. Struensee, R.T. Pack and J.S. Cohen; *Boundary-Value Approach to Nuclear Effects in Muon-Catalyzed d-t Fusion*; this volume.

[9] M. Danos, L.C. Biedenharn and A. Stahlhofen; *Comprehensive Theory of Nuclear Effects on the Intrinsic Sticking Probability: I*; this volume.

[10] M. Danos, L.C. Biedenharn and A. Stahlhofen; *Comprehensive Theory of Nuclear Effects on the Intrinsic Sticking Probability: II*; this volume.

[11] L.C. Biedenharn, private communication.

[12] H.E. Rafelski and B. Müller; *Density Dependent Stopping Power and Muon Sticking in Muon Catalyzed Fusion*; this volume.

[13] R.M. Kulsrud; *A Proposed Method for Reducing the Sticking Constant in M.C.F*; this volume.

[14] T. Tajima; *A New Concept for Muon Catalyzed Fusion*; this volume.

[15] L.I. Menshikov; *Muon Catalytic Processes in a Dense Low-Temperature Plasma*; preprint IAE-4589/2, Institut Atomnoj Energii im. I.B. Kurchatova March 1988. L.I. Menshikov and L.I. Ponomarev; Pisma ZhETF 46, 246 (1987).

[16] I know of several colleagues who recognized independently that the calculations in above reference are in error: M. Jändel, private communication; D.M. Eardley, private communication.

[17] S.A. Alexander, H.J. Monkhorst and K. Szalewicz; *A Comparison of Muonic Molecular Calculations*; this volume.

[18] K. Szalewicz, S.A. Alexander, P. Froelich, S.E. Haywood, B. Jeziorski, W. Kolos, H.J. Monkhorst, A. Scrinzi, C. Stodden, A. Velenik and X. Zhao; *Theoretical Description of Muonic Molecular Ions*; this volume.

[19] J.S. Cohen, M. Leon and N.T. Padial; *Pressure Broadening of the* $[(dt\mu)dee]^*$ *Formation Resonances*; this volume.

[20] V.Yu. Petrov and Yu.V. Petrov; *Formation of Hydrogen Mesic Molecules at Moderate Gas Densities*; this volume.

[21] A.N. Anderson; *Investigation of* Q_{1s}; this volume.

[22] L.I. Menshikov and L.I. Ponomarev, Pisma ZhETF 39, 542 (1984) [JETP Lett. 39, 663 (1984)] and Z. Phys. D2, 1 (1986).

[23] B. Müller, J. Rafelski, M. Jändel and S.E. Jones; *Possible Influence of Vacuum Polarization on* Q_{1s} *in Muon Catalyzed dt Fusion*; this volume.

[24] S. Barnett and A.M. Lane; *Increase in Meso-Molecular Formation by Laser Induced Resonances*; this volume.

[25] S. Eliezer and Z. Henis; *Laser Induced* $\mu-$*molecular Formation*; this volume.

[26] H. Takahashi; *Muonic Molecular Formation under Laser Irradiation and in the Clustered Ion Molecule*; this volume.

[27] P. Froelich, K. Szalewicz, H.J. Monkhorst, W. Kolos and B. Jeziorski; *Some properties of Three Body Resonances of* $dt\mu$ *Related to Muon Catalyzed Fusion*; this volume.

LIST OF PARTICIPANTS

Gerard Aissing
Dept. of Physics
University of Florida
Gainseville, FL 32611

Steven A. Alexander
Dept. of Physics
University of Florida
Gainsville, FL 32611

Alan N. Anderson
2602 Lemp St.
Boise, ID 83702

Konrad Aniol
Physics & Astronomy
California State Univ.
Los Angeles, CA 90032

Olive Baker
MPDO MSH844
Los Alamos Natl. Lab.
Los Alamos, NM 97545

A. Bertin
Dept. of Physics
University of Bologna
40126 Bologna, Italy

Anand Bhatia
NASA/Goddard Space Fl.Ctr
Lab for Astronomy
Greenbelt, MD 20771

Larry Biedenharn
Physics Dept.
Duke University
Durham, NC 27706

Helmut Bossy
Dept. 50-205
Lawrence Berkeley Lab
Berkeley, CA 94720

James Bradbury
MS H844
Los Alamos Natl. Lab
Los Alamos, NM 87545

Wolfgang H. Breunlich
Inst. for Medium Energy Physics
Vienna, Austria A-1020

Aurel Bulgac
Dept. of Physics
Michigan State Univ.
East Lansing,MI 48824

A.J. Caffrey
Idaho Natl. Eng Lab
P.O. Box 1625
Idaho Falls, ID 83415

Michael Cargnelli
Inst. for Med. Energy Phys
Czerningasse 17/14
Vienna, Austria

Lali Chatterjee
Physics Dept.
Jadavpur University
Calcutta, India

John W. Clark
Physics Dept.
Washington University
St. Louis, MO 63130

James Cohen
Dept. T-12, MS J569
Los Alamos Natl. Lab
Los Alamos, NM 87545

Mike Danos
National Bureau of Standards
Washington, DC 20234

John D. Davies
Physics Dept.
Univ. of Birmingham
Chilton, England

Stephen Dean
Fusion Power Associates
2 Professional Dr. Ste 248
Gaithersburg, MD 20879

V.P. Dzhelepov
Joint Inst. Nucl Res
PO Box 79, Head P.O.
101000 Moscow, USSR

Shalom Eliezer
Plasma Physics Dept.
Soreq Nucl. Res. Ctr.
Yavne 70600, Israel

V.V. Filchenkov
Joint Inst. Nucl. Res.
P.O. Box 79, Head Post Off.
101000 Moscow, USSR

Robert L. Forward
Forward Unlimited
P.O. Box 2783
Malibu, CA 90265

Piotr Froelich
Quantum Chemistry
Uppsala University
Uppsala, Sweden

Kenji Fukushima
Med. Energy Phys. Dvn.
Los Alamos Natl. Lab
Los Alamos, NM 87545

M.P. Goldsworthy
Univ. of New So Wales
Kensington 2033
New So Wales, Australia

Ryszard Gajewski
U.S. Dept. of Energy
Dvn. of Adv Eng. Projects
Washington, DC 20545

Gerald Hale
Group T-2, MS B243
Los Alamos Natl. Lab
Los Alamos, NM 87545

David Harley
Physics Bldg. 81
University of Arizona
Tucson, AZ 85721

F. Joachim Hartmann
Physik Dept. TU Munchen
James-Franck-Str.
D-8046 Garching, W. Ger.

Susan Haywood
Quantum Theory Project
University of Florida
Gainsville, FL 32611

Zohar Henis
Plasma Physics Dept.
Soreq NRC
Yavne, Israel 70600

Arnold Honig
Physics Dept.
Syracuse University
Syracuse, NY 13244

Dezso Horvath
Nuclear Physics Dept.
Central Res. Just. Physics
Budapest, Hungary H-1525

C.Y. Hu
Physics Dept.
California State Univ.
Long Beach, CA 90840

Tsutomu Ishigami
Dept. of Nucl. Energy
Brookhaven Natl. Lab
Upton, NY 11973

Magnus Jandel
Siegbhan Inst.
Roslagsvagen 100
S-104 05 Stockholm,SWEDEN

Herbert Jones
Physics Dept.
Florida A&M Univ.
Tallahassee, Fl 32307

Steven Jones
Physics & Astronomy
Brigham Young Univ.
Provo, Utah 84602

Masayasu Kamimura
Dept. of Physics
Kyushu University
Fukuoka, Japan 812

Grigory Korenman
Inst. of Nucl. Physics
Moscow State Univ.
Moscow 119899, USSR

Russell Kulsrud
PPPL, P.O. Box 451
Plasma Physics Lab
Princeton, NJ 08544

Nai-hang Kwong
Physics Dept.
University of Arizona
Tucson, Az 85721

Sigurd Larsen
Dept. of Physics
Temple University
Philadelphia, PA 19122

Everett G. Larson
Physics & Astronomy
Brigham Young University
Provo, Utah 84602

Guy Larson
Quantum Chemistry Dept.
Uppsala University
Uppsala, Sweden

Melvin Leon
Los Alamos Natl. Lab
MP-DO, MS H844
Los Alamos, NM 87545

Ping Li
Physics & Astronomy
Brigham Young University
Provo, Utah 84602

Chii-Dong Lin
Physics Dept.
Kansas State University
Manhattan, KS 66506

B. Maglich
AE Labs, Inc. Bldg 6
14 Washington Rd.
Princeton, NJ 08550

Johann Marton
Inst. for Med.Energ. Phys.
Czerningasse 17/14
Vienna, Austria A-1020

Hendrik J. Monkhorst
Quantum Theory Project
University of Florida
Gainesville, FL 32611

John D. Morgan III
Physics & Astronomy
University of Delaware
Newark, DE 19716

Berndt Mueller
Theoretical Physics
University of Frankfurt
Frankfurt, W Ger D-6000

Kanetada Nagamine
Meson Science Lab
University of Tokyo
Tokyo, Japan 113

Michael Paciotti
Los Alamos Natl. Lab
Mail Stop H809
Los Alamos, NM 87545

Paolo Pasini
Istituto Nazionale
di Fiska Nucleare
40138 Bologna, Italy

Arnold Perlmutter
Physics Dept.
University of Miami
Coral Gables, FL 33124

Claude Petitjean
Paul Scherrer Inst.
CH-5234 Villigen
Switzerland

Yuri Petrov
Academic of Science
Leningrad Nuc.Phys.Inst.
Gatchina (P/I 188350) USSR

Walter Polansky
Dvn. of Adv. Energy Proj.
U.S. Dept. of Energy
Washington, DC 20545

Leonid I. Ponomarev
I.V. Kurchatov Atomic
Energy Inst.
Moscow 123182, USSR

Johann Rafelski
Physics Dept.,Bldg. 81
University of Arizona
Tucson, AZ 85721

J.J. Reidy
Physics Dept.
University of Mississippi
University, MS 38677

Subhash Saini
Lawrence Livermore Lab
L-417, P.O. Box 808
Livermore, CA 94550

Armin Scrinzi
Physics Dept.
University of Delaware
Newark, DE 19716

Eugene Sheely
Brigham Young University
284E 300 St.
Lehi, UT 84043

Janine Shertzer
Physics Dept.
Holy Cross College
Worcester, MA 01610

James Shurtleff
Chemistry Dept.
Brigham Young University
Provo, UT 84602

Robert Siegel
Physics Dept.
William & Mary College
Williamsburg, VA 23185

Gerald A. Smith
Physics Dept.
Penn. State Univ.
Univ. Park, PA 16802

Vedene Smith
Chemistry Dept.
Queen's University
Kingston, Canada
K7L 3N6

Alfons Stahlhofen
Physics Dept.
Duke University
Durham, NC 27706

Donald Steiner
Dept. of Nucl. Eng.
Rensselaer Poly. Inst.
Troy, NY 12180

Clifford Stodden
Quantum Theory Project
University of Florida
Gainesville, FL 32611

Szymon Suckewer
Plasma Physics Lab
Princeton University
Princeton, NJ 08543

Krzysztof Szalewicz
Physics Dept.
University of Delaware
Newark, DE 19716

Toshi Tajima
Inst. for Fusion Studies
University of Texas
Austin, TX 78712

Hiroshi Takahashi
Dept. of Nucl. Energy
Brookhaven Natl. Lab
Upton, NY 11973

David Taqqu
Paul Scherrer Inst.
CH-5234 Villigan
Switzerland

Stuart Taylor
Physics Dept.
Brigham Young Univ.
Provo, UT 84604

A. Vitale
Physics Dept.
University of Bologna
40126 Bologna, Italy

Alvin M. Weinberg
Inst for Eng. Analysis
Oak Ridge Assoc. Univ.
Oak Ridge, TN 37831

David Worledge
Elec. Power Res. Inst.
3412 Hillview Ave.
Palo Alto, CA 95120

Xiaolin Zhao
Dept. of Physics
University of Delaware
Newark, DE 19716

AIP Conference Proceedings

		L.C. Number	ISBN
No. 1	Feedback and Dynamic Control of Plasmas – 1970	70-141596	0-88318-100-2
No. 2	Particles and Fields – 1971 (Rochester)	71-184662	0-88318-101-0
No. 3	Thermal Expansion – 1971 (Corning)	72-76970	0-88318-102-9
No. 4	Superconductivity in *d*- and *f*-Band Metals (Rochester, 1971)	74-18879	0-88318-103-7
No. 5	Magnetism and Magnetic Materials – 1971 (2 parts) (Chicago)	59-2468	0-88318-104-5
No. 6	Particle Physics (Irvine, 1971)	72-81239	0-88318-105-3
No. 7	Exploring the History of Nuclear Physics – 1972	72-81883	0-88318-106-1
No. 8	Experimental Meson Spectroscopy –1972	72-88226	0-88318-107-X
No. 9	Cyclotrons – 1972 (Vancouver)	72-92798	0-88318-108-8
No. 10	Magnetism and Magnetic Materials – 1972	72-623469	0-88318-109-6
No. 11	Transport Phenomena – 1973 (Brown University Conference)	73-80682	0-88318-110-X
No. 12	Experiments on High Energy Particle Collisions – 1973 (Vanderbilt Conference)	73-81705	0-88318-111–8
No. 13	π-π Scattering – 1973 (Tallahassee Conference)	73-81704	0-88318-112-6
No. 14	Particles and Fields – 1973 (APS/DPF Berkeley)	73-91923	0-88318-113-4
No. 15	High Energy Collisions – 1973 (Stony Brook)	73-92324	0-88318-114-2
No. 16	Causality and Physical Theories (Wayne State University, 1973)	73-93420	0-88318-115-0
No. 17	Thermal Expansion – 1973 (Lake of the Ozarks)	73-94415	0-88318-116-9
No. 18	Magnetism and Magnetic Materials – 1973 (2 parts) (Boston)	59-2468	0-88318-117-7
No. 19	Physics and the Energy Problem – 1974 (APS Chicago)	73-94416	0-88318-118-5
No. 20	Tetrahedrally Bonded Amorphous Semiconductors (Yorktown Heights, 1974)	74-80145	0-88318-119-3
No. 21	Experimental Meson Spectroscopy – 1974 (Boston)	74-82628	0-88318-120-7
No. 22	Neutrinos – 1974 (Philadelphia)	74-82413	0-88318-121-5
No. 23	Particles and Fields – 1974 (APS/DPF Williamsburg)	74-27575	0-88318-122-3
No. 24	Magnetism and Magnetic Materials – 1974 (20th Annual Conference, San Francisco)	75-2647	0-88318-123-1
No. 25	Efficient Use of Energy (The APS Studies on the Technical Aspects of the More Efficient Use of Energy)	75-18227	0-88318-124-X

No. 26	High-Energy Physics and Nuclear Structure – 1975 (Santa Fe and Los Alamos)	75-26411	0-88318-125-8
No. 27	Topics in Statistical Mechanics and Biophysics: A Memorial to Julius L. Jackson (Wayne State University, 1975)	75-36309	0-88318-126-6
No. 28	Physics and Our World: A Symposium in Honor of Victor F. Weisskopf (M.I.T., 1974)	76-7207	0-88318-127-4
No. 29	Magnetism and Magnetic Materials – 1975 (21st Annual Conference, Philadelphia)	76-10931	0-88318-128-2
No. 30	Particle Searches and Discoveries – 1976 (Vanderbilt Conference)	76-19949	0-88318-129-0
No. 31	Structure and Excitations of Amorphous Solids (Williamsburg, VA, 1976)	76-22279	0-88318-130-4
No. 32	Materials Technology – 1976 (APS New York Meeting)	76-27967	0-88318-131-2
No. 33	Meson-Nuclear Physics – 1976 (Carnegie-Mellon Conference)	76-26811	0-88318-132-0
No. 34	Magnetism and Magnetic Materials – 1976 (Joint MMM-Intermag Conference, Pittsburgh)	76-47106	0-88318-133-9
No. 35	High Energy Physics with Polarized Beams and Targets (Argonne, 1976)	76-50181	0-88318-134-7
No. 36	Momentum Wave Functions – 1976 (Indiana University)	77-82145	0-88318-135-5
No. 37	Weak Interaction Physics – 1977 (Indiana University)	77-83344	0-88318-136-3
No. 38	Workshop on New Directions in Mossbauer Spectroscopy (Argonne, 1977)	77-90635	0-88318-137-1
No. 39	Physics Careers, Employment and Education (Penn State, 1977)	77-94053	0-88318-138-X
No. 40	Electrical Transport and Optical Properties of Inhomogeneous Media (Ohio State University, 1977)	78-54319	0-88318-139-8
No. 41	Nucleon-Nucleon Interactions – 1977 (Vancouver)	78-54249	0-88318-140-1
No. 42	Higher Energy Polarized Proton Beams (Ann Arbor, 1977)	78-55682	0-88318-141-X
No. 43	Particles and Fields – 1977 (APS/DPF, Argonne)	78-55683	0-88318-142-8
No. 44	Future Trends in Superconductive Electronics (Charlottesville, 1978)	77-9240	0-88318-143-6
No. 45	New Results in High Energy Physics – 1978 (Vanderbilt Conference)	78-67196	0-88318-144-4
No. 46	Topics in Nonlinear Dynamics (La Jolla Institute)	78-57870	0-88318-145-2
No. 47	Clustering Aspects of Nuclear Structure and Nuclear Reactions (Winnepeg, 1978)	78-64942	0-88318-146-0
No. 48	Current Trends in the Theory of Fields (Tallahassee, 1978)	78-72948	0-88318-147-9

No. 49	Cosmic Rays and Particle Physics – 1978 (Bartol Conference)	79-50489	0-88318-148-7
No. 50	Laser-Solid Interactions and Laser Processing – 1978 (Boston)	79-51564	0-88318-149-5
No. 51	High Energy Physics with Polarized Beams and Polarized Targets (Argonne, 1978)	79-64565	0-88318-150-9
No. 52	Long-Distance Neutrino Detection – 1978 (C.L. Cowan Memorial Symposium)	79-52078	0-88318-151-7
No. 53	Modulated Structures – 1979 (Kailua Kona, Hawaii)	79-53846	0-88318-152-5
No. 54	Meson-Nuclear Physics – 1979 (Houston)	79-53978	0-88318-153-3
No. 55	Quantum Chromodynamics (La Jolla, 1978)	79-54969	0-88318-154-1
No. 56	Particle Acceleration Mechanisms in Astrophysics (La Jolla, 1979)	79-55844	0-88318-155-X
No. 57	Nonlinear Dynamics and the Beam-Beam Interaction (Brookhaven, 1979)	79-57341	0-88318-156-8
No. 58	Inhomogeneous Superconductors – 1979 (Berkeley Springs, W.V.)	79-57620	0-88318-157-6
No. 59	Particles and Fields – 1979 (APS/DPF Montreal)	80-66631	0-88318-158-4
No. 60	History of the ZGS (Argonne, 1979)	80-67694	0-88318-159-2
No. 61	Aspects of the Kinetics and Dynamics of Surface Reactions (La Jolla Institute, 1979)	80-68004	0-88318-160-6
No. 62	High Energy e^+e^- Interactions (Vanderbilt, 1980)	80-53377	0-88318-161-4
No. 63	Supernovae Spectra (La Jolla, 1980)	80-70019	0-88318-162-2
No. 64	Laboratory EXAFS Facilities – 1980 (Univ. of Washington)	80-70579	0-88318-163-0
No. 65	Optics in Four Dimensions – 1980 (ICO, Ensenada)	80-70771	0-88318-164-9
No. 66	Physics in the Automotive Industry – 1980 (APS/AAPT Topical Conference)	80-70987	0-88318-165-7
No. 67	Experimental Meson Spectroscopy – 1980 (Sixth International Conference, Brookhaven)	80-71123	0-88318-166-5
No. 68	High Energy Physics – 1980 (XX International Conference, Madison)	81-65032	0-88318-167-3
No. 69	Polarization Phenomena in Nuclear Physics – 1980 (Fifth International Symposium, Santa Fe)	81-65107	0-88318-168-1
No. 70	Chemistry and Physics of Coal Utilization – 1980 (APS, Morgantown)	81-65106	0-88318-169-X
No. 71	Group Theory and its Applications in Physics – 1980 (Latin American School of Physics, Mexico City)	81-66132	0-88318-170-3
No. 72	Weak Interactions as a Probe of Unification (Virginia Polytechnic Institute – 1980)	81-67184	0-88318-171-1
No. 73	Tetrahedrally Bonded Amorphous Semiconductors (Carefree, Arizona, 1981)	81-67419	0-88318-172-X

No. 74	Perturbative Quantum Chromodynamics (Tallahassee, 1981)	81-70372	0-88318-173-8
No. 75	Low Energy X-Ray Diagnostics – 1981 (Monterey)	81-69841	0-88318-174-6
No. 76	Nonlinear Properties of Internal Waves (La Jolla Institute, 1981)	81-71062	0-88318-175-4
No. 77	Gamma Ray Transients and Related Astrophysical Phenomena (La Jolla Institute, 1981)	81-71543	0-88318-176-2
No. 78	Shock Waves in Condensed Matter – 1981 (Menlo Park)	82-70014	0-88318-177-0
No. 79	Pion Production and Absorption in Nuclei – 1981 (Indiana University Cyclotron Facility)	82-70678	0-88318-178-9
No. 80	Polarized Proton Ion Sources (Ann Arbor, 1981)	82-71025	0-88318-179-7
No. 81	Particles and Fields –1981: Testing the Standard Model (APS/DPF, Santa Cruz)	82-71156	0-88318-180-0
No. 82	Interpretation of Climate and Photochemical Models, Ozone and Temperature Measurements (La Jolla Institute, 1981)	82-71345	0-88318-181-9
No. 83	The Galactic Center (Cal. Inst. of Tech., 1982)	82-71635	0-88318-182-7
No. 84	Physics in the Steel Industry (APS/AISI, Lehigh University, 1981)	82-72033	0-88318-183-5
No. 85	Proton-Antiproton Collider Physics –1981 (Madison, Wisconsin)	82-72141	0-88318-184-3
No. 86	Momentum Wave Functions – 1982 (Adelaide, Australia)	82-72375	0-88318-185-1
No. 87	Physics of High Energy Particle Accelerators (Fermilab Summer School, 1981)	82-72421	0-88318-186-X
No. 88	Mathematical Methods in Hydrodynamics and Integrability in Dynamical Systems (La Jolla Institute, 1981)	82-72462	0-88318-187-8
No. 89	Neutron Scattering – 1981 (Argonne National Laboratory)	82-73094	0-88318-188-6
No. 90	Laser Techniques for Extreme Ultraviolt Spectroscopy (Boulder, 1982)	82-73205	0-88318-189-4
No. 91	Laser Acceleration of Particles (Los Alamos, 1982)	82-73361	0-88318-190-8
No. 92	The State of Particle Accelerators and High Energy Physics (Fermilab, 1981)	82-73861	0-88318-191-6
No. 93	Novel Results in Particle Physics (Vanderbilt, 1982)	82-73954	0-88318-192-4
No. 94	X-Ray and Atomic Inner-Shell Physics – 1982 (International Conference, U. of Oregon)	82-74075	0-88318-193-2
No. 95	High Energy Spin Physics – 1982 (Brookhaven National Laboratory)	83-70154	0-88318-194-0
No. 96	Science Underground (Los Alamos, 1982)	83-70377	0-88318-195-9

No. 97	The Interaction Between Medium Energy Nucleons in Nuclei – 1982 (Indiana University)	83-70649	0-88318-196-7
No. 98	Particles and Fields – 1982 (APS/DPF University of Maryland)	83-70807	0-88318-197-5
No. 99	Neutrino Mass and Gauge Structure of Weak Interactions (Telemark, 1982)	83-71072	0-88318-198-3
No. 100	Excimer Lasers – 1983 (OSA, Lake Tahoe, Nevada)	83-71437	0-88318-199-1
No. 101	Positron-Electron Pairs in Astrophysics (Goddard Space Flight Center, 1983)	83-71926	0-88318-200-9
No. 102	Intense Medium Energy Sources of Strangeness (UC-Sant Cruz, 1983)	83-72261	0-88318-201-7
No. 103	Quantum Fluids and Solids – 1983 (Sanibel Island, Florida)	83-72440	0-88318-202-5
No. 104	Physics, Technology and the Nuclear Arms Race (APS Baltimore –1983)	83-72533	0-88318-203-3
No. 105	Physics of High Energy Particle Accelerators (SLAC Summer School, 1982)	83-72986	0-88318-304-8
No. 106	Predictability of Fluid Motions (La Jolla Institute, 1983)	83-73641	0-88318-305-6
No. 107	Physics and Chemistry of Porous Media (Schlumberger-Doll Research, 1983)	83-73640	0-88318-306-4
No. 108	The Time Projection Chamber (TRIUMF, Vancouver, 1983)	83-83445	0-88318-307-2
No. 109	Random Walks and Their Applications in the Physical and Biological Sciences (NBS/La Jolla Institute, 1982)	84-70208	0-88318-308-0
No. 110	Hadron Substructure in Nuclear Physics (Indiana University, 1983)	84-70165	0-88318-309-9
No. 111	Production and Neutralization of Negative Ions and Beams (3rd Int'l Symposium, Brookhaven, 1983)	84-70379	0-88318-310-2
No. 112	Particles and Fields – 1983 (APS/DPF, Blacksburg, VA)	84-70378	0-88318-311-0
No. 113	Experimental Meson Spectroscopy – 1983 (Seventh International Conference, Brookhaven)	84-70910	0-88318-312-9
No. 114	Low Energy Tests of Conservation Laws in Particle Physics (Blacksburg, VA, 1983)	84-71157	0-88318-313-7
No. 115	High Energy Transients in Astrophysics (Santa Cruz, CA, 1983)	84-71205	0-88318-314-5
No. 116	Problems in Unification and Supergravity (La Jolla Institute, 1983)	84-71246	0-88318-315-3
No. 117	Polarized Proton Ion Sources (TRIUMF, Vancouver, 1983)	84-71235	0-88318-316-1

No. 118 Free Electron Generation of Extreme Ultraviolet Coherent Radiation (Brookhaven/OSA, 1983)	84-71539	0-88318-317-X
No. 119 Laser Techniques in the Extreme Ultraviolet (OSA, Boulder, Colorado, 1984)	84-72128	0-88318-318-8
No. 120 Optical Effects in Amorphous Semiconductors (Snowbird, Utah, 1984)	84-72419	0-88318-319-6
No. 121 High Energy e^+e^- Interactions (Vanderbilt, 1984)	84-72632	0-88318-320-X
No. 122 The Physics of VLSI (Xerox, Palo Alto, 1984)	84-72729	0-88318-321-8
No. 123 Intersections Between Particle and Nuclear Physics (Steamboat Springs, 1984)	84-72790	0-88318-322-6
No. 124 Neutron-Nucleus Collisions – A Probe of Nuclear Structure (Burr Oak State Park - 1984)	84-73216	0-88318-323-4
No. 125 Capture Gamma-Ray Spectroscopy and Related Topics – 1984 (Internat. Symposium, Knoxville)	84-73303	0-88318-324-2
No. 126 Solar Neutrinos and Neutrino Astronomy (Homestake, 1984)	84-63143	0-88318-325-0
No. 127 Physics of High Energy Particle Accelerators (BNL/SUNY Summer School, 1983)	85-70057	0-88318-326-9
No. 128 Nuclear Physics with Stored, Cooled Beams (McCormick's Creek State Park, Indiana, 1984)	85-71167	0-88318-327-7
No. 129 Radiofrequency Plasma Heating (Sixth Topical Conference, Callaway Gardens, GA, 1985)	85-48027	0-88318-328-5
No. 130 Laser Acceleration of Particles (Malibu, California, 1985)	85-48028	0-88318-329-3
No. 131 Workshop on Polarized ^{3}He Beams and Targets (Princeton, New Jersey, 1984)	85-48026	0-88318-330-7
No. 132 Hadron Spectroscopy–1985 (International Conference, Univ. of Maryland)	85-72537	0-88318-331-5
No. 133 Hadronic Probes and Nuclear Interactions (Arizona State University, 1985)	85-72638	0-88318-332-3
No. 134 The State of High Energy Physics (BNL/SUNY Summer School, 1983)	85-73170	0-88318-333-1
No. 135 Energy Sources: Conservation and Renewables (APS, Washington, DC, 1985)	85-73019	0-88318-334-X
No. 136 Atomic Theory Workshop on Relativistic and QED Effects in Heavy Atoms	85-73790	0-88318-335-8
No. 137 Polymer-Flow Interaction (La Jolla Institute, 1985)	85-73915	0-88318-336-6
No. 138 Frontiers in Electronic Materials and Processing (Houston, TX, 1985)	86-70108	0-88318-337-4
No. 139 High-Current, High-Brightness, and High-Duty Factor Ion Injectors (La Jolla Institute, 1985)	86-70245	0-88318-338-2

No. 140	Boron-Rich Solids (Albuquerque, NM, 1985)	86-70246	0-88318-339-0
No. 141	Gamma-Ray Bursts (Stanford, CA, 1984)	86-70761	0-88318-340-4
No. 142	Nuclear Structure at High Spin, Excitation, and Momentum Transfer (Indiana University, 1985)	86-70837	0-88318-341-2
No. 143	Mexican School of Particles and Fields (Oaxtepec, México, 1984)	86-81187	0-88318-342-0
No. 144	Magnetospheric Phenomena in Astrophysics (Los Alamos, 1984)	86-71149	0-88318-343-9
No. 145	Polarized Beams at SSC & Polarized Antiprotons (Ann Arbor, MI & Bodega Bay, CA, 1985)	86-71343	0-88318-344-7
No. 146	Advances in Laser Science–I (Dallas, TX, 1985)	86-71536	0-88318-345-5
No. 147	Short Wavelength Coherent Radiation: Generation and Applications (Monterey, CA, 1986)	86-71674	0-88318-346-3
No. 148	Space Colonization: Technology and The Liberal Arts (Geneva, NY, 1985)	86-71675	0-88318-347-1
No. 149	Physics and Chemistry of Protective Coatings (Universal City, CA, 1985)	86-72019	0-88318-348-X
No. 150	Intersections Between Particle and Nuclear Physics (Lake Louise, Canada, 1986)	86-72018	0-88318-349-8
No. 151	Neural Networks for Computing (Snowbird, UT, 1986)	86-72481	0-88318-351-X
No. 152	Heavy Ion Inertial Fusion (Washington, DC, 1986)	86-73185	0-88318-352-8
No. 153	Physics of Particle Accelerators (SLAC Summer School, 1985) (Fermilab Summer School, 1984)	87-70103	0-88318-353-6
No. 154	Physics and Chemistry of Porous Media—II (Ridge Field, CT, 1986)	83-73640	0-88318-354-4
No. 155	The Galactic Center: Proceedings of the Symposium Honoring C. H. Townes (Berkeley, CA, 1986)	86-73186	0-88318-355-2
No. 156	Advanced Accelerator Concepts (Madison, WI, 1986)	87-70635	0-88318-358-0
No. 157	Stability of Amorphous Silicon Alloy Materials and Devices (Palo Alto, CA, 1987)	87-70990	0-88318-359-9
No. 158	Production and Neutralization of Negative Ions and Beams (Brookhaven, NY, 1986)	87-71695	0-88318-358-7

No. 159	Applications of Radio-Frequency Power to Plasma: Seventh Topical Conference (Kissimmee, FL, 1987)	87-71812	0-88318-359-5
No. 160	Advances in Laser Science–II (Seattle, WA, 1986)	87-71962	0-88318-360-9
No. 161	Electron Scattering in Nuclear and Particle Science: In Commemoration of the 35th Anniversary of the Lyman-Hanson-Scott Experiment (Urbana, IL, 1986)	87-72403	0-88318-361-7
No. 162	Few-Body Systems and Multiparticle Dynamics (Crystal City, VA, 1987)	87-72594	0-88318-362-5
No. 163	Pion–Nucleus Physics: Future Directions and New Facilities at LAMPF (Los Alamos, NM, 1987)	87-72961	0-88318-363-3
No. 164	Nuclei Far from Stability: Fifth International Conference (Rosseau Lake, ON, 1987)	87-73214	0-88318-364-1
No. 165	Thin Film Processing and Characterization of High-Temperature Superconductors	87-73420	0-88318-365-X
No. 166	Photovoltaic Safety (Denver, CO, 1988)	88-42854	0-88318-366-8
No. 167	Deposition and Growth: Limits for Microelectronics (Anaheim, CA, 1987)	88-71432	0-88318-367-6
No. 168	Atomic Processes in Plasmas (Santa Fe, NM, 1987)	88-71273	0-88318-368-4
No. 169	Modern Physics in America: A Michelson-Morley Centennial Symposium (Cleveland, OH, 1987)	88-71348	0-88318-369-2
No. 170	Nuclear Spectroscopy of Astrophysical Sources (Washington, D.C., 1987)	88-71625	0-88318-370-6
No. 171	Vacuum Design of Advanced and Compact Synchrotron Light Sources (Upton, NY, 1988)	88-71824	0-88318-371-4
No. 172	Advances in Laser Science–III: Proceedings of the International Laser Science Conference (Atlantic City, NJ, 1987)	88-71879	0-88318-372-2
No. 173	Cooperative Networks in Physics Education (Oaxtepec, Mexico 1987)	88-72091	0-88318-373-0
No. 174	Radio Wave Scattering in the Interstellar Medium (San Diego, CA 1988)	88-72092	0-88318-374-9
No. 175	Non-neutral Plasma Physics (Washington, DC 1988)	88-72275	0-88318-375-7

No. 176	Intersections Between Particle Land Nuclear Physics (Third International Conference) (Rockport, ME 1988)	88-62535	0-88318-376-5
No. 177	Linear Accelerator and Beam Optics Codes (La Jolla, CA 1988)	88-46074	0-88318-377-3
No. 178	Nuclear Arms Technologies in the 1990s (Washington, DC 1988)	88-83262	0-88318-378-1
No. 179	The Michelson Era in American Science: 1870–1930 (Cleveland, OH 1987)	88-83369	0-88318-379-X
No. 180	Frontiers in Science: International Symposium (Urbana, IL, 1987)	88-83526	0-88318-380-3